Table of Contents

To the Student ix

Chapter 1 – Introduction to Toxicology 1
    1-1   Historical Background 3
    1-2   Sources of Chemical Information 5
            Environmental Protection Agency (EPA) 5
            Occupational Safety and Health Administration (OSHA) 6
            National Institute for Occupational Safety and Health (NIOSH) 6
            Centers for Disease Control and Prevention (CDC) 6
            Nuclear Regulatory Commission (NRC) 7
            Agency for Toxic Substances and Disease Registry (ATSDR) 7
            American Conference of Governmental Industrial Hygienists (ACGIH) 7
            Electronic Databases 7
    1-3   Role of the Toxicologist 11
            Descriptive Toxicology 11
            Mechanistic Toxicology 11
            ✷Forensic Toxicology 13
            Environmental Toxicology 13
            Regulatory Toxicology 13
            Food Toxicology 13
            Clinical Toxicology 15
    Summary 16
    Application and Critical Thinking Activities 16

Chapter 2 – Epidemiological and Toxicological Studies ... 17
- 2-1 Introduction ... 18
- 2-2 Epidemiological Studies ... 19
  - Retrospective and Prospective Epidemiological Studies ... 21
  - Risk ... 25
  - Bias ... 26
- 2-3 Toxicological Studies ... 29
- 2-4 Epidemiological Studies versus Toxicological Studies ... 30
  - Establishing a Cause-Effect Relationship ... 30
  - Controlled Study versus Noncontrolled Study ... 30
  - Exposure: Routes, Concentration, Duration, and Frequency ... 31
  - Exposed Individuals ... 31
  - Extrapolation of Data ... 31
- Summary ... 32
- Application and Critical Thinking Activities ... 32

Chapter 3 – Routes of Exposure ... 35
- 3-1 Introduction ... 36
- 3-2 Basic Cell Structure and Function ... 37
  - Cell Membrane Permeability ... 38
  - Endoplasmic Reticulum ... 38
  - Mitochondria ... 38
  - Passive and Active Transport Mechanisms ... 39
- 3-3 Pathways of Exposure ... 43
  - Dermal Absorption ... 43
  - Inhalation ... 45
  - Ingestion ... 47
  - Eye ... 52
  - Other Pathways of Exposure ... 53
- 3-4 Local versus Systemic Effects ... 54
- Summary ... 55
- Application and Critical Thinking Activities ... 55

Chapter 4 – The Dose-Response Relationship ... 57
- 4-1 Introduction ... 58
- 4-2 Acute and Chronic Toxicological Studies ... 58
  - Acute Toxicity Studies ... 58
  - Chronic Toxicity Studies ... 61
- 4–3 Dose-Response Curves ... 65
  - Relative Toxicity ... 69
- 4-4 Variables Influencing Toxicity ... 72
  - Dose-Time Relationship ... 72
  - Routes of Exposure and Physical and Chemical Factors ... 73
  - Interspecies and Intraspecies Variability ... 76
- Summary ... 83
- Application and Critical Thinking Activities ... 83

Basic ...egacy

Chris Kent

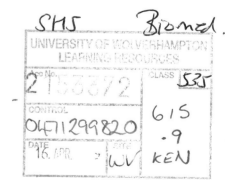

JOHN WILEY & SONS, INC.

New York / Chichester / Weinheim / Brisbane / Singapore / Toronto

Technical Illustration: Richard J. Washichek, Graphic Dimensions, Inc.
Photo Research: Beatrice Hohenegger

This book is printed on acid-free paper. ∞

This material is based on work supported by the National Science Foundation under Grant No. DUE94-54521. Any opinions, findings, and conclusions or recommendations expressed in this material are those of the author(s) and do not necessarily reflect those of the National Science Foundation.

Copyright ©1998 by INTELLICOM Intelligent Communications and
The Partnership for Environmental Technology Education

Published by John Wiley & Sons, Inc. All rights reserved.
Published simultaneously in Canada.

No part of this publication may be reproduced, stored in a retrieval system or transmitted in any form or by any means, electronic, mechanical, photocopying, recording, scanning or otherwise, except as permitted under Sections 107 or 108 of the 1976 United States Copyright Act, without either the prior written permission of the Publisher, or authorization through payment of the appropriate per-copy fee to the Copyright Clearance Center, 222 Rosewood Drive, Danvers, MA 01923, (978) 750-8400, fax (978) 750-4744. Requests to the Publisher for permission should be addressed to the Permissions Department, John Wiley & Sons, Inc., 605 Third Avenue, New York, NY 10158-0012, (212) 850-6011, fax (212) 850-6008, E-Mail: PERMREQ @ WILEY.COM.

This publication is designed to provide accurate and authoritative information in regard to the subject matter covered. It is sold with the understanding that the publisher is not engaged in rendering professional services. If professional advice or other expert assistance is required, the services of a competent professional person should be sought.

Library of Congress Cataloging-in-Publication Data:

0-471-29982-0

Printed in the United States of America.

10 9 8 7 6 5 4 3 2 1

Chapter 5 – Absorption, Distribution, Metabolism, and Elimination of Toxics 85
    5-1   Introduction 86
    5-2   Absorption of Toxic Substances 86
           Mechanisms of Absorption 87
           Skin Absorption 89
           Lung Absorption 90
           Gastrointestinal Absorption 90
    5-3   Distribution of Toxic Substances 91
           Anatomy of the Circulatory System 91
           Binding of Plasma Proteins 94
           Storage of Toxic Substances 95
           Specialized Barrier: Blood-Brain Barrier 102
           Redistribution of Toxic Substances 103
    5-4   Metabolism (Biotransformation) 105
           Phase I Reactions 110
           Phase II Reactions 111
           Deactivation versus Bioactivation 114
           Factors Affecting Metabolism 117
    5-5   Excretion 120
           Excretion by the Liver 120
           Excretion by the Kidneys 120
           Excretion by the Lungs 122
    Summary 122
    Application and Critical Thinking Activities 123

Chapter 6 – Target Organ Effects 125
    6-1   Introduction 126
    6-2   Neurotoxicity 127
           General Organization of the Nervous System 127
           Cells of the Nervous System 127
           General Physiology of the Nervous System 130
           Neurotoxic Effects 133
    6-3   Hematoxicity 139
           Blood Cell Hematopoiesis 140
           Hematoxic Effects 141
           Chemically Induced Hypoxia 142
    6-4   Immunotoxicity 144
           Structure and Function of the Immune System 144
           Symptoms of Immunotoxicity 147
    6-5   Cardiotoxicity 150
           Structure and Function of the Heart 150
           Cardiotoxic Effects 152
    6-6   Pulmonary Toxicity 155
           Functions of the Respiratory System 155
           Pulmonary Toxic Effects 156
    6-7   Hepatotoxicity 160
           Functions of the Liver 160
           Hepatotoxic Effects 161

6-8 Nephrotoxicity 162
     Functions of the Kidneys 162
     Nephrotoxic Effects 162
6-9 Reproductive Toxicity 164
     Anatomy and Physiology of the Male Reproductive System 165
     Toxic Effects in the Male Reproductive System 166
     Anatomy and Physiology of the Female Reproductive System 167
     Toxic Effects in the Female Reproductive System 168
6-10 Toxicity of the Eye 170
     Toxic Effects to the Eye 171
Summary 173
Application and Critical Thinking Questions 175

Chapter 7 – Survey of Toxic Substances 177

7-1 Introduction 178
7-2 Outdoor and Indoor Air Pollution 184
     Outdoor Air Pollution 184
     Indoor Air Pollution 189
7-3 Petroleum Distillates 191
     Halogenated Hydrocarbons 191
     Aromatic Hydrocarbons 193
     Other Petroleum Distillates 194
7-4 Metals and Metalloids 196
7-5 Other Important Chemicals 207
7-6 Pesticides 215
     Organophosphate and Carbamate Insecticides 215
     Organochlorine Insecticides 218
     Herbicides 220
     Fungicides 222
     Fumigants 223
     Rodenticides 223
Summary 224
Application and Critical Thinking Activities 226

Chapter 8 – Radiation, Pathogens, and Naturally Occurring Toxins 229

8-1 Introduction 230
8-2 Radiation, Radioactive Materials, and Mixed Wastes 231
8-3 Chemical and Biological Warfare Agents 236
     Chemical Warfare Agents 236
     Biological Warfare Agents 239
8-4 Blood-borne Pathogens 243
8-5 Infectious and Medical Wastes 247
8-6 Naturally Occurring Toxins 249
     Bacteria-Derived Toxins 249
     Virus-Caused Illnesses 250
     Animal-Derived Toxins 251
     Plant-Derived Toxins 251

Summary 252
Application and Critical Thinking Activities 253

Chapter 9 – Mutagens, Teratogens, and Carcinogens 255
   9-1 Introduction 256
   9-2 General Chromosome Structure and Function 256
   9-3 Mutagens 260
        Categories of Mutations 260
        DNA Repair and Mutations 265
   9-4 Teratogens 267
        Mechanisms of Action 267
        Teratogenic Agents 271
   9-5 Carcinogens 273
        Definitions 273
        Mechanisms of Action 274
        Sources of Carcinogens 276
        Identification and Classification of Carcinogens 279
   9-6 Radiation 280
        Ionizing Radiation 280
        Nonionizing Radiation 283
Summary 285
Application and Critical Thinking Questions 286

Chapter 10 – Risk Assessment and Acute Exposure Treatment 287
   10-1 Introduction 288
   10-2 Risk Assessment 291
        Hazard Identification 291
        Hazard Evaluation and Dose-Response Assessment 293
        Exposure Assessment 297
        Risk Characterization 304
        Risk Management 305
   10-3 Administering First Aid 306
        Assessing Nature and Severity of Circumstances 306
        Basic First Aid Steps 308
   10-4 Toxic Substance Exposure, Symptoms, and Treatment 313
Summary 338
Application and Critical Thinking Activities 338

Glossary 341

Bibliography 373

Index 387

To the Student

The study of toxicology focuses on the factors in our environment that have the potential to cause adverse health effects. These factors may occur naturally in a variety of plants and animals or as a result of human activities. Understanding the nature of these factors and their actions on living organisms contributes to our making more informed decisions in environmental matters.

Basics of Toxicology has been written to provide environmental technology students with a broad overview of the discipline of toxicology. The first chapters are intended as an introduction to the history, the terminology, and the basic concepts of toxicology. Explanations on how exposures occur and how toxicological data are obtained are also provided. The subsequent chapters discuss absorption, distribution, metabolism, and elimination of toxic substances as well as their effect on target organs. These are followed by two chapters that survey man-made and naturally occurring toxic substances and that discuss the effects of radiation on the body. Mutagens, teratogens, and carcinogens are treated in a separate chapter. The last chapter discusses risk assessment and basic first aid procedures. Numerous illustrations, diagrams, and tables are included to facilitate the understanding of the various concepts. A glossary containing nearly 800 terms as well as a comprehensive index are provided as valuable reference and research tools.

Although the text was written for the beginning student, the thoroughness of the approach to the subject makes this a valuable reference book for a professional library. Basic anatomy, physiology, and chemistry concepts and descriptions are included throughout the text; they provide the needed background information and illustrate how various toxicological agents interact with target tissues.

It is my hope that this text will provide the students with a basic understanding of the subject of toxicology and stimulate their interest and curiosity to continue the study of this fascinating topic.

(Please note that the bold terms throughout the text are glossary terms.)

Acknowledgments

Any endeavor of this nature cannot occur without the help and assistance of many people. I would like to express my thanks particularly to:

INTELECOM, Intelligent Telecommunications; The Partnership for Environmental Technology (PETE); and the National Science Foundation (NSF) for making the *Preserving the Legacy* Project and this book possible.

The members of the National Academic Council for their reviews and helpful suggestions during the initial development of the manuscript: Ann Boyce, Bakersfield College; Doug Feil, Kirkwood Community College; A. J. Silva, Eastern Idaho Technical College; Dave Boon, Front Range Community College; Bill Engel, University of Florida; Eldon Enger, Delta College; Doug Nelson, SUNY Morrisville; Stephen Onstot, Esq. Burk, Williams, and Sorensen; and Ray Seitz, Columbia Basin College.

Howard Guyer, Academic Team Leader and A. J. Silva, member of the National Academic Council for their unrelenting efforts to get the chapters written, reviewed, rewritten, and edited.

Beatrice Hohenegger, INTELECOM Senior Editor, for her attention to detail, her suggestions, and particularly her efforts and patience during the months of editing and rewriting the chapters.

Graphic Dimensions for their creative talent in converting the manuscript into a readable text.

Debbie Davis who typed and retyped the initial versions of the text and was able to decipher my ancient cryptic scrawlings and put them to paper.

And to my family who has had to endure my absence both mentally and physically during the preparation of this text. Words can never express my appreciation of their understanding and patience.

Chris Kent, Ph.D.
Idaho National Engineering
and Environmental Laboratory
Idaho Falls, Idaho

1

Introduction to Toxicology

Chapter Objectives

Upon completing this chapter, the student will be able to:

1. **Define** the term toxicology.
2. **Identify** sources of information for chemicals.
3. **Define** some of the principal terminology used in toxicology.
4. **Identify** various disciplines of toxicology.

Chapter Sections

1-1 Historical Background

1-2 Sources of Chemical Information

1-3 Role of the Toxicologist

Box 1-1 ■ Chemical Accident: Bhopal, India

Site of Union Carbide facility. *Copyright Jay Ullal/Das Fotoarchiv/Black Star 1994.*

On December 3, 1984 a chemical, **methyl isocyanate (MIC)**, was dicharged from a Union Carbide facility into the surrounding community of Bhopal, India with a population of 800,000. Water leaked into a tank containing MIC. The leak occurred as a result of human error and non-operational safety equipment. Mixing of methyl isocyanate with water produced heat and caused an increase in tank pressure, eventually resulting in the release of an estimated 20 to 30 tons of methyl isocyanate into the air. The released gas was heavier than air and formed a deadly cloud close to the ground. A warning alarm was initiated an hour after the leak began. However, by then most of the toxic effects had occurred. It is estimated that there were 2,500 deaths caused immediately and up to 300,000 were injured. An additional 8,000 have died since the initial exposure. Reports indicate that up to 40,000 people still had respiratory problems three months after exposure.

MIC is a reactive chemical used in the manufacture of **carbaryl**, an insecticide sold under the trade name **Sevin**. Sevin is used to control fleas on dogs and cats, and in mosquito control programs. It is extremely irritating to **mucous membranes** and can be acutely toxic when inhaled. Exposure results in coughing, increased saliva production, and effects similar to exposure to tear gas (irritation of the mucous membrane of the eye and excess tear formation). Damage to the **cornea** can occur. At high concentrations breathing difficulties occur characterized by a feeling of pressure on the chest and chest pains. Severe exposure results in the development of pulmonary edema, which can cause death.

Initially the Indian government sought $3 billion in liabilities related to the release, but it eventually settled for $470 million and agreed not to recognize any future civil claims. A report issued in December of 1996 stated that as many as 50,000 people continue to be affected by the events associated with the Bhopal accident. These individuals suffer from a combination of physical and psychological problems. Physical disorders include loss of memory, speech, or fine motor skills. Psychological disorders include **post-traumatic stress disorder**, a condition similar to the one suffered by individuals who served in the Vietnam war. In addition, reports indicate that as many as 250,000 more claims are expected to be filed with the Indian government with respect to injuries associated with the chemical release.

MIC effects included cows and other animals. *Copyright Pablo Bartholomew/Gamma Liaison 1984.*

1-1 Historical Background

Identification and use of **toxic** substances have occurred throughout recorded history. Records from as early as 1500 B.C. describe plant poisons such as hemlock, aconite, and opium, as well as toxic metals such as antimony (Sb), copper (Cu), and lead (Pb). A Greek physician, Dioscorides (circa A.D. 50-90), was the first to attempt the classification of poisons. Through the use of drawings and written descriptions, he divided the substances into three categories: plant poisons, animal poisons, and mineral poisons. This work was the standard for sixteen centuries.

As early as the 15th century it was recognized that individuals associated with certain occupations were being exposed to toxic substances. It was during this time that a more structured approach was taken to identifying the appearance of various illnesses and diseases with the exposure to toxic substances. Workers involved in occupations such as goldsmithing, mining, printing, and pottery-making were exposed to toxic substances such as mercury (Hg) and lead (Pb). This information was documented in two major publications: "On the Miners' Sickness and Other Diseases of Miners" (Paracelsus, 1567) and "Discourse on the Diseases of Workers" (Ramazzini, 1700). The Swiss-German physician Paracelsus, who is often referred to as the father of **toxicology**, developed many of the ideas that comprise the major toxicological principles used today. He believed that the toxic agent, which he referred to as the **toxicin**, was a single chemical found in a toxic substance. In order to understand the effects of "toxicins," Paracelsus believed that: 1) it was necessary to use experimentation to identify and understand responses to chemicals; 2) there is a difference between toxic and therapeutic properties of chemicals; 3) the **dose** of the chemical is important in making the distinction between therapeutic effects and harmful effects; and 4) it is possible to identify – to some extent – a degree of chemical specificity. These ideas subsequently became an integral part of the science of toxicology, which is the study of the adverse effects of chemicals on living organisms.

In the 18th century the Industrial Revolution led to a significant increase in the development and use of chemicals. By 1880 ten thousand organic chemicals such as **chloroform, carbon tetrachloride,** and **carbonic acid** had been created. Poison gases, used during World War I, had been synthesized by the early 19th century. In the 20th century organochlorine pesticides, such as **DDT (dichloro-diphenyltrichloroethane)**, and organophosphates, such as **malathion**, were manufactured to eliminate a variety of unwanted insects. **Sulfanilamide** was developed to fight bacterial infections. Derivatives of **quinine** were produced to fight symptoms of **malaria**. Now many new chemicals are developed daily. Many of them are petroleum-based products used in the development of plastics. New antibacterial and antiviral drugs, such as those used to combat the **AIDS (Acquired Immunodeficiency Syndrome)** virus, are being developed to improve health care throughout the world.

Along with the many benefits brought by industrialization and chemical use came an increase in the number of occupational diseases and illnesses being observed. An increased rate of scrotal cancer among chimney sweeps was observed by Percival Pott in 1775. This appears to be the first documented evidence of polyaromatic hydrocarbon-induced cancer. More obvious and direct effects of toxic substances were on display during World War I when chemical warfare agents – such as **phosgene** and **mustard gas** – were used to injure and kill thousands of soldiers. It was not until the 1960s that the long-term effects of pesticides were documented in Rachel Carson's book *Silent Spring*. Furthermore, dropping the atomic bomb during World War II unveiled to the world, in a dramatic fashion, the lethal effects of **acute exposure** to **radiation**.

> Acute exposure is exposure to a substance over a short period of time, usually 24 hours.

Ether, chloroform, **thalidomide**, and sulfanilamide were developed for medicinal purposes. However, the unintended side effects were at times disastrous. Ether and chloroform were used as anesthetics. Their use during pregnancy resulted in numerous fetal deaths. Thalidomide was administered during the early stages of pregnancy for "morning sickness." This drug caused several thousand babies to be born with serious birth defects, such as problems with arm and leg development. Sulfanilamide, used to fight bacterial infections, was dissolved in ethylene glycol (a component used in modern day antifreeze). Since ethylene glycol affects the kidneys, many patients died from kidney dysfunction.

With the increasing awareness of the damaging effects chemicals may have on individuals and the environment, various laws and government agencies were established to control the manufacturing and use of chemicals. The harmful effects associated with the use of sulfanilamide led to the formation of the U.S. Food and Drug Administration (FDA) in 1938. Today the FDA regulates food additives, cosmetics, medical devices, and drugs that are made available to the public. In 1947 the Federal Insecticide, Fungicide, and Rodenticide Act (FIFRA) was passed. This law requires manufacturers of pesticides to prove that the pesticides can be manufactured and used safely.

Events at Love Canal during the late 1970s and early 1980s focused additional attention on even more harmful effects of toxic substances. From 1942 to 1952, 21,000 tons of various types of chemical wastes were disposed of at the Love Canal landfill. The site was covered and subsequently sold to the city of Niagara Falls for one dollar. It was not until the late 1970s – after years of complaints from local residents – increased numbers of illnesses, and visible signs of chemicals appearing in yards and homes, that the magnitude of the problem was discovered. As a result, residents were moved from the Love Canal area, and cleanup and removal of the buried chemicals was initiated.

The 1970s and 1980s also saw the most significant increase in the passage of laws and the growth of government agencies to address the manufacture, use, and disposal of chemicals. It was during this time that the **Environmental Protection Agency (EPA)**, the **Occupational Safety and Health Administration (OSHA)**, the **Nuclear Regulatory Commission (NRC)**, and the **Agency for Toxic Substances and Disease Registry (ATSDR)** were established. The **Centers for Disease Control and Prevention (CDC)** was established initially in 1946 as the Communicable Disease Center, which focused primarily on preventing the spread of diseases such as smallpox, poliomyelitis, and malaria. The CDC's role was expanded to include hazards associated with chemical exposure; it still remains in the forefront in addressing "modern-day" diseases such as AIDS, **Legionnaire's disease,** and **Lyme disease**. All of these agencies are involved in gathering and evaluating information on chemicals and infectious agents, as well as evaluating the **risk** from exposure in order to set public policy protective of both people and the environment.

Checking Your Understanding

1. What is meant by the phrase "a substance is toxic"?
2. Who was Paracelsus? What was his contribution to the science of toxicology?
3. What is the relationship between the terms toxicology and toxicin?
4. Define the term "dose."
5. What is meant by the term "risk"?

1-2 Sources of Chemical Information

Environmental Protection Agency (EPA)

The U.S. Environmental Protection Agency (EPA) was established in 1970 and is responsible for formulating and implementing programs as well as enforcing federal laws designed to protect human health and the environment. There are ten regional offices throughout the United States, with EPA headquarters located in Washington, D.C. To help meet its objectives EPA has developed and promulgated regulations that implement the requirements of several comprehensive environmental laws (Table 1-1). These laws and regulations set standards and requirements for the manufacturing, use, and disposal of thousands of toxic chemicals. The specific chemicals regulated by EPA and the associated standards are found in the **Code of Federal Regulations** or **CFR** (Table 1-1). The CFR contains the implementing language and regulatory requirements for all federal laws that are passed. There are several different sections of the CFR called Titles; those that apply to EPA are in Title 40.

In many cases environmental laws require extensive evaluation of information about various toxic substances. Most of the information is derived from two major sources: toxicological studies (animal experimentation) and epidemiological studies. The data from these types of studies are reviewed to determine the potential risk associated with exposure.

> **Epidemiology** is the study of the causative factors that are associated with the incidence and distribution of disease or illness in a specific population.

Name	Acronym	Date Established	Purpose	CFR
Clean Air Act	CAA	1970	Implements regulations that control and abate air emissions from stationary and mobile sources.	40 CFR 50-93
Clean Water Act	CWA	1972	Regulates discharge of pollutants to surface waters.	40 CFR 100-149
Safe Drinking Water Act	SDWA	1974	Establishes standards for contaminants in drinking water; regulates discharges to underground injection wells, sole source aquifers, and public drinking water systems.	40 CFR 141-143
Comprehensive Environmental Response, Compensation, and Liability Act (Superfund)	CERCLA	1980	Cleanup of hazardous waste sites and definition of requirements for response to hazardous waste spills.	40 CFR 300-306
Resource Conservation and Recovery Act	RCRA	1976	Identification and regulation of hazardous waste treatment, storage, and disposal.	40 CFR 260-266, 268, 270-272, 279, 280-282, 148
Federal Insecticide, Fungicide, and Rodenticide Act	FIFRA	1948	Requires registration and testing of pesticides; regulates their sale, distribution, and use.	40 CFR 152-180
Toxic Substances Control Act	TSCA	1976	Requires testing and reporting of chemicals prior to manufacturing, distribution, and use; restricts the use of chemicals that pose a threat to human health and the environment.	40 CFR 701-720, 761
Emergency Planning and Community Right to Know Act	EPCRA	1986	Requires companies to report inventories of hazardous chemicals and toxic releases; requires state and local governments to develop plans for responding to emergency releases.	40 CFR 355, 370, 372

Table 1-1: A selection of EPA laws that regulate chemicals in the environment.

This information is then used to establish acceptable exposure limits. The limits may be incorporated directly into an environmental law; or they may be used to establish regulatory requirements for handling toxic substances, which would then result in a decreased chance of harmful exposure. The **Office of Research and Development (ORD)** within EPA studies the effects of toxic exposure on people and the environment.

Occupational Safety and Health Administration (OSHA)

In 1970 the Occupational Safety and Health Act was passed. The Act is enforced by the Occupational Safety and Health Administration. OSHA's stated mission is to save lives, prevent injuries, and protect the health of America's workers. Exposure to toxic substances in the workplace is one of the major areas of interest to OSHA. Like EPA, OSHA is involved in the accumulation and evaluation of toxic chemical information in order to establish acceptable exposure levels for chemicals found in the workplace. OSHA regulations – which establish standards for a variety of hazardous and toxic chemicals – are found in 29 CFR 1910. Much of the information that is utilized to establish acceptable workplace standards is derived from research and investigations performed by the **National Institute for Occupational Safety and Health (NIOSH)**.

National Institute for Occupational Safety and Health (NIOSH)

NIOSH is a federal agency established by the Occupational Safety and Health Act of 1970. NIOSH is a part of the Centers for Disease Control and Prevention (CDC). It has several functions, such as investigating potentially hazardous work conditions, evaluating chemical hazards in the workplace, and conducting research on chemicals. Information derived from these efforts is used to help reduce disease, injury, and disability in the workplace. The information is assembled into criteria documents, which OSHA utilizes to establish standards. NIOSH performs a similar function in support of TSCA, CERCLA, and the CAA.

Centers for Disease Control and Prevention (CDC)

The Centers for Disease Control and Prevention (CDC) is an agency of the U.S. Public Health Service, in the Department of Health and Human Services. The CDC's stated mission is to promote health and quality of life by preventing and controlling disease, injury, and disability. The CDC is composed of eleven different organizations (Table 1-2). Historically the CDC has focused on the study and prevention of infectious diseases such as malaria and smallpox. However, its responsibilities have expanded to include environmental and occupational hazards.

Organizations within the CDC – such as NIOSH and the **National Center for Environmental Health (NCEH)** – address hazards associated with chemical exposure inside and outside the workplace. Both organizations are involved in establishing standards, guidelines, and recommendations for toxic chemicals.

National Center for Chronic Disease Prevention and Health Promotion
National Center for Environmental Health
National Center for Health Statistics
National Center for Infectious Disease
National Center for Injury Prevention and Control
National Center for Prevention Services
National Institute for Occupational Safety and Health
Epidemiology Program Office
International Health Program Office
Public Health Practice Program Office
National Immunization Program

Table 1-2: Organizations within the Centers for Disease Control and Prevention (CDC).

Nuclear Regulatory Commission (NRC)

The Nuclear Regulatory Commission (NRC) was established in 1974 and assumed the roles and responsibilities of the Atomic Energy Commission. The NRC is responsible for regulating the use of nuclear materials for commercial, industrial, academic, and medical purposes. This includes regulating commercial nuclear power plants, nuclear materials used in the diagnosis and treatment of cancer, and nuclear materials used in smoke detectors. The NRC is also responsible for regulating non-power research, test and training reactors; nuclear fuel cycle facilities (the production of nuclear fuel); and the transport, storage, and disposal of nuclear materials and waste. Like OSHA and EPA, the NRC is involved in obtaining and evaluating information concerning acceptable exposure levels for workers handling nuclear materials.

Agency for Toxic Substances and Disease Registry (ATSDR)

The Agency for Toxic Substances and Disease Registry is part of the U.S. Public Health Agency. It was established in 1980 to provide health-based information in support of the cleanup of chemical waste disposal sites mandated by CERCLA. The ATSDR is primarily concerned with the health effects associated with exposure to toxic chemicals. This information is maintained in a database. The ATSDR publishes *Toxicological Profiles, Case Studies in Environmental Medicine,* and *Public Health Statements* that contain information relative to toxic chemical exposure.

American Conference of Governmental Industrial Hygienists (ACGIH)

The **American Conference of Governmental Industrial Hygienists (ACGIH)** is a professional organization that produces a listing of **Threshold Limit Values (TLV)** and **Biological Exposure Indices (BEI)** for several hundred chemicals, updating them every year. The TLVs are used as guidelines for occupational exposure to hazardous chemicals. This information is published in a booklet called *Threshold Limit Values (TLVs) for Chemical Substances and Physical Agents and Biological Exposure Indices.*

Biological exposure indices are recommended maximum concentrations of various types of toxic substances; they are used as guidelines to evaluate the potential health hazards associated with exposure. The maximum concentrations may be measured in the blood, urine, or exhaled air. Threshold limit values are established primarily as a result of airborne exposure. Biological exposure indices are a measure of exposure from all pathways such as inhalation, ingestion, or absorption through the skin. Both TLVs and BEIs are based on exposure during a normal working week; eight hours a day, five days a week. The indices are indicators of potential health problems. Exceeding these values does not necessarily mean individuals will develop diseases or illnesses.

Electronic Databases

The Internet has made it possible for anyone with a computer and modem to access information on toxic chemicals. The sources of information and means of accessing them are numerous; only a few will be mentioned here. The **National Library of Medicine (NLM)** has an electronic network called the **Toxicology Data Network (TOXNET)**. Several databases such as the **Hazardous Substances Data Bank (HSDB)** and the **Registry of Toxic Effects of Chemical Substances (RTECS)** can be found in the TOXNET database. The HSDB has toxicological information on more than 4,000 chemicals. It also contains information on emergency handling procedures, environmental data, regulatory status, and human exposure. The RTECS are maintained by NIOSH and contain information on the health effects for more than 90,000 chemicals.

CHEMTREC (Chemical Transportation Emergency Center) and MEDTREC (Medical Transportation Emergency Center) are databases used to obtain information concerning chemical exposure. CHEMTREC is a service provided by the Chemical Manufacturer's Association; it supplies basic information about chemicals currently produced. In addition, a caller can be put into direct contact with the manufacturer. MEDTREC is an

Box 1-2 ■ Agency for Toxic Substances and Disease Registry – Public Health Statement

Lead
ATSDR Public Health Statement, June 1990

What is lead?
Lead is a naturally occurring bluish-gray metal found in small amounts in the earth's crust. Lead and its compounds can be found in all parts of the environment, for example, in plants and animals used for food, in air, drinking water, rivers, lakes, oceans, in dust, and soil. Lead in the air attaches to dust and can be carried for long distances from where it is released. The lead-containing dust is removed from the air by rain. Lead stays in soil for many years. However, heavy rain can cause lead-containing soil to move into water.

At this time, lead and lead compounds have been found at 635 out of 1,177 sites on the National Priorities List of hazardous waste sites in the United States. As more sites are evaluated by EPA, this number may change.

Lead used by industry comes from mined ores or from recycled scrap metal. In 1985, 20 percent of the world's total production of lead was in the United States. Lead has a wide range of uses. Its main use is in the manufacture of storage batteries. Other uses include the production of chemicals, including paint, gasoline additives, various metal products (for example, sheet lead, solder, and pipes), and ammunition.

How might I be exposed to lead?
You may be exposed to lead and lead compounds from breathing air, drinking water, and eating soil or foods that contain lead. Breathing in air with lead-containing dust or swallowing lead-containing soil – such as might be found at a hazardous waste site or near areas with heavy automobile traffic – are also sources of exposure. Another source of lead exposure for children is the swallowing of non-edible items such as chips of lead-containing paint. This activity is known as **pica** (an abnormal eating habit). Children who put toys or other items in their mouths may also swallow lead if lead-containing dust and dirt are on these items. Touching dust and dirt containing lead can happen every day, but not much lead passes through the skin. During normal use of lead-containing products, very little skin contact takes place.

The burning of gasoline has been the single largest source (90 percent) of lead in the atmosphere since the 1920s. Today less than 35 percent of the lead released to the air comes from gasoline because EPA reduced the amount of lead allowed in it.

Other sources of release to the air may include emissions from iron and steel production, smelting operations, municipal waste incinerators, and lead-acid-battery manufacturers. Lead is released to the air from active volcanoes, windblown dust, and the burning of lead-painted surfaces. Cigarette smoke is a source of lead; people who smoke tobacco or who breathe in tobacco smoke may be exposed to more lead than people who are not exposed to it.

The major sources of lead released to water are lead plumbing and solder in houses, schools, and public buildings; lead-containing dust and soil carried into water by rain and wind; and wastewater from industries that use lead. Lead is released to dust and soil from such sources as lead-containing waste in municipal and hazardous waste dumps; or when fertilizers that contain sewage sludge are used; and from automobile exhaust.

Food and beverages may contain lead if lead-containing dust deposits onto crops while they are growing, and during food processing. Plants may take up lead from contaminated soil at a hazardous waste site or near areas with heavy automobile traffic.

Workers may be exposed to lead in a wide variety of occupations including smelting and refining industries, steel welding and cutting operations, battery manufacturing plants, gasoline stations, and radiator repair shops.

How does lead get into my body?
Lead can enter your body when you breathe air containing lead dust or particles of lead. Almost all of the lead in the lungs enters the blood and moves to other parts of the body.

When lead-containing substances are ingested much more of the lead enters the blood – and moves to other body parts – in children than it does in adults.

Much less lead enters the body through the skin than through the lungs or gastrointestinal tract. Regardless of how lead enters your body, most of it is stored in bone. Because some lead is stored in the body each time you are exposed, the levels of lead in bone and teeth get higher as a person gets older. Lead that is not stored in the body is removed in the urine and feces.

How can lead and its compounds affect my health?
Once lead is in the body, its effects are the same no matter how it entered. However, exposure to lead is especially dangerous for unborn children, whose bodies can be harmed while they are being formed. Exposure to lead by a pregnant woman can be carried over to the unborn child and cause premature birth, low birth weight, or even abortion. Children's bodies absorb lead more easily and are more sensitive to its effects. For infants or young children, lead exposure has been shown

to decrease intelligence (IQ) scores, to slow growth, and to cause hearing problems. Lead exposure may have lasting effects and interfere with successful performance in school. Adverse health effects can happen at exposure levels once thought to be safe.

The ability of lead to cause cancer in humans has not been shown. To date, workplace studies do not provide enough information to determine the risk of cancer for workers exposed to lead. However, tumors have developed in rats and mice given large doses of lead. The U.S. Department of Health and Human Services has determined that lead acetate and lead phosphate may reasonably be anticipated to be carcinogens.

Exposure to high levels of lead may cause the brain and kidneys of adults and children to be badly damaged. Lead exposure may increase blood pressure in middle-aged men. It is not known if lead increases blood pressure in women. Also, high levels of lead may affect the sperm or damage other parts of the male reproductive system.

Is there a medical test to determine if I have been exposed to lead?

It is possible to determine lead exposure by measuring the amount of lead in the blood. Lead in bone and teeth can be measured using X-ray techniques; however, this test is no longer commonly performed.

Exposure to lead can also be tested by measuring the amount of a substance in red blood cells called **erythrocyte protoporphyrin (EP)**. This method is commonly used to test children for lead poisoning. The level of EP in red blood cells is high when the amount of lead in the blood is high. However, high EP levels are not necessarily due to lead exposure; they may be caused by other diseases that affect red blood cells, such as anemia.

What levels of exposure have resulted in harmful health effects?

The graph below (Health Effects from Ingesting and Breathing Lead) shows the relationship between the amount of lead in the body and known health effects.

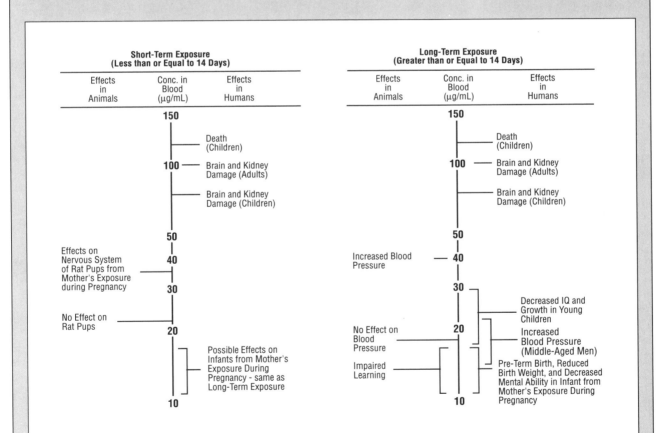

Health effects from ingesting and breathing lead. *Obtained from the Agency for Toxic Substances and Disease Registry Internet site.*

> The common way of measuring the amount of lead in the blood is by micrograms per deciliter (μg/dL). The column on the left side describes known health effects in laboratory animals and people for exposures of 14 days or less; the column on the right side is for exposures beyond 14 days. The levels of lead exposure that cause damage to a man's reproductive system and the levels that cause abortion in women are not well known; they are therefore not included in the graph.
>
> **What recommendations has the federal government made to protect human health?**
> The Centers for Disease Control (CDC) recommends that screening (testing) for lead poisoning be included in health care programs for children under 72 months of age, and especially for those under 36 months of age. If many children in a community have blood lead levels ≥ 10 μg/dL, lead poisoning prevention activities throughout the community should begin.
>
> To help protect small children – who might swallow chips of paint – the **Consumer Product Safety Commission (CPSC)** does not allow the amount of lead in most paints to be more than 0.06 percent. CDC suggests that indoor and outdoor paint used in residential buildings be tested for lead. If the level of lead is high, the paint should be removed and replaced with a paint that contains an allowable level of lead.
>
> CPSC, EPA, and the states are required by the 1988 Lead Contamination Control Act to test drinking water in schools for lead and to remove the lead if levels are too high. Drinking water coolers must also be lead-free, and any that contain lead have to be removed. EPA regulations limit lead in drinking water to 0.015 milligrams per liter (mg/L).
>
> To limit the amount of lead people are exposed to in the air, EPA does not allow the amount of lead that a person breathes over 3 months to be more than 1.5 micrograms of lead per cubic meter of air (1.5 μg/m^3). The National Institute for Occupational Safety and Health (NIOSH) recommends that workers not be exposed to levels of more than 100 μg/m^3 for up to 10 hours. EPA also limits the amount of lead in leaded gasoline to 0.1 gram of lead per gallon of gasoline (0.1 g/gal), and in unleaded gasoline to 0.05 g/gal.
>
> **Where can I get more information?**
> If you have more questions or concerns, please contact your state health or environmental department or: Agency for Toxic Substances and Disease Registry, Division of Toxicology, 1600 Clifton Road, E-29, Atlanta, Georgia 30333.
>
> This information was obtained from the Agency for Toxic Substances and Disease Registry Internet site.

extension of CHEMTREC and provides information relative to medical treatment in response to chemical exposures.

Material Safety Data Sheets (MSDS) are also found on the Internet. MSDSs provide information such as physical and chemical properties of a substance, first aid information, emergency response, and disposal information. Paper copies of MSDSs can be obtained from chemical manufacturers.

Checking Your Understanding

1. What is the Code of Federal Regulations (CFR)?
2. Which of the laws listed in Table 1-1 regulate the cleanup of a hazardous waste site?
3. Identify the function of each agency or organization listed in this chapter. Are you aware of any other agencies or organizations that are involved in evaluating and regulating toxic substances? If so, what are they and what is their function?
4. Define the terms Biological Exposure Indices (BEI) and Threshold Limit Value (TLV). How are they measured?
5. List two organizations that are responsible for developing regulations establishing acceptable exposure levels for chemicals in the air. Research the literature and list at least 10 chemicals regulated by each of the agencies and give exposure levels for each. Are there any substances that are regulated by both agencies?

1-3 Role of the Toxicologist

The number of toxic chemicals and their possible effects are so numerous and diverse that it would be difficult for any one individual to know what all the effects are and to evaluate the possible risk associated with exposure. As a result, toxicologists may focus on one or two specific areas. The following descriptions identify some of the various toxicological disciplines, providing a brief explanation of the types of activities and terminology associated with the various disciplines.

As discussed previously, evaluation of information concerning chemical exposure is important in the development and implementation of various rules and regulations administered by the different government agencies. The information used to establish acceptable exposure levels is obtained from several sources such as chemical analogies, epidemiological studies, or animal experimentation (toxicological studies). These sources are used to identify possible effects associated with chemical exposure, to assess the risk as a result of exposure, and to develop exposure limits. Information obtained from these studies includes determining the effective dose, differentiating between **acute** or **chronic** exposure, and determining the **dose-response relationship**. Each of these terms will be briefly described below. More detailed discussions will be found in subsequent chapters.

The dose is the actual amount of a chemical that enters the body. The dose received may occur as a result of either acute or chronic exposure. An acute exposure occurs over a very short period of time, usually 24 hours. Chronic exposures occur over long periods of time such as weeks, months, or years. In either case, a particular response is measured to determine the effects of chemical exposure. A dose-response relationship can be established by measuring the response relative to an increasing dose. The dose-response relationship is important in determining the toxicity of a particular substance. In general, the larger the dose the greater the response. Information such as the dose, exposure time, and observed response, is important in determining the risk associated with chemical exposure in a particular environment. This information is then used by the regulatory agencies to establish acceptable exposure levels.

Descriptive Toxicology

Descriptive toxicologists are involved in obtaining toxicological information from animal experimentation. These types of experiments are used to establish dose-response relationships and may involve either acute or chronic exposure to a toxic substance. A particular response is measured in order to identify and quantify the toxic effect. Changes in the route of exposure (inhalation, dermal contact, ingestion), species of test animal (rats, mice, guinea pigs, etc.), age, sex, and other variables may be evaluated to determine their effect on toxicity of a particular substance. Government agencies, such as the Environmental Protection Agency (EPA), the Occupational Safety and Health Administration (OSHA), and the Food and Drug Administration (FDA), use information derived from these studies to establish regulatory exposure limits.

Mechanistic Toxicology

Observations on how toxic substances cause their effects form another specialized area of toxicology, **mechanistic toxicology**. The effects of exposure are dependent on a variety of factors such as the solubility of the substance in water or fatty tissues, the size of the molecule, and the specific tissue type and cellular components affected. Understanding the properties of the toxic substance is important for the understanding of the mechanism of action and for determining if the effects observed in one species of test animal are applicable to others, such as humans. Perhaps one of the best examples of this principle is given by exposure to organochlorine and organophosphate pesticides. The mechanism of action for both of these substances is similar for insects, mice, rats, and humans. As a result, the toxicity to humans can be predicted, and safe levels of exposure can be developed based upon how these pesticides produce their effects in other organisms. Similarly, this type of information has been used to create harmful substances such as the nerve gases **soman** and **tabun** (See Chapter 8).

Box 1-3 ■ Minamata Disease: Mercury Poisoning

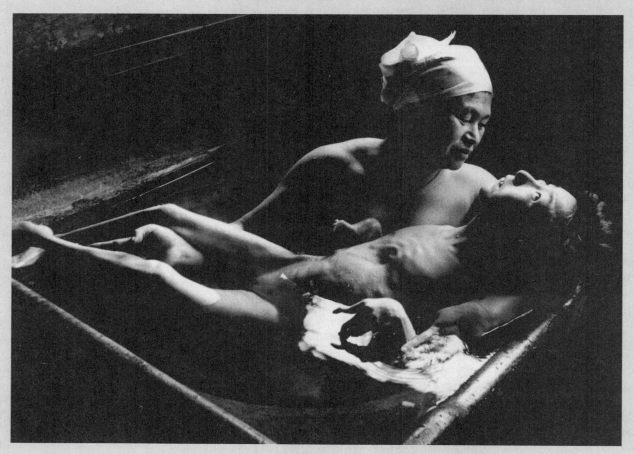

"Tomoko in her Bath." *Copyright W. Eugene Smith/Black Star 1972-75.*

While Rachel Carson's book *Silent Spring* chronicled the environmental effects of pesticide use in the United States, environmental disasters in other areas of the world were being identified as well. A "strange disease" appeared in the mid-1950s in and around a small Japanese village called Minamata. In general, victims demonstrated symptoms consistent with nervous system degeneration. Symptoms included numbness in the extremities of the hands and feet as well as the lips. The victims developed slurred speech and impaired vision. In more severe cases individuals developed severe brain damage, uncontrolled muscle spasms, and – in some cases – they became unconscious. In addition, cats demonstrated strange behavior and birds seemed to "drop from the sky."

It was not until 1959 that a correlation was made between the occurrence of the "strange disease" and the presence of mercury in the diet of individuals who demonstrated these symptoms. The source of the mercury was a factory located in Minamata. The factory manufactured plastics, drugs, and perfumes using the chemical acetaldehyde. Acetaldehyde is produced using mercury. From 1932 to 1968, 27 tons of mercury compounds were dumped into Minamata Bay. In the bay, bacteria converted the mercury to methylmercury, which was subsequently concentrated in the fish and shellfish of Minamata Bay. The chief source of food for many of the residents was fish and shellfish found in the bay. Consumption of the contaminated food supply resulted in methylmercury poisoning and development of what has been called Minamata Disease.

Mercury is no longer being dumped into the bay. The Japanese courts were still resolving claims as late as 1993. More than 3,000 people had been identified as having developed symptoms associated with Minamata Disease.

Forensic Toxicology

Forensic toxicologists are often involved in providing toxicological information in support of legal activities. Drug testing and blood alcohol determinations, in support of **DUI (driving under the influence)** cases, are two areas where forensic toxicologists can be involved. In more severe circumstances, forensic toxicologists are involved in determining whether a toxic chemical played a part in the injury or death of an individual. This type of work not only requires a knowledge of the effects associated with the parent compound; it also requires an understanding of how the substance is metabolized and of the toxic effects associated with the metabolites.

Environmental Toxicology

Environmental toxicologists are concerned with the interaction of toxic substances with all components of the ecosystem. Various aspects of mechanistic toxicology, descriptive toxicology, and forensic toxicology are utilized to understand how various plants, animals, and humans are affected by exposure to toxic substances. For example, in the 1960s a decline in the population of bald eagles and other raptors was observed. The decline eventually led to the bald eagle being listed as an endangered species. Environmental toxicologists attributed the cause of the decline primarily to the widespread use of DDT after World War II. DDT bioaccumulates or is concentrated in the food chain. Predatory birds such as the bald eagle are at the top of the food chain. They eat fish that in turn have eaten smaller fish, insects, and larvae that, in this case, had been exposed to DDT. The end result was that the bald eagles ate fish with high concentrations of DDT. Environmental toxicologists determined that the high concentration of DDT found in the birds was correlated with a decrease in the thickness of the eggshell resulting in eggs that broke more readily during incubation. The thinner eggshell was caused by the interaction of DDT with an enzyme involved with calcium deposition in the eggshell. As a result, the eggshell was thinner, more fragile, and easier to break. This discovery eventually led to a ban on the use of DDT in the United States. In the last twenty years, since the use of DDT was banned, the bald eagle population has rebounded. Similarly, mercury-contaminated shellfish were identified as the causative agent of Minamata Disease, a severe, debilitating neurological disorder (see Box 1-3 Minamata Disease: Mercury Poisoning).

Regulatory Toxicology

Regulatory toxicologists are involved in compiling and evaluating existing toxicological information to establish concentration-based standards of "safe" exposure. In general, the standard implies that an individual may be exposed to a chemical at the concentration established without adverse effects. As will be discussed in later chapters, there are several variables, such as duration of exposure, age, and health of an individual, that affect the response of an individual exposed to a particular chemical.

The information provided by regulatory toxicologists is utilized by a number of government agencies such as EPA, OSHA, and the FDA. Based upon the information obtained, EPA establishes standards and regulations associated with the production, use, and distribution of chemicals and pesticides; the identification and handling of hazardous waste; and acceptable levels of chemicals in the air, surface water, and drinking water. OSHA uses the information to establish acceptable levels of exposure in the workplace for various types of toxic chemicals. The FDA utilizes the information to control the amount of chemical substances in food as well as the production and use of drugs and cosmetics.

Food Toxicology

Food toxicologists are involved in delivering a safe and palatable supply of food to the consumer. During processing, a variety of different substances may be added to the food supply to make it look, taste, or smell better (see Food Additives: Safe or Harmful?). Fats, oils, fatty acids, sugar, protein, and starch may be added to modify the texture and the taste of the food. Various types of color additives such as FD&C Blue No. 1, FD&C Green No. 3 or FD&C Red No. 4 give color to the food to make it more appealing. All these additives are evaluated to determine if, and at what quantities, they will produce any adverse effects.

Another area of interest includes the study of food allergies. It is estimated that up to 30 percent of the American population has some degree of food allergy. Many people have trouble digesting milk.

Box 1-4 ■ Food Additives: Safe or Harmful?

Additives are used to appeal to the consumer's preference for a food supply that not only tastes good but looks and feels good as well. Oranges are colored to produce a brighter color, food coloring is added to icing used on cakes or in various liquid substances, such as sodas and syrup. Red dye is added to various types of meat to make it more appealing. The health and fitness craze has led the food industry to develop various types of additives designed to reduce calorie intake and "make people thin." Artificial sweeteners such as **saccharin**, **cyclamates**, and more recently, **aspartame**, were developed and manufactured to replace sugar. Saccharin and aspartame are used in commercial foods and the diet drink industry. Cyclamates are not currently approved for use. In the 1990s further efforts to produce "health food" additives resulted in the use of **olestra** as a substitute for fat in various products. The FDA approved its use in products such as potato, corn and tortilla chips, and crackers.

The issue of food additives has not been without controversy. Research and discussions concerning the safety of new substances being developed and those currently in use are ongoing. The following information addresses a few of the additives currently found in various foods and some of the issues associated with their use.

Artificial Sweeteners: Saccharin, Cyclamate, Aspartame

Saccharin – Saccharin was first used in the food industry in the 1960s. It can be found in artificial sweeteners. It is thought to be about 300 times sweeter than sugar with fewer calories. Saccharin contains only 1 to 8 calories per teaspoon. Controversy exists concerning the **carcinogenic** properties of saccharin. In 1977 reports indicated that saccharin caused cancer in the urinary bladder. Other studies have suggested that it may cause cancer of the stomach and intestine. This information and other research have been evaluated by the FDA, which has concluded that saccharin is considered "safe" for use. However, a warning message, *"Use of this product may be hazardous to your health. This product contains saccharin, which has been determined to cause cancer in laboratory animals,"* is required on diet beverage cans and packets containing the artificial sweetener.

Cyclamates – Consumption of saccharin-containing products leaves a bitter aftertaste. To address this problem, cyclamates were added to these products. Cyclamates appear to be about 30 times sweeter than sugar. However, in the 1980s studies indicated that the combination of saccharin and cyclamates caused cancer in lab rats. As a result cyclamate use in combination with saccharin was banned.

Aspartame – Aspartame is one of the more recent additions to the list of artificial sweeteners. It is 180-200 times as sweet as sugar and contains only four calories per gram. It is formed by combining two amino acids, aspartic acid, and phenylalanine. There are several issues associated with the use of this product. Once ingested, aspartame is broken down in the digestive tract, releasing phenylalanine, which is absorbed into the bloodstream. One in 20,000 newborn babies have difficulty metabolizing phenylalanine. As a result, elevated blood levels of phenylalanine may occur, which may be toxic and cause mental retardation. Other symptoms that have been associated with the use of this product are headaches, dizziness, and epileptic-like seizures. More recent reports have suggested that there may be an association between an increase in the occurrence of brain tumors and the use of aspartame.

Fat Substitute: Olestra

Ironically, olestra was first developed for the purpose of increasing fat intake in premature babies. The olestra molecule is composed of sucrose and fatty-acid residues. It can be used to fry certain food types, as in the production of potato and corn chips. The olestra molecule is not broken down by enzymes or bacteria present in the digestive tract. Therefore, the molecule remains too large to be absorbed, and it passes out of the digestive tract. Although olestra is not absorbed across the walls of the small intestine, it does bind with other vital nutrients, such as fat soluble vitamins A, D, E, and K, and removes them from the body. Adverse effects at this time have been limited to the gastrointestinal tract, with diarrhea being the primary symptom. The FDA requires all olestra-containing materials to carry a label stating, *"This product contains olestra. Olestra may cause abdominal cramping and loose stools. Olestra inhibits the absorption of some vitamins and other nutrients. Vitamins A, D, E, and K have been added."*

> ## Box 1-5 ■ It Only Takes A Little
>
> It only took a little! This is the message behind the story of Loretta Boberg, a 62-year old woman from Wisconsin who always tastes food before serving it to company. In this case the company can be very thankful she did.
>
> When Mrs. Boberg opened a jar of home-canned carrots last January, she dipped in a finger to taste the juice. Not liking the taste, she served home-canned beans to her guests instead. Within two days Mrs. Boberg became dizzy and had difficulty walking. At first, hospital staff thought she had suffered a stroke because of her slurred speech and muscle weakness. The doctor did ask her if she had eaten any spoiled food lately. However, too weak to speak, Mrs. Boberg wrote "carrots" on a piece of paper.
>
> If this physician had not suspected botulism, even though he had seen only a few cases, Mrs. Boberg would probably have died. The toxin moved through the respiratory system, paralyzing her muscles. A sample from the jar was fed to a laboratory mouse and it died instantly. A sample of Mrs. Boberg's blood was given to another mouse and it, too, died instantly.
>
> The road to recovery for this lady was very slow. Six months later, she remained in the hospital on a respirator, still being fed intravenously. She had stood for only three minutes since the incident, and talked through a tube in her trachea when not out of breath. Muscle movement was returning slowly with the help of physical therapy. Hospital officials estimated that her bill was running about $200,000. These results are a terrible tragedy, but they could be even worse. Botulism is fatal in many cases.
>
> Mrs. Boberg used a boiling water canner for the carrots that gave her botulism. Yes, this was the same method she had used – and only by luck had gotten away with – for the past 44 years. This year she was not so lucky. If, like Mrs. Boberg, you are canning low-acid foods such as vegetables (except tomatoes), red meats, seafood, and poultry in a boiling water canner or by the open kettle method, you may wish to think twice before taking another chance.
>
> Because native spores of *Clostridium botulinum*, the bacteria that cause the potentially fatal botulism, are extremely difficult to destroy at boiling water temperatures, all low-acid foods should be processed at the much higher temperatures achieved with pressure canners.
>
> (This case history obtained from the Institute of Food and Agricultural Sciences at the University of Florida.)

Others have been shown to be allergic to cheese, wheat, tomatoes, peanuts, and eggs.

Often toxic substances such as pesticides may be applied to a food crop in the field. Toxic substances such as cadmium, lead, and arsenic are naturally present in soil and water and may be absorbed by plants. Toxicologists must determine the acceptable daily intake levels for these substances.

Genetic engineering of food plants has opened a new field of study for toxicologists. Generally, if the new plant's composition has not been altered, or if it is the same or similar to the composition of the original plant, then further study is not undertaken. However, changing the genetic composition of plants to be used as food by consumers must be evaluated in a manner similar to introducing other types of food additives.

Food toxicologists also deal with food safety relative to the presence of microorganisms in the food supply and their possible effects. Food-borne illnesses can be caused by a variety of bacteria such as *Escherichia coli*, *Clostridium botulinum* (see Box 1-5 It Only Takes A Little) and *Staphylococcus aureus*.

Clinical Toxicology

Clinical toxicologists are concerned with diseases and illnesses associated with acute or chronic exposure to toxic chemicals. They include emergency-room physicians who must be familiar with the symptoms associated with exposure to a wide variety of toxic substances in order to administer the appropriate treatment. The type of symptom demonstrated helps to narrow the type of toxic agent involved. Clinical toxicologists rely on a variety of laboratory tests to identify the causative agent and determine whether treatment is successful. In addition, clinical toxicologists contribute to the development of the treatment methods used for individuals exposed to toxic substances.

Checking Your Understanding

1. What is meant by the phrase "dose-response relationship"?
2. Distinguish between acute and chronic exposure.
3. List the various disciplines of toxicology and provide a brief description of the areas of interest associated with each.

Summary

Interest in toxic substances has occurred throughout history. As the world became more industrialized, exposure to chemicals increased along with the number of occupational diseases. New government agencies and laws were created to regulate the manufacturing, use, and disposal of chemicals. Toxicologists play an important role in obtaining and evaluating information relative to chemical exposure and effects. This information is utilized to establish safe exposure levels that are protective of human health and the environment as well as in support of various law enforcement activities.

Application and Critical Thinking Activities

1. As a class project, access some of the chemical databases mentioned in this chapter. What type of information is available for a given chemical on each of the databases? Identify at least three other databases and the types of information available on a specific chemical.

2. Obtain Material Safety Data Sheets for the same chemical from different manufacturers. Does the information vary from one sheet to the other and, if so, how? Compare the information contained in the chemical MSDS with the chemical information contained in one of the electronic databases mentioned in the chapter.

3. Choose ten consumer products and identify what may be considered a food additive, other than those mentioned in this chapter.

4. Identify a chemical accident you may be familiar with – other than the one mentioned in this chapter – and discuss the toxic substance involved and any actual or potential effects as a result of exposure.

5. Much debate has occurred concerning the number of government agencies and the regulations they impose. Organize a class discussion on the various regulations you are familiar with and their impact on you and your classmates. In your opinion, which ones are perceived to be helpful? Which ones may be considered a burden?

6. Choose a food additive, other than those mentioned in this chapter, which has been or still is controversial. Discuss the controversy associated with the additive.

Epidemiological and Toxicological Studies

Chapter Objectives

Upon completing this chapter, the student will be able to:

1. **Explain** the difference between retrospective and prospective epidemiological studies.
2. **Define** and calculate the prevalence rate and the incidence rate.
3. **Explain** the difference between relative risk and attributable risk.
4. **Describe** selection and observational bias.
5. **Differentiate** between epidemiological and toxicological studies.

Chapter Sections

2-1 Introduction

2-2 Epidemiological Studies

2-3 Toxicological Studies

2-4 Epidemiological versus Toxicological Studies

2-1 Introduction

Information concerning the effects of toxic substances is obtained for a variety of reasons. Many government agencies such as EPA and OSHA, discussed in Chapter 1, use the information to set acceptable exposure levels for toxic substances found in the workplace or to set allowable levels in the food we eat, the water we drink, or the air we breathe. Additionally, the information may be utilized to address public health issues and establish public health policy. Information is obtained for these purposes by performing epidemiological (human) or toxicological (animal) studies.

Epidemiological and toxicological studies seek to determine the cause of a particular illness or disease. **Etiology** is the study of causes; in this case, the causes of a disease or illness. The etiological or causative agents of **infectious** diseases are relatively easy to establish. The etiological factor for the common flu is a virus, for pneumonia it is a bacterium, and for athlete's foot it is a fungus. However, it is often difficult to identify the noninfectious factors that may cause a particular illness or adverse effect.

For noninfectious diseases the etiological factor could be any one of a number of toxic substances, such as toxic metals, pesticides, radiation, or some other type of chemical or physical agent, all of which are found in the environment. Only after extensive study and investigation can an association between a noninfectious agent and an observed effect be established. This is particularly true if the information is being derived only from epidemiological studies.

Epidemiological studies focus on "real-world" situations and attempt to determine if there is an association between a specific factor and an observed effect. Toxicological studies occur in a controlled laboratory environment where a known toxic substance is introduced and a specific effect is measured. In each type of study an attempt is made to determine the effects of exposure to a specific etiological factor. The information then may be used to determine a "safe" exposure level for the factor in question, which ultimately reduces risk to the worker in the workplace.

2-2 Epidemiological Studies

Epidemiology is the study of the causative factors that are associated with the incidence and distribution of disease and illness in a specific population. The population may be selected so that it consists of individuals who have a common occupation, are located in a specific community, have the same hobbies, are of the same ethnic group, or have some other common or shared experience. Epidemiological studies have been useful in establishing those factors that can cause cancer, heart disease, stroke, and other illnesses. They are particularly useful because large numbers of people can be studied, and the data obtained are representative of exposures and effects that reflect daily reality. There is less reliance on extrapolating information from toxicological (animal) studies, which use high doses of a toxic substance and various types of experimental animals.

Modern epidemiology had its origin in the study of the outbreak of diseases such as **cholera, bubonic plague,** and **typhus** that occurred in Europe and Asia. John Snow, an English physician, was one of the early pioneers of modern epidemiology. During an outbreak of cholera in 1849, Snow observed that the occurrence of disease was higher in some areas of London than in others. The higher rate was associated with households that obtained their drinking water from two companies, the Lambeth Company and the Southwark and Vauxhall Company. Both companies provided drinking water from a section of the Thames River heavily polluted with sewage. During a second cholera outbreak in 1854, Snow was able to ascertain that a major factor in determining the number of individuals who developed cholera was the source of their drinking water. The Lambeth Company had its water intake moved to a less polluted portion of the Thames prior to 1854. From historical data Snow determined the number of households supplied by each company and calculated the number of deaths from cholera during the first seven weeks of the epidemic, comparing it to the rest of London. His findings indicated a higher mortality rate for individuals obtaining their water from the Southwark and Vauxhall Company (315 deaths per 10,000 households) than individuals who obtained their water from the Lambeth Company (37 deaths per 10,000 households). As a result of the information obtained from this and other investigations on cholera, Snow was able to demonstrate that there was an association between cholera outbreaks and water polluted with sewage.

The causes of noninfectious diseases may also be the subject of epidemiological studies. The relationship between cigarette smoking and the development of lung cancer is well known and controversial. An investigation by E. C. Hammond (1966) was one of the earliest efforts to study the relationship between smoking and the development of lung cancer. More than 1,000,000 participants were involved in the study. Figure 2-1 illustrates some of the results of the study and suggests that the higher the number of cigarettes smoked per day the higher the risk of dying from lung cancer.

More focused epidemiological studies can be performed in relation to certain types of industries. For example, a study was performed by Rinsky and his colleagues (1987) to determine the relationship between benzene exposure from manufacturing rubber hydrochloride and death caused by **leukemia**. Rubber hydrochloride is manufactured by dissolving natural rubber in benzene and spreading it on a conveyor. The benzene is then evaporated and recovered. The thin rubber film is peeled from the conveyor and milled. The group under study consisted of 1,165 men who were involved in the manufacture of rubber hydrochloride from January 1, 1940 to December 31, 1965. Results (Table 2-1) from the study indicated a strong association be-

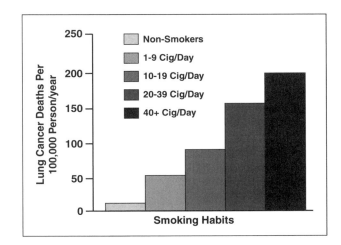

Figure 2-1: Death rates from lung cancer for men smoking different numbers of cigarettes per day. *Adapted from Hammond E.C., "Smoking in relation to the death of one million men and women" in Epidemiological Studies of Cancer and Other Chronic Diseases, National Cancer Institute (NCI) Monograph no. 19.*

Latency (yr)	Exposure (ppm/yr)				
	0.001 - 40	40 - 200	200 - 400	> 400	Total*
< 5	2/0.10	0/0.02	–	–	2/0.12
5 - 10	0/0.16	0/0.05	0/0.01	–	0/0.22
10 - 15	0/0.22	1/0.07	1/0.02	0/0.00	2/0.31
15 - 20	0/0.27	1/0.09	1/0.03	2/0.01	3/0.39
20 - 25	0/0.32	0/0.10	0/0.03	1/0.01	1/0.46
25 - 30	0/0.37	0/0.12	0/0.04	0/0.01	0/0.54
> 30	0/0.40	0/0.16	1/0.04	0/0.01	1/0.62
Total*	2/1.83	2/0.62	2/0.17	3/0.04	9/2.66

a There are two numbers in each column. The first number represents observed deaths from leukemia and the second represents the expected number of deaths from leukemia.
* The numbers of expected deaths have been rounded.

Table 2-1: Observed and expected deaths[a] from leukemia in 1,165 white men with at least one day of exposure to benzene from January 1, 1940 through December 31, 1965 according to cumulative exposure and years of latency. *Obtained and modified with permission from Rinsky R. A. et al., "Benzene and Leukemia: An Epidemiological Risk Assessment," 1987, New England Journal of Medicine 316: 1044-1050.*

tween occupational exposure to benzene and leukemia mortality.

These examples illustrate several important attributes of epidemiological studies.

— First, epidemiological studies usually involve a comparison between two groups of individuals. The groups can be determined in two ways. They may be divided based upon an observed response. One group may exhibit the symptoms of an illness while the other has fewer or no individuals with symptoms of the illness. Or, they may be divided based upon exposure or nonexposure to a particular etiological factor; in this case, the observed effects are recorded for each group.

— Second, epidemiological studies often do not provide information in which a clear and specific cause-effect relationship can be established. Instead, the results often suggest an association between some causative factor or factors and the observed effect. Establishing a direct cause-effect relationship is easier when the disease or illness is caused by an infectious agent, such as the one associated with cholera. In the cholera example there was a strong association between individuals who had cholera and the drinking of sewage-contaminated water. It was relatively easy to conclude that the sewage-contaminated water was the direct cause of cholera. Subsequent investigations led to the discovery of the microorganisms that cause cholera as well as the establishment of a direct cause-effect relationship for this illness. It is more difficult to establish a definitive cause-effect relationship for toxic substances using epidemiological studies. Often there is a long period of time between exposure to the substance and the development of disease, as in the development of cancer or heart disease. In some situations, however, some diseases are more closely associated with a specific type of toxic substance. This is the case with a form of lung cancer called **mesothelioma**. Although this type of cancer is rare, it occurs more frequently in individuals who have been exposed to asbestos. There is a strong association between mesothelioma and exposure to asbestos, but not everyone exposed to asbestos develops mesothelioma. The same is true for individuals exposed to benzene. Although persons exposed to benzene may have higher inci-

> Epidemiological studies are designed to establish an association between an etiological factor and an observed effect.

dences of leukemia, not all individuals exposed to benzene develop leukemia.

— Third, information obtained from epidemiological studies, which attempt to compile information related to past exposures, may be incomplete and inaccurate. Completeness and accuracy of the information depends on the type of epidemiological study being performed. These types of studies rely on the ability of individuals being studied to recall situations where exposure to a causative factor occurred. Information obtained from this method can be highly variable since everyone's memory of past events is different. Also, the level of exposure is either not known or its measurement uncertain. As a result, nothing more than an association between the illness or disease and the causative factor can be established.

Retrospective and Prospective Epidemiological Studies

There are two major types of epidemiological studies, **retrospective** and **prospective**. Briefly, retrospective studies attempt to retrieve information from the past. Often the information is incomplete because of the uncertainties associated with the data collection methods. As a result, it is sometimes difficult to determine if a relationship exists between the observed effect and a specific factor. Prospective studies obtain information from current, ongoing investigations and therefore yield more complete and accurate data than retrospective studies. Both types of studies, however, provide a useful mechanism for identifying adverse effects associated with a given toxic substance.

Retrospective Epidemiological Studies

Retrospective epidemiological studies compare the incidence of illness or disease in two groups of individuals by reconstructing information related to exposure to a toxic substance. These types of studies look at the effects after exposure has occurred and the disease or illness has already developed. It is a look back into the past to determine which factor may be responsible for the observed illness. Another way to describe this type of epidemiological study is, "Why did this happen to me?" What caused the occurrence of cancer, or liver failure, or heart disease, or birth defects? These types of studies are performed more often because they are quicker and cheaper than prospective studies. The group with the disease or illness is compared with a group without it. Each group consists of individuals who are

> Retrospective epidemiological studies evaluate individuals who currently have a specific illness or disease and try to determine the cause by collecting information from past activities.

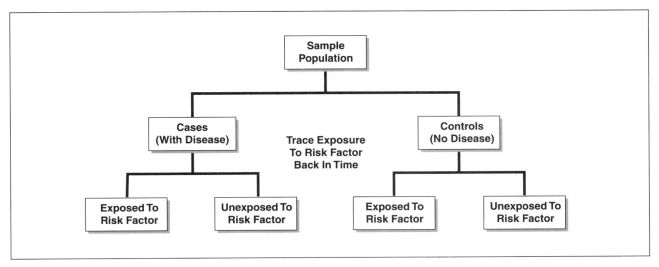

Figure 2-2: A generalized schematic illustrating the experimental design of a typical retrospective epidemiological study. *Obtained from the U. S. Environmental Protection Agency,* EPA Toxicology Handbook, *Rockville, MD Government Institute, 1994.*

very similar in attributes such as sex, age, or race, thereby ideally eliminating variables unrelated to the occurrence of the disease. Figure 2-2 is a generalized schematic representing how retrospective studies are performed. A study population is identified and consists of two groups of individuals. One group has the disease or illness (cases) being studied, the other group does not (controls). These two groups are further divided to evaluate the causative agent of concern. In each group there will be some individuals who were exposed to the causative agent and some who were not. Comparison of the rate of occurrence in each group will determine whether an association exists between the illness or disease and the causative factor being evaluated.

The role of oral contraceptives in causing an increase in the number of women developing blood clots, for example, was determined by performing a retrospective study. A group of 175 women hospitalized with this illness was compared with another group of 175 women hospitalized for different reasons. More women who developed blood clots used oral contraceptives than the group that did not use them. Consequently, it was determined that the use of oral contraceptives may increase the risk of developing blood clots.

There are disadvantages associated with using retrospective studies. Information for these types of

> Information collected during retrospective epidemiological studies is often incomplete because of the variety and types of methods that were used to collect the data and the added issue of confounding factors.

studies depends on reconstructing past events (e.g., working conditions, lifestyle, diet, etc.) that occurred over a prolonged period of time prior to the occurrence of the disease. Information is also collected from documented sources such as medical records, employment histories, or questionnaires. Obtaining information by these methods may be incomplete, inaccurate, and **confounding factors** may not be identified. This is particularly true when the information is obtained by using questionnaires. Individuals who have a particular illness are more likely to remember possible exposures than individuals who do not have the illness. Exposure data may have been collected over a number of years with different standards and methods of collection. Therefore, the data may not be reliable or accurate. Confounding factors are those attributes associated with the individuals being studied that may affect the association between the etiological factor (causative agent) and the measured effect. Confounding fac-

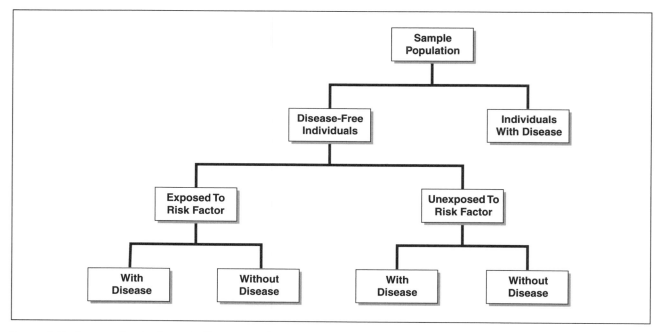

Figure 2-3: A generalized schematic illustrating the experimental design of a typical prospective epidemiological study. *Obtained from the U. S. Environmental Protection Agency*, EPA Toxicology Handbook, *Rockville, MD Government Institute, 1994.*

tors include, but are not limited to, sex of the individuals, ethnic groups, personal habits, specific work conditions, source and form of the etiological factor, and the duration of exposure. For example, if you were studying the effects of a drug on blood pressure, you would want to consider the weight, age, and race of the individuals. Each of these attributes has an effect on blood pressure, independently from the drug's effects, and is considered a confounding factor. In general, the older and heavier an individual the higher the blood pressure. African-Americans, though, typically show a higher incidence of elevated blood pressure, which is associated with genetics.

Exposure to asbestos is another example of a confounding factor. It should be considered when determining the effects of smoking on the rate of lung cancer in a specific population. Individuals who smoke and are exposed to asbestos have a higher occurrence of lung cancer than individuals who only smoke cigarettes. However, exposure to asbestos alone also increases the rate of lung cancer versus nonexposed individuals.

In summary, the disadvantages of retrospective epidemiological studies include: 1) incomplete records of exposure, 2) inability to completely identify or control confounding factors, 3) higher susceptibility to bias due to data obtained primarily from recall, and 4) the fact that only a weak to moderate association between the illness or disease and the causative effect may be established. Its advantages include: 1) more rapid performance of the study due to the fact that the illness or disease already exists, and 2) less costly because it doesn't require following the individuals until the illness or disease occurs.

Prospective Epidemiological Studies

Prospective studies are forward-looking studies. They are designed to determine what will happen to a group of healthy individuals when exposed to the etiological agent of concern. The group is studied for a defined period of time to determine the effects of exposure. Figure 2-3 is a generalized illustration representing how a prospective study is performed. Similarly to a retrospective study, a sample population is identified and divided into an illness or disease-free group, and a group with the illness or disease of concern. However, unlike the retrospective study, only the individuals without the illness or disease are divided into subgroups. One group has been exposed to the causative factor of

> Prospective epidemiological studies evaluate the health of individuals who are exposed to a causative agent to determine if the agent will increase the occurrence of an illness or disease in exposed versus nonexposed individuals.

concern and one has not. Comparison of the rate of occurrence in each group with or without the disease or illness will determine if an association exists between the factor being studied and the observed effect.

Usually a single disease or illness is chosen in order to evaluate the effect of exposure to the agent of concern. For example, there has been considerable interest in whether **electromagnetic fields (EMF)** associated with power lines can cause leukemia. A prospective study could be performed by gathering information concerning the occurrence of leukemia in individuals who live near power lines and those who do not. Both groups would consist of healthy individuals who do not exhibit any symptoms associated with leukemia. The rate of occurrence of leukemia in the group living near the power lines would be compared with the rate of occurrence of leukemia of those not living near power lines to determine if the electromagnetic fields have an effect.

Another hypothetical example of a prospective study would be to determine the influence of drug use on the development of Acquired Immunodeficiency Syndrome (AIDS). Exposure to the AIDS virus as a result of drug use does not generally result in the immediate development of symptoms associated with the disease. Symptoms usually develop several years after exposure. A group of individuals who take drugs by a variety of pathways could be observed over a period of time to determine if AIDS symptoms develop, and if so, which pathway is more likely to result in the occurrence of the symptoms.

Individuals accidentally exposed to a toxic substance would be good candidates for a prospective study. For instance, individuals receiving a single, accidental exposure to benzene could be evaluated to determine if there is an increased incidence of leukemia in the exposed population versus a nonexposed group. Inhabitants of Bhopal, India were accidentally exposed to methyl isocyanate. This chemical can cause severe lung irritation. A prospective study could be performed to evaluate how many exposed individuals develop respiratory diseases versus nonexposed individuals.

Prospective studies have several advantages over the retrospective method: 1) the exposure data are measurable and more reliable, 2) they can identify and attempt to control confounding factors, 3) they are able to measure the time between initial exposure and occurrence of disease or illness, 4) they may be able to establish a stronger association between the illness or disease and the causative agent, and 5) they offer a better control of bias.

The disadvantages of prospective epidemiological studies are: 1) the results are not available immediately – this is particularly true in illnesses such as cancer, which does not appear until many years after exposure has occurred, 2) the cost is higher since study may occur over a period of five, ten, or even twenty years, and 3) due to the duration of the study it is difficult to obtain and maintain information from individuals in the population – individuals may move and not be counted in the final evaluation.

Prevalence Rates and Incidence Rates

One of the primary objectives of epidemiological studies is to determine if there is a cause-effect relationship or an association between a causative agent and some observed effect. The causative agent may be some infectious organism such as a bacterium or a virus, or it may be some type of toxic substance. In order to determine if there is an association between the causative factor and the observed effect, there must be a means to measure the number of individuals who demonstrate the effect and those who do not. Two of the most often used measurements are the **prevalence rate** and the **incidence rate**.

The prevalence rate is a measure of the number of individuals who have a particular attribute relative to the total number of individuals in the specific population being observed. Attribute in this case means a particular illness, disease, or characteristic that is measurable. This relationship can be represented by the formula:

$$\text{Prevalence rate (P.R.)} = \frac{\text{number of people with attribute during a specified time interval}}{\text{total number of people in specific population during the same time interval}}$$

> The prevalence rate measures the number of individuals who demonstrate a particular attribute relative to a given population at a specific point in time.

The prevalence rate is expressed in terms of the number of individuals who exhibit the attribute in a population of 1,000, 10,000, or even 100,000 if occurrence of the attribute is rare. Generally the prevalence rate is expressed as the number of cases per 1,000. For instance, suppose you wanted to identify the prevalence of heart disease in a "community" of 500 people at the end of the year. The "community" may be defined as the population of a small town, or individuals in the workplace, or vegetarians. At the end of the year you identified five people with heart disease. The prevalence rate would be P.R. = 5/500 = 0.01. In terms of the number of people who would have heart disease per 1,000 individuals, the P.R. of 0.01 is multiplied by 1,000. Therefore, you would expect 10 cases of heart disease per 1,000 people.

It is uncertain if there is a higher incidence of cancer in workers at nuclear power plants. Determining the prevalence rate may provide some information concerning the rate of occurrence of cancer in nuclear power plant workers versus workers in other industries. Suppose the number of workers developing cancer was 10 for nuclear power plants and 11 for other industries, and the total population being studied was 500 for each group. The prevalence rate for the nuclear power workers would be:

$$\text{P.R.} = \frac{10}{500} = .02 \text{ or } 20/1,000$$

The prevalence rate for workers in other industries would be:

$$\text{P.R.} = \frac{11}{500} = .025 \text{ or } 25/1,000$$

In this example, comparison of the two rates would suggest that working in the nuclear power industry does not increase the cancer occurrence rate relative to other industries. It is important to remember that the prevalence rate for nuclear power workers does not imply that you will not develop cancer. It simply means that, in this hypothetical example, the occurrence of cancer in the nuclear power plant industry is less than in other industries.

The incidence rate is the number of new cases (individuals) developing the particular attribute being studied – which is identified during a fixed period of time – relative to the number of people at risk. The incidence rate can be represented by the formula:

$$\text{Incidence rate (I.R.)} = \frac{\text{number of new cases in fixed time period}}{\text{total number of people at risk during the same time period}}$$

> The incidence rate is the number of new cases that occur in a fixed period of time relative to the number of people at risk.

There are two major differences between the incidence rate and the prevalence rate. The incidence rate is measured over a specified period of time such as a week, month, or year. The prevalence rate is a measurement at a specific point in time i.e., it is a snapshot in time. Secondly, the incidence rate is a measure of the number of new cases that occur during the specified period of time. The prevalence rate is a measure of the total of all individuals who have the attribute being studied. For example, if in year one, 5 of 500 people developed heart disease, and the number of people who develop heart disease during the second year is 10, then the incidence rate for year 2 is: I.R. = 10/495 = .02 or = 20/1,000. Notice that the number of people at risk is 495, not 500. Since five individuals already had heart disease, they are not included in the total number of people who could possibly develop heart disease during the second year. The total number of people at risk would be 500 if the five individuals did not have the disease at the beginning of the study period. The number with the disease at the end of the second year is 15. Since five individuals had heart disease at the end of the first year, and ten more developed heart disease the second year the prevalence rate is:

$$\text{P.R.} = \frac{\text{number with heart disease after second year (15)}}{\text{number of people who could have disease (500)}} = \frac{15}{500} = .03 \text{ or } 30/1{,}000.$$

Risk

The prevalence and incidence rates are simply measures of the frequency of occurrence of some observed effect that may be associated with a particular etiological factor. Often it is desirable to know the risk of developing a particular illness or symptom associated with the particular etiological factor. As discussed previously, not all individuals exposed to asbestos and benzene developed mesothelioma and leukemia, respectively. However, some individuals do develop these diseases indicating that there is some degree of risk associated with exposure to these substances.

Relative Risk

Both prevalence rate and incidence rate measure the number of exposed individuals who develop a measurable response following exposure to some etiological factor. Not all individuals who are exposed to the etiological factor develop the response being studied. Conversely, individuals who are not exposed to the same etiological factor may develop the symptoms being studied. A comparison of the incidence rates between those who have been exposed and those who have not been exposed provides useful information concerning the probability of an individual developing a response to a given etiological factor. The ratio of the incidence rate for exposed individuals to the incidence rate for nonexposed in-

> Relative risk is an estimate of the probability that a given individual will develop a particular effect when exposed to a specific etiological factor compared to a nonexposed individual.

dividuals is indicative of the chance an individual may develop symptoms associated with exposure and is referred to as the **relative risk**. For example, Hammond's study on lung cancer and smoking – comparing the incidence rates of lung cancer between smokers and nonsmokers – shows that 188/1,000 smokers develop lung cancer while only 19/1,000 nonsmokers do. To determine the relative risk (or the chance of developing lung cancer from smoking cigarettes), the number of individuals who smoke and develop lung cancer is divided by the number of individuals who do not smoke and develop lung cancer, 188/19 = 9.9. Therefore – based on this study – a person who smokes is approximately 10 times more likely to develop lung cancer "relative" (compared) to a nonsmoker.

Similarly, the risk of developing heart disease as a result of eating red meat or a high fat diet versus a vegetarian diet could be evaluated. The relative risk of developing heart disease could be measured by comparing the incidence rates for heart attacks in the two groups. If, for example, the heart attack incidence rates were 5/1,000 for the vegetarians and 35/1,000 for the nonvegetarian group, the relative risk would be calculated as follows: 35/5 = 7. Therefore, the nonvegetarian group would have 7 times the chance of having a heart attack. In general, the larger the ratio, the stronger the association between the observed effect and the suspected etiological factor.

Attributable Risk

Although the relative risk can identify the increased chance of developing a particular symptom associated with exposure to an etiological factor, it is often desirable to know how many individuals will actually exhibit an attribute in direct response to exposure to the etiological factor. The number of individuals who will develop the symptoms as a result of exposure to the etiological factor can be determined by calculating the **attributable risk**. The attributable risk (A.R.) is defined as follows:

A. R. = Incidence in exposed group − Incidence in unexposed group

Using the examples discussed above, the attributable risk from smoking cigarettes would be: incidence of lung cancer in smokers (188) minus the incidence rate in nonsmokers 188 − 19 = 169 per

> Attributable risk is defined by the number of cases that develop a particular illness or disease and that is associated with a specific etiological factor.

1,000. Therefore, based upon these data, we can expect 169 more cases of lung cancer per 1,000 as a result of smoking than we would expect from other causes.

Calculation of the attributable risk can be used to establish safe working conditions. For instance, comparison of the incidence of respiratory illness in dusty and not dusty workplaces could be used to determine if respiratory protective equipment is necessary, or if a new ventilation system is needed. Suppose the incidence of respiratory illness in the exposed group (dusty workplace) is 10/1,000 and in the nonexposed (not dusty workplace) group it is 9/1,000. The attributable risk is calculated as follows: 10 − 9 = 1 per 1,000. This means that we could expect one more person to develop a respiratory illness working in a dusty environment than working in an environment that is not dusty. The company or regulatory agency may decide that the additional cost of providing respiratory protection is not justified if only one person is expected to develop a respiratory illness from working in dusty conditions.

Bias

One of the major difficulties associated with epidemiological studies is the high degree of variability associated with human subjects. Attempts are made to compare two groups of individuals that are identical or very similar; however, this is not always possible. As a result, some **bias** may be introduced into the study, which could potentially result in a false association being developed between the etiological factor and the measured response. Although there are many different types of bias, only two of them, **selection** and **observational bias,** will be discussed here.

The previously discussed confounding factors can also introduce bias, particularly if they are unknown. These factors can cause a weaker or stronger association to be drawn between the causative agent and the illness or disease than really exists, if they are not identified and accounted for when performing the study and evaluating the results.

Selection Bias

Usually two groups of individuals are selected to determine if a particular attribute is more prevalent in the group exposed to a particular etiological factor than in the nonexposed (control) group. The choice of the two groups must be done with care to assure that the individuals studied are truly representative of the population of "exposed" and "nonexposed" individuals. For example, suppose a study was undertaken to determine if stressful jobs, such as doctors or company presidents, resulted in a higher incidence of heart disease. One group would consist of individuals from one of those occupations, the other group would consist of individuals that represent a cross-section of other occupations. However, if the occupations chosen are sedentary in nature, then there is an increased probability that the individuals in the control group may be overweight. Therefore, because of the choice of occupations, there is a greater chance (bias) that an overweight individual will be selected. There is an increased chance of developing heart disease if an individual is overweight regardless of occupation. In this case, the difference between the number of individuals who develop the disease in each group will be smaller; as a consequence, the association between stress and heart disease would appear less significant. If the difference is small enough, then it may be falsely concluded that stress does not result in an increased incidence of heart disease.

Similarly, suppose you wanted to study the effects of nutrition on the occurrence of birth defects in newborns. Selection bias would be introduced if the individuals in the two groups had a large number of single teenage mothers. Many of the individuals may not have the resources to adequately obtain proper nutrition during pregnancy. The incidence rates would be similar and, therefore, the relative risk would be low, suggesting there is a low risk of developing birth defects as a result of poor nutrition. In reality, several studies have shown that poor nutrition does cause birth defects. For instance, lack of the B vitamin, folic acid, can increase the chance of developing spinal cord malformations. Decreased intelligence has also been associated with poor nutrition during pregnancy.

Selection bias would also be introduced if we were studying the relationship between asbestos exposure and lung cancer and only surviving individuals were included in the study. Individuals exposed to asbestos and no longer living would need to be included as well. The cause of their death would have to be known in order to determine if asbestos-induced cancer was the cause.

Observation Bias

Observation bias is concerned with the means by which information is obtained and analyzed. Retrospective epidemiological studies are particularly susceptible to this type of bias. As mentioned previously, these types of studies are performed by obtaining information from the past. It is likely that an individual who develops a particular illness or disease is going to give more thought to the cause than someone who did not develop the illness or disease. However, both individuals may have been exposed in a similar manner. There is also a chance the investigator may place one of the individuals in the exposed group and the other in the nonexposed group based only upon the presence or absence of the illness without evaluating the exposure data.

A good example of this type of bias is using death certificates as a source of data for identifying the group of individuals who die of cancer. There is no doubt that cancer causes death. However, death may occur not from cancer itself but rather as a result of heart or respiratory failure. Therefore, examination of death certificates can be misleading. Although some individual certificates may indicate that the cause of death is cancer, others may indicate heart or respiratory failure as the cause and not mention cancer. As a result, these individuals would not be part of the study group.

In both situations described above, the introduced bias could lead to a false association. In the first situation, because those individuals who develop the disease are more likely to recall exposure, the association could appear to be stronger because those individuals who do not develop the illness or disease may not be included in the exposed group. This would result in a large proportion of individuals who

---

Selection bias occurs when groups of individuals are not comparable in all respects except the attribute being studied.

Observation bias occurs as a result of the misidentification of individuals based upon the presence or absence of the attribute being studied.

have the illness or disease being included in the exposed group to the exclusion of individuals who had equal exposure but no illness or disease. The second case is very similar to the first except that the investigator has introduced the bias by assigning individuals to a group based solely on the presence or absence of the illness or disease, which in turn is data derived from incomplete or misleading documentation.

Checking Your Understanding

1. Define etiology and explain its relationship to epidemiological and toxicological studies.
2. What is an infectious disease? Give an example.
3. What substances might cause noninfectious diseases or illnesses?
4. Define the term epidemiology.
5. What characteristics of a population may be used to identify individuals to be included in an epidemiological study?
6. What was the contribution of John Snow to the field of epidemiology?
7. What are three attributes that characterize epidemiological studies?
8. What are some of the advantages and disadvantages of retrospective and prospective epidemiological studies?
9. What is the relationship between relative risk, attributable risk, prevalence rate, and incidence rate?
10. Define selection bias and observation bias. Explain how these factors can affect the cause-effect association identified by an epidemiological study.

2-3 Toxicological Studies

The high degree of variability associated with human subjects and the uncertainty associated with the data collection process make it difficult, at times, to clearly establish a relationship between an etiological factor and a specific illness or disease when performing epidemiological studies. The relative risk and the attributable risk are indicators of the probability that an individual exposed to a given etiological factor may develop the specific effect being studied. However, the information obtained from these types of measurements does not clearly define a specific cause-effect relationship. To determine if an etiological factor is directly responsible for an observed response, it is important to be able to demonstrate that the factor produces the specific effect observed at the specific concentration of concern. Also, it must be demonstrated that there is a significant increase in the number of individuals who develop the response as a result of exposure compared to those who develop the response and are not exposed. Toxicological studies are designed to establish a cause-effect relationship between a specific toxic substance and a measured response. In general, these types of studies are performed in a controlled environment using a specific type of test animal. The test animal is exposed to a range of chemical doses and the observed effect is measured. The toxicological study may be short term (**acute toxicity** study), which is usually less than 14 days in length; or it may be of longer duration (**chronic toxicity** study), lasting more than 14 days, perhaps weeks, months, or years. The route of exposure may be by dermal exposure, inhalation, ingestion, or injection. The frequency of exposure varies. The test animal may be exposed to a single, high dose of the toxic substance (acute exposure) or to continuous or multiple doses over the time of the study (chronic exposure).

The design of the toxicological study should mimic as closely as possible the conditions identified during the epidemiological study. As such, comparison of the toxicological data with the epidemiological data will have a higher degree of validity and the conclusions concerning cause-effect will be more supportable. This information can then be utilized to help establish "acceptable" exposure levels for the toxic substance being studied. A more detailed discussion of the methodology used in toxicological studies occurs in Chapter 4.

> Toxicological studies are controlled investigations on animals designed to establish a direct cause-effect relationship.

Checking Your Understanding

1. Why are toxicological studies performed?
2. List some of the characteristics of toxicological studies.
3. Differentiate between chronic and acute toxicity.

2-4 Epidemiological Studies versus Toxicological Studies

Information derived from both epidemiological and toxicological studies is important in determining exposure levels. It is important to evaluate the information based upon the method used and the reliability of the data. Combining information from both types of studies for a given etiological factor will provide a more realistic understanding of the potential effects associated with exposure.

According to A. W. Hayes in *Principles and Methods of Toxicology*, the advantages of epidemiological studies are: 1) the exposure conditions are realistic, 2) there is occurrence of interactive effects among individual chemicals, 3) the effects are measured in humans, and 4) the full range in human susceptibility is frequently expressed. The disadvantages are: 1) they are costly and time consuming, 2) they are post facto and not protective of health, 3) they pose a difficulty in defining exposure and pose problems with confounding factors, and 4) the effects measured are often relatively crude, i.e. mortality.

The advantages of animal toxicological studies are: 1) the exposure conditions are controlled, 2) they give the ability to measure many types of responses, 3) they give the ability to assess effect of host characteristics (e.g., gender, age, genetics) and other modifiers (e.g., diet) of response, and 4) there is the potential to evaluate mechanisms associated with the development of the symptoms. The disadvantages are: 1) there are uncertainties in the relevance of animal response to human response, 2) controlled housing, diet, etc. is of questionable relevance to humans, and 3) the exposure concentrations and time frames are often very different from those experienced by humans.

Establishing a Cause-Effect Relationship

Epidemiological studies only identify an association or correlation between a toxic substance and some observed effect. Although the association may be strong, a definitive link between cause and effect is difficult to establish, usually because of the number of confounding factors associated with the study. These factors can result in incomplete and misleading information. Toxicological studies are designed

> Epidemiological studies establish an association between an etiological factor and the observed effect. Toxicological studies establish a direct cause-effect relationship.

to establish a direct cause-effect relationship. Usually the test animal is exposed to a single toxic substance and a specific effect recorded. Since most of the other environmental variables are controlled, it is easier to establish the direct cause-effect relationship.

Controlled Study versus Noncontrolled Study

Toxicological studies are performed in such a manner that the confounding factors that could potentially affect the toxicity of the etiological agent are controlled. By this we mean that all the test animals are maintained in the same environment, i.e., they are all in the same area, eat the same food, breathe the same air, and are exposed to the same toxic substance by the same route. In addition, the test animals are selected in a way that other variables, such as age, sex, health, and type, are the same.

Confounding factors in epidemiological studies for the most part are not controlled and, as discussed previously, make it difficult to establish a definite cause-effect relationship. The population is usually much more diverse, made up of individuals of different ages, sex, or ethnic background; also, some are in good health while others are not. In addition, their diets are varied and they may be exposed to the toxic substance by a variety of different pathways. All of these factors can have an effect on toxicity and, consequently, on the observed response.

> Identification and control of confounding factor allows establishment of a cause-effect relationship.

Exposure: Routes, Concentration, Duration, and Frequency

Exposure of the test animals during a toxicological study may occur as a result of inhalation, dermal **absorption**, ingestion, or as the result of injection. Usually only one route is chosen. Regardless of which route is chosen, the variability associated with the concentration of the toxic substance, duration of exposure, and frequency of exposure are all eliminated by strict and controlled laboratory protocol. Control of these factors is important in establishing a direct cause-effect relationship.

In many cases, particularly in retrospective epidemiological studies, the concentration, frequency, and duration of exposure are either not known or are estimated. Although an apparent route of exposure may be easily identified, it is not always clear that it is the only one. The performance of prospective epidemiological studies allows some information concerning routes of exposure, concentration, frequency, and duration of exposure to be documented accurately and controlled to some extent. However, some of the confounding factors that could affect the chemical toxicity may not be known.

Exposed Individuals

Many different types of test animals are used in toxicological studies. The most frequently used animals are rats and mice. Many of the confounding factors such as sex, age, and health – related to the test animal – can be controlled. This helps minimize the number of variables that could affect chemical toxicity as well as the ability to accurately extrapolate animal data to humans. The human population is highly variable, and a direct correlation between the results obtained from animal studies and the human population is therefore not always possible. Also, the selection of mice or rats may not be the best model to indicate the effects that may occur in humans because these animals may respond differently than humans. This is discussed in more detail in Chapter 4.

Epidemiological studies focus on a specific human population. Although there is a high variability in the type of individuals studied, it is more representative of the "real world." If a particular etiological factor can be associated with a specific effect, the data from this type of study is useful in developing a cause-effect relationship and in determining exposure levels that are reflective of the "real world."

Extrapolation of Data

Most toxicological studies are performed in a controlled environment and are not very representative of exposure to toxic substance in everyday life. In addition, the dose to which a test animal is exposed is usually much higher than the one encountered in the real world. In many situations it is not clear whether the toxic substance causes the same response both at low and high exposure levels.

Conversely, toxic effects observed in epidemiological studies are usually those associated with the lower doses. However, the presence of confounding factors and incomplete information concerning exposure make it difficult to draw definitive conclusions concerning the effects resulting from exposure at these levels.

Epidemiological and toxicological studies have their advantages and disadvantages, and useful information can be derived from each type. However, this information must be evaluated carefully. The circumstances under which the data are obtained can affect the interpretation of the results.

Checking Your Understanding

1. Explain what is meant by the phrases: "Epidemiological studies establish an association between an etiological factor and an observed response," and "Toxicological studies establish a cause-effect relationship."

2. What are some of the differences and similarities between epidemiological and toxicological studies?

3. Under what circumstances would effects from a toxicological study be predictive of effects that you would see as a result of exposure to a toxic substance in the "real world"?

Summary

Information obtained from epidemiological and toxicological studies is used to establish exposure levels for toxic substances. Epidemiological studies establish an association between the causative factor and the observed effect. Retrospective epidemiological studies reconstruct information from the past. The objective of these studies is to try to determine the causative factor by evaluating individuals who have developed a particular illness. Prospective epidemiological studies obtain information from current, ongoing investigations. Healthy individuals are studied to determine if they develop a specific illness as a result of exposure.

The prevalence rate is a measure of the number of individuals who have a particular attribute relative to the total number of individuals in the specific population being studied. The incidence rate is the number of new cases having a particular attribute relative to the total number of people who could develop the attribute. Relative risk is an estimate of the probability that a given individual will develop a particular illness or disease when exposed to a specific etiological factor compared to a nonexposed individual. Attributable risk is the incidence rate of the exposed group minus the incidence rate of the nonexposed group. It represents the number of illnesses associated with exposure to the causative agent. Selection bias occurs when individuals are not representative of the exposed and nonexposed populations. Individuals are chosen with an attribute that affects the illness or disease being studied but is not the specific attribute being evaluated. Observational bias occurs when errors are made concerning the presence or absence of symptoms or data, resulting in the selection or elimination of individuals who may or may not be representative of the populations being studied. This may be caused by incorrect classification of individuals based on incomplete or inaccurate data.

Toxicological studies are designed to establish a direct cause-effect relationship. They are performed in a controlled environment where all individuals are exposed to the same environmental conditions. In toxicological studies, confounding factors are identified and accounted for; only the etiological factor under investigation is changed. Epidemiological studies are performed in a noncontrolled environment where confounding factors may not be identified; this may affect interpretation of the results. Consequently, data concerning exposure frequencies, duration, and concentration may be incomplete or inaccurate. Therefore, a definitive cause-effect relationship may not be established. Neither study by itself is fully representative of actual exposures. Collectively the information from both types of studies must be evaluated to set exposure levels.

Application and Critical Thinking Activities

1. A group of 50 individuals on a cruise ship contracted a respiratory disease called Legionnarie's disease; 20 eventually died and the rest recovered. The remaining 1,150 passengers did not contract the disease. In order to determine the cause would you perform a retrospective or prospective epidemiological study? Why? What confounding factors might you have to consider when determining why some individuals contracted the disease and others did not?

2. Sick-building syndrome is a relatively new illness associated with workers who occupy newly constructed and furnished office buildings. How would you determine whether worker illnesses were associated with sick-building syndrome or not? Include in your answer a discussion on how incidence rate and attributable risk were used to answer the question.

3. You have been given the assignment to try and determine the causes for two separate illnesses: heart disease and open sores that develop on the skin of individuals who swim in the waters of the southeast coast of the United States. For each illness indicate whether you would perform an epidemiological or a toxicological study. Briefly explain your answer.

4. It has been speculated that spontaneous abortions in women might occur as a result of exposure to a wide variety of environmental factors such as lead, smoking, alcohol, halogenated hydrocarbons, or photographic developers. Develop a table that shows the number of individuals included in your study, the number of successful pregnancies, and the number of spontaneous abortions for each of the environmental factors listed above. Based upon the numbers you provided, calculate the incidence rate for

each environmental factor. Exposure to which of the environmental factors has a greater chance in causing a spontaneous abortion?

5. Suppose you were assigned the task of determining whether a specific illness is caused by an infectious or a noninfectious agent. Briefly explain how you might perform this task.

6. Explain the differences and similarities between retrospective and prospective epidemiological studies. Give an example of each.

7. What are confounding factors? How can these factors affect the cause-effect association established by an epidemiological study?

8. Define the terms prevalence rate and incidence rate. Can the prevalence rate and the incidence rate be the same? Why or why not? Give an example to support your answer. Include calculations to support your position.

9. You and your partner are conducting a study to determine how skin pigmentation affects the development of skin cancer. Both of you are choosing individuals to be included in the study. The individuals will be placed in separate groups based on the color of their skin. Would results of this study be more likely affected by selection bias or observation bias? Explain your choice.

10. Exposure to asbestos can result in fibrosis, a lung disease. Given the following information calculate the prevalence rate and incidence rate of fibrosis after 4 years, 14 years, and 20 years. What is the relative risk of developing this disease?

Years Employed	Number of Individuals Examined	Cases of Fibrosis
0 - 4	89	0
5 - 9	141	36
10 - 14	84	27
15 - 19	28	15
20 and over	21	17
TOTAL	363	95

3

Routes of Exposure

Chapter Objectives

Upon completing this chapter, the student will be able to:

1. **List** and explain the factors that affect absorption of toxic substances across the cell membrane.
2. **Compare** and contrast passive and active transport processes.
3. **Identify** the major routes of exposure.
4. **Describe** the general anatomy of skin, gastrointestinal tract, and lungs.
5. **Explain** the factors for each of the major pathways of exposure that reduce or increase the rate of absorption of toxic substances.
6. **Explain** the difference between local and systemic effects.

Chapter Sections

3–1 Introduction

3–2 Basic Cell Structure and Function

3–3 Pathways of Exposure

3–4 Local versus Systemic Effects

3-1 Introduction

It is important to distinguish between exposure to and absorption of a toxic substance. Exposure to toxic substances occurs on a daily basis through a variety of pathways. The most common routes of exposure are inhalation, ingestion, and skin contact. However, exposure alone does not determine whether the substance will have toxic effects. In general, for any substance to have its effects in an organism it must first be absorbed. **Absorption** occurs when a substance passes into a cell or into the circulatory system regardless of the route of exposure. It is entry into the cell or circulatory system that initiates those processes associated with distribution, metabolism, and elimination of the substance and the development of any toxic effects.

In order to understand how a toxic substance may exert its effects, it is important to have a basic understanding of cellular structure and function.

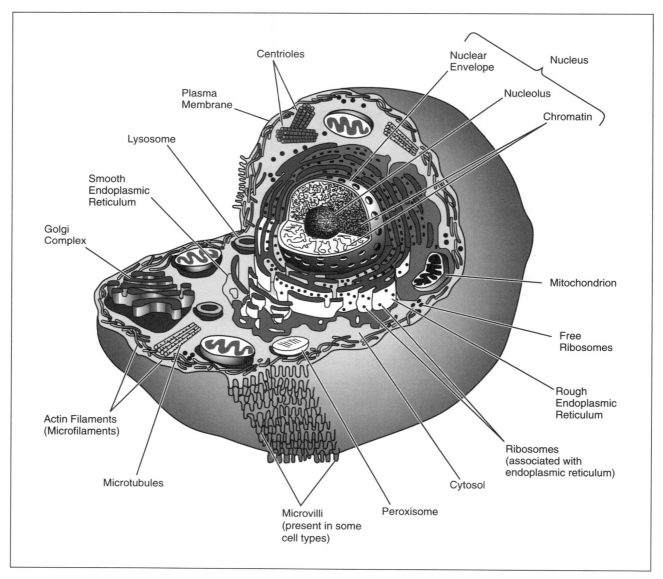

Figure 3-1: Three-dimensional model illustrating some of the basic cell structures of a "typical" animal cell. Depending on cell function, the number and extent of the structures will vary. *From* Principles of Cell and Molecular Biology, *2nd edition, by Lewis J. Kleinsmith and Valerie M. Kish. Copyright © 1995 by HarperCollins College Publishers. Reprinted by permission of Addison Wesley Educational Publishers.*

Both the structure and the functions of many different kinds of cells influence how quickly toxic substances are absorbed and what their effects will be.

The following is a general discussion of cell structure and function focusing primarily on the **plasma cell membrane**, the **endoplasmic reticulum (ER),** and the **mitochondrion**. Figure 3-1 is a generalized illustration of the cell. These structures are important in absorption, metabolism, and elimination of toxic substances.

> Exposure to toxic substances occurs daily; however, absorption of the toxic substance must occur in order for toxic effects to be observed.

3-2 Basic Cell Structure and Function

The plasma cell membrane is a thin structure that separates one cell from another as well as the inside of the cell from its external environment. As we can see in Figure 3-2, the membrane is composed primarily of **phospholipid** molecules (75 percent), **proteins** (20 percent), and **carbohydrates** (5 percent). The phospholipid molecules are arranged in two

> Plasma membrane is composed primarily of phospholipids in parallel layers. Proteins are found throughout the cell membrane and on both surfaces. The proteins are important in regulating movement of substances across the cell membrane.

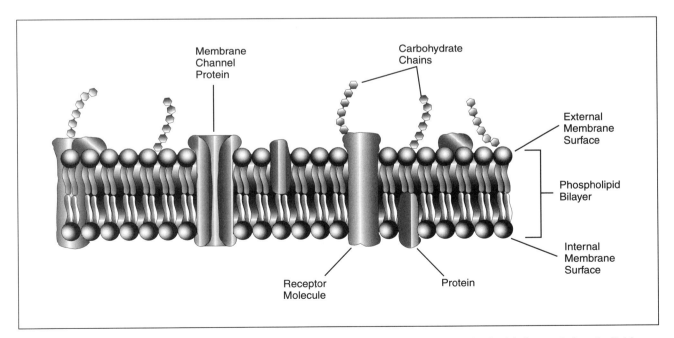

Figure 3-2: Typical structure of the cell membrane. The cell membrane is composed of a double layer of phospholipid molecules with proteins "floating" in the membrane. Proteins in the membrane can form membrane channels that act as carrier molecules or as receptors.

parallel layers. The proteins appear to "float" within the double layer of phospholipid molecules, either singly or in pairs. Some proteins are also located on both surfaces of the membrane. These proteins can act as cell receptors, or as **carrier molecules** that attract and transport substances across the cell membrane. They may also form "channels" through which water-soluble molecules and electrically charged **ions** may pass. The size of the channels varies and determines which substances will cross the cell membrane.

Cell Membrane Permeability

The plasma membrane regulates the entrance and exit of the materials into and out of the cell. It is freely permeable to some substances such as water, but it prevents passage of other materials. The membrane is therefore referred to as being **selectively permeable**.

> Selective permeability is a property of the cell membrane that allows some substances to cross the cell membrane and prevents the crossing of others.

There are four major factors affecting the ease with which a substance will move across the cell membrane:

1. Size of the molecule: Large-sized molecules, such as proteins, cannot readily enter or leave the cell. The molecules may be too large to "squeeze" between the molecules that compose the cell membrane.

2. Solubility of molecule in **lipids** (fats): As discussed previously, the cell membrane is composed of parallel layers of phospholipids. Therefore, molecules that easily dissolve in lipids readily move through the cell membrane. For example, ethanol, a lipid-soluble molecule, is easily absorbed by cells lining the stomach and the intestine. Xylene (paint thinner) is easily absorbed by the skin.

3. Electrical charge of ions and molecules: Many substances have electrical characteristics. The cell membrane, through which the substances are trying to enter, also has an electrical charge associated with it. These charges affect the ease with which a substance can move across a membrane. For instance, a positively charged portion of the membrane will repel a positive ion, decreasing the rate of absorption of that ion; however, an ion with a negative charge would be more readily absorbed.

4. Carrier molecules: Special molecules (proteins) in the cell membrane can attract and transport substances across the cell membrane. This occurs regardless of size, lipid solubility, or electrical charge.

> The selectively permeable nature of the cell membrane is influenced by: the size, solubility, charge of the material being transported, and the presence of carrier molecules within the membrane.

Endoplasmic Reticulum

The endoplasmic reticulum is a membranous system located within the cell and continuous with the cell membrane (Figure 3-1). The ER is associated with the **intracellular** (within the cell) **transport**, storage, and **secretion** of molecules. Many chemical reactions occur on the surface of these membranes. In most cells the ER is associated with other cellular structures called the **ribosomes**. Ribosomes are the site of protein synthesis. **Enzymes** and other proteins synthesized by the ribosomes may play an important role in the active and passive transport processes associated with the cell membrane as well as in the biotransformation of toxic substances.

Mitochondria

The mitochondrion is variable in shape. There are spherical and rod-shaped mitochondria (Figure 3-1). The mitochondrion is a double-membrane structure; its inner membrane is the site of many important chemical reactions, the most important of which are those associated with the production of energy in the form of molecules called **Adenosine Triphosphate (ATP)**. The energy associated with these molecules is important in the functioning of cellular transport mechanisms and protein synthesis associated with the ribosomes.

> ATP is the primary source of energy required by living things.

Passive and Active Transport Mechanisms

All the factors discussed above will determine collectively if a substance will be absorbed by a cell. The actual movement of the substance across the cell membrane will be the result of either **passive transport** or **active transport** mechanisms. Because the cell cannot always distinguish between the substances needed to function normally and those that may be potentially harmful, toxic substances may be transported into the cell by the same cellular mechanisms.

Passive Transport

Passive transport does not require the expenditure of energy by the cell to move a substance across the cell membrane. The substance moves in response to a **concentration gradient** that exists across the cell membrane. **Diffusion, osmosis,** and **facilitated diffusion** are all examples of passive transport.

> Concentration gradients result in passive movement of substances across the cell membrane.

Diffusion is the result of constant, random movement of molecules, atoms, or ions. This motion occurs in all directions regardless of the concentration. However, because there are more molecules in an area of higher concentration than in an area of lower concentration, the chances are

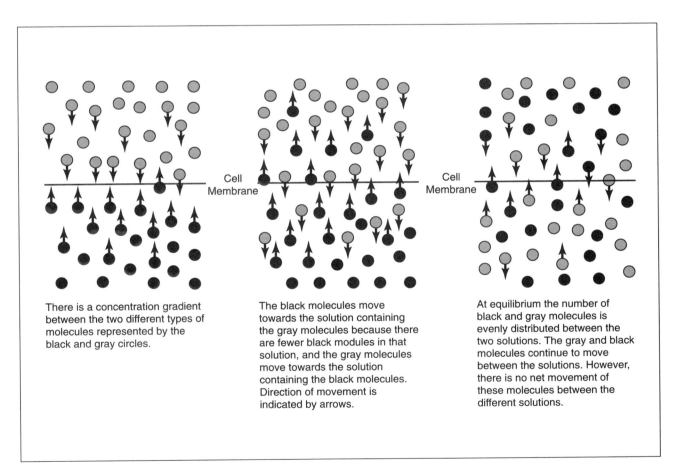

Figure 3-3: Diffusion.

greater that there will be a net movement of molecules from the higher to the lower area of concentration. At **equilibrium**, the net movement of molecules stops, and the movement of molecules in any one direction is equal to the movement of molecules in the opposite direction (Figure 3-3). An example of diffusion is the movement and distribution of perfume throughout a room, or the movement of oxygen from the lungs to the blood and the movement of carbon dioxide from the blood to the lungs. Inhaled toxic substances such as carbon monoxide and diethyl ether pass from the lungs to the bloodstream as a result of diffusion. In each example the molecules are moving from an area of higher concentration to an area of lower concentration.

> Diffusion is the net movement of molecules from an area of higher concentration to an area of lower concentration.

Osmosis is the passage of water (a solvent) across a selectively permeable membrane, from a region of higher water concentration to a region of lower water concentration. The selectively permeable membrane prevents passage of – or is less permeable to – substances dissolved in the water. Figure 3-4 illustrates this principle. Sugar molecules are dissolved in the water contained inside the funnel. There are no dissolved sugar molecules in the water contained in the beaker. Therefore, there are "fewer" water molecules inside the funnel than outside. As a result, the water moves freely from the beaker into the funnel. The water is moving from an area of higher water concentration to an area of lower water concentration. The selectively permeable membrane does not allow the sugar to move into the container with pure water.

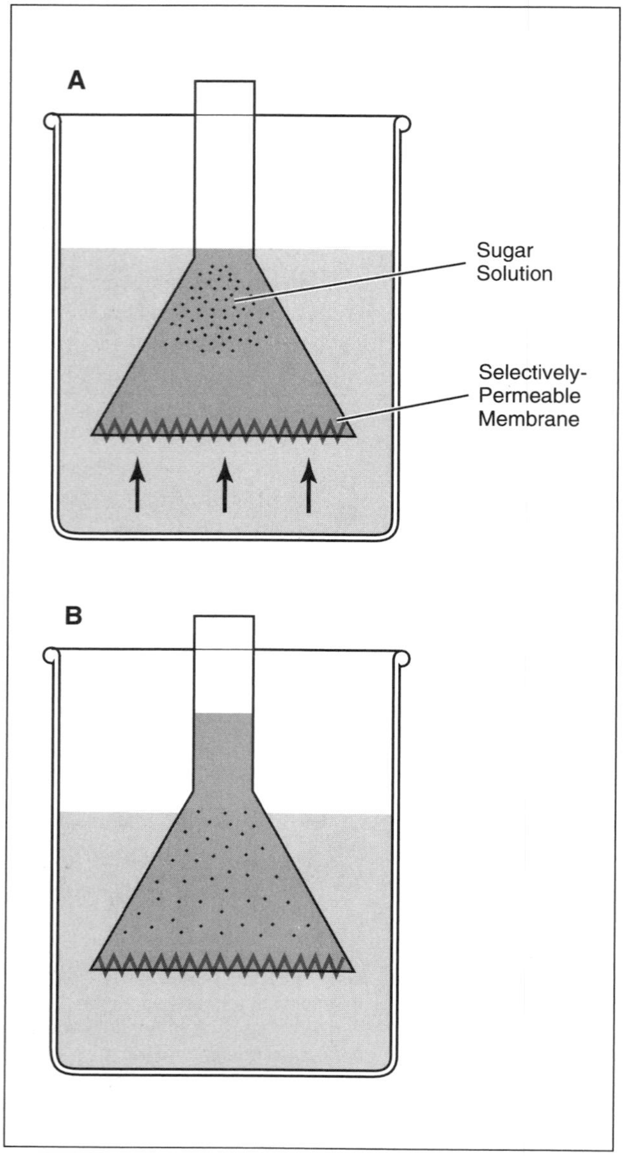

Figure 3-4: Osmosis is the net movement of water from an area of higher water concentration to an area of lower water concentration. The water inside the beaker contains no dissolved sugar, while the water inside the funnel does. Therefore more water moves into the funnel than out.

> Osmosis is the net movement of water (a solvent) molecules from a region of higher water concentration to a region of lower water concentration.

Facilitated diffusion is characterized by the movement of substances across the cell membrane using protein carrier molecules that are embedded in the cell membrane. How this mechanism works is uncertain, but it is believed that a carrier protein attaches to a substance on one side of the cell membrane and then undergoes a change in shape, thus allowing transport of the substance to the other side of the cell membrane (Figure 3-5). As in the case with diffusion and osmosis, the movement of substances by facilitated diffusion also takes place from an area of higher to an area of lower concentration. Glucose moves across the cell membrane by facilitated diffusion.

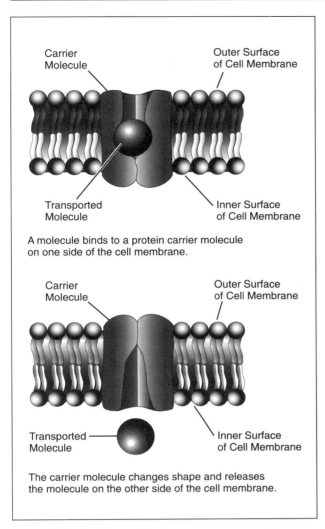

Figure 3-5: Carrier-mediated transport across cell membrane.

> Facilitated diffusion is the net movement of a substance, using a carrier molecule, across a cell membrane.

Active Transport

Active transport consists of several steps in which a carrier molecule attaches to a substance located outside the cell. The carrier, with the attached substance, then moves across the cell membrane and releases the attached substance on the other side of the cell membrane. In some cells the protein carrier, after releasing the substance, may attach to a different type of molecule inside the cell, move it across the cell membrane, and release it on the outside of the cell. This movement requires the expenditure of energy (Figure 3-6) and allows the cell to move substances across the cell membrane against a concentration gradient (from lower concentrations to higher concentrations). This mechanism is also thought to be important in the absorption of metals in the digestive tract.

> Active transport mechanisms expend energy and move substances against a concentration gradient.

Phagocytosis is another process where a particle, located outside the cell, can be transported into the cell. This process also requires the expenditure of energy. The particle is surrounded by extensions of the cell called **pseudopodia**. These cellular extensions fuse, forming a membranous sac – called a **vesicle** – around the particle and bring it inside the cell (Figure 3-7). Phagocytosis is an important mechanism used by white blood cells to remove or destroy harmful substances, such as asbestos or silica, which may enter the respiratory system.

Pinocytosis is similar to phagocytosis in that the cell membrane is involved in moving substances from the outside to the inside of the cell, also with an expenditure of energy. However, pinocytosis is primarily used to move liquids. Pseudopodia do not form. The cell membrane simply folds inward in response to contact with liquids, forming a vesicle that subsequently releases the liquid into the cell. This is a common transport mechanism in the small intestine. Solutions containing metals such as cadmium and lead may be absorbed by pinocytosis.

> Phagocytosis and pinocytosis are mechanisms that transport substances located outside the cell to inside the cell with expenditure of energy.

In addition to the structural and functional properties of individual cells affecting absorption, the arrangement of the cells in the various tissues can also affect absorption. The cells lining the interior cavities of the lungs, the digestive tract, the blood vessels, the exterior surface of organs, the interior walls of the body cavity, as well as the cells that compose the outer layer of the skin, are referred to as **epithelial cells**. These cells are closely packed with little or no intercellular space between them, thereby decreasing the potential for absorption between the

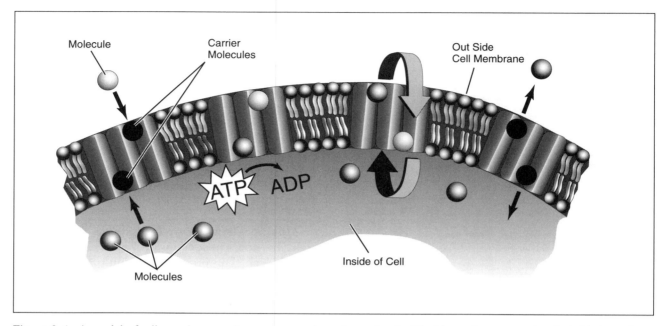

Figure 3-6: A model of cell membrane active transport. A carrier molecule (black) attaches to a molecule (white) on the outside of the cell membrane and then releases the molecule on the inside of the cell. The carrier molecule may attach to a molecule inside the cell and transport it to the outside of the cell. This mechanism requires the expenditure of energy – which is represented by the conversion of ATP to ADP – transporting substances against a concentration gradient. *From Biology: The World of Life, 6th edition, by R. A. Wallace, Copyright © 1992 by HarperCollins College Publishers Inc. Reprinted by permission of Addison-Wesley Educational Publishers.*

cells. In some cases the epithelial cells are arranged into several layers, such as in the skin and parts of the digestive tract, making absorption of a toxic substance into the bloodstream more difficult. In other parts of the digestive tract and in the lungs the epithelial cells are arranged in a single layer, and absorption occurs readily. These characteristics, along with those associated with the cell membrane, will collectively determine the rate and amount of absorption of a toxic substance and its subsequent toxic effects.

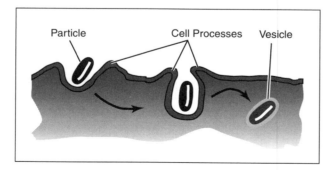

Figure 3-7: Phagocytosis. The particle to be taken into the cell is surrounded by cellular processes (pseudopodia). The processes fuse to form a vesicle that contains the particle.

Checking Your Understanding

1. List the factors that affect the absorption of toxic substances across the cell membrane.

2. What are two major differences between active and passive processes involved in movement of substances across the plasma membrane?

3. Define the term equilibrium. How does equilibrium affect the transport of substances across the cell membrane?

4. Define and give an example of each of the following: diffusion, facilitated transport, osmosis, active transport, phagocytosis, and pinocytosis.

5. Name and describe three cell structures that may be involved in the transport of toxic substances into the cell. How are the functions of these three structures interrelated in terms of the transport of toxic substances?

6. What are epithelial cells? How might they affect absorption of toxic substances?

3-3 Pathways of Exposure

As mentioned earlier, exposure to toxic substances usually occurs by one of three major pathways: dermal absorption, inhalation, and ingestion. The eye is a particularly vulnerable organ that is often exposed to various types of substances in the workplace and in the home. Exposure to toxic substances can damage the eye causing impaired vision or blindness. Therefore, a general discussion of the structure of the eye is included. Specific effects as a result of exposure will be discussed in Chapter 6.

Dermal Absorption

The most common path of exposure to toxic substances is by contact with the skin. The skin is composed of three main parts: 1) the outer layer composed of several cell layers, called the **epidermis**; 2) an inner and thicker layer, the **dermis**, which contains a number of structures including blood vessels, hair follicles, sweat, and **sebaceous** (oil) **glands**; and 3) the fatty, **subcutaneous** layer (Figure 3-8).

> Skin is the most common route of exposure.

Anatomy of the Skin

Epidermis – The outermost layer of the epidermis is called the **stratum corneum** (cornified layer). It consists of several layers of cells that are derived from the underlying basal layer (Figure 3-8). Cells in the basal layer continuously divide, with the older cells migrating toward the stratum corneum, where they become flattened and scale-like. The cells of the epidermis do not contain any blood vessels; they must therefore obtain their oxygen and nutrients from the blood vessels located in the underlying dermis. The lack of a direct blood supply is a contributing factor to the death of the cells as they migrate to the surface of the epidermis where they begin to degenerate and die. As the cells die, a specialized protein called **keratin** forms from proteins interacting with enzymes located in the cells. Keratin gives the epidermis its rough, impermeable characteristic. It is the most insoluble of all proteins and is extremely resistant to a variety of substances and environmental conditions such as heat, cold, and changes in pH. The ability of a toxic substance to penetrate the stratum corneum determines the rate at which the substance will be absorbed by the skin. For example, an organophosphate pesticide such as malathion, which easily penetrates the stratum corneum, moves easily through the remaining layers in the skin and is rapidly absorbed into the bloodstream. On the other hand, an organochlorine pesticide such as DDT does not easily penetrate the stratum corneum; the rate of absorption is therefore much slower.

> Stratum corneum is the structure that determines the rate of absorption of substances through the epidermis.

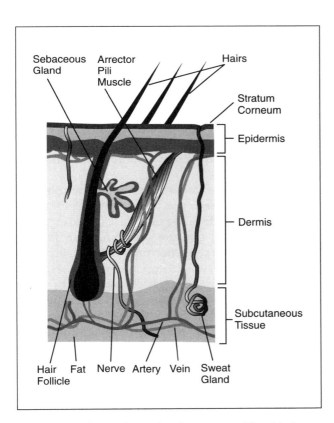

Figure 3-8: Skin and associated structures. The skin is composed of several layers: the epidermis, the dermis, and the subcutaneous fatty tissue. Blood vessels that supply the nutrients to the skin are found in the dermis along with a number of other accessory structures such as sweat glands, sebaceous glands, and hair follicles.

Dermis – The second layer of the skin is the dermis, sometimes referred to as the true skin. The dermis is composed primarily of connective tissue. This tissue binds the skin to the bone and muscle. In animal hides, this is the layer that turns into leather when chemically processed. Blood vessels are distributed throughout the dermis and are responsible for supplying nutrients to the cells of the dermis and the epidermis. Also found in the dermis are other structures such as the hair follicles, sweat glands, and sebaceous glands. Associated with the hair follicles are smooth muscles (arrector pili muscles). It is the contraction of these muscles that results in "goose bumps." These structures play a limited role in the absorption of substances across the skin.

> The dermis is the source of oxygen and of nutrients for the epidermis.

Subcutaneous fatty tissue – This layer provides a cushion for the underlying structures and allows the skin to move to some extent. The subcutaneous fatty tissue contributes to the connection of the dermis to the underlying tissues such as the muscle.

Factors Affecting Absorption

The ability of the skin to absorb a toxic substance is dependent upon two major factors: 1) the condition of the skin, and 2) the chemical nature of the toxic substance. The outer portion of the epidermis is composed of several layers of thin flat cells. The outermost cells are dead and are continuously replaced from below, approximately every two weeks. This outer layer serves as an effective barrier. Because of its relative impermeability it retards the diffusion of gases, water, and chemicals, preventing many types of toxic substances from entering the body.

> The outer layer of the epidermis is an effective barrier to absorption of some toxic substances.

Physical damage to the protective barrier, such as a cut or an abrasion, allows toxic substances to easily penetrate the epidermis and enter the dermis, where they may be more readily absorbed into the bloodstream and carried to other parts of the body to exert their effects. Acids, such as sulfuric and nitric acid, as well as strong bases, such as sodium hydroxide, are examples of toxic substances that are capable of breaking down the outer layers of the epidermis. When chemical burns occur, toxic substances – which are usually poorly absorbed – are more readily absorbed through the skin and have a greater potential to exert their toxic effects. Repetitive exposure to soaps, detergents, and organic solvents may result in water loss and dryness of the skin; this can enhance absorption of toxic substances.

> Damage to the epidermis can facilitate absorption.

In general, inorganic chemicals are a class of toxic substances that is not readily absorbed through intact, healthy skin. Examples of inorganic, toxic substances are metals such as cadmium (Cd), lead (Pb), and chromium (Cr). However, metallic mercury is absorbed across the skin and into the bloodstream, where it can produce systemic effects. The effects of exposure to mercury are discussed in Chapter 7.

The absorption of organic chemicals is affected by the form of the substance to which the skin is exposed. Organic chemicals dissolved in water do not easily penetrate the skin because the skin is impermeable to water. On the other hand, organic solvents, such as paint thinner or gasoline, are easily absorbed through the epidermis because they are lipid-soluble and therefore penetrate the cell membrane readily.

Other factors can influence the amount of absorption across the skin. Increasing the concentration of the toxic substance or the exposure time can increase the rate or amount of material absorbed. Conversely, concentration of and exposure time to a toxic substance can be decreased by "flooding" the point of contact, usually with water. This method is normally used when an individual is exposed to an acid such as HCl (hydrochloric acid). Copious amounts of water are used, diluting the concentration of HCl and decreasing the exposure time by flushing the acid from the skin's surface.

Although some toxic substances may not be absorbed through the skin, they still are capable of producing local effects. Acids, bases, certain metals (chromium), organic solvents, and naturally occurring toxins (poison ivy) exert local effects that may be characterized by reddening of the skin, appearance of a rash, blisters, or destruction of skin tissue resulting in scarring. These symptoms occur because of the effect of the toxic substances on various intracellular components of skin cells. For example, bases damage keratin in cells; organic solvents dissolve skin surface lipids and remove lipids in the cells; and metals, such as arsenic and chromium, alter cell protein structure. These effects are localized, which means that symptoms appear only at the point of contact between the toxic substance and the skin.

Toxic substances that pass through the skin may also cause local effects. In addition, they are absorbed by the bloodstream and transported to other tissues of the body where they can produce systemic effects. For example, dermal contact with carbon disulfide can affect the nerve tissue and the heart. Absorption of carbon tetrachloride will affect the liver, the kidneys, and the nervous system, resulting in kidney and liver damage as well as depression of the nervous system. Dermal exposure to organophosphate pesticides can produce symptoms associated with the dysfunction of the digestive tract, nerve-muscle functioning, and the lungs. Systemic effects as a result of exposure to toxic substances are described in Chapters 6, 7, 8, and 9.

Inhalation

Perhaps the easiest, most rapid, and most direct route of exposure is by inhalation. The lining of the respiratory tract is not effective in preventing absorption of toxic substances into the body. The lungs are designed to make transport of molecules easy.

> Toxic substances are readily absorbed in the respiratory tract.

Anatomy of the Respiratory Tract

The respiratory tract is composed of the **nasal cavity**, **pharynx**, **larynx**, **trachea**, **primary bronchi**, **bronchioles**, and **alveoli** or air sacs (Figure 3-9). Coarse hairs found in the nose filter out and remove particulate matter thereby cleaning the air entering the respiratory tract. In addition, **mucous cells** and

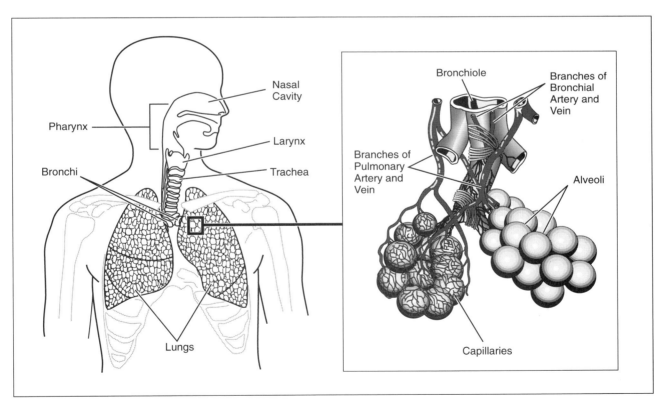

Figure 3-9: General representation of the organs that compose the human respiratory system. The respiratory pathway continues to divide and terminates in the small, sac-like alveoli. *From* Biology: The World of Life, *6th edition, by R. A. Wallace, Copyright © 1992 by HarperCollins College Publishers Inc. Reprinted by permission of Addison-Wesley Educational Publishers. Inset recreated with permission from Dorothy D. Schottelius,* Textbook of Physiology, *1978.*

ciliated epithelial cells are found along the entire length of the respiratory tract except in the alveoli. The mucous cells produce a thick, sticky mucous layer that lines the respiratory tract. Most of the particulates within the inhaled air are trapped in the mucus. The cilia are small hair like structures that move the mucus with trapped particulates from the respiratory tract toward the throat where it is swallowed. Various activities or diseases can affect this cleansing mechanism. Smoking decreases the ciliary activity thereby affecting the cleansing mechanism. As a result, airborne particulates are not removed as efficiently; they reside longer in the lungs, and may begin to affect normal respiratory functions and produce other toxic effects throughout the body.

> Mucous cells and cilia are the primary protective barriers to particulates in the respiratory system.

The alveolus is the functional unit of the respiratory tract. Each alveolus is composed of a single layer of thin, flat cells that are surrounded by a thin layer of connective tissue (Figure 3-10). It is estimated that each lung contains 150 million alveoli. Surrounding the alveoli is an extensive capillary bed that carries blood to and from these structures. It is in the alveoli that exchange of oxygen and carbon dioxide occurs between the bloodstream and the lungs. It is also where toxic substances are absorbed into the bloodstream, or where toxic particulates may accumulate.

> Alveoli are the primary site of exchange of toxic substances in the respiratory tract.

Factors Affecting Absorption

Both the entry of airborne toxicants into the circulatory system and the distribution to the body tissues depend upon several factors: 1) concentration of the toxicant in the air; 2) solubility of the toxicant in the blood and tissue; 3) the toxicant's molecular or particulate size; 4) respiration rate; 5) duration of exposure; and 6) functional integrity of the respiratory tract.

Many of the toxic substances are inhaled gases such as carbon monoxide (CO) and nitrogen dioxide, or vapors from volatile substances such as benzene or carbon tetrachloride. These toxic substances diffuse from the lungs into the bloodstream. The rate of diffusion is dependent on the concentration gradient between the lungs and the blood. Initially the diffusion is rapid because the concentration gradient between the lungs and the blood is high. As the concentration in the blood increases the rate of diffusion decreases until an equilibrium is established. In general, at equilibrium the concentration of the toxic substance in the blood is equal to the concentration in the lungs. If the concentration of the toxic substance in the bloodstream becomes greater than in the lungs, diffusion will occur in the opposite direction.

Solubility is an important factor in determining how readily a toxicant is absorbed into the bloodstream and into the tissue of the respiratory tract. Many organic solvent vapors such as benzene and chloroform are very soluble in the bloodstream and rapidly diffuse from the lungs into the blood. Ethylene is less soluble in the bloodstream and diffuses more slowly from the lungs.

Many types of particulates such as silica, fiberglass, or asbestos, are either almost or completely insoluble. Particulates also demonstrate the effect

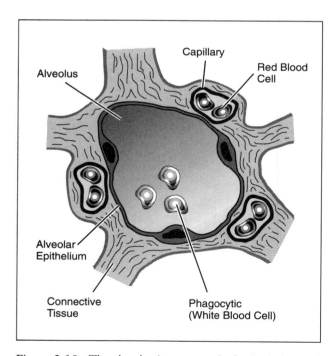

Figure 3-10: The alveolus is composed of a single layer of cells and is surrounded by an extensive network of capillaries. The close association facilitates the absorption of materials to and from the lungs.

of size on the rate of absorption. They do not easily pass through the cell membrane; rather, they may become embedded in the surface of the alveoli. In some cases the particles are phagocytized or engulfed (see Figure 3-7) by white blood cells found in the respiratory tract. After engulfing the particle, these cells may pass through the cell membrane; or they may migrate toward the cilia and mucous-lined portions of the respiratory tract, which then transport the particle-containing cells to the back of the throat where they are either expectorated or swallowed. Active and passive transport mechanisms – discussed previously in this chapter – are necessary to move large molecules, which normally diffuse very slowly (or not at all) across the cell membrane.

An individual's respiratory rate can affect the amount of toxic substance absorbed, especially when the substance is readily soluble in the blood. For example, due to its high solubility, virtually all the chloroform in one breath is removed from the lungs and diffuses rapidly into the bloodstream. Therefore, the quantity of chloroform that can be dissolved in the blood before equilibrium is reached is greater than that for low-solubility substances such as ethylene. Increasing the respiration rate will then increase the amount of chloroform absorbed, while it would have little effect on absorption of ethylene because of its low solubility.

Ingestion

Ingestion of toxic substances other than those contained in food is uncommon; when it does occur it is usually done accidentally or unknowingly. As discussed in the previous section, ingestion also takes place indirectly as a result of swallowing materials removed from the respiratory tract by ciliary activity.

The digestive tract is essential in the digestion and absorption of the food we eat. Various nutrients and substances are absorbed into the bloodstream along the digestive tract and transported throughout the body. The same actions that allow food to be used by the body also result in the absorption of many toxic substances.

Anatomy of the Digestive Tract

The digestive tract is composed of five major areas: 1) mouth and pharynx. 2) esophagus; 3) stomach; 4) small intestine; and 5) large intestine (Figure 3-11). From the **esophagus** to the large intestine, the digestive tract is composed of four basic layers of tissue (Figure 3-12). The outermost layer is referred to as the **serosa** and is composed of a thin sheet of cells that covers the organs. The second layer is a muscular layer composed of an outer, longitudinal layer and an inner, circular layer of smooth muscle. In the mouth, pharynx, and esophagus, skeletal muscle comprises part of the muscular layer. Contraction of the skeletal muscle in the pharynx and esophagus initiates the swallowing reflex. Collectively, this layer is referred to as the **muscularis** layer. The third layer is referred to as the **submucosa** and is composed of loose connective tissue as well as blood and lymph vessels. The fourth layer is the **mucosa** (mucous membrane). The epithelial cells that line the **lumen** of the digestive tract are found in this layer. The cells are nonkeratinized and stratified in the mouth and esophagus. However, the epithelial layer in the rest of the digestive tract is composed of a single layer of columnar cells.

Ninety per cent of all absorption in the digestive tract occurs in the small intestine, including absorption of many types of toxic substances. The

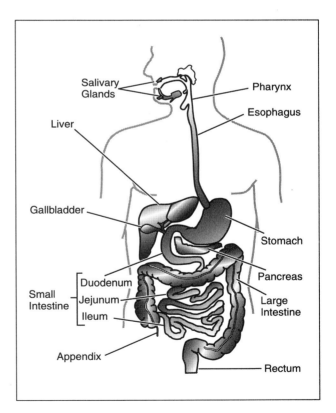

Figure 3-11: Generalized representation of the human digestive tract and some associated structures. *Recreated with permission from Dorothy D. Schottelius,* Textbook of Physiology, *1978.*

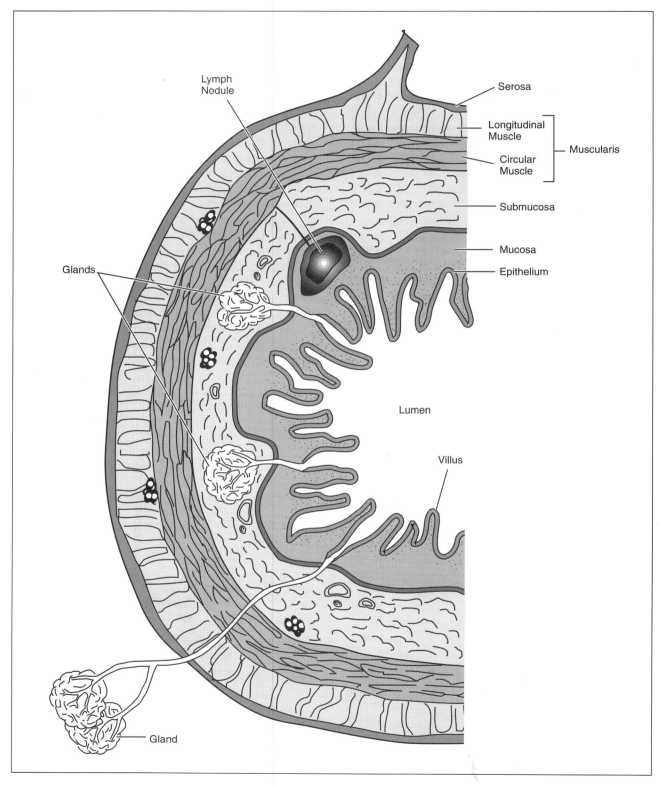

Figure 3-12: General structure of the digestive tract. From the esophagus to the end of the large intestine, the walls of the digestive tract have the same basic structure. The tissues are arranged in four basic layers: the serosa, the muscularis, the submucosa, and the mucosa. The mucosa lines the lumen of the digestive tract and is involved in absorption. *From* Biology: The World of Life, *6th edition, by R. A. Wallace, Copyright © 1992 by HarperCollins College Publishers Inc. Reprinted by permission of Addison-Wesley Educational Publishers.*

remaining 10 percent occurs in the stomach and large intestine.

> Absorption of toxic substances in the digestive tract occurs primarily in the small intestine.

The small intestine is divided into three segments: the **duodenum**, the **jejunum**, and the **ileum**. The duodenum is the first portion of the small intestine and is approximately 10 to 12 in long. The jejunum is the section of the small intestine between the duodenum and the ileum and is 3 to 4 ft long. The ileum is the portion of the small intestine between the jejunum and the large intestine and is 6 to 7 ft long.

The mucosa and submucosa in the small intestine form folds (Figure 3-13) that extend around the entire circumference of the small intestine. These folds are referred to as the **plicae circulares**. The mucosa also forms projections in the small intestine called **villi** (Figure 3-14). There are approximately 4 to 5 million villi in the small intestine. Blood vessels embedded in the villi carry substances – absorbed by the epithelial cells – throughout the body. The epithelial layer is composed of a single layer of cells. The plasma membranes of most of the epithelial cells are modified and form finger-like projections called **microvilli** (Figure 3-15).

The plicae circulares, villi, and microvilli increase the surface area (200 m²) in the small intestine, which facilitates the absorption of substances into the bloodstream. Absorption through the epithelial cells depends on the various passive and active transport mechanisms discussed earlier. These same mechanisms affect the rate of absorption of toxic substances as well.

> Absorption is enhanced in the small intestine due to presence of increased surface area associated with the microvilli, villi, and plicae circulares.

Very little absorption of toxic substances occurs in the large intestine. The large intestine is shorter than the small intestine and its lining varies significantly from the small intestine. There are no circular folds (plicae circulares), villi, or microvilli. Therefore, there is less surface area than in the small intestine. The epithelium is a single layer of cells, some of which are modified and secrete mucose; this facilitates the passage of material along the length of the large intestine and protects the epithelial lining. The large intestine absorbs primarily water and some inorganic salts.

The amount of toxic substance absorbed as a result of ingestion varies. For example, only 10 percent of the lead ingested is absorbed. Ingestion of cadmium or chromium results in 1.5 percent and 1.0 percent being absorbed, respectively.

> The absorption rates vary according to the different toxic substances.

Factors Affecting Absorption

Absorption of toxicants in the digestive tract depends on several physical and chemical factors. As discussed previously, the presence of villi, microvilli – and the fact that epithelial lining is only one cell thick – enhances the absorption rate of toxicants in

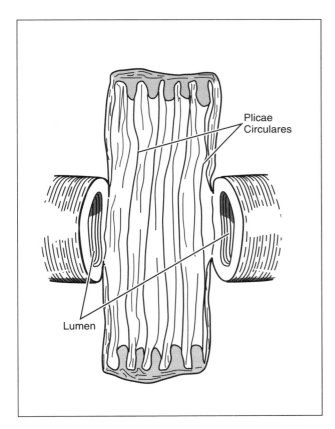

Figure 3-13: Folds in the mucosa and submucosa of the small intestine form the plicae circulares, which extend around the entire circumference of the small intestine.

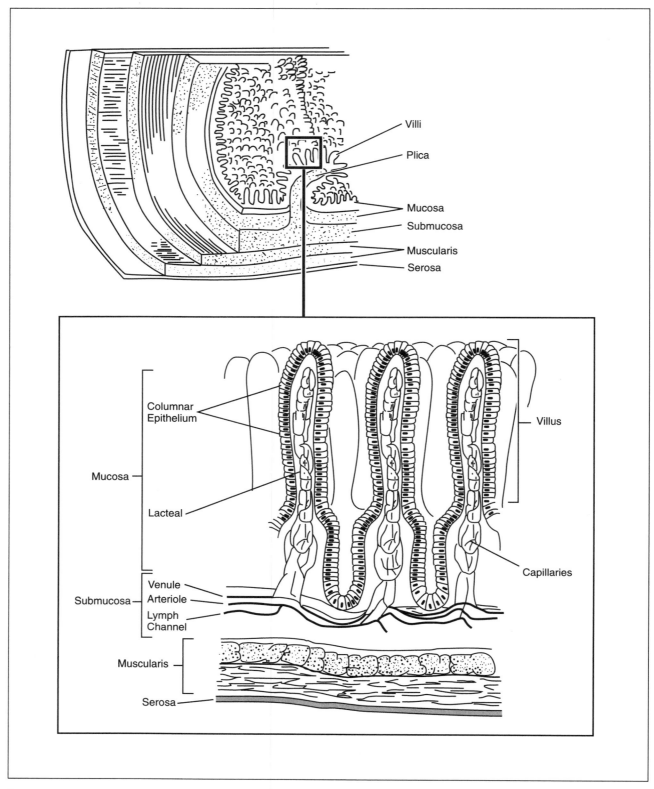

Figure 3-14: Cross section of the small intestine illustrating the relationship between the villi and the plicae circulares. The insert is a magnification of the villi showing their relationship to the tunics, the circulatory system, and other related structures. *From* Biology: The World of Life, *6th edition, by R. A. Wallace, Copyright © 1992 by HarperCollins College Publishers Inc. Reprinted by permission of Addison-Wesley Educational Publishers.*

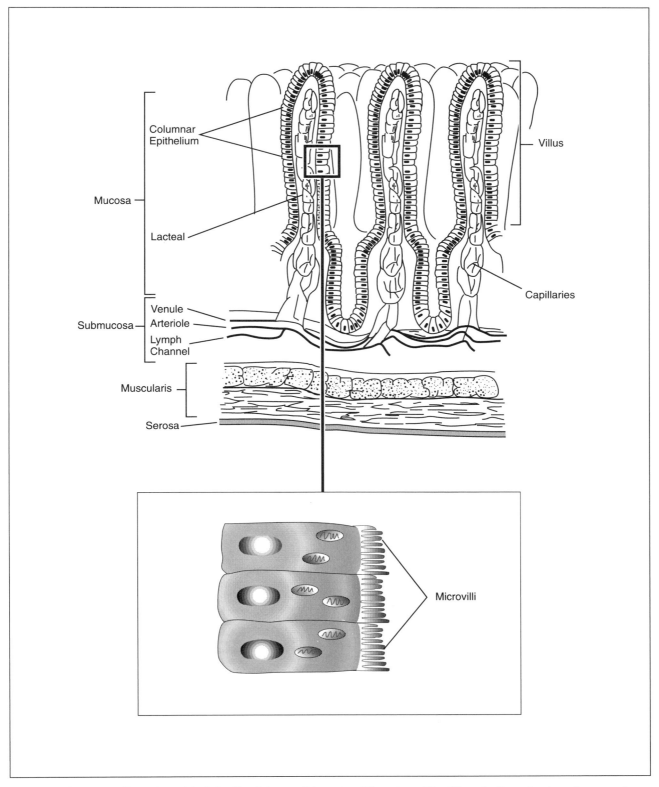

Figure 3-15: Microvilli on the epithelial cells of the small intestine. The microvilli, villi, and plicae circulares increase the surface area in the small intestine; this results in greater absorption of substances in this region of the digestive tract. *From Biology: The World of Life, 6th edition, by R. A. Wallace, Copyright © 1992 by HarperCollins College Publishers Inc. Reprinted by permission of Addison-Wesley Educational Publishers.*

the small intestine. Lipid solubility and molecular size are also important in determining how efficiently toxic substances are transported across the digestive tract epithelium and into the bloodstream. Toxicants that are lipid-soluble, such as organic solvents like methylene chloride, are easily absorbed as a result of diffusion. Similarly, toxicants that have a small molecular size will diffuse more readily through the cell membrane pores.

Active and passive transport mechanisms facilitate absorption of essential nutrients such as amino acids, calcium, sodium, and iron. Competition may occur between the essential nutrients and various types of toxicants for the carrier molecule used during active transport and facilitated diffusion. Lead may compete for the same carrier molecule used by calcium. Similar competition may be involved in the absorption of other toxicants such as cadmium and mercury.

Another important factor affecting absorption is the length of time food containing the toxicant remains in the small intestine. Generally, the longer the food remains in the intestine the more toxicant will be absorbed.

Eye

Anatomy of the Eye

The eye is composed primarily of three layers of tissue (Figure 3-16). The outer layer consists of dense connective tissue that gives shape to the eye. This tissue is referred to as the **sclera** and – in the front

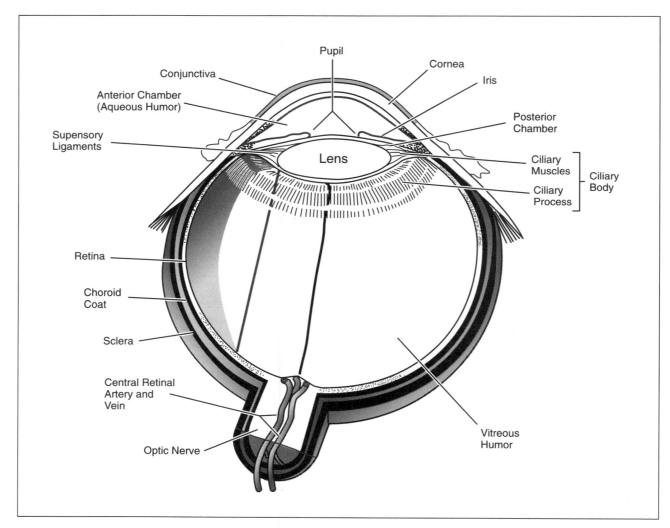

Figure 3-16: Cross section of the eyeball illustrating some important anatomical structures. *Recreated with permission from Dorothy D. Schottelius,* Textbook of Physiology, *1978.*

of the eyeball – forms into the transparent cornea. Often it is the cornea that first comes into contact with toxic substances and is damaged. The middle layer contains the blood vessels that supply the various types of eye tissue with oxygen and nutrients. This middle layer is called the **choroid**; in the front of the eye it forms the **ciliary bodies** as well as the pigmented **iris**, which gives color to the eyes. The innermost layer, the **retina**, is composed of the sensory tissues of the eye that are involved in producing the images we see.

> The eye is a common point of contact for toxic substances.

Enclosed within these tissues are two compartments that contain an aqueous fluid-like substance called the **aqueous humor** and a transparent jelly-like substance called the **vitreous humor**. Located between the aqueous and vitreous humors, and forming part of the barrier between these two substances, is a transparent oval structure called the **lens.** The cornea, the humors, and the lens are all transparent and allow light from the exterior to be focused on the retina to form images.

The primary point of contact of toxic substances to the eye is the cornea. Acids and alkalis are probably the most common types of exposures that do damage, generally they cause damage to the corneal epithelium. Depending on the amount of damage done, the cornea may be able to repair itself. This process is slow because the cornea does not have a direct blood supply. All nutrients and oxygen must diffuse through the aqueous humor in the **anterior chamber** to the corneal epithelium in order for the damaged tissue to be repaired. More detailed information concerning specific toxic effects will be discussed in Chapter 6.

> The eye can be easily damaged and will heal slowly or not at all.

Other Pathways of Exposure

In toxicology studies the administration of the toxic substance may sometimes occur by one of the three major pathways described in this chapter. However, most commonly it occurs by injecting the substance into the experimental animal. The injection may occur **intravenously** (into a vein), **intramuscularly** (into the muscle), **intraperitoneally** (into the peritoneal cavity), **intradermally** (into the skin), and subcutaneously (under the skin). This is due primarily to standard medical practices.

In the general population this type of exposure may occur as a result of accidents or acts of nature as, for example, bee stings, spider bites, snakebites, or by accidental penetration by puncture wounds or splinters. If the object is composed of or contaminated with a toxic substance, then these types of exposure may introduce the substance into the body, however, these nonroutine exposures play a relatively minor role in introducing toxic substances. The primary means of toxic substances entering the body is by dermal exposure, inhalation, or ingestion.

Checking Your Understanding

1. List the three primary routes of exposure. Which route is the most common?
2. Describe the various layers of the skin.
3. What attribute of the skin limits the rate of absorption?
4. What types of substances are easily absorbed through the skin? Why?
5. List three factors that increase skin permeability.
6. Describe the components of the respiratory tract.
7. What is the primary site of absorption of toxic substances in the respiratory tract?
8. What characteristics of the small and large intestine account for their different rates of absorption?
9. Which part of the eye is most susceptible to injury as a result of exposure to a toxic substance?

3-4 Local versus Systemic Effects

Once exposure to a toxicant occurs, effects may be observed locally or systemically. Some substances may demonstrate both.

Local effects are those observed at or near the site of exposure. The cells and tissue are directly affected resulting in a localized effect that does not spread much beyond the contact point. For example, exposure to caustic substances such as acids or alkaline (bases) produce a localized effect that can be characterized by pain and reddening of the tissue at the point of contact. Inhalation of particulates can result in irritation of the lungs characterized in some cases by chest pains and difficulty in breathing.

Many substances are absorbed into the bloodstream at the site of exposure and transported to other parts of the organism where they cause their effects. These effects are referred to as systemic effects. Systemic toxicants produce most of their effects in one or two target organs.

> Systemic effects occur in target organs as a result of absorption at the site of exposure and transport by the bloodstream.

Whether systemic effects occur or not is dependent on the concentration of the toxic substance in the target organ, which in turn is dependent on the absorption, distribution, **biotransformation**, and **excretion** of the toxic substance.

> Biotransformation is the conversion of a substance from one form to another, which may result in a change of the toxic properties.

As discussed earlier in the chapter, absorption of a toxic substance is affected by several factors such as the route of exposure, cell membrane permeability, and the physical and chemical properties of the toxic substance. Toxic metals like cadmium and lead do not readily penetrate the skin because of the cornified outer layer of the epidermis. However, these substances are absorbed in the small intestine by active and passive transport mechanisms. Therefore, systemic effects are more likely to be observed if these types of toxic substances are ingested. Lipid-soluble substances such as benzene and xylene are easily absorbed by the skin, in the lungs, and by the cells lining the digestive tract because of the high lipid content of the cell membrane. Exposure to these toxic substances by either of these routes may result in systemic effects. In general, the factors that result in greater absorption of the toxic substance increase the chance that systemic effects will be observed.

Distribution of a toxic substance is performed primarily by the circulatory system. Once in the bloodstream the toxic substance may either undergo biotransformation, or be stored in various tissues, or be excreted from the body. Each of these processes can further affect the concentration of the substance in the target organ, potentially decreasing its systemic toxicity.

Biotransformation usually occurs in several steps during which the original substance is chemically modified to form a new, chemically distinct substance. Biotransformation occurs primarily in the liver, but may also occur in other tissue such as the kidneys, lungs, and digestive tract. This process is important in transforming lipid-soluble substances – such as DDT and organophosphate pesticides – to water-soluble substances that easily pass through the cell membrane of the kidney tissues and are excreted from the body. Biotransformation is important in decreasing the concentration of toxic substances in the body. However, the rate at which biotransformation occurs does not continue to increase as the concentration of the toxic substance increases. Depending upon the toxic substance, there is a concentration at which the rate of accumulation in the tissues is faster than the rate of biotransformation and subsequent excretion. When this occurs systemic effects begin to be observed.

Many toxic substances are absorbed from the bloodstream and stored in tissues in which they have little or no toxic effects. These tissues – fat, bone, liver, and kidneys – can serve as a buffer by decreasing the availability of the toxic substance to the target organ, thereby decreasing the systemic toxicity of the substance. For example, the highest concentration of chlorinated hydrocarbons, such as DDT and chlordane, are found in the fatty tissue of the body because they are very lipid-soluble; they do not produce toxic effects while stored in this tissue type. Lead is stored primarily in the long bones of the body, but it is toxic primarily to the nervous system.

Of course, the substances that accumulate in their **target tissues** result in systemic effects. For example, organic solvents are readily absorbed and are concentrated in the liver, the primary target organ for these types of toxic substances. High enough concentrations will result in liver damage.

The rate at which a toxic substance is excreted will also determine if systemic effects will occur. Excretion results in the discharge of a substance from the body, from an organ, or from a cell. There are several important organs that are involved in the excretion of toxic substances. The kidney eliminates more toxicants from the body than any other organ. Biotransformation of toxicants to water-soluble substances is necessary in order to eliminate them in the urine. The lungs are important in eliminating substances that are in the gas phase such as carbon dioxide. The liver removes substances, such as DDT and lead, by excreting them into the **bile** (produced by the liver), which subsequently enters the small intestine. Once in the small intestine, the toxic substance can be eliminated in the feces.

Checking Your Understanding

1. What are the differences between local and systemic effects?
2. What are target tissues?

Summary

Toxic substances can enter the body by three major routes: dermal absorption, inhalation, and ingestion. The rate of absorption by each of these routes varies. It is dependent on the structure and function of the epithelial cells at the point of contact as well as the chemical and physical properties of the toxic substance. Although contact with the skin is the most common route of exposure, many substances are not easily absorbed. Absorption of many toxic substances is easier and faster if it is inhaled. In the digestive tract most absorption occurs in the small intestine. Exposure can result in either local or systemic effects. Local effects are those that are observed at or near the site of exposure and do not spread much beyond the point of contact. Systemic effects occur when the toxic substance is absorbed into the bloodstream and transported to other parts of the organism, where its effects are observed in the target tissues. The concentration of the toxic substance in the target tissue will determine whether systemic effects are observed. The concentration is affected by the absorption, distribution, biotransformation, and the rate of excretion of the toxic substance.

Application and Critical Thinking Activities

1. Identify at least five different toxic substances you may be exposed to at home or in your workplace. What are the routes of exposure? As a result of exposure, did you observe any effects? Were they local or systemic effects?

2. Often individuals who smoke cigarettes are more susceptible to toxic substances and respiratory diseases. Why is this true? Relate your answer to information discussed in this chapter.

3. Acceptable levels of exposure have been identified for a variety of toxic substances. For each of the major pathways of exposure make a list of these substances and the acceptable exposure levels. (References produced by the Environmental Protection Agency and the Occupational Safety and Health Administration will be useful.)

4. For each of the three major pathways of exposure, identify toxic substances that can cause local and systemic effects.

4

The Dose-Response Relationship

Chapter Objectives

Upon completing this chapter, the student will be able to:

1. **Describe** the two major types of toxicological studies, their attributes as well as their advantages and disadvantages.
2. **Define** terminology associated with acute and chronic toxicological studies.
3. **Explain** and illustrate how dose-response curves are constructed.
4. **Explain** relative toxicity of chemicals using information derived from dose-response curves.
5. **List** and explain how various factors affect chemical toxicity.
6. **Describe** types of chemical interactions and their effects on toxicity.

Chapter Sections

4–1 Introduction

4–2 Acute and Chronic Toxicological Studies

4–3 Dose-Response Curves

4–4 Variables Influencing Toxicity

4-1 Introduction

Chemicals are part of our everyday life; therefore, exposure to them is unavoidable. We are exposed to them at home and in the workplace. Given these circumstances, it is desirable to establish safe levels of exposure for chemicals in order to protect us from possible harmful effects. The effect of multiple chemical exposures, as well as individual variations, are but two of the factors that make it difficult to establish a single, safe level for all individuals. Both of these factors, along with several others, will be discussed later in the chapter.

Nonetheless, the need to establish some criteria to base the relative safety of chemicals is essential. Several methods are used to obtain data in order to establish safe levels and provide information about the relative toxicity of chemicals. The methods used include epidemiological studies, chemical analogy, and animal experimentation. Most of the information concerning chemical toxicity is derived from animal experimentation.

4-2 Acute and Chronic Toxicological Studies

Information concerning the toxicity of a substance is obtained primarily from either acute or chronic toxicity studies. These types of studies entail exposure of test animals – usually rats or mice – to a range of chemical doses. Effects as a result of exposure are recorded over a period of time. The information obtained is utilized to establish a safe level of exposure for that particular substance.

Acute Toxicity Studies

Acute toxicity studies are the most commonly performed studies for obtaining information on the effects of chemical exposure. They are short-term, relatively inexpensive tests with death of the test animal being the usual observed effect. However, other effects may be considered to evaluate the toxicity of a substance. Table 4-1 lists various types of responses that can be observed or measured when performing acute toxicity tests. Information obtained from acute toxicity tests can be used to 1) determine the relative toxicity of different chemicals by comparing the respective **LD_{50}** values, which are the doses that prove to be lethal for 50 percent of the test animals; 2) identify the target organs (heart, liver, kidneys, etc.) that are affected as a result of exposure; and 3) determine the appropriate doses for long term, chronic studies. In addition, some of the parameters and observations usually associated with chronic toxicity tests (see Tables 4-4 and 4-5) may, under certain circumstances, also be useful in evaluating exposure associated with acute toxicity tests.

To perform acute toxicity studies the test animal population is divided into several groups, with an equal number of individuals (usually 10 to 50) in each group. One group of test animals – referred to as the control group – is exposed to all the same experimental conditions except exposure to the toxic substance. The control group is used to determine the number of individuals that will die as a result of being exposed to the various experimental conditions associated with the study and not due to exposure of the toxic substance. For example, rats caged in groups of ten had an LD_{50} for **isoproterenol** of 80 mg/kg while rats caged singly had an LD_{50} of 50 mg/kg. In this example, it was the num-

Physical and Observational Examination of Animals in Toxicity Studies		

1. MOTOR ACTIVITY
    Dec Locomotor Activity
    Inc Locomotor Activity
    Jumping
	12. **SOMATIC** (Body) RESPONSE
	    Inc **Preening**
	    Dec Preening
	    Rubbing Nose
	    Inc Scratching
	    Dec Scratching
		23. URINATION
		    Inc Urination
		    Dec Urination
		    Hematuria (blood in urine)

2. SENSITIVITY TO PAIN
    Inc Sensitivity
    Dec Sensitivity
    Analgesia (no pain)
	13. POSTURAL REFLEXES
	    Grasping Reflex Depr
	    Grasping Reflex Abs
	    Righting Reflex Depr
	    Righting Reflex Abs
		24. **APNEA** (absence of breathing)

3. SENSITIVITY TO SOUND
    Inc Sensitivity
    Dec Sensitivity
	14. **PROSTRATION**
	    Prostration (exhaustion)
	    Loss of Consciousness
		25. **DYSPNEA** (labored breathing)

4. SENSITIVITY TO TOUCH
    Inc Sensitivity
    Dec Sensitivity
	15. TREMORS
	    Tremors - During rest and movement
	    Tremors - Movement
		26. CARDIAC
		    Heart Rate
		    Pulse

5. SOCIAL INTERACTION
    Inc Exploratory Behavior
    Dec Exploratory Behavior
    Inc Rearing (Standing) Frequency
    Inc Speed of Rearing
    Dec Rearing Frequency
	16. EYE IRRITATION
	    Eye **Opacity**
	    Blinking - Excessive
		27. RESPIRATION
		    Inc Respiration Rate
		    Dec Respiration Rate
		    Inc Respiration Depth
		    Dec Respiration Depth
		    Irregular Respiration

6. Abnormal Tail
    Rigid Tail
    Limp Tail
	17. **CORNEAL** (eye) **REFLEX** (closing eyelid)
	    Corneal Reflex Abs
	    Corneal Reflex Depr
		28. NASAL DISCHARGE

7. AGGRESSIVE BEHAVIOR
    Inc - Species toward same species
    Dec - Species
	18. **LACRIMATION**
	    (tear formation)
		29. BODY TEMP
		    Inc Rectal Temp
		    Dec Rectal Temp

8. **ATAXIA** (loss of coordination)	19. **PUPILLARY LIGHT REFLEX**
	    Pupillary Reflex Abs
	    Pupillary Reflex Depr
		30. **CYANOSIS**

9. CONVULSIONS
    Tonic (continuous) Convulsions
    Clonic (spasmodic) Convulsions
	20. **PHOTOPHOBIA**
	    (Light avoidance)
		31. DEATH

10. MUSCLE TONE
    Inc Muscle Tone - Trunk
    Dec Muscle Tone - Trunk
    Inc Muscle Tone - Limbs
    Dec Muscle Tone - Limbs
	21. DEFECATION
	    Inc Defecation
	    Dec Defecation
	    Diarrhea
	    Bloody Stool
		32. NO EFFECT

11. PARALYSIS	22. SALIVATION
	    Salivation
	    Dry Mouth	

ABBREVIATIONS Dec - Decreased Inc - Increased Depr - Depressed Abs - Absent		

Table 4-1: Responses that can be measured or observed when performing acute toxicity animal studies. *Obtained and modified with permission from Loomis T. A. and Hayes A.W.,* Essentials of Toxicology, *Academic Press/Harcourt Brace, New York, 1996.*

ber of animals in a cage that affected toxicity. Strychnine, nicotine, malathion (an organophosphate pesticide), and sarin (a nerve gas) toxicity is increased at lower temperatures. However, other organophosphate pesticides such as parathion are less toxic at lower temperatures. Other factors, such as handling the test animals or injury associated with the route of exposure (e.g. injection), may also contribute to their death.

> Isoproterenol is a drug used to relieve breathing difficulties associated with asthma.

Each of the remaining test groups are exposed to a specific dose of the toxic chemical. One group will receive the lowest while another will receive the highest dose. Each of the remaining groups will receive one of the doses between the two extremes. The groups are observed for 14 days, and the number of deaths in each group is recorded. Some acute toxicity tests may be performed for shorter periods of time such as 24 or 96 hours (four days). Figure 4-1 illustrates the standard response associated with an acute or chronic toxicity test. Each bar repre-

Toxic Substance	Species	LD_{50} (mg/kg of body weight)
Ethanol	Mouse	10,000
Sodium chloride	Mouse	4,000
Ferrous sulphate	Rat	1,500
Morphine sulphate[a]	Rat	900
Phenobarbital, sodium[b]	Rat	150
DDT	Rat	100
Picrotoxin[c]	Rat	5
Strychnine sulphate[d]	Rat	2
Nicotine	Rat	1
d-Tubocurarine[e]	Rat	0.5
Tetrodotoxin[f]	Rat	0.1
Dioxin[g]	Guinea-pig	0.001
Botulinus toxin[h]	Rat	0.00001

[a] A drug used to relieve pain or as a sedative.
[b] A drug used as a sedative or as an anticonvulsant (stops uncontrolled muscle spasms).
[c] A toxic substance obtained from the seeds of a shrub, *Anamirta coccolas*, which can produce uncontrolled muscle spasms.
[d] A toxic substance that can cause uncontrolled contraction of muscles and death.
[e] Drug used to treat poisoning due to black widow spider bite; used on tips of poison arrows; can produce skeletal muscle relaxation; at certain doses may cause death.
[f] A nerve poison found in eggs of California newt and certain puffer fish; in concentrated form more toxic than cyanide.
[g] Toxic substance that is a contaminant in widely used herbicides.
[h] Toxic substance produced by bacterium *Clostridium botulinum*, which is associated with food poisoning.

Table 4-2: Acute LD_{50} values for a variety of chemical agents. *Obtained and modified with permission from Loomis T. A. and Hayes A.W., Essentials of Toxicology, Academic Press/Harcourt Brace, New York, 1996.*

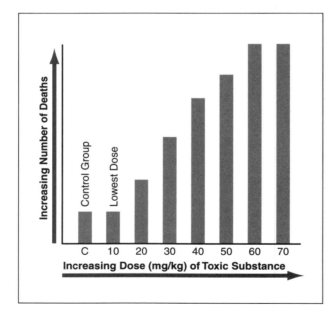

Figure 4-1: a) Simplified graph illustrating the standard response associated with an acute or chronic toxicity study. Each bar represents a group of test animals (usually 10 to 50) exposed to the toxic substance.
b) The number of deaths in the control group equals the number of deaths in the group exposed to the lowest dose of the toxic substance.

sents a group of test animals exposed to a specific dose of the toxic substance. The number of deaths in the control group and the group receiving the lowest dose are the same indicating – in this particular example – that the death of the test animals at the lowest dose may not be related to exposure to the toxic substance. As the dose increases, so does the number of deaths.

> Acute toxicity tests are short-term, usually lasting no longer than 14 days, with death as the observed effect.

Results from acute toxicity studies are often expressed as LD_{50} or **LC_{50}** values. LD means "lethal dose," and the subscript means that the dose was lethal to 50 percent of the test animals. A subscript of 10 would mean that, at a given dose, 10 percent of the animals died. The unit of measure for the LD_{50} is usually expressed as milligrams per kilogram (mg/kg), which means milligrams of chemical per kilogram body weight of the test animal, or parts per million (ppm).

Test	Cost, $
General acute toxicity	
Acute toxicity (rat; two routes)	6,500
Acute dermal toxicity (rabbit)	3,000
Acute inhalation toxicity (rat)	6,500
Acute dermal irritation (rabbit)	900
Acute eye irritation (rabbit)	500
Skin sensitization (guinea pig)	700
Repeated dose toxicity	
14-day exposure (rat)	
90-day exposure (rat)	100,000
1-year (diet; rat)	225,000
1-year (oral; rat)	275,000
2-year (diet; rat)	625,000
2-year (oral; rat)	800,000
Genetic toxicology tests	
In vivo (living) chromosome aberration (rat)	26,500
Mammalian bone marrow cytogenetics (in vivo; rat)	25,000
Other tests	
Acute toxicity in fish (LC_{50})	1,750
Daphnia reproduction study	1,750
Algae growth inhibition	1,750

Table 4-3: Typical costs of various types of toxicity tests. *Obtained and modified with permission from Klaasen C. D. et al.,* Casarett & Doull's Toxicology, The Basic Science of Poisons, *5th edition, McGraw Hill, New York, 1996.*

> One milligram per kilogram is equal to one part per million.

The route of exposure is either by ingestion, dermal absorption, or injection. Table 4-2 lists the LD_{50} values for a variety of toxic substances. Results from acute toxicity tests lasting 24 or 96 hours may be expressed as: 24-h LD_{50} or 96-h LD_{50}.

> The term lethal dose (LD) with its subscript is used to express the percent of deaths as a result of either ingestion, dermal absorption, or injection of the toxic substance under study.

LC_{50} is a measure of chemical toxicity resulting from inhalation. Experimental animals are usually exposed continuously to a toxic substance for four hours. LC means "lethal concentration" and the subscript has the same meaning as described previously for LD. The unit of measure for the LC_{50} is usually expressed as part per million (ppm) when the toxic chemical in the air is in the gas or vapor form, or as milligrams per cubic meter of air (mg/m^3) if in particulate form.

> The term lethal concentration (LC) and its subscript is used to express the percent of deaths as a result of inhalation of the toxic substance under study.

Chronic Toxicity Studies

Chronic toxicity studies are performed to obtain information on effects that develop over a long period of time as a result of continuous exposure to non-lethal doses of a toxic substance. These types of studies are useful in determining which toxic substances cause diseases such as cancer. For these studies, the test animals are divided into groups as described before. However, instead of a single exposure, the toxic substance may be administered daily or intermittently by incorporating it in the food, air, or water. Instead of lasting 14 days, chronic toxicity tests may be performed over a period of months, years, or the lifetime of the test animal. Doses of the toxic substance are selected to assure that most of the animals will survive the entire time the study is performed. Maintaining a viable (living) test population over a long period of time is costly. Therefore, chronic tests are not performed as often as acute toxicity tests. Table 4-3 lists the approximate cost of performing various types of chronic and acute toxicity tests. Responses that may be measured or observed in association with chronic toxicity testing are listed in Tables 4-4 and 4-5. Table 4-5 lists changes in various tissue types that can occur as a result of chronic exposure.

Table 4-6 identifies various protocols (test plans) that are considered when performing acute and chronic toxicity tests. In general, chronic and acute toxicity tests are performed to determine what effect dose, route of exposure, species, sex of the test animal, and duration of exposure will have on the toxicity of a chemical. Each of these factors will impact the observed effects. Different species of test animals may be more or less sensitive to the same toxic chemical. This is related to the ability of some species to detoxify the substance more efficiently. Therefore, different species are used to evaluate toxic substances. In addition, males and females of the same species may respond differently to the same substance. This is related to the differences in **metabolism** between the two groups. The effects of various factors on toxicity are discussed in more detail later in this chapter.

Although acute and chronic studies provide useful information in evaluating chemical toxicity, it is important to understand that they are not truly representative of the environment to which people are exposed in everyday life. Normally, acute and chronic toxicity studies are performed in an environment where most of the variables influencing chemical toxicity are controlled. This is done because of the need to establish a cause-effect relationship. For instance, the test animals are generally exposed to only

Test	Parameters
Hematology (blood)[a]	**erythrocyte** (red blood cell) **count**[b] **total leucocyte** (white blood cell) **count**[c] **differential** (different types) **leucocyte count**[d] **hematocrit**[e] **hemoglobin**[f]
Blood Chemistry	sodium potassium chloride calcium carbon dioxide serum glutamate-pyruvate transaminase* serum glutamate-oxalacetic transaminase** serum alkaline phosphatase* fasting blood sugar total serum protein
Urine Analyses	pH total protein glucose

[a] Hematology is the science concerned with blood and the blood forming tissues.
[b] Erythrocyte (red blood cell) count is a test to determine the number of red blood cells present in the circulatory system. Calculated by counting the number of red blood cells in a milliliter of blood.
[c] Total leucocyte (white blood cell) count is a test to determine the number of white blood cells present in the circulatory system. Calculated by counting the number of white blood cells in a milliliter of blood.
[d] There are different types of white blood cells in the circulatory system. The differential leucocyte count is a test to determine the number of each type present in the circulatory system. Calculated by counting the number of white blood cells in a milliliter of blood.
[e] Hematocrit is the percentage of the volume of a blood sample composed of cells.
[f] Hemoglobin is the iron-containing pigment of the red blood cells; its function is to carry oxygen from the lungs to the tissues.
* = enzymes measured to evaluate liver function
** = enzyme to test for injured tissue cells

Table 4-4: Various parameters associated with a normal functioning individual that can be tested to evaluate the effects of exposure to toxic substances during chronic toxicity testing. *Obtained and modified with permission from Loomis T. A. and Hayes A.W., Essentials of Toxicology, Academic Press/Harcourt Brace, New York, 1996.*

Weight Determination of	
body **thyroid**[a] heart liver	spleen kidneys **adrenals**[b] testes
Tissue Examination of	
adrenals heart liver large intestine small intestine spleen ovary **lymph nodes**[c] all tissue lesions (injured, altered tissue)	**pituitary**[d] thyroid kidneys stomach **pancreas**[e] urinary bladder testes

[a] The thyroid gland is located on both sides of the neck just below the voice box. It is involved in regulating normal body metabolism.
[b] The adrenal glands are located on top of the kidneys. They secrete several chemical substances that affect various tissues of the body.
[c] Lymph nodes are round, oval or bean-shaped bodies located throughout the body. They produce white blood cells that fight bacteria, and filter the blood.
[d] The pituitary gland is a small, gray, rounded body located at the base of the brain. It secretes a number of chemical substances (hormones) that regulate body processes such as growth, reproduction and metabolism.
[e] The pancreas is a gland located near the stomach and small intestine that regulates blood sugar levels.

Table 4-5: Tissue examinations commonly perfomed in chronic toxicity tests. *Obtained and modified with permission from Loomis T. A. and Hayes A.W., Essentials of Toxicology, Academic Press/Harcourt Brace, New York, 1996.*

Type of Study	Test Species	Number of Dose Levels [a]	Dosing Regimen	Approximate Number of Animals of each Sex per Dose Level	Observation Period	Typical Observations
ACUTE						
Oral	Rats, mice, guinea pigs	4-6	Single dose	5-10	14 days	Survivors, body weight changes (day 14), gross changes in tissue, changes in toxicity, blood chemistry
	Dogs	–	Single dose	2-3	14 days	
Dermal	Rabbits	3-4	Single application for 24 hours	5	14 days (evaluated 24 hours, 7 days, 14 days)	Survivors, body weight changes (day 14), toxicity and gross changes in tissue, especially of skin
Inhalation	Rats	4-5	4-hour exposure	5	14 days	Survivors, body weight changes (day 14), toxicity and gross changes in tissue, especially of lungs
SUBCHRONIC [b]						
Oral	Rats	3-4	Daily doses	20	90 days	Survivors, body weight changes, diet consumption, urinalysis, changes in blood composition and chemistry; gross and microscopic examination of major tissues and organs
	Dogs	3-4	Daily doses	6	90 days	
Dermal	Rabbits	3-4	Daily application	10	90 days	
Inhalation	Rats	3-4	Daily doses [c]	10	90 days	
CHRONIC						
Oral	Rats	3-4	Daily doses	50	2 years	Survivors, body weight changes, diet consumption, urinalysis, changes in blood composition and chemistry, gross and microscopic examination of major tissues and organs
	Dogs	3-4	Daily doses	6	2 years	
Inhalation	Rats	3-4	Daily doses [c]	50	2 years	

[a] Includes a dose level of zero (control)
[b] Subchronic refers to toxicity studies with a duration of 5 to 90 days
[c] Minimum of 5 days per week

Table 4-6: Typical testing protocols for acute and chronic toxicity studies. *Obtained and modified from EPA Toxicology Handbook, Government Institute, Rockville, MD, 1986.*

one chemical and, in acute studies – from which most toxicity information is derived – they are usually exposed to a single dose. In addition, there usually is a single route of exposure. The number of animals exposed is small and in many instances – when rats and mice are used – the genetic make-up of the population is not very diverse. Generally, only individuals in good health and/or of the same sex are selected. Collectively, these factors can have an impact on the observed response.

In reality, people are exposed to toxic substances through several different routes. Toxic substances may be in the air we breathe, the food we eat, or the water we drink. In addition, exposures often occur over a prolonged period of time and, usually, there is exposure to more than one chemical at a time. The healthy and unhealthy are both exposed and – because of the greater genetic variability in a larger population – the range of responses to a given chemical is more diverse. The collective interactive effect of all these factors on toxicity makes it difficult to establish a cause-effect relationship; further, attempts to project information based on extrapolation from animal studies is affected by these differences.

> Cause-effect relationships observed as a result of performing chronic and acute studies are useful in establishing safe levels of chemical exposure; however, the results must be evaluated in terms of all the factors affecting the information.

Checking Your Understanding

1. Why are acute and chronic toxicity studies performed?
2. What are the major differences and similarities between acute and chronic toxicity studies?
3. Explain the differences between LD_{50} and LC_{50}.
4. What are the advantages and disadvantages of using acute or chronic toxicity studies in determining chemical safety?

4-3 Dose-Response Curves

The major purpose for performing acute and chronic toxicity studies is to establish a cause-effect relationship between exposure to a toxic substance and an observed effect in order to determine a safe exposure level. In general, as the dose increases, so does the number of individuals in each group exhibiting the measured response. By plotting this information on a graph, with the horizontal axis representing increasing doses and the vertical axis representing an increasing response, a curve can be drawn that illustrates the relationship between the dose administered and the observed response. This curve is referred to as the **dose-response curve** (Figure 4-2). Dose-response curves for ethyl alcohol, methyl mercury, and actinomycin D, are illustrated in Figures 4-3, 4-4, and 4-5 respectively. In each graph the threshold can be determined for the response being measured. A dose-response curve can be developed for most chemicals. From these curves the threshold level and the **relative toxicity** of chemicals can be obtained to help establish safe levels of chemical exposure.

Dose-response curves illustrate the relationship between dose and observed effects.

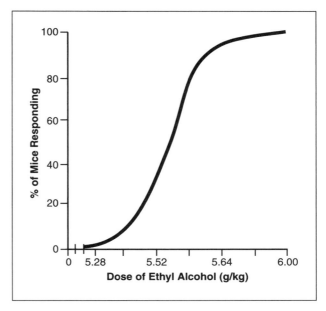

Figure 4-3: Response of mice to various doses of ethyl alcohol. The response measured is death. *Recreated with permission from Levine R. R.*, Pharmacology, Drug Action, and Reactions, *Lippincott Raven, Philadelphia, 1973.*

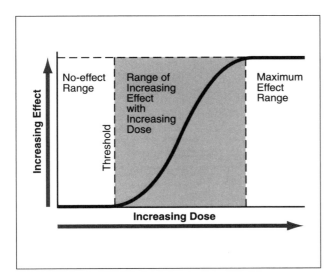

Figure 4-2: Dose-response curve. *Recreated with permission from Ottoboni M. A.*, The Dose Makes the Poison, *Van Nostrand Reinhold, New York, 1991.*

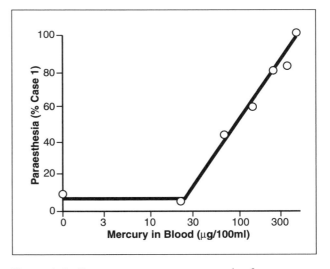

Figure 4-4: Dose response curve as a result of exposure to methyl mercury. The response is measured in the percent of cases (individuals) that experience paraesthesia, which is a burning, tingling sensation associated with the skin. *Recreated with permission from Lu F. C.*, Basic Toxicology: Fundamentals, Target Organs, and Risk Assessment, *Taylor and Francis, New York, 1985.*

66 ■ Basics of Toxicology

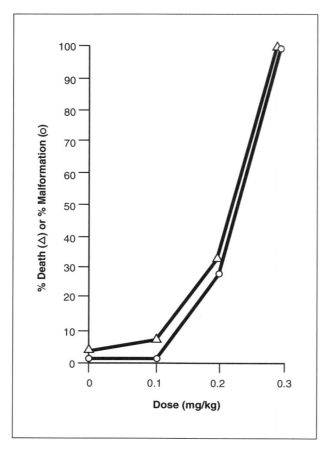

Figure 4-5: Two different responses of the developing fetus to exposure to Actinomycin^D (an antibiotic). The observed response is dependent on when exposure occurs during development. *Recreated with permission from Timbrell J. A.*, Principles of Biochemical Toxicology, *London, Taylor and Francis, 1982.*

The **threshold** is the dose below which no effect is detected or above which an effect is first observed. (See Box on Terminology Associated with Dose-Response Curves). The threshold information is useful in extrapolating animal data to humans and calculating what may be considered a safe human dose for a given toxic substance.

The threshold dose ($ThD_{0.0}$) is measured as mg/kg/day. It is assumed that humans are as sensitive as the test animal used. To determine the equivalent dose in man the $ThD_{0.0}$ is multiplied by the average weight of a man, which is considered to be 70 kg. The calculation used to determine the safe human dose is as follows:

$$SHD = \frac{ThD_{0.0} \times 70 \text{ kg}}{SF} = \text{Amount mg/day of toxic substance}$$

Where

SHD = Safe Human Dose

$ThD_{0.0}$ = Threshold dose at which no effect is observed

70 Kg = Average weight of a man

SF = Safety Factor (ranges from 10 to 1,000), which varies according to the type of test and data used to obtain the $ThD_{0.0}$

The safety factor chosen is dependent on the slope of the dose-response curve, the type of experimental animal used, and the availability of data from human exposure. In general, the lower the LD_{50} or LC_{50} the larger the safety factor used. The lower LD_{50} or LC_{50} implies a more toxic substance, and a higher safety factor is chosen to ensure that a safe human dose is established. For example, suppose the $ThD_{0.0}$ for substance A has an LD_{50} of 0.5 mg/kg and the $ThD_{0.0}$ for substance B has an LD_{50} of 5 mg/kg. If all other test protocols are the same, substance A is 10 times more toxic than substance B. Therefore, to determine a safe human dose the safety factor chosen for substance A will be larger (100 or 1,000)

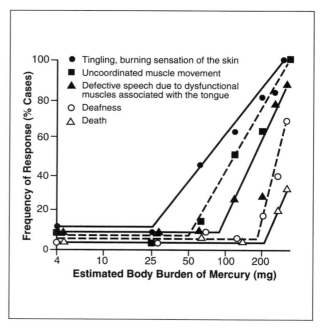

Figure 4-6: Dose-response curves for various responses as a result of exposure to methyl mercury. *Recreated with permission from Lu F. C.*, Basic Toxicology: Fundamentals, Target Organs, and Risk Assessment, *Taylor and Francis, New York, 1985.*

Box 4-1 ■ Terminology Associated with Dose-Response Curves

There are several terms associated with the dose-response curve that are utilized to describe in a quantitative manner the level of response observed as a result of exposure to a toxic substance (see Figure to the right). Depending upon the publication, these terms may be used interchangeably, which sometimes creates confusion. The term threshold is defined as the lowest dose of a toxic substance at which a specific, measurable effect is observed and below which it is not observed. Two other terms, **NEL** and **NOEL** are often used interchangeably with this term. NEL is the acronym for **No-Effect Level**; and NOEL is the acronym for **No-Observed-Effect Level**. Both describe the fact that there is no observed effect. Once effects are observed several terms are used. The level of exposure where some minor effects are observed but are not directly related to the response being measured is referred to as the **LOEL, Lowest-Observed-Effect Level**. The level where some toxicant-related effects are observed but the observed effect seen at higher levels is still absent, is referred to as the **NOAEL, No-Observed-Adverse-Effect Level**. The level where the first signs of adverse effects are seen similar to those seen at higher doses is the **LOAEL, Lowest-Observed-Adverse-Effect Level**. The level where overt effects are observed is referred to as the **Frank-Effect Level (FEL)**.

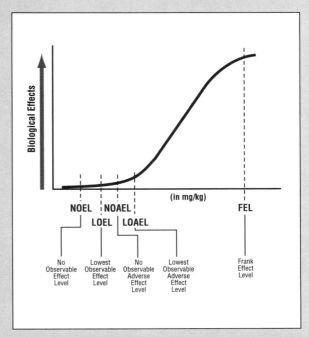

Illustration of the locations of various effect-no effect levels on a dose-response curve. *Reprinted with permission from Ecobiochen D., The Basis of Toxicity Testing, Copyright CRC Press, Boca Raton, FLA 1990/1992.*

than the safety factor chosen for substance B, which may be 10. Performing the calculations using these data will result in a safe human dose for substance A being smaller than for substance B. Calculation of the **Acceptable Daily Intake (ADI)** is another means of establishing safe levels of chemical exposure. The factors involved in calculating these levels are described in more detail in Chapter 10.

Each threshold or NOEL boundary is based upon a specific effect that manifests itself when the test subjects are exposed to different concentrations (dosage) of the chemical. Only one effect is usually chosen as a measure of response and, therefore, the threshold applies only to the response being measured. In many acute toxicity studies death of the organism is the observed and measured response. However, exposure to toxic substances may result in a wide array of effects. Other effects (Tables 4-1, 4-4, and 4-5) may occur at doses lower than the ones causing death. Figures 4-5 and 4-6 illustrate several different effects that can occur when exposed to Actinomycin D and methyl mercury. The threshold will vary depending on the type of response chosen to measure the effects of exposure to the toxic substance.

Exposure to nitrates and nitrites also produces a variety of effects. Nitrates and nitrites are added to several types of food, particularly meats such as ham, bacon, and sausages. These substances, along with many others, are added to "enhance" the flavor, color, and texture of the food. In addition, nitrites – and to a lesser extent nitrates – inhibit the growth of the bacterium *Clostridium botulinum*, which is responsible for causing **botulism** (food poisoning). Other sources of nitrates include vegetables and drinking water (see Box 4-2 on Nitrates Awareness: A Health Issue). It is estimated that the average total daily intake from all sources is 75 mg.

Box 4-2 ■ Nitrate Awareness: A Health Issue

Q: What are nitrogen compounds and where are they?

A: Nitrates and nitrites are essential nutrients for plants and have been commonly used in lawn treatments and fertilizers for many years. Frequently they find their way into groundwater supplies. Shallow wells and groundwater often contain higher levels of these substances than deeper waters not in contact with the surface soil. However, an Iowa State University Special Report found that old or depleted wells – often just abandoned and not filled with concrete as most water specialists recommend – become readily available reservoirs for runoff and excessive groundwater. As they refill with surface water, they may become concentrated with potentially toxic lawn care and agricultural chemicals, contaminating new wells dug in the same area.

Q: Why should I be concerned?

A: Nitrates may cause problems in young children and farm animals, as they bind very strongly to hemoglobin affecting the blood's ability to carry and release oxygen. This is how it happens: nitrates are ingested – through water or food – and nitrate-reducing bacteria in an infant's digestive tract convert these nitrates to nitrites. These nitrites find their way to the circulatory system, and bind very tightly with hemoglobin, which is the component of the blood that attaches to oxygen in the lungs releasing it to the body tissues that need it. If nitrites bind to hemoglobin, the latter becomes practically useless for oxygen transfer, causing shortness of breath, increased susceptibility to illness, heart attacks, and even death by asphyxiation in extreme cases. Nitrates are not usually a problem for older children and adults. As a child develops, the acidity of the stomach becomes stronger, and nitrate-reducing bacteria are killed.

Q: How can I tell if my water has nitrates in it?

A: The surest way to tell is by having your water tested. Use a reputable lab – either run by your County Health Department, or certified by state authorities or the EPA. The EPA has also published an extensive, sophisticated national survey of nitrate and pesticide levels in drinking water wells. The National Pesticide Survey tested 1,349 community and domestic rural wells in every state. Nitrates were detected in some 57 percent of domestic wells, and 52 percent of community wells. In 1992, when the survey was released, some 22,500 infants drinking domestic well water were estimated to be exposed to levels of nitrate exceeding the EPA safe drinking water limits. For community systems, the number of infants was estimated to be 43,500.

Q: What can I do about it?

A: Water found to contain excessive nitrates can be treated by a variety of methods. Point-of-Use systems reduce the levels by either reverse osmosis filtration, distillation, or a disposable mixed-bed deionizer – and can remove the nitrates (and other contaminants) for water specifically to be used for drinking and cooking. Another option is a system very much like a water softening system, using a strong anionic exchange resin bed rather than the cationic exchange resin bed commonly used for water softening. It is regenerated in a similar way to conventional softeners. This system provides a whole-house removal solution and is most effective in tandem with a water-softening system.

Many communities with a municipal water treatment system split off a portion of the water and treat it to remove nitrates. This purified water is blended into the general water supply, effectively diluting the nitrate levels to below EPA standards. Consumers especially concerned with nitrate levels may still prefer to use a Point-of-Use system to remove remaining nitrates.

This information is modified from articles published by the Water Quality Association, Water Review Technical Briefs, Internet Site http://www.wqa.org/WQIS/wqb-nitrates.

When nitrates are ingested they can be converted to nitrites by bacteria located in the digestive tract. Nitrites in the blood stream oxidize the iron atom (Fe^{2+}) of hemoglobin in the red blood cells to Fe^{3+} to form **methemoglobin**.

> Methemoglobin is a compound formed when iron in the hemoglobin molecule is oxidized from the ferrous (Fe^{2+}) to the ferric state (Fe^{3+}).

Methemoglobin does not combine with oxygen. If enough nitrates are consumed, **cyanosis** and **anoxia** (lack of oxygen) may occur as a result of decreased circulating oxygen concentrations.

> Cyanosis is a bluish discoloration of the skin as a result of decreased oxygen in the blood.

Usually the body has the ability to cope with elevated levels of nitrites in the blood. However, baby formula using water high in nitrates, high nitrate content of spinach, and in some cases, meat with high nitrate and nitrites, have resulted in life-threatening **methemoglobinemia**.

> Methemoglobinemia is the condition in which more than 1 percent of the iron in hemoglobin has been oxidized and is therefore unable to carry oxygen.

Under normal circumstances, it is estimated that the amount of hemoglobin that has been oxidized to methemoglobin is one percent in adults and two percent in children, with no observable effects. Cyanosis usually appears when 10 percent of the hemoglobin has been oxidized, and cerebral anoxia occurs around 20 percent. Coma and death usually result when 60 percent of the hemoglobin has been oxidized to methemoglobin. Plotting dose-response curves on the same graph for each of these effects would look analogous to the graphs for methyl mercury (Figure 4-6). Depending on the response chosen, the threshold will change.

In the examples given above a threshold concentration can be determined. However, some chemicals are believed to produce adverse effects at any level of exposure. Therefore, no threshold would exist (Figure 4-7). Certain carcinogenic substances are thought to be characterized by a dose-response

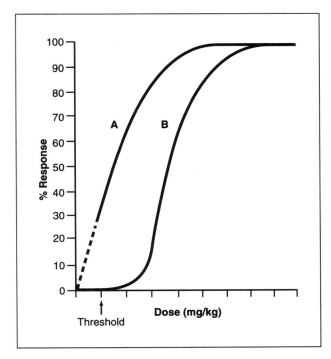

Figure 4-7: Two dose-response curves, one illustrating a chemical that does not have a threshold (A) and one that does (B). Chemicals with no threshold often produce their effects at any concentration. The lower concentrations cannot be easily quantified; therefore, the lower portion of the curve is extrapolated and passes through 0. *Recreated with permission from Timbrell J. A.,* Introduction to Toxicology, *Taylor and Francis, New York, 1989.*

curve with no threshold. This type of dose-response curve suggests that there is a response at any concentration and, therefore, no concentration is safe. This is often controversial since the response at the lower doses tends to be extrapolated on the basis of the responses at the high doses. Extrapolation is necessary because of the difficulty of accurately quantifying the lower doses. The dotted line in Figure 4-7 illustrates the extrapolated portion of the dose-response curve. Lead, which causes nerve damage in children and adults, may also be characterized as having no threshold dose.

Relative Toxicity

A comparison of the dose-response curves for two or more different chemicals – administered to the same type of test animal – can be used to determine the relative toxicity of the chemicals. This concept is illustrated in Figure 4-8. The LD_{50} for chemical A is less than the LD_{50} for chemical B. Therefore, chemical A is considered more toxic than chemical B. Although this may be true over the entire range of doses illustrated in Figure 4-8, it may not be true for other chemicals. Figure 4-9 illustrates a situation where chemical A is more toxic than chemical B at the higher doses, but at the lower doses Chemical B is more toxic than chemical A. Therefore, it is important to evaluate the relative toxicity across the entire range of doses. Table 4-7 lists the relative toxicity of various substances administered to a variety of species.

> Relative toxicity information can be used to help determine if a given chemical is safer than others.

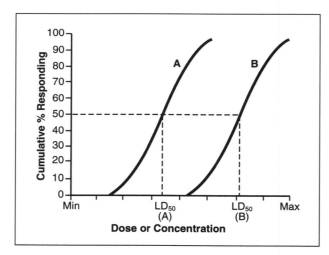

Figure 4-8: Dose-response curves for two chemical agents. Chemical A is more toxic over the entire range of doses. *Recreated with permission from Loomis T. A. and Hayes A.W.,* Essentials of Toxicology, *Academic Press/Harcourt Brace, New York, 1996.*

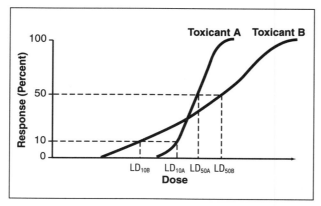

Figure 4-9: Dose-response curves for two chemicals. Chemical A is more toxic at higher doses. Chemical B is more toxic at lower doses. *Recreated and modified with permission from Williams P. L. and Burson J. L,.* Industrial Toxicology: Safety and Health Applications in the Workplace, *Van Nostrand Reinhold, New York, 1985.*

Compound	Susceptibility Factor[a]			
	Mouse	Guinea Pig	Rabbit	Dog
Chlorobenzilate	1.4	–	–	–
DDT	0.3-0.8	0.3	0.3-0.5	–
Methoxychlor[b]	2.7-3.8	–	–	–
Lindane	1.0	0.7-0.9	0.4-1.5	–
Aldrin	0.9-1.2	1.2-1.6	0.5-1.1	0.4-0.8
Chlordane	0.8	–	1.1-3.4	–
Dieldrin	1.2	0.9-1.0	0.9-1.0	0.7-0.8
Endrin	–	0.5-2.7[c]	1.8-6.2	–
Heptachlor	0.6-1.5	0.8-0.9	–	–
Azinphosmethyl	1.6	0.2	–	–
Chlorthion®	0.7	–	–	–
Diazinon	0.9-1.4	0.3	0.8	–
Dimethoate	3.0-5.4	0.5-0.9	–	–
Dioxathion	–	–	–	1.1-11.8
Malathion	1.6-1.9	–	–	–
Methyl parathion	0.4	–	–	–
Mevinphos	0.9-1.6	–	–	–
Oxydemeton-methyl	1.0-2.5	0.3-0.6	–	–
Parathion	0.2-5.0	1.6-3.2	0..5-3.0	–
Phosphamidon	1.3	–	–	–

[a] A factor of less 1.0 indicates less susceptibility than that of the male rat; a factor greater than 1.0 indicates greater susceptibility.
[b] Both sexes.
[c] Approximate

Table 4-7: Relative susceptibility of different species to pesticides. Based on oral LD_{50} values using the male rat as a standard. *Obtained with permission from Hayes W. J. Jr. and Laws E. R.,* Handbook of Pesticide Toxicology Vol. 1 General Principles, *Academic Press/Harcourt Brace, New York, 1991.*

Relative toxicity can also be evaluated by comparing the slope of each dose-response curve. A steep dose-response curve is indicative of a chemical that is rapidly and extensively absorbed and slowly detoxified; therefore, the chemical is more toxic. A small change in the dose results in a large change in the observed response. A dose-response curve that is relatively flat is indicative of a chemical that is poorly absorbed and detoxified readily; therefore, the chemical is less toxic. A small change in the dose results in a smaller change in the observed response. Figure 4-10 illustrates the difference in observed response for two chemicals having different slopes. Notice the change in response for chemical A – having the steeper slope – is greater than for chemical B, for which the slope is less steep.

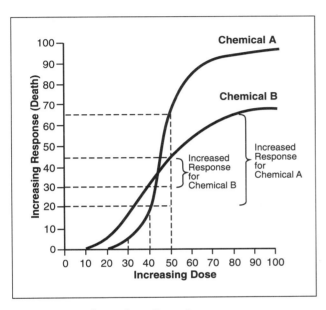

Figure 4-10: Illustration of two dose-response curves with different slopes. The slope for chemical A is steeper than the one for chemical B. A change in dose (40 to 50 mg/kg) will result in a greater response for chemical A than for chemical B.

Checking Your Understanding

1. What is a dose-response curve?

2. How are dose-response curves used to evaluate toxicity?

3. Define the term threshold and explain its significance in establishing chemical toxicity.

4. What other terms are used to describe doses where 1) no effects are observed and 2) effects are observed. Define each term.

5. What is the difference between a toxic substance that has a threshold and one that does not?

6. Identify the factors involved in calculating a safe human dose. How does the threshold dose affect the choice of a safety factor used in the calculation?

7. Define relative toxicity. How can dose-response curves be used to predict relative toxicity of various substances?

4-4 Variables Influencing Toxicity

Relative toxicity and threshold levels are attributes of a particular chemical and are indicative of its toxicity. These attributes are affected by a variety of factors. These factors include the dose-time relationship, the route of exposure, the physical and chemical properties of the toxic substance, the chemical interactions, the interspecies and intraspecies variability, genetics, age, sex, and health of the individual collectively. These factors will affect the absorption, distribution, metabolism, and elimination of the toxic substance. The toxicity of a given substance is the net result of the complex interaction between all of these various factors.

Dose-Time Relationship

Every day we ingest a number of synthetic and natural substances that have the potential to be lethal. For example, one hundred cups of strong coffee contain enough caffeine to be lethal. There is a lethal dose of oxalic acid in 10 to 20 pounds of spinach or rhubarb. And there is a lethal dose of aspirin in a bottle of 100 tablets. Most individuals are obviously unaffected by consuming small amounts of these items. In fact, people are encouraged to consume leafy, green vegetables such as spinach because of their nutritional value; and aspirin is sold to relieve symptoms of illness.

The examples discussed above demonstrate several important points concerning chemical exposure and toxicity. Not only is the dose of a toxic substance important in determining toxicity, but so is the frequency and length of exposure. For each of the toxic substances mentioned there is a specific concentration in the blood (threshold) that is associated with the occurrence of a toxic effect. This is true for any chemical, and can occur as a result of chronic or acute exposure.

Consumption (exposure) of the substances mentioned above occurs in small amounts (doses) and usually over a prolonged, infrequent period of time (chronic exposure). The exposure and absorption will result in increased levels of these substances in the blood. How long the toxicant levels remain elevated in the blood is dependent on how fast the substance is metabolized and removed from the body. For instance, daily doses of saccharin result in elevated blood levels (Figure 4-11). However, saccharin is easily metabolized and excreted from the body. The blood levels of saccharin therefore return to the baseline level. The rapid elimination of the substance from the bloodstream decreases the probability of an adverse effect.

Other substances are not as readily metabolized and excreted from the blood. Continuous exposure to these compounds will result in elevated blood levels. Relative to other chemicals, such as saccharin, methyl mercury is not rapidly removed from the bloodstream. Repeated exposure over 270 days

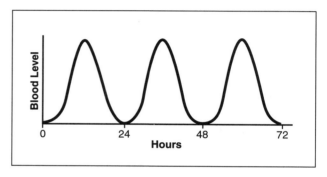

Figure 4-11: Blood levels of saccharin as a result of daily dosing. The saccharin is easily metabolized and excreted resulting in blood levels returning to zero. *Recreated with permission from Lu F. C.,* Basic Toxicology: Fundamentals, Target Organs, and Risk Assessment, *Taylor and Francis, New York, 1985.*

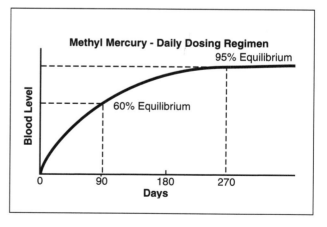

Figure 4-12: Blood level of methyl mercury as a result of daily dosing. The methyl mercury is not as readily metabolized and excreted as saccharin therefore it accumulates in the blood. *Recreated with permission from Lu F. C.,* Basic Toxicology: Fundamentals, Target Organs, and Risk Assessment, *Taylor and Francis, New York, 1985.*

results in a gradual accumulation of methyl mercury (Figure 4-12). Gradual accumulation increases the probability that an adverse effect will occur.

Most individuals do not experience any lethal effects from the routine consumption of aspirin, coffee, or spinach. The body is able to lower the circulating levels of the toxic substances between exposures. Therefore, the lethal threshold is not exceeded. A lethal dose would require an acute exposure to the toxic substance i.e., drinking 100 cups of coffee or consuming a bottle of aspirin over a short period of time.

Acute effects are more readily apparent and observed after exposure to high doses of a toxic substance. Usually, different target tissues are affected, and the observed effects are therefore different than observed effects of chronic exposure. Take examples such as chlorinated solvents, lead, or sodium fluoride: symptoms of acute exposure to sublethal quantities of chlorinated solvents, such as chloroform or carbon tetrachloride, are primarily associated with the nervous system, such as excitability, dizziness and narcosis. Chronic exposure to these solvents is primarily associated with liver damage. Symptoms of chronic lead exposure include damage to the nervous and muscular systems as well as to the blood-forming tissues such as bone marrow. Symptoms of acute exposure to ingested lead are primarily associated with the gastrointestinal tract. Frequent exposure to large quantities of lead can produce symptoms associated with both acute and chronic exposure. Acute exposure of sodium fluoride – oral $LD_{50} \approx 35$ mg/kg – is also highly toxic. However, chronically, oral ingestion in very small amounts (1-2 mg daily), is essential to good dental health.

Routes of Exposure and Physical and Chemical Factors

Exposure to toxic substances occurs primarily by inhalation, ingestion, or by contact with the skin. Toxicity is determined, in part, by the quantity ab-

Compound	LD_{50} in Male Rats (mg/kg)		"No Effect Level"* (mg/kg/day)	ADI** (mg/kg/day)
	Oral	Dermal		
DDT	113 (p,p'-DDT) 217 (technical)	–[a] 2,510	Rat - 0.05	0.005
DDE	880	–[a]	–[a]	–[a]
DDA	740	–[a]	–[a]	–[a]
Methoxychlor	5,000-7,000	–[a]	Rat - 10	0.1
Aldrin	39	98	Rat - 0.025 Dog - 0.025	0.0001
Dieldrin	46	90	Rat - 0.025 Dog - 0.025	0.0001
Endrin	18	18	Rat - 0.05 Dog - 0.025	0.0002
Heptachlor	100	195	Rat - 0.25 Dog - 0.06	0.0005
Chlordane	335	840	Rat - 1.0 Dog - 0.06	0.001
Lindane	88	1,000	Rat - 1.25	0.0125
Mirex	740	>2,000	–[a]	–[a]

* Threshold
** Acceptable Daily Intake
[a] No Data Available

Table 4-8: Toxicity of some organochlorine insecticides. *Obtained with permission from, Fitzgerald J. J., Toxicology for Scientists and Engineers, 1985.*

sorbed into the bloodstream and distributed to the target tissue. As we saw in Chapter 3, regardless of the route of exposure, the amount absorbed is dependent on the structural and chemical characteristics of the cells at the point of contact as well as the chemical and physical characteristics of the toxicant.

In general, absorption through the skin is more difficult for some types of toxic substances because of the relatively nonpermeable nature of the outer layer of skin, the epidermis. The outer, keratinized layers of the skin are designed to prevent excessive evaporation of water from the body. Conversely, it is difficult for water-soluble substances such as salts of toxic metals to be readily absorbed. Substances that are fat-soluble (lipid), such as organic solvents like xylene (paint thinner), acetone, and trichloroethylene, readily penetrate the skin. Other substances, such as nicotine, strychnine, and "nerve gas" (sarin), can also penetrate the skin. Most toxic metals and particulates are not absorbed through the skin. An exception is metallic mercury. Particulates may be absorbed if the skin is damaged and if the particulate is soluble in the fluids present in the exposed area.

Compound	LD_{50} in Male Rats (mg/kg)		"No Effect Level"* (mg/kg/day)	ADI** (mg/kg/day)
	Oral	Dermal		
TEPP (tetraethyl pyrophosphate)	1.1	2.4	–a	–a
Mevinphos	6.1	4.7	–a	–a
Disulfoton	6.8	15	–a	–a
Azinphosmethyl	13	220	Rat - 0.125 Dog - 0.125	0.0025
Parathion	13	21	Rat - 0.05 Man - 0.05	0.005
Methylparathion	14	67	–a	–a
Chlorfenvinphos	15	31	Rat - 0.05 Dog - 0.05	0.002
Dichlorvos	80	107	Rat - 0.5 Dog - 0.37 Man - 0.033	0.004
Diazinon	108	200	Rat - 0.1 Monkey - 0.05 Dog - 0.02 Man - 0.02	0.002
Dimethoate	215	260	Rat - 0.4 Man - 0.04	0.02
Trichlorfon	630	>2,000	Rat - 2.5 Dog - 1.25	0.01
Chlorothion	880	1,500-4,500	–a	–a
Malathion	1,375	>4,444	Rat - 0.5 Man - 0.2	0.02
Ronnel	1,250	>5,000	Rat - 0.5 Dog - 1.0	0.01
Abate	8,000	>4,000	–a	–a

* Threshold
** Acceptable Daily Intake
a No Data Available

Table 4-9: Toxicity of some organophosphate insecticides. *Obtained with permission from, Fitzgerald J. J.,* Toxicology for Scientists and Engineers, *1985.*

Amine	Oral LD$_{50}$ (mg/kg)	Skin LD$_{50}$ (mg/kg)	Dermal Effects
Methylamine	0.02	0.04	Necrosis[a]
Ethylamine	0.4	0.4	Necrosis[a]
Propylamine	0.4	0.4	Necrosis[a]
Butylamine	0.5	0.5	Necrosis[a]
Hexylamine	0.7	0.4	Slight necrosis[a]

[a] Death of cells

Table 4-10: Toxicity of some amines. *Obtained with permission from Williams P. L. and Burson J. L.* Industrial Toxicology: Safety and Health Applications in the Workplace, *Van Nostrand Reinhold, New York, 1985.*

Numerous toxic substances may enter the body via the respiratory system. Small, lipid-soluble substances such as organic solvent vapors (ie., methylene chloride) are easily absorbed through the alveoli into the bloodstream. Toxic gases such as carbon monoxide and hydrogen cyanide pass through the alveoli by diffusion. Asbestos and lead, adsorbed to particulates, can be engulfed by phagocytic cells and transported to other portions of the body.

Small amounts of toxic substances enter the body as a result of the food we eat and the water we drink. In general, ingested toxic substances such as heavy metals are poorly absorbed from the digestive tract and are readily removed in the feces. Also, the heavy metals must compete with essential nutrients such as calcium and iron for the carrier molecules involved in absorption. Organic solvents such as benzene and gasoline are most likely to enter the digestive tract as a result of accidental exposure. These types of toxic substances are fat-soluble and readily diffuse into the bloodstream. Symptoms associated with accidental poisonings as well as intentional ingestion (suicide) are generally the result of ingestion of larger amounts of toxic substances. Increased absorption then results in elevated levels of circulating toxins, which cause the observed effects associated with exposure.

Because of the differences in the characteristics of cells at the point of contact – as well as the physicochemical differences associated with chemicals – the toxicity of various substances may be different. Tables 4-8 and 4-9 list the LD$_{50}$ values for various organochlorine and organophosphate pesticides resulting from ingestion and skin exposure. In addition, the threshold ("no-effect level") values are

Site of Application	Hydrocortisone[b]		Parathion[c]	
	Percent Absorbed	Compared with Forearm (Ratio)	Percent Absorbed	Compared with Forearm (Ratio)
Ventral forearm[a]	1.04	1	8.6	1
Foot arch, plantar	0.17	0.14	–	–
Ball of foot	–	–	13.5	1.6
Palm of hand	0.78	0.83	11.8	1.3
Back	1.26	1.7	–	–
Abdomen	–	–	18.5	2.1
Axilla	3.7	3.6	63	7.4
Scalp	4.41	3.5	32	3.7
Forehead	7.65	6.0	36	4.2
Jaw angle	12.85	13	33	3.9
Ear canal	–	–	46	5.4
Scrotum	36.2	42	101	11.8

[a] Many drug absorption studies are done on the skin of the ventral forearm, since this is a part that everybody will expose readily.
[b] C^{14} hydrocortisone (hormone) in acetone on surface of 13 cm^2 for 24 hr.
[c] C^{14} parathion in acetone, 4 mg/cm^2 for 24 hr.

Table 4-11: Regional differences of dermal absorption in human male. *Obtained with permission from Sperling F.,* Toxicology Principles and Practices, Volume 2. *Copyright © 1984 by John Wiley and Sons, Inc. Reprinted by permission.*

provided along with a calculated acceptable daily intake. (Table 4-10 lists the LD_{50} values for various amines from oral and skin exposures.)

Absorption of toxic substances in the digestive and respiratory tract occurs primarily in the small intestine and alveoli, respectively. In both instances the toxic substance is transported across a single layer of cells. As a result, toxic substances will enter the bloodstream readily. Absorption of toxic substances through the skin requires movement of the toxicant through several layers of cells. The thickness of the layers, in particular the keratinized stratum corneum, varies in the different regions of the body. As a result, the rate and amount of toxic substance absorbed into the bloodstream will be different. Table 4-11 illustrates the different absorption rates for various regions of the skin in the human male.

Chemical toxicity can be affected by the arrangement of the atoms in a molecule. For instance, 1,1-dichloroethylene, cis-1,2-dichloroethylene, and trans-1,2-dichlorethylene all have the same number of carbon, chlorine, and hydrogen atoms, but their orientation in the molecule is different (Figure 4-13). The chlorine atoms of 1,1-dichloroethylene are bound to the same carbon; in 1,2-dichloroethylene they are bound to different carbon atoms. There are two **isomers** of 1,2-dichloroethylene: the cis and the trans form. Cis indicates that the chlorine atoms are on the same side of the double bond between the two carbons. Trans means they are on opposite sides.

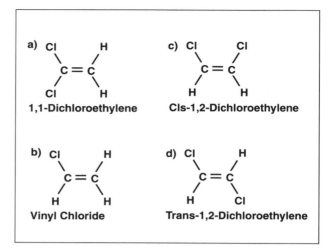

Figure 4-13: Structural formulae for dichloroethylene and vinyl chloride.

> An isomer is one of two or more chemical substances that have the same molecular formula but a different arrangement of the atoms in the molecule resulting in different chemical and physical properties.

These structural differences result in different toxic effects being observed as a result of exposure. The LD_{50} for 1,1-dichloroethylene is 5,750 mg/kg and the LD_{50} for the 1,2-dichloroethylene isomers is 770 mg/kg. However, based on animal studies, 1,1-dichloroethylene is a suspected human carcinogen. Its structure is similar to a known human carcinogen, vinyl chloride (Figure 4-13 b). The 1,2-dichloroethylene isomers are not carcinogenic, but they do have different effects. The cis isomer is an irritant and acts as a narcotic (causes drowsiness). The trans isomer affects the central nervous system (brain and spinal cord) and the gastrointestinal system causing weakness, tremors, cramps, nausea; it may also cause liver and kidney damage.

The differences in effects appear to be related to the rate of metabolism and the chemical intermediates formed. Cis-1,2-dichloroethylene is metabolized faster than trans-1,2-dichloroethylene. Both are converted to the chemical intermediates dichloroacetic acid and dichloroethanol, which are further metabolized and excreted. Metabolism of 1,1-dichloroethylene results in the formation of a toxic intermediate that causes cancer.

Interspecies and Intraspecies Variability

Most toxicological studies are performed on various species of animals such as rats, mice, dogs, rabbits, and monkeys. These animals are used because they are either plentiful, and/or because they represent the best model for predicting the response to a toxic substance in humans. It is assumed that humans, when exposed to the same toxic substance, will have the same response as the test species. Depending on the species, however, the response to a particular toxic substance may or may not be the same; some responses are species-specific as illustrated in Table 4-12. It is necessary to know and understand the importance of species variation in response to toxic chemical exposure in order to properly extrapolate results from animal toxicity studies to man.

Compound	Toxicity	Reactive Species[a]	Non-Reactive Species[b]
Beta-naphthylamine	carcinogen	dog, man	rat
Aflatoxin B1	carcinogen	rat, trout	monkey, mouse
Phenylthiourea	**pulmonary edema**[c]	rat	rhesus monkey
Norbormide	respiratory failure	rat	cat, dog, mouse
DDT	weak carcinogen	mouse	hamster
3-methylcholantrene	carcinogen	mouse, rat	rhesus monkey
Thalidomide	teratogen	rabbit, man	hamster, rat, mouse

[a] Reactive species are those in which an effect is observed.
[b] Non-reactive species are those in which an effect is not observed.
[c] Pulmonary edema is the excessive accumulation of fluid in the lungs.

Table 4-12: Toxicity of various compounds to different test species. *Obtained from Galli C. L. et al.*, The Principles and Methods in Modern Toxicology, *1980.*

Interspecies Variability

The variability in response between species is primarily the result of the evolution of different biological mechanisms involved in the metabolism of toxic substances. Each species may absorb, metabolize, or excrete the substance at different rates, or not at all. A comparison of LD_{50} values for different species exposed to the same toxic compound is indicative of the inherent capability to metabolize toxic substances at different rates. Table 4-7 lists the LD_{50} values for several different species.

The differences in toxicity may be due to the rate at which metabolism occurs, or whether metabolism produces chemical intermediates more toxic than the parent compound. For example, the pesticide malathion is metabolized at a different rate in mammals than in insects. In mammals, the chemical intermediates are rapidly metabolized to deliver end products that are excreted from the body. Insects metabolize malathion more slowly and produce a chemical intermediate – malaoxon – that is toxic to the insect. The differences in the rate of metabolism as well as the production of the toxic chemical intermediate both account for the selective toxicity of this compound.

Similarly, the pesticide 2-acetylaminofluorene may cause cancer in some species (rat, rabbit, hamster) and not in others (Table 4-13). The pesticide undergoes a chemical reaction, N-hydroxylation, which produces the cancer-causing intermediate. This reaction does not occur in the guinea pig, and cancer does not occur.

Other chemicals also demonstrate species variability. Ethylene glycol, for example, is more toxic to cats than rats. Cats produce more of the toxic intermediate oxalic acid than rats do. Aniline is more toxic to cats and dogs than to rats and hamsters. Cats and dogs produce more o-aminophenol, which is more toxic. Rats and hamsters convert aniline to p-aminophenol, which is less toxic.

Both chronic and acute exposure to methanol by means of ingestion or inhalation are toxic to man and other primates such as monkeys. Methanol intoxication results in ocular damage and blindness. However, in other species methanol is less toxic and the symptoms associated with human intoxication are not observed.

Acute and chronic exposure to an industrial chemical named **tri-ortho-cresyl phosphate (TOCP)**

Species	Carcinogenicity	N-hydroxylation[b] % of the Dose
Rat	+	8
Rabbit	+	21
Hamster	+	17
Dog	+	5
Guinea Pig	–	0
Steppe Lemming	–	trace
Man[c]	?	9

[a] Pesticide
[b] Represents the percent of 2-acetylaminofluorene that underwent the chemical reaction N-hydroxylation.
[c] Effects unknown in man.

Table 4-13: Appearance of cancer in various species exposed to 2-acetylaminofluorene[a]. *Obtained from Galli C. L. et al.*, The Principles and Methods in Modern Toxicology, *1980.*

causes nerve damage in man and chickens. Similar symptoms are not observed in dogs and rats when they are exposed to TOCP.

Even humans and monkeys – who are very similar to each other physiologically – do not always respond in the same manner. They respond the same way to methanol intoxication but – when humans are exposed to nitrobenzene – the hemoglobin in the blood is converted to methemoglobin, which is incapable of carrying oxygen. Nitrobenzene does not affect monkey hemoglobin; however, dogs are affected in the same manner as humans. In this situation, dogs would be a better model to predict the possible toxic effects of nitrobenzene to humans than monkeys. The effects of metabolism on toxicity are discussed in more detail in Chapter 5.

Other interspecies differences can play a role in affecting toxicity. Rats do not have a vomiting reflex. Rodenticides, such as squill, are more effective if they are ingested because the rat can not discard the substance by vomiting. Chronic exposure of rats and mice to 15 ppm formaldehyde results in fewer nasal tumors in mice than in rats. When exposed to formaldehyde the rate and depth of breathing declined in both species. However, the decline was greater in the mice. Therefore, less formaldehyde came in contact with the cells lining the nose resulting in fewer mice developing cancer.

Absorption differences between species also play an important role in affecting toxicity. DDT, for instance, is more readily absorbed across the skeletal chitin of insects than across the skin of humans. Table 4-14 illustrates how the skin from different species can affect the penetration rate of phenoxy-herbicides. Differences in the amount and the rate at which these substances penetrate the skin will play a role in determining the toxicity of these substances.

It is clear from these examples that no one species handles all chemicals in the same manner as other species; and there is no one animal model that can be used to study the toxicity of all chemicals for humans. Therefore, extrapolation of information concerning the toxicity of chemical substances to humans must be performed with the utmost care.

Intraspecies (Individual) Variability

Typically, individuals within a population – exposed to a single dose – will demonstrate a wide array of responses. For example, 10 mg/kg of the pesticide parathion will not kill every rat. It will kill approximately 50 percent of the rats and a larger dose will be required to kill the rest over the same period of time. Of the 50 percent killed initially, several may have been killed by a smaller dose.

This type of variability can be caused by differences associated with individual genetics and sex of the individuals within the population as demonstrated by the results in Table 4-15. Different strains of rats were exposed to a chemical intermediate (N-hydroxy-AAF) of the pesticide 2-acetylamino-

Strains	Source	LD_{50}	
		mg/kg	M/F Ratio[c]
Sprague-Dawley			
Female	Holtzman	315 (241-413)	6.3
Male	Holtzman	50 (42-58)	
Sprague-Dawley			
Female	Carworth CFE	506 (372-689)	7.4
Male	Carworth CFE	68 (54-86)	
Wistar			
Female	Carworth CFN	116 (76-175)	2.2
Male	Carworth CFN	52 (32-83)	
Fischer 344			
Female	Charles River	52 (34-80)	0.9
Male	Charles River	61 (38-97)	

[a] Pesticide
[b] Strain refers to a different variety or type of rat.
[c] Male/Female LD_{50} ratio

Species	Dose (mg/kg of body weight)	Penetration (%)
Rat	1	5
	10	2
	100	1
Rabbit	1	18
	10	13

Table 4-14: Dermal penetration of phenoxy-herbicides in the rat and rabbit. *Obtained with permission from Volans G. N. et al. editors,* V International Congress of Toxicology, Basic Science in Toxicology, *Taylor and Francis, New York, 1990.*

Table 4-15: Acute toxicity (LD_{50}) of N-hydroxy acetylaminofluorene[a] in various strains[b] of rats. *Obtained from Galli C. L. et al.,* The Principles and Methods in Modern Toxicology, *1980.*

fluorene. The table illustrates a couple of important points relative to factors that affect intraspecies toxicity. First, some strains are more sensitive to the toxic substance than others. The LD_{50} values for male Fischer 344 rats (a specific strain or type of rat) are less than the LD_{50} values for other male rats. Second, there is a difference in the toxicity of the substance between males and females on average. Therefore, the strain and the sex of the test animal can significantly affect the individual's response to a toxic substance. No one factor by itself will necessarily result in a differing response, although that is sometimes possible.

Genetics

The differences in genetic makeup both within species and between species markedly influence the disposition and the metabolism of toxic substances. This in turn affects the susceptibility of the species to toxic chemicals. For example, some individuals have an enzyme (glucose-6-phosphate dehydrogenase) deficiency resulting in their red blood cells being more fragile than normal. Consequently, red blood cells of these individuals are more susceptible to chemicals such as phenylhydrazine and primaquine, which cause **hemolysis** (breakup of red blood cells). Therefore, these types of chemicals are more toxic to those individuals who have a deficiency of glucose-6-phosphate dehydrogenase.

The enzyme **aryl hydrocarbon hydroxylase (AHH)** transforms the chemical benzo[a]pyrene to its carcinogenic **metabolite**. Some individuals produce lower amounts of AHH; benzo[a]pyrene is therefore less toxic to these individuals. Similarly, low AHH individuals may be less likely to develop cancer from cigarette smoke – a source of benzo[a]-pyrene. Benzo[a]pyrene is one of many **polycyclic aromatic hydrocarbons (PAH)** contained in cigarette smoke; many PAH are potential carcinogens.

Age

In general, younger and developing individuals are more sensitive to toxic substances than older individuals. The difference can be primarily attributed to differences in metabolism; specifically to differences in the functioning of the liver. The liver is the primary site for transformation of toxic substances that enter the bloodstream. In younger individuals the mechanisms for detoxifying substances are less developed than in older individuals. However, some

Pesticide	Age	LD_{50} (mg/kg) with 95% Confidence Limits
Malathion	Newborn	134.4 (94.0-190.8)
	Pre-weaning	925.5 (679.0-1,261.0)
	Adult	3,697.0 (3,179.0-4,251.0)
DDT	Newborn	>4,000.0
	Pre-weaning	437.8 (346.3-553.9)
	Weaning	355.2 (317.2-397.8)
	Adult	194.5 (158.7-238.3)
	Middle-aged	235.8 (208.0-267.4)
Dieldrin	Newborn	167.8 (140.8-200.0)
	Pre-weaning	24.9 (19.7-31.5)
	Adult	37.0 (27.4-50.1)

Table 4-16: Effect of age on acute toxicity of malathion, DDT, and dieldrin in rats. *Obtained with permission from Lu F. C., Basic Toxicology: Fundamentals, Target Organs, and Risk Assessment, Taylor and Francis, New York, 1985.*

substances are less toxic to the young than to adults (Table 4-16). DDT is relatively nontoxic to newborn rats, but has an LD_{50} of 200 to 300 mg/kg in adult rats. Conversely malathion, parathion, and boric acid are acutely toxic to immature individuals and less toxic to adults. Certain heavy metals, such as lead and cadmium, are more toxic in young children than in adults. This can be attributed to the greater absorption rate of these substances from the intestinal tract. The lead absorption rate is four to five times greater in young individuals than in adults; the cadmium absorption rate is 20 times greater than in adults.

Sex

The difference in toxicity between genders seems to be primarily influenced by the sex hormones (estrogen and testosterone), which can affect metabolism. As with other factors discussed previously, the toxicity varies between and within classes of toxic substances. For example, some organophosphate pesticides are more toxic to females than males. Parathion is metabolized more rapidly in females resulting in a higher concentration of the more toxic intermediate, paraoxon. Female rats are also more susceptible to strychnine and warfarin. Male rats are 10 times more susceptible to liver damage than female rats as a result of chronic oral exposure to DDT.

Exposure to dinitrotoluene results in a greater incidence of liver cancer in the male than in the female. Dinitrotoluene is metabolized in the liver and excreted into the small intestine. Bacteria in the small intestine metabolize the dinitrotoluene intermediate, and the cancer-producing substance is reabsorbed into the liver. The female excretes dinitrotoluene-metabolic products through the kidneys. Similarly, kidney damage caused by exposure to chloroform occurs more readily in male mice. This effect can be reversed by castration, which removes the male sex hormones. Table 4-15 depicts differences in susceptibility between sexes of the experimental animals. In another study several compounds (unleaded gasoline, D-Lomolene, isophorone, and 1,4-dichlorobenzene) produced cancer in the kidney of male rats only, and not in female rats or mice exposed to the same chemicals (see Volans G.N et al., *Basic Science in Toxicology*, Taylor And Francis, New York, 1989).

State of Health

The liver and the kidney are important organs for detoxifying and removing toxic substances. Therefore, conditions that lead to liver or kidney disease enhance the toxic effects of substances normally detoxified by these organs. Excessive consumption of alcohol, for example, may cause liver disease, which decreases the rate at which a toxic substance is detoxified. Individuals who smoke cigarettes are more susceptible to toxic substances that affect the respiratory system such as asbestos. The number of individuals who develop lung cancer as a result of asbestos exposure is higher for smokers than for nonsmokers. Active individuals are generally better equipped than inactive individuals to handle exposure to a toxic substance. For instance, obese individuals may develop heart disease. This can affect circulation to the kidney and liver decreasing the rate of detoxification and excretion of toxic substances.

Chemical Interactions

Most toxicological studies look at the effects as a result of exposure to a single substance, usually over a short period of time. In reality, as discussed earlier, we are exposed to a wide variety of chemicals in our daily activities. The interaction between two or more chemicals and their subsequent effects have not been widely studied; not because these interactions are not important, but rather because the results are difficult to interpret and extrapolate, and also because the studies are longer and costlier.

The presence of other chemicals at the time of exposure or as a result of previous exposure can affect the response to the specific toxicant being studied. These effects can be **additive**, **synergistic**, **antagonistic**, or they may demonstrate the characteristics associated with **sensitization** or **potentiation**. In general, the presence of other chemicals tends to either enhance or decrease the toxicity of the chemical being studied.

Additive Effects – The most common effect observed when two chemicals are given together is additive. An additive effect occurs when the combined effect of the two chemicals is equal to the sum of the effects of each individual chemical. For example, chemical A produces 25 percent mortality in 14 days when administered alone, and chemical B also produces 25 percent mortality in the same time period, when administered alone. The same type of test animal is used in each study. If chemical A and chemical B are given together, the additive effect would be 50 percent mortality. Organophosphate pesticides generally have the same effect on organisms. Investigation of the acute toxicity of 43 pairs of organophosphate pesticides demonstrated that 21 pairs had additive properties. The others had less than additive effects, and a small number displayed some level of synergism. In general, it may be assumed that exposure to two organophosphate pesticides may result in an additive effect.

Synergistic – A synergistic effect occurs when the observed effects are greater than what would be predicted by simply summing the effects of each individual chemical. For example, chemical A produces 25 percent mortality, and chemical B produces 25 percent mortality. However, if the effects are synergistic, when chemicals A and B are administered together the mortality is greater than 50 percent. As an example, both carbon tetrachloride and ethanol are toxic to the liver, but together they produce much more liver injury than the sum of their individual effects.

When two organophosphate pesticides – malathion and **o-ethyl o-(p-nitrophenyl) phenylphosphorothionate** (**EPN**) – were added to the diets of dogs and rats, synergism was observed. The two toxic chemicals were fed to the rats and dogs in doses equal to 1/12th the LD_{50} of both pesticides. There was a 10-fold increase in toxicity for both agents in the dog. Other combinations of organophosphate pesticides that demonstrate synergistic effects are malathion-dipterex and dipterex-guthion.

Antagonistic – Antagonistic effects are the basis on which antidotes are developed to counteract the effects of toxic substances that enter the body. Antagonistic effects occur when the sum of the effects of each individual chemical is less than the sum of the effects for both chemicals. For example, chemical A produces 25 percent mortality and chemical B produces 25 percent mortality. If the chemicals are antagonistic, when administered together, the mortality is less than 50 percent. Mathematically this could be represented, for example, as 25 percent + 25 percent = 40 percent.

There are four major types of antagonism: functional, chemical, dispositional, and receptor antagonism:

1. **Functional antagonism** refers to a type of reaction two chemicals have on a specific physiological function in the test organism, usually producing counterbalancing or opposite effects. For example, convulsions are often produced when toxic doses of chemicals are administered. Sufficient amounts of anticonvulsants, such as benzodiazepines, are given as an antidote to counterbalance the effects of the convulsion-inducing substance.

2. **Chemical antagonism** may be referred to as chemical inactivation. This type of interaction is the result of a reaction between two chemicals to produce a less toxic product. For example, arsenic, mercury, and lead ions can be made less toxic by reacting with a chelating (binding) agent such as dimercaprol. Animal toxins, such as snake venom, are treated with antitoxins that function as chemical antagonist.

3. As discussed previously, the toxicity of a given substance is affected by the distribution, the metabolism, and the elimination of the toxicant. If these mechanisms can be altered in such a way that the toxic substance is either transformed into a less toxic substance or it is eliminated from the body more rapidly, then the observed effects will be less than the expected ones. This is referred to as **dispositional antagonism**. Dispositional antagonism has been demonstrated between organophosphate and organochlorine insecticides. A single, oral dose of the organochlorine pesticide aldrin significantly reduced the toxicity of several organophosphate pesticides in mice. For parathion, **TEPP** (tetraethyl pyrophosphate), and EPN the lethalities were reduced from 35, 95, and 50 percent, respectively, to zero in the aldrin-pretreated groups. Similar effects have been observed in parathion-exposed rodents pretreated with DDT, DDE, and chlordane. The organochlorine pesticides provided a protective effect by increasing the activity of the organophosphate detoxification mechanism.

4. Many toxic chemicals exert their effects by first binding to a target tissue receptor. **Receptor antagonism** occurs when two chemicals require binding to the same receptor in order to exert their effect. The result is that when the two chemicals are administered together the observed effect is less than the sum of the effects for each chemical. Chemicals that are receptor antagonists are often referred to as blockers. Oxygen and carbon monoxide bind to the same receptor in the red blood cell. Carbon monoxide poisoning is treated by administering oxygen. The increased oxygen level in the blood results in a decreased amount of carbon monoxide bound to the red blood cell receptor.

Sensitization – Initial exposure to toxic chemicals does not always result in an observable or measurable effect. However, subsequent exposure to the same chemical may result in an observable or measurable effect. This response is the result of the individual becoming sensitized to the chemical with the initial exposure. Sensitization occurs as a result of the interaction of the body's immune system with the toxic chemical during the initial exposure. When the chemical enters the body it is transformed, usually by interacting with a protein within the body. This newly created "protein-toxic chemical" structure causes the immune system to produce **antibodies** that will interact with the transformed chemical. When subsequent exposures occur, the antibody levels increase dramatically and produce an observable effect. The effect can range from a slight reddening or irritation of the skin to constriction of breathing, which may lead to death.

Potentiation – At first glance it would appear that synergism and potentiation are the same. As described, each synergistic chemical has an effect when administered alone; however, in a chemical interaction that demonstrates potentiation, only one of the chemicals has an observable effect when administered alone. Chemical A has no effect on mor-

Methylcarbamate	Common Name or Commercial Designation	Topical LD$_{50}$ (µg per Female Fly)		Degree of Synergism[a] (A/B)
		Alone (A)	With Pbx (B)	
1-napthyl	carbaryl	>100	0.25	>400
4-dimethylamino-3,5-xylyl	Zectran	1.2	0.27	4.4
4-(methylthio)-3,5-xylyl	methiocarb	0.48	0.25	1.9
4-dimethylamino-m-tolyl	aminocarb	1.6	0.50	3.2
m-isopropylphenyl	UC 10854 AC 5727	2.0	0.20	10.0
o-isopropylphenyl	PPC-3	2.1	0.48	4.4
o-isopropoxyphenyl	propoxur	0.47	0.14	3.4
2,3-dihydro-2,2-dimethyl-7-benzofuranyl	carbofuran	0.13	0.05	3.0
2-methyl-2-(methylthio)-propionaldehyde O-oximyl	aldicarb	0.11	0.067	1.6
3,4,5-trimethylphenyl + 2,3,5-trimethylphenyl	Landrin	1.3	0.27	4.8
3,5-di-tert-butylphenyl	butacarb	0.78	0.12	6.5
m-tert-butylphenyl	RE 5030	>100	0.16	>625
3,4-xylyl	Meobal	2.4	0.57	4.2
3-methyl-5-isopropyl(m-cym-5-yl)	promecarb	0.58	0.11	5.3

[a] Ratio of the response to substance A over the response to substance B.

Table 4-17: Synergism of some commercially important phenyl methylcarbamates with piperonyl butoxide against susceptible houseflies. *Reprinted with permission from Calabrese E. J.,* Multiple Chemical Interactions, *Copyright CRC Press, Boca Raton, FLA 1990/1992.*

tality (0 percent). Chemical B causes 25 percent mortality. When the two chemicals are administered together the effect is greater than the sum of their individual effects (0 percent + 25 percent = 100 percent). For example, when administered alone isopropanol is not toxic to the liver. When isopropanol is administered in addition to carbon tetrachloride, the toxicity of carbon tetrachloride is much greater than when given alone.

As shown in Table 4-17, carbamate insecticides have a lower LD$_{50}$ when administered with the nontoxic piperonyl butoxide (Pbx). The LD$_{50}$ for the carbamate insecticide by itself is provided in Column A. Column B represents the LD$_{50}$ for each carbamate when it is combined with piperonyl butoxide. Pbx enhances carbamate toxicity by inhibiting the enzymes involved in carbamate metabolism.

Checking Your Understanding

1. Explain the relationship between frequency of exposure, dose, and the toxicity of a substance.

2. How does the route of exposure affect toxicity? Why?

3. What factors affect the toxicity of a substance in different species?

4. Intraspecies variability in response to a given toxic substance may be attributed to what factors?

5. What specific factors associated with the age and sex of the test animal affect toxicity?

6. List and define the various types of chemical interactions that affect toxicity.

7. Many people are allergic to a given chemical; however, symptoms do not occur until after two or more exposures. What type of chemical interaction does this represent?

Summary

Information concerning the toxicity of a substance is obtained primarily from either acute or chronic toxicity studies. The information obtained is utilized to establish a safe level of exposure for that particular substance. Acute toxicity studies are the most common types of toxicological studies performed. They are short-term studies usually lasting no more than 14 days. Chronic toxicity studies occur over a longer period of time, usually months, years, or the lifetime of the test animal. Results from these tests are often expressed as either LD_{50} or LC_{50} values. These values represent the concentration at which 50 percent of the exposed individuals die. A relationship exists between the dose administered and the effect observed, and is represented by the dose-response curve. This curve provides information concerning the threshold and relative toxicity of a given chemical.

Chemical toxicity is affected by the dose-time relationship, route of exposure, interspecies and intraspecies variability, genetics, age, sex, health of the individual, physical and chemical properties and chemical interactions. These factors may increase or decrease chemical toxicity. Therefore, they must be considered when performing animal toxicity tests to establish safe levels of exposure for a toxic substance.

It is important to take into consideration all the factors that affect the toxicity of a given substance. The extrapolation of data from animal studies to humans must be tempered and evaluated in light of all these variables. They can affect the toxicity of a substance and result in the establishment of exposure levels that may not be safe.

Application and Critical Thinking Activities

1. Make a list of some chemicals found in products in your home. Review the literature and create a table that shows the LD_{50}/LC_{50} for the chemical, the amount of chemical in your product, and the volume you would have to be exposed to during a given time to reach the LD_{50} or LC_{50}.

2. Some individuals develop contact dermatitis, which is a common skin disease found in the workplace after repeated exposure to a toxic substance. Explain how and why this may occur based on information provided in this chapter.

3. Snakebites are often treated by administering a specific "anti-venom." Based upon information in this chapter explain how "anti-venom" may work.

4. Suppose a safe pain relief medication is available to the general public. After many years it is suggested that it has an adverse effect on the nervous system of young children. First, given the information in this chapter, how might you determine if the medication affected children? Second, speculate what factors you might have to address that may not have been considered during the initial testing of the product. Why?

5. What effect may the chemical structure of a toxic substance have on its toxicity? Give an example.

5

Absorption, Distribution, Metabolism, and Elimination of Toxics

Chapter Objectives

Upon completing this chapter, the student will be able to:

1. **Identify** the cell membrane transport mechanisms involved in absorption.
2. **List** and explain various factors that affect skin, lungs, and digestive tract absorption of toxic substances.
3. **Describe** the general anatomy and function of the circulatory system.
4. **Explain** the role of the circulatory system in the distribution of toxic substances.
5. **Identify** types of plasma proteins and their effect on absorption.
6. **Identify** the major storage sites of toxicants.
7. **Describe** attributes of the major storage sites that facilitate storage of toxic substances.
8. **Describe** the physico-chemical characteristics and function of the blood-brain barrier.
9. **Describe** the different phases of toxic substance metabolism.
10. **Identify** the different types of Phase I and Phase II metabolic reactions.
11. **Define** the difference between deactivation and bioactivation.
12. **Identify** the two major groups of factors that affect toxic substance metabolism.
13. **Identify** the major sites of excretion.

Chapter Sections

5-1 Introduction

5-2 Absorption of Toxic Substances

5-3 Distribution of Toxic Substances

5-4 Metabolism (Biotransformation)

5-5 Excretion

5-1 Introduction

People are exposed to toxic substances on a daily basis. For an effect from exposure to be observed, the toxicant must reach the target tissue in a high enough concentration to cause a response. The amount toxicant that actually reaches the target tissue is dependent upon the amount absorbed, the distribution of the substance throughout the body, the metabolism (change in the chemical and physical characteristics) of the substance, and the rate of excretion. Although we will discuss each of these separately, it is important to understand that these activities occur simultaneously: the amount of a substance that actually reaches the target tissue is the net result of the interaction between absorption, distribution, metabolism, and excretion.

5-2 Absorption of Toxic Substances

In order for a toxicant to undergo distribution, metabolism, and excretion in an organism it must be absorbed through the membrane of cells that comprise the various layers of the skin and that line the digestive and respiratory tracts. The cell membrane is selectively permeable; therefore, only certain substances are able to pass through it. The selective permeability is determined by molecular size, lipid (fat) solubility, and electrical charge associated with the toxic molecule, as well as by cell membrane active and passive transport mechanisms.

The toxic substance must also pass through the membrane of the cells that comprise the capillaries as well as the cell membranes of the target tissues involved in metabolism, excretion, and storage. Figure 5-1 is a generalized illustration representing the movement of a toxic substance through these structures. The amount and rate of toxic substance transported across these barriers is determined by the characteristics associated with cell membrane transport and the physico-chemical attributes of the toxic substance described above.

Recall that the cell membrane is composed of a bilayer of phospholipids interspersed with proteins and carbohydrates (Figure 5-2). The type of phospholipids and proteins varies with the cell type. The proteins may be structural or they may function as a carrier molecule involved in either active transport or facilitated diffusion of toxic substances across the cell membrane. Toxic substances may also be trans-

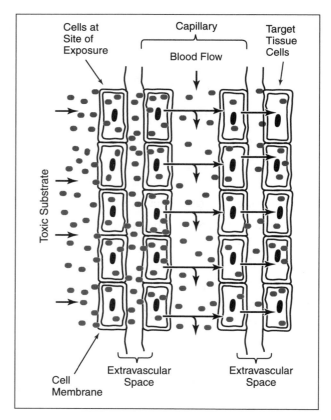

Figure 5-1: General schematic illustrating the movement of a toxic substance from the point of contact to the target tissue. The toxic substance must pass through several layers of cells to reach the target tissue.

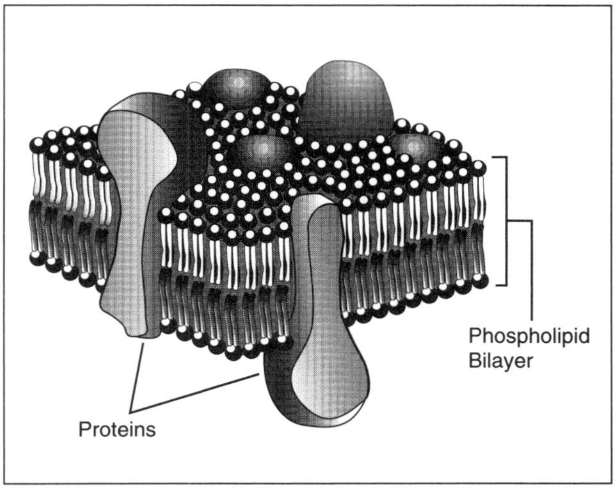

Figure 5-2: Generalized representation of the cell membrane illustrating the lipid bilayer and proteins. *Recreated with permission from Timbrell J. A.*, Introduction to Toxicology, *Taylor and Francis, New York, 1989.*

ported across the cell membrane by passive diffusion, phagocytosis, or pinocytosis.

Mechanisms of Absorption

Passive Diffusion

Most toxic substances move across the cell membrane by diffusion (Figure 5-3). Diffusion is dependent on three basic factors: 1) the concentration gradient across the cell membrane, 2) lipid solubility, and 3) the electrical charge associated with the molecule.

The rate of diffusion is primarily based on the differences in the concentration of the substance across the cell membrane, the surface area and thickness of the cell membrane. For instance, the walls of the alveoli – where gas exchange occurs in the lungs – are only one cell thick (Figure 5-4) and provide a large surface area (about 40 times the surface area of the outer body); this allows for rapid diffusion of toxic substances such as carbon monoxide and ether. Diffusion will continue until an equilibrium is established across the cell membrane. In many cases, once the substance moves across the cell membrane it is either transformed or absorbed into the bloodstream, so the concentration inside the cell continually changes, decreasing in amount. Therefore, transformation and absorption help maintain the concentration gradient, which facilitates movement of the toxic substance across the cell membrane.

Lipid solubility is also an important factor in determining the diffusion rate of a toxic substance since 75 percent of the cell membrane is composed of lipids. Solvents – such as carbon tetrachloride and chloroform – are lipid-soluble and are therefore readily absorbed. Lipid solubility results in the **bioaccumulation** (concentration of a substance) of some types of toxic substances such as DDT in fatty tissues. As a result, small amounts of a toxic substance that normally would produce no adverse effects become concentrated to a level that may cause detrimental effects.

The electrical charge associated with a toxic substance molecule or atom will also affect the rate of diffusion. In general, nonionized toxic substances diffuse more readily across the cell membrane than ionized substances. Organic substances such as nitrous oxide, ethylene, and divinyl ether diffuse across the cell membrane of the alveoli easily because they do not have an electrical charge and are lipid-soluble.

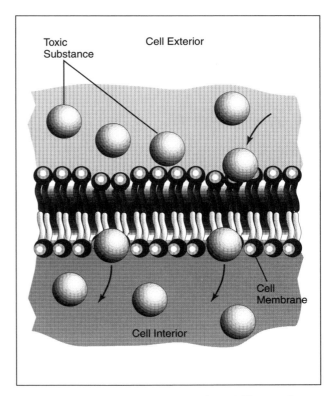

Figure 5-3: Simplified illustration of the diffusion of a toxic substance across the cell membrane. Movement of liquids and gases generally occur by diffusion.

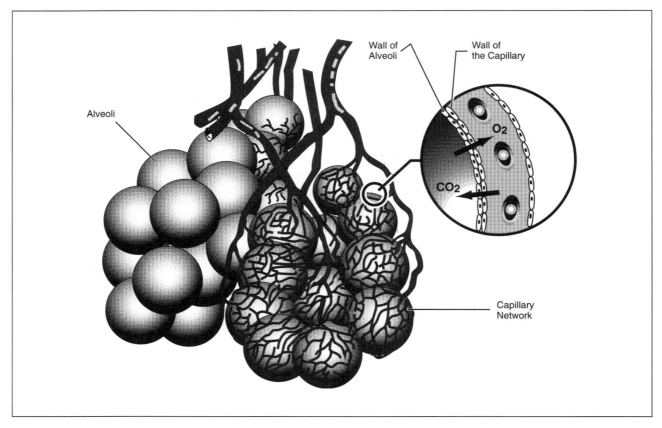

Figure 5-4: Illustration of the structure of the wall of the alveoli, which is only a single cell thick. The thin barrier between inhaled substances and the bloodstream increases the chance of toxic substances being absorbed into the bloodstream and then transported to the target tissue.

Carrier-Mediated Transport

Facilitated diffusion (transport) and active transport are carrier-mediated mechanisms associated with the cell membrane. These transport mechanisms are important when: 1) moving molecules across the cell membrane against a concentration gradient; 2) the size of the toxic molecule is too large to diffuse through the pores of the cell membrane; and 3) there is low lipid solubility and the substance has an electrical charge associated with it.

Active transport moves substances across the cell membrane against a concentration gradient. Transporting the substance requires expenditure of energy in the form of ATP (adenosine triphosphate) produced by the mitochondria. Essential nutrients such as calcium and amino acids are absorbed from the small intestine by active transport.

> ATP is the energy source for cellular activities.

When toxic substances are present in the small intestine, in some cases they will compete with the nutrients for the binding sites associated with the active transport mechanism. For instance, calcium is actively transported from the small intestine across the cell membrane by a **calcium-binding protein (CaBP)**. The CaBP binds to the calcium in the small intestine, moves it across the cell membrane and releases it inside the cell. Toxic substances such as cadmium, lead, or strontium will also bind to the CaBP because their physico-chemical characteristics are similar to calcium (they all have two positive charges associated with the ion form, Ca^{2+}, Pb^{2+}, Cd^{2+}, Sr^{2+}). The amount of the toxic substance absorbed will depend on the concentration of the toxic substance versus the concentration of calcium in the small intestine. Low circulating amounts of calcium due to a calcium-deficient diet causes increased production of the CaBP which can result in an increased absorption of lead from the small intestine. High protein diets appear to reduce the toxicity of lead and cadmium, because amino acids derived from proteins compete for the same protein-carrier molecules. These same toxic substances will also bind to the cell membrane protein-carrier molecule responsible for transporting iron from the small intestine, thereby affecting iron absorption.

Facilitated diffusion (see Chapter 3) is characterized by a cell membrane protein attaching to a substance outside the cell and transporting it across the cell membrane (Figure 5-5). This is similar to the active transport mechanism. However, no energy is expended and transport of the substance is from an area of high concentration to an area of low concentration. This mechanism is used to transport substances that are too large to diffuse through the cell membrane pores or that have an electrical charge associated with them.

Phagocytosis and Pinocytosis

Phagocytosis plays an important role in the disposition of particulates that enter the respiratory tract. Particulates such as asbestos, silica dust, or uranium dioxide are engulfed by white blood cells found throughout the respiratory tract, particularly the alveoli. These particulate-containing cells may then remain in the respiratory system or be transported out of the system by the ciliary-clearance mechanism discussed in Chapter 3. Pinocytosis occurs frequently in the lining of the lumen in the small intestine. Water-soluble substances and particulates appear to be transported via this mechanism.

Skin Absorption

All toxicants that pass through the skin do so as a result of passive diffusion. The rate at which a toxic substance is absorbed is determined by its ability to penetrate the keratinized outer layer of the epidermis; it is also dependent on the physico-chemical properties of the toxic substance. In general, gases penetrate the epidermis freely; liquids less freely; and solids that are not water soluble – as well as other solids – are for the most part incapable of penetrating the skin.

Organic solvents like carbon tetrachloride and hexane are lipid-soluble and are easily absorbed through the skin, entering the bloodstream. Small water-soluble toxic compounds such as hydrazine are absorbed through the skin and may cause local and systemic effects. Organophosphate pesticides such as parathion are more readily absorbed through the skin than some organochlorine pesticides such as DDT. The nerve gas sarin is easily absorbed through the skin.

Absorption is enhanced as a result of damage to the skin's outer layer. Abrasions as well as chemicals – such as acids (sulfuric, nitric), alkalis (sodium hydroxide), and mustard gas – cause damage to the structural integrity of skin cells allowing toxic substances to pass to the underlying tissue. Application of certain chemicals such as methyl and ethyl alcohol, hexane, and acetone are lipid-soluble and can be used to alter skin permeability by degrading the lipid barrier of the cell membrane.

The blood flow to the skin will also influence the rate of absorption. Absorption of toxic substances is enhanced in areas of the skin with a well developed blood supply. The epidermis is several cells thick and contains no blood vessels, thus slowing the diffusion rate of some substances. Once the substance diffuses through the epidermis it is absorbed into the bloodstream, which will carry it to other parts of the body. Distribution and removal of the toxic substance by the circulatory system helps main-

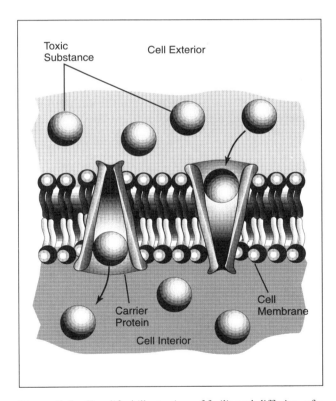

Figure 5-5: Simplified illustration of facilitated diffusion of a toxic substance across the cell membrane. The substance attaches to a cell membrane protein and is transported across the cell membrane. Movement is from an area of high concentration to an area of low concentration.

tain the concentration gradient necessary for diffusion to occur.

Lung Absorption

Toxic substances in the form of gases, vapors, particulates, or aerosols are present in the air we breathe. Gases (i.e., carbon monoxide, sulfur dioxide, and nitrogen oxides) and vapors of volatile compounds (benzene, methylene chloride) diffuse across the alveolar cell membrane and are absorbed into the bloodstream. Although lipid solubility is important in determining the rate of absorption, an even more important factor is the solubility of the toxic substance in the blood and its interaction with components of the blood. For example, chloroform is very soluble in the bloodstream and is rapidly removed from the site of absorption by the blood in the capillary system surrounding the alveoli. Due to its lipid solubility, chloroform is carried throughout the body by the blood and is readily deposited in tissues with a high fat content. Its high solubility in blood, its rapid removal from the site of absorption, and its deposition in fatty tissues throughout the body help establish and maintain the concentration gradient between the lungs and the blood. This results in enhanced removal of chloroform from inhaled air.

Ethylene, on the other hand, is not very soluble in the blood. Ethylene is absorbed into the blood and – because of its low solubility – the blood becomes saturated, quickly establishing an equilibrium between the lungs and the blood. Because the removal rate from the blood is slow, ethylene remains in the inhaled air for a longer period of time when compared to chloroform.

Particulates and aerosols of various sizes are removed and deposited along various regions of the respiratory tract. Particles 5 μm or larger in diameter are usually trapped in the nose and upper portions of the respiratory tract. They are removed from the respiratory tract by nose wiping, blowing, or by sneezing. Soluble particles may dissolve and be absorbed into the bloodstream as a result of diffusion. Particles 2-5 μm in diameter are trapped in the trachea and bronchial region of the respiratory tract. These particles are primarily removed by the **mucociliary escalator** (discussed in Chapter 3), which transports these substances to the back of the throat where they are swallowed or expectorated. Particles 1 μm in diameter and smaller reach the alveoli. These particulates may be dissolved and absorbed into the bloodstream.

Dissolved compounds and particulates may also be removed by pinocytosis and phagocytosis. Uranium dioxide particles are removed by pinocytosis; lead particles are removed by phagocytosis. Once engulfed, the substance is removed by the mucociliary escalator or carried to other parts of the body. Some particulates (asbestos, fiberglass) that are phagocytized remain in the lungs where they may have adverse effects and result in the development of respiratory disease.

Gastrointestinal Absorption

Most of the absorption in the digestive tract occurs in the small intestine. Aspirin and alcohol are absorbed in the stomach, and some medications are absorbed in the mouth, usually by placing the substance under the tongue.

Absorption in the small intestine is enhanced by the presence of villi and microvilli (see Chapter 3), which increase its surface area. The primary mechanism of absorption is by diffusion but, as we already know, facilitated and active transport also occurs. In addition to the previously mentioned factors that affect absorption, increased absorption will occur if food containing the toxic substance remains in the digestive tract for a longer period of time. Generally, about 10 percent of the lead, 4 percent of the manganese, 1.5 percent of the cadmium, and 1 percent of chromium salts is absorbed after ingestion.

Checking Your Understanding

1. Define diffusion.
2. What three factors affect the rate of diffusion?
3. What is bioaccumulation?
4. What is the difference between simple diffusion and facilitated diffusion?
5. What types of toxic substances are most likely to be transported by phagocytosis? And by pinocytosis?
6. Identify the mechanisms involved in the absorption of toxic substances as a result of ingestion, inhalation, and skin exposure.

5-3 Distribution of Toxic Substances

Once the toxic substance enters the bloodstream it is transported and distributed by the circulatory system to the various tissues of the body where it is stored, metabolized, or excreted. Absorption into and out of the bloodstream occurs in only one part of the circulatory system: the capillaries. Inhaled toxicants diffuse across the alveoli and are absorbed into the **pulmonary circulation** (Figure 5-6).

> The pulmonary circulatory system consists of blood vessels associated with the lungs.

Skin and oral exposure (ingestion) result in absorption and transport of the toxicant via the **systemic (body) circulation** (Figure 5-6).

> The systemic circulation consists of the circulatory system not associated with the lungs.

Anatomy of the Circulatory System

The circulatory system is composed of the heart along with an extensive system of blood vessels that transport oxygen and nutrient-rich blood to the cells. The blood picks up waste products of cellular metabolism and distributes them to various organs, which subsequently eliminate them from the body. Movement of oxygen, waste products, and nutrients – and by extension, toxic substances – into and out of the bloodstream occurs by a variety of active and passive transport mechanisms.

Anatomy of the Heart

The heart is a thick, muscular organ containing four chambers: the right and left **atria** and the right and left **ventricles** (Figure 5-7). The right and left atria are small, thin-walled chambers. The right atrium receives low-oxygenated blood from the systemic (body) circulation and the left atrium receives oxygen-rich blood from the lungs. The ventricles are thick, muscular chambers. The right ventricle pumps blood through the **pulmonary artery** to the extensive vascular network (pulmonary circulation) located in the lungs. The pulmonary capillaries are in close association with the alveoli of the lungs; this facilitates the diffusion of substances – such as oxygen, carbon dioxide, or airborne toxic substances (methylene chloride, hydrogen sulfide) – between the circulatory system and inhaled air. Oxygenated blood returns to the left ventricle via the **pulmonary veins** and left atria. It is pumped from the left ventricle via the aorta through the systemic circulation, which supplies blood to various tissues and organs of the body.

Anatomy of the Vascular System

Once the blood leaves the ventricles it enters either the systemic circulation via the aorta or the pulmonary circulation via the pulmonary artery. The arteries slowly change in structure and size, eventually forming small-diameter **arterioles.**

> Arterioles are small-diameter arteries; the blood flows from arterioles to capillaries.

Arterioles supply blood to the capillaries that have walls composed of a single layer of thin, flattened cells. The capillaries, in turn, evolve into small-diameter **venules,** which eventually develop into large veins that return blood to the heart (Figure 5-8).

> Venules are small veins that collect blood from capillaries; several venules unite to form larger veins.

There are 10 billion capillaries in the body. They are composed of a single layer of flat endothelial cells, having a diameter of 4-8 μm, and are the principal site of exchange of substances between the blood and the surrounding tissues (Figure 5-9). Nutrients, gases, and toxic substances enter or leave the bloodstream by active or passive transport mechanisms associated with the capillary endothelial cells. Diffusion is the primary mechanism responsible for the movement of gases, lipid-soluble

92 ■ Basics of Toxicology

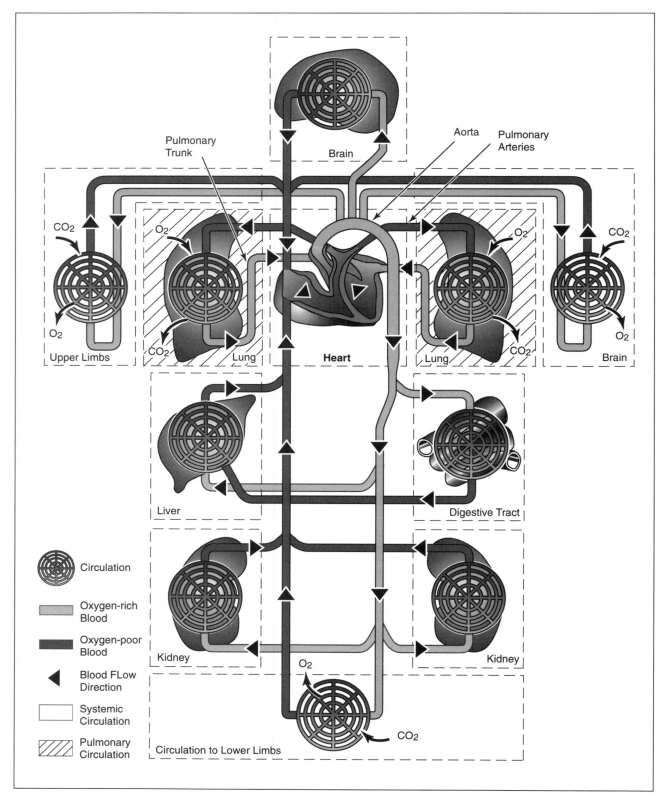

Figure 5-6: Generalized blood flow through the circulatory system. Oxygen-poor blood is returned to the right side of the heart where it is pumped through the pulmonary circulation. Oxygen-rich blood is returned to the left side of the heart where it is pumped through the systemic circulation, delivering oxygen-rich blood to tissues and organs of the body. The arrows in the illustration indicate the direction of the blood flow.

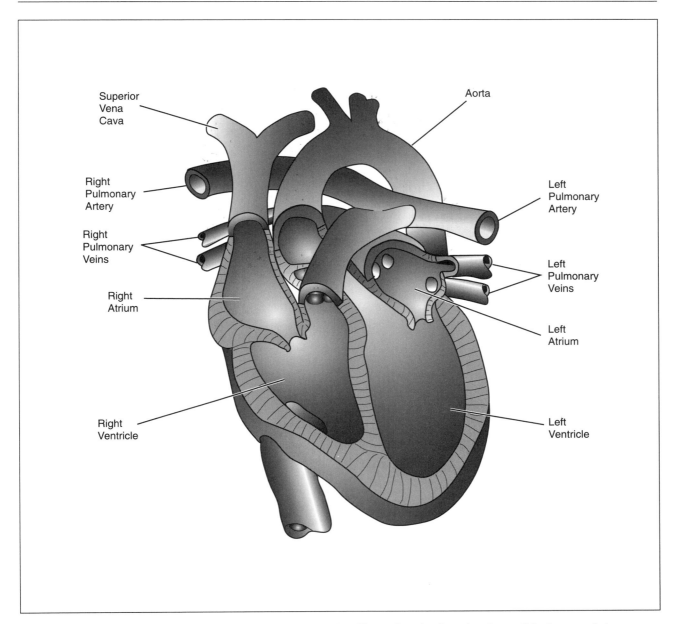

Figure 5-7: General anatomy of the heart. Frontal plane section illustrating the four chambers of the heart and the associated major blood vessels. The superior vena cava returns blood to the right atrium, and the pulmonary veins return blood from the lungs to the left atrium. The left ventricle pumps oxygenated blood to the body via the aorta. The pulmonary artery carries low-oxygenated blood to the lungs.

substances, and water-soluble molecules. Nutrients, nongaseous and nonlipid-soluble substances will be transported through the walls of the capillaries by either active transport or facilitated transport mechanisms. The limiting factor – in terms of transporting toxic substances from the capillaries – appears to be the size of the molecule. Pinocytosis seems to be the major mechanism for transporting these types of substances. Pinocytosis is a relatively slow process; therefore, large toxic molecules will remain in the blood for a longer period of time.

The competition for protein-carrier molecules may affect the rate of absorption and subsequent toxicity of a substance. For example, a toxic substance with a similar molecular size, lipid solubility, and electrical charge as an essential nutrient – such as an amino acid – will successfully compete with the amino acid for the protein-carrier molecule in the cell membrane. If the concentration of amino acids is low, then the toxic substance will be absorbed more readily from the bloodstream.

> Absorption of toxic substances into tissues occurs in the capillaries and occurs as a result of active and passive transport mechanisms.

Toxic substances that are readily absorbed are more likely to reach target tissues in high enough concentrations to exert a toxic effect. However, there are several mechanisms that oppose distribution to the target tissue: 1) binding to **plasma** proteins, 2) distribution to storage sites; and 3) specialized barriers.

> Plasma is the liquid portion of blood minus the blood cells.

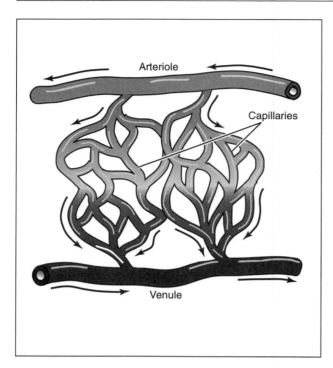

Figure 5-8: Illustration of the structural relationship between arterioles, capillaries, and venules. Arterioles give rise to capillaries, which in turn converge and empty into venules. The arrows in the illustration indicate the direction of the blood flow.

Binding of Plasma Proteins

When toxic substances enter the bloodstream they may bind with plasma proteins such as **albumin**, transferrin, **globulin**, and **lipoproteins**. Most toxic substances will bind with the plasma protein albumin.

> Albumin, transferrin, and lipoproteins are blood proteins that transport toxic substances through the bloodstream.

However, there is always some portion of the toxic substance that is not bound and is in equilibrium with the bound portion. Equilibrium in this case does not mean there are equal amounts of bound and free toxicant. The chemical characteristics of a toxic substance and of the blood (i.e., pH) will determine how much of the toxicant will be bound and/or free. For example, up to 99 percent of the pesticides dieldrin and warfarin are bound to a lipoprotein and one percent is nonbound. The toxicant-plasma protein structure has a high molecular weight

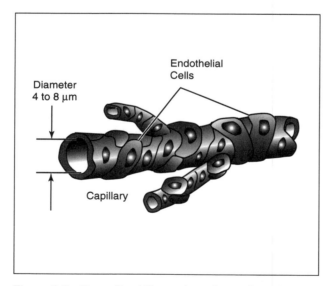

Figure 5-9: Generalized illustration of a capillary showing the single-cell wall comprised of endothelial cells. Capillaries are the principal sites of exchange of substances between blood and surrounding tissue.

and does not easily pass through the capillary walls to the surrounding tissue.

> Many toxic substances bind to plasma proteins; this affects the rate of absorption through capillaries.

The length of time a toxic substance remains bound to the plasma protein is dependent on the amount of toxicant that exists free in the blood (not bound to plasma proteins), and the rate at which the nonbound toxicant passes through the capillaries to the surrounding tissues. As the unbound toxicant passes through the endothelial cells of the capillary into the **extravascular space** (a space between the wall of the capillary and the tissue cell membrane, filled with fluid), the bound toxicant disassociates (separates) from the protein to maintain an equilibrium between the bound and unbound toxicant in the blood. The amount of free toxicant in the extravascular space is in equilibrium with the amount of free toxicant in the blood. The previously mentioned factors (i.e. lipid solubility, size of toxicant molecule, etc.) determine the rate of transport out of the capillary and into the extravascular space. These same factors will also determine how fast the toxicant is absorbed from the extravascular space into the various tissues of the body.

> A dynamic equilibrium exists between the bound and unbound forms of a toxicant. The equilibrium will be determined by factors that affect absorption, subsequently affecting toxicity.

Storage of Toxic Substances

Toxic substances are stored in several different tissue types found throughout the body. The toxicant may be stored in the target tissue, possibly resulting in an adverse response, or it may be stored in other tissue types, which may not be readily affected. For instance, organochlorine pesticides like DDT, dieldrin, and chlordane exert their effects primarily on nerve tissue. However, the toxicants are fat-soluble and are easily absorbed into fatty deposits throughout the body where they may be stored for long periods of time. Similarly, lead is readily absorbed and stored in bone. Lead also affects nerve tissue. In general, storage of these substances in these tissue types will have no effect on their primary target tissue, the nerve tissue.

Major tissue storage sites include the liver, kidneys, fat, and bone. Many toxicants are stored in these tissues for several years. A brief description of their normal structure and function is discussed below. This information will be useful in understanding how toxic substances reach these tissues and the factors that affect the distribution of toxic substances within them.

Storage in the Liver and Kidneys

The liver and the kidneys are involved in the detoxification and removal of toxic substances circulating in the bloodstream. They have a high affinity for toxic substances and store more toxicants than any other tissue in the body. The reason for the high affinity is not clearly understood but may be attributable to anatomical and physiological characteristics associated with both organs. The ability to concentrate toxic substances in these organs is advantageous for the detoxification and the excretion of toxic substances; however, it may also increase the occurrence of adverse effects in the liver and kidneys.

Structure of the Liver – The liver is the single largest organ of the body and can weigh as much as 1.40 kg (3 lbs.). It is comprised of two lobes, each of which is divided into numerous functional units called **lobules**. A lobule consist of cords of liver cells arranged around a **central vein** (Figure 5-10).

Most venous blood passes through the liver before it reaches the heart. The liver receives blood from the lower extremities, kidneys, **spleen,** and gastrointestinal tract. The hepatic portal vein carries nutrient-rich, but oxygen-poor blood from the digestive tract to the liver. The blood from the branches of the hepatic portal vein passes into spaces called **hepatic sinusoids** located between the lobular cords. In the sinusoids, venous blood is mixed with oxygen-rich and nutrient-poor blood from the **hepatic artery**, the second major source of blood to the liver. The blood from the sinusoids drains toward the center of the lobule and into the central vein and then exits the liver through the hepatic veins.

96 ■ Basics of Toxicology

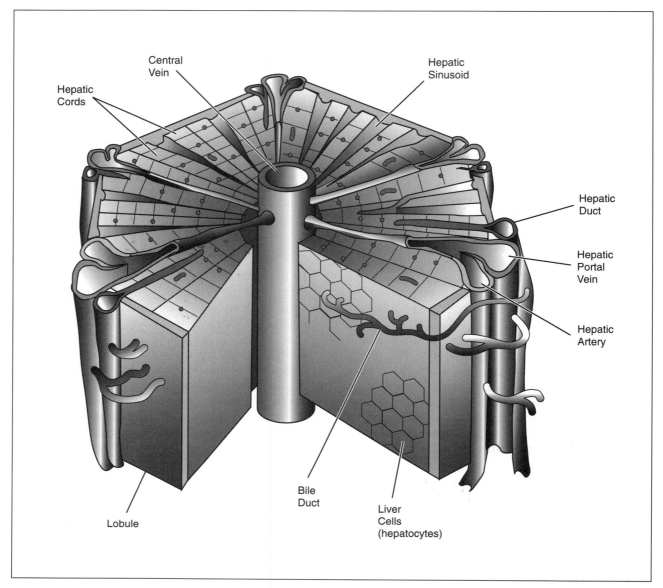

Figure 5-10: General anatomy of the liver.

Storage in the Liver – Absorption of toxic substances in the liver is affected by the same factors that regulate absorption in other tissue types. Lipophilic substances, such as organochlorine pesticides and organic solvents (trichloroethane, methyl chloroform), are readily absorbed. If they are not biotransformed into water-soluble substances they will be retained in the liver for long periods of time. Active transport mechanisms facilitate concentration of toxicants in the liver, subsequently causing liver damage. The **hepatocytes** absorb copper (Cu^{2+}) and iron (Fe^{2+}) from the blood by protein-carrier molecules in the cell membrane.

Hepatocytes are cells of the liver.

These metals are important in maintaining normal cell function. Some toxic metals are absorbed and concentrated in a similar manner. They have similar physical and chemical properties as iron and copper and will compete with these essential metals for the protein-carrier molecule in the cell membrane. Therefore, substances such as mercury and lead are easily absorbed in the liver.

> The liver is a major storage site for water-insoluble toxic substances such as heavy metals.

In addition to the cell membrane transport mechanisms, there are two other features associated with the liver, which facilitate the absorption and concentration of toxicants. As it passes through the liver, the blood enters the sinusoids located between strands of hepatocytes. The layer of endothelial cells lining the sinusoids is discontinuous with small and large **fenestrae** (pores) (Figure 5-11). The fenestrae allow larger, blood-borne molecules to pass through the endothelial lining. Therefore, it is easier for toxic substances that bind with blood proteins to pass through the endothelial lining and to be absorbed by the hepatocytes. The liver cells also have a high concentration of the intracellular protein **metallothionein**.

> Metallothionein is an intracellular protein with a high affinity for heavy metals such as lead, cadmium, and mercury.

Metallothionein has a high affinity (binding up to 99 percent) for many different kinds of toxic metals such as cadmium, mercury, and lead. When bound to metallothionein the protein-metal complex does not readily pass through the cell membrane. Therefore, the affinity of metals for the cell membrane transport mechanisms, the presence of fenestrae, and the high affinity of metals for intracellular metallothionein facilitate the storage of some toxicants in the liver at higher concentrations than in the surrounding tissue. For instance, given a single dose, the concentration of lead in the liver – after 30 minutes – is 50 times higher than in the blood.

> Cell membrane transport mechanisms, fenestrae, and metallothionein facilitate absorption and storage of toxic substances in the liver.

Structure and Function of the Kidneys – The kidneys are two bean-shaped organs located on the posterior wall of the abdominal cavity. Each kidney is divided into two parts, the outer portion referred to as the **cortex** and the inner portion referred to as the **medulla**. Within the medulla are cone-shaped structures called **renal pyramids**. Urine passes from the tips of the pyramids to a large funnel-like structure called the **renal pelvis**. The renal pelvis narrows to form the **ureter,** which connects the kidney with the urinary bladder (Figure 5-12).

The functional unit of the kidney is the **nephron**, parts of which are found in the renal cortex and the medulla. There are approximately 1.3 million nephrons in each kidney. Each nephron is composed of the following structures: a **renal corpuscle**, a **proximal convoluted tubule**, a **Loop of Henle**, and a **distal convoluted tubule**. The distal convoluted tubule empties into a **collecting duct,** which transports urine toward the renal pelvis area where it enters the ureter (Figure 5-13).

The renal corpuscle is composed of **Bowman's capsule** and a ball-like arrangement of capillaries referred to as the **glomerulus**. Substances carried in the blood are transported to the renal corpuscle of the nephron. Liquid from the blood containing dissolved materials is filtered from the glomerulus into Bowman's capsule; from there it moves into the proximal convoluted tubule and into the Loop of Henle. The latter is composed of a descending and an ascending limb. These structures are surrounded by an extensive blood supply referred to as the **peritubular capillaries**. The ascending limb gives rise to the distal convoluted tubule, which empties into the collecting duct (Figure 5-13).

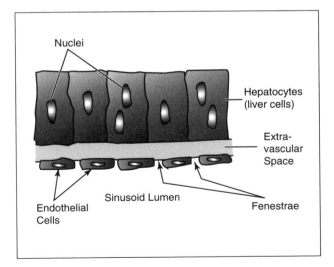

Figure 5-11: Fenestrae of the hepatic sinusoid endothelium. The endothelial cells lining the sinusoids are discontinuous and have small and large openings (fenestrae) that allow large molecules to pass out of the bloodstream and to be absorbed by the hepatocytes.

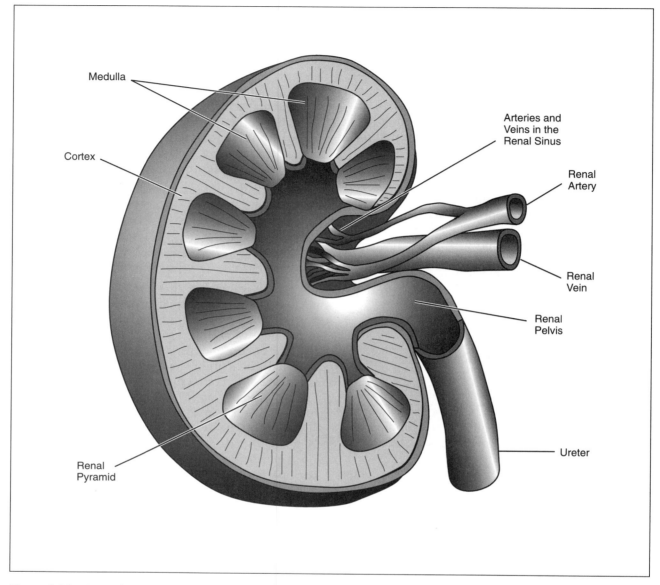

Figure 5-12: General anatomy of the kidney. The cortex forms the outer part of the kidney, the medulla forms the middle part. Urine flows from the tip of the renal pyramid through the renal pelvis and into the ureter.

The kidneys receive approximately 25 percent of the total cardiac (heart) output, which is about 1.2-1.3 liters of blood per minute. The **perfusion rate** for the kidneys is greater than that for the heart, brain, or liver.

As a result, any toxic chemical present in the blood is delivered to the kidneys in large quantities. The primary functions of the kidneys are to remove metabolic waste from the body and help maintain **homeostasis** of the body.

> The perfusion rate is the amount of blood delivered to tissues or organs over a specific period of time.

> Homeostasis is the existence and maintenance of a relatively constant environment within the body with respect to functions and composition of fluids and tissues.

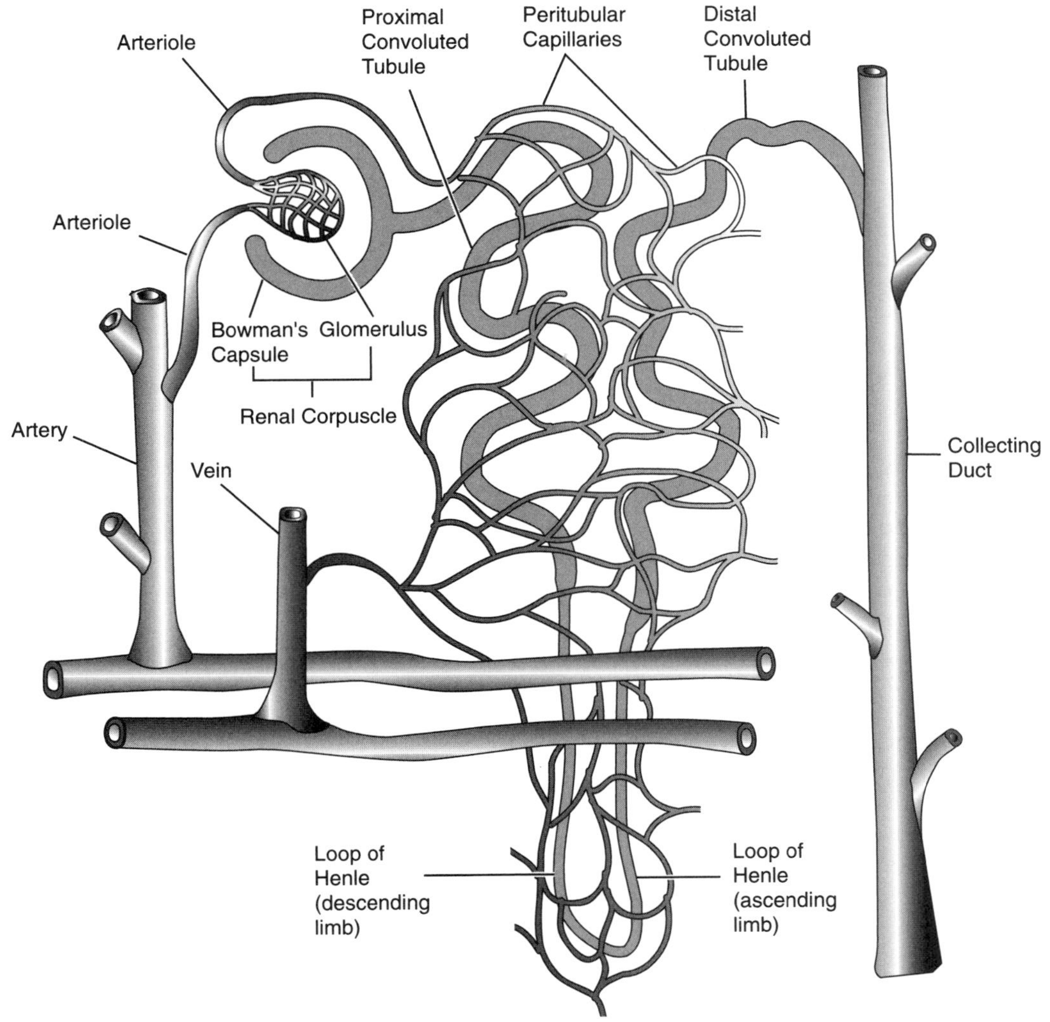

Figure 5-13: The nephron of the kidney. It consists of the Bowman's capsule, the proximal convoluted tubule, the Loop of Henle, and the distal convoluted tubule. Toxic substances enter the nephron primarily via the glomerulus. They may be reabsorbed in the proximal convoluted tubule or excreted in the urine.

These functions are performed by the nephrons and involve the absorption and excretion of solutes and water from the bloodstream to the nephrons and vice versa.

A high perfusion rate may result in larger quantities of toxic substances being delivered to the kidneys than to other tissues.

Factors such as molecular size, electrical charge, lipid-water solubility, and active and passive transport mechanisms will determine which substances will be transported from the blood into the kidneys and reabsorbed from the kidneys back into the blood. The cells that compose the walls of the glomerulus and Bowman's capsule allow water and solutes of small molecular size to pass into the nephron. However, large molecules such as proteins (i.e., albumin) do not easily pass through the walls of

Bowman's capsule. The negative charge associated with both albumin and the cell membrane of capillaries repel each other, which further hinders the movement of albumin from the blood into the nephron. Therefore, only a small amount of protein enters the kidneys.

> Toxic substances leave the bloodstream and enter the nephron primarily through Bowman's capsule.

Once a substance enters the kidneys it may be reabsorbed by either active or passive transport mechanisms along the various parts of the nephron. The proximal convoluted tubule is the primary site of reabsorption of water and solutes. Proteins, amino acids, glucose molecules, as well as sodium, potassium, and chloride ions are actively transported from the proximal convoluted tubule to the peritubular capillaries. Water from the tubule also enters the capillaries as a result of osmosis. Movement of water and solutes in the descending limb of the Loop of Henle is primarily by passive processes. Active transport mechanisms predominate in the ascending limb of the Loop of Henle. Passive processes exist primarily in the distal convoluted tubule (Figure 5-13).

Secretion of various substances from the peritubular capillaries to the nephron can also occur. Ammonia, for example, diffuses from the capillaries to the nephron tubules. Substances such as hydrogen and potassium ions, as well as penicillin, are actively transported to the lumen of the nephron.

Storage in the Kidneys – Toxic substances that enter the bloodstream are eliminated from the body by the same mechanisms involved in excretion of normal waste products. Lipid-soluble, water-soluble, and small size toxicants easily pass into the nephron. If they are not reabsorbed as they pass through the nephron they will be excreted in the urine. Lipid-soluble toxicants will be reabsorbed readily. Similarly to the liver, unbound metals like cadmium and mercury can be reabsorbed by active transport mechanisms in the cells of the proximal convoluted tubule. Once in the cell they bind to metallothionein resulting in concentration of these toxicants in the kidneys, possibly causing adverse effects.

For instance, the kidneys store 10 times the amount of cadmium found in the liver, and it can be stored for 10 years or more. Cadmium concentrations as high as 200 mg/g may result in renal tubular damage. Trimethylpentane, which is present in gasoline, is adsorbed to a blood protein called globulin. In this form it is reabsorbed in the proximal convoluted tubule. It is known to produce renal **tumors** in rats. The kidneys have the greatest concentration of mercury after exposure to inorganic salts of mercury and mercury vapor.

Similarly to storage in the liver, storage of toxicants in the kidneys decreases their toxicity to other target tissue, i.e., the nerve tissue. However, bioaccumulation in the kidneys may ultimately cause kidney damage and, if severe enough, result in complete renal (kidney) failure.

Storage in Fat

Lipid-soluble substances such as the organochlorine pesticide DDT, polychlorinated biphenyls (PCBs), and organic solvents are stored in body fat. Biotransformation and degradation (breakdown) occurs slowly, thereby allowing bioaccumulation and the long retention of these types of toxicants in the body.

> Bioaccumulation is the retention and concentration of a substance by an organism.

Bioaccumulation results in the concentration of non-toxic quantities of a substance to a toxic level as small quantities are accumulated over time.

Storage of toxicants in the fat initially results in a lower concentration of the toxicant reaching the target tissue. Like storage in the plasma protein, liver, and kidneys the toxic substance in the fat is in equilibrium with the toxicant in the blood. Larger amounts of the toxicant are released when fat reserves are mobilized in response to increased metabolism of fats, such as during periods of fasting or starvation. Therefore, signs of intoxication may occur under starvation conditions even though they may not have appeared during the initial exposure.

Storage in Bone

Bone is the major storage site for calcium in the body. It is composed of bone cells called **osteocytes** and an extracellular matrix arranged in sheets called **lamellae**.

> Lamellae are thin sheets or layers of bone.

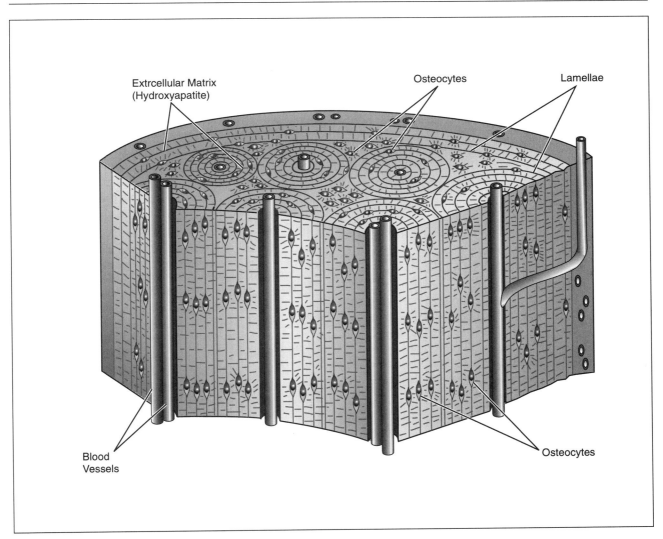

Figure 5-14: General structure of the bone illustrating the bone layers (lamellae), the extracellular matrix composed of hydroxyapatite, and bone cells (osteocytes).

The extracellular matrix is composed of a mineral complex called **hydroxyapatite** (Figure 5-14), which is composed primarily of calcium and phosphate ions.

> Hydroxyapatite is the complex crystal structure that makes up the lamellae of the bone.

When blood levels of calcium are low, the hydroxyapatite is broken down, and calcium is released from the bone into the bloodstream. When the calcium levels increase in the blood, calcium is absorbed into the hydroxyapatite of the bone. It is during the exchange between blood and bone that calcium can be replaced by toxic substances such as lead and strontium. These substances have a similar size and electrical charge as calcium. Ninety percent of the total body burden of lead is in the bone. Significant amounts can remain in the bone for 10-20 years.

Storage of toxic substances in the bone may or may not have a detrimental effect. Lead is not toxic to bone. However, it may be released from the bone resulting in nerve damage. Mobilization of the lead can occur in older individuals suffering from **osteoporosis**.

102 ■ Basics of Toxicology

> Osteoporosis is a bone disorder in which the bone becomes brittle and breaks easily.

Since osteoporosis is associated with the mobilization and excretion of bone calcium when the bone is degraded, the lead stored in the degraded bone is mobilized as well. Increased blood levels of lead then can cause nerve damage. Mobilized lead in the bone of expectant mothers may be passed on to the fetus. The nervous system of the fetus is very susceptible to the toxic effects of substances such as lead and may result in birth defects associated with the nervous system.

The hydroxide ion (OH$^-$) is also a component of the metabolic process associated with bone metabolism. Chronic exposure to elevated levels of fluoride may result in the fluoride ion (F$^-$) replacing the hydroxide ion during normal bone metabolism. Increased deposition of fluoride in the bone can result in **fluorosis,** which is characterized by weakening of the bone. **Radioactive** strontium stored in bone may cause certain types of bone cancer.

> A substance capable of giving off radiant energy in the form of particles or rays.

Specialized Barrier: Blood-Brain Barrier

The **blood-brain barrier** is created by the close association of brain capillaries with specialized cells formed in the nervous system. This barrier effectively decreases the type and amount of toxic substances that are transferred from the blood to the brain tissue. Reduced absorption of toxicants is attributed to several unique characteristics associated with the blood-brain barrier: 1) the closely packed endothelial cells of the capillaries; 2) the **astrocytes**, and 3) the low protein content of the interstitial fluid.

Normally, capillary endothelial cells are loosely joined to each other with pores, about 4 **nm (nanometers)** in diameter, located between the cells. The pores are one of the major mechanisms that facilitate the movement of substances into and out of the capillaries. The capillary endothelial cells of the blood-brain barrier are more closely arranged to each other; therefore, few or no pores exist. This anatomical arrangement of the cells restricts the movement of substances through the capillaries. Lipid-soluble substances like ethanol easily diffuse through the phospholipid cell membrane and into the fluid surrounding the brain. Water-soluble substances such as glucose – which is necessary for nerve cell functioning – must be transported across the blood-brain barrier by carrier-mediated mechanisms.

The capillaries are surrounded by cellular processes (extensions) from astrocytes (Figure 5-15). Astrocytes are specialized cells found in the nervous system.

> Astrocytes are specialized, star-shaped cells in the nervous system that form part of the blood-brain barrier.

The cell membrane of the astrocyte has a high lipid content that helps to form an effective barrier,

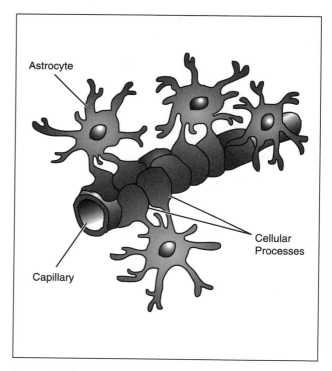

Figure 5-15: General illustration of the blood-brain barrier. Astrocyte processes surround capillaries that supply blood to the brain tissue. The cell membrane of the astrocyte has a high lipid content and forms an effective barrier slowing the movement of substances from the bloodstream into the brain tissue.

slowing the rate of movement of water-soluble molecules.

The interstitial fluid is the liquid between the external wall of the capillary and the cell membrane of the surrounding tissue; it has a low protein content. Once water-insoluble toxic substances leave the bloodstream, they bind with the interstitial proteins. Because of the low protein content of the fluid, reduced amounts of the protein-toxic substance complex reaches the brain tissue.

> The blood-brain barrier reduces the effects of most toxic substances on the brain by limiting their movement from the blood to the brain.

The same physical and chemical factors that affect the rate of absorption of toxicants in other areas of the body apply to absorption across the blood-brain barrier. For example, the small pores associated with the capillary endothelial cells make it more difficult for large water-soluble toxicant molecules to pass from the blood to the surrounding tissue. Small and medium sized water-soluble molecules can pass through the pores. However, size is not the only issue: the **hydrophobic** cell membrane of the astrocyte processes slows or inhibits their passage.

> A hydrophobic substance does not have a high affinity to water.

Absorption of **lipophilic** toxicants is slowed because the substance must pass through the phospholipid cell membrane of both the endothelial cells and the astrocyte.

> A lipophilic substance has an affinity for lipids and is easily absorbed in fats.

If toxic substances are able to overcome these barriers, then further transport to the brain tissue may be diminished because of the low protein content in the interstitial fluid. The proteins are necessary for water-insoluble substances to be carried through the aqueous interstitial fluid to the brain tissue.

Some toxic substances do penetrate the blood-brain barrier. Methyl mercury, for example, enters the brain through a carrier-mediated process. It appears to first combine with the amino acid cysteine; then the metal-amino acid complex thus formed is absorbed and transported across the blood-brain barrier. Inorganic mercury compounds such as mercuric chloride are not lipid-soluble and unable to readily cross the blood-brain barrier. Therefore, they have no effect on the brain, but are concentrated in the kidneys where they may produce toxic effects. Due to their lipophilic nature, organophosphate pesticides also penetrate the blood-brain barrier readily exerting their effects on the nervous system. On the other hand, dioxins – which are very fat-soluble – are not easily transported to the brain, possibly because of their affinity for plasma proteins.

Redistribution of Toxic Substances

Initially a toxic substance may be concentrated and stored for a short period of time in a particular tissue type. However, over time the toxicant may be transported to a different tissue where it will be stored for a more prolonged period of time. Toxic substances have a tendency to concentrate initially in well perfused tissues (those that receive large amounts of blood) such as the liver and kidneys. However, other tissue types may have a greater affinity for the toxicant and over time – because of the dynamic equilibrium discussed previously – the toxicant will eventually be transported and stored in these tissues. For example, 50 percent of the lead dose is found in the liver two hours after administration. One month later 90 percent of the remaining dose is found in the bone.

Checking Your Understanding

1. Describe the difference between the systemic and pulmonary circulation.
2. Where is the primary site of exchange of toxic substances between the blood and the surrounding tissue?
3. What characteristics of capillaries facilitate the exchange of toxic substances between the blood and surrounding tissues?
4. What three factors may affect the distribution of toxic substances to the target tissues?

5. Identify the types of blood proteins that bind to toxic substances. Why does binding to these proteins affect distribution and toxicity?

6. Identify the major storage sites for toxic substances.

7. What are fenestrae?

8. What is metallothionein?

9. How do the fenestrae and metallothionein facilitate absorption and storage of lipid-soluble toxicants and heavy metals in the liver?

10. Describe the structure and function of the nephron.

11. Explain the mechanisms involved in the storage of toxic substances in the kidneys. Where does most absorption of heavy metals occur in the kidneys? Why?

12. Identify the types of toxic substances stored preferentially in the fatty tissue of the body. Why are these types of toxic substances stored more readily?

13. Identify what types of toxic substances may be stored in the bone. Why might this occur?

14. What effect would high levels of calcium (Ca^{2+}) in the blood have upon cadmium (Cd^{2+}) absorption?

15. What are the components of the blood-brain barrier?

16. What characteristics associated with the blood-brain barrier affect the absorption of toxic substances? How does this affect toxicity?

5-4 Metabolism (Biotransformation)

Biotransformation of toxicants occurs in various tissues, resulting in a change in the chemical structure of the toxic compound. These changes occur in a variety of tissues such as the liver, kidney, lung, skin, nasal mucosa, eye, and gastrointestinal tract. The liver is the primary organ where biotransformation occurs. The transformation process begins when the toxic substance is absorbed across the cell membrane, where a variety of metabolic processes take place.

The transformation process may take place as a result of the interaction of the toxic substance with enzymes found primarily in the cell endoplasmic reticulum, **cytoplasm**, and mitochondria.

> Cytoplasm is the fluid, translucent substance located outside of the cell nucleus and bound by the cell membrane; it is composed of fats, proteins, and other types of molecules. Intracellular structures such as ribosomes and endoplasmic reticulum are found in this substance.

Interaction with these enzymes may change the toxicant to either a less or a more toxic form. The metabolic processes also may change the chemical nature of the toxicant from a water-insoluble chemical to a water-soluble chemical that can be excreted by the kidneys. In addition, the biotransformation can result in the production of chemical intermediates that are more toxic than the original compound.

Generally, biotransformation occurs in two phases. Phase I involves **catabolic** reactions that break down the toxicant into various components. Catabolic reactions include **oxidation**, **reduction**, and **hydrolysis**.

> Oxidation occurs when a molecule combines with oxygen, loses hydrogen, or loses one or more electrons.

> Reduction occurs when a molecule combines with hydrogen, loses oxygen, or gains one or more electrons.

> Hydrolysis is the process in which a chemical compound is split into smaller molecules by reacting with water.

In most cases these reactions make the chemical less toxic, more water soluble, and easier to excrete. Phase II reactions involve the binding of molecules to either the original toxic molecule or the toxic molecule metabolite derived from Phase I reactions. The final product is usually water soluble and, therefore, easier to excrete from the body. Phase II reactions include **glucuronidation**, **sulfation**, **acetylation**, **methylation**, **conjugation** with glutathione, and conjugation with amino acids (such as glycine, taurine, and glutamic acid).

> Phase I and II reactions result in the breakdown and synthesis of compounds that are generally less toxic, water soluble, and therefore easier to excrete.

Although individual types of reactions will be discussed in the following sections, it is important to understand that these reactions may occur simultaneously or sequentially. Figure 5-16 is a generalized illustration of the sequential metabolism of benzene with a single intermediate substance being produced. Also, biotransformation of a single toxic substance can result in multiple intermediate products being produced. Figure 5-17 illustrates the intermediates that can be produced as a result of oxidation of benzene.

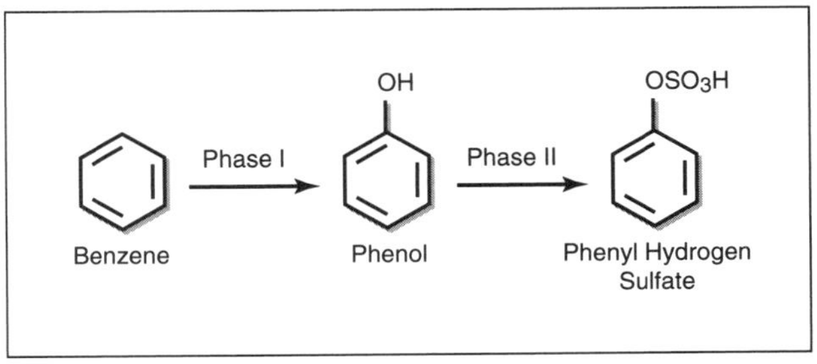

Benzene →(Phase I)→ Phenol (OH) →(Phase II)→ Phenyl Hydrogen Sulfate (OSO_3H)

Figure 5-16: Generalized illustration showing the sequential metabolism of benzene.

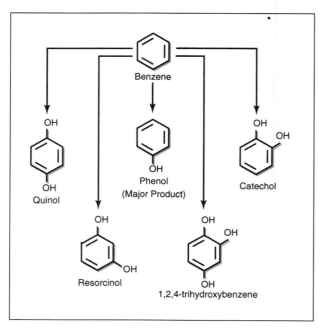

Figure 5-17: Oxidation of benzene resulting in several different chemical intermediates. The major product – phenol – will be further metabolized during Phase II to produce phenyl sulfate.

It is important to understand that chemical intermediates may or may not be more toxic than the parent compound. Whether the more toxic intermediate produces an effect is dependent on 1) how rapidly the intermediate undergoes further metabolism to less toxic substances, 2) how much of it is produced and accumulated in cells, 3) what type of cellular damage is caused by the toxic intermediate, and 4) what factors may affect excretion of the toxic material. Therefore, the production of toxic chemical intermediates does not necessarily mean that an adverse effect will occur. However, in some cases, toxic intermediates are produced and do cause adverse effects. In the case of parathion, the production of a toxic intermediate is desired in order to kill the pest. On the other hand, toxic chemical intermediates associated with alcohol consumption can lead to liver disease if consumption is excessive.

Table 5-1 lists the various types of Phase I and Phase II chemical reactions. Notice that there are several different types of chemical reactions associated with each of the major Phase I and Phase II reactions.

PHASE I	
I. OXIDATIONS	
(a) Microsomal oxidations	
Aliphatic oxidation	$RCH_3 \longrightarrow RCH_2OH$
Aromatic hydroxylation	benzene → phenol
Epoxidation	$R-CH_2-CH_2-R \longrightarrow R-CH-CH-R$ (epoxide)
Oxidative deamination	$R-CH(NH_2)-CH_3 \longrightarrow [R-C(OH)(NH_2)-CH_3] \longrightarrow R-CO-CH_3 + NH_2$

Table 5-1: Metabolic Transformations. *Recreated with permission from Lu F. C.,* Basic Toxicology: Fundamentals, Target Organs, and Risk Assessment, *Taylor and Francis, New York, 1985.*

PHASE I (continued)	
N-dealkylation	$R-N(CH_3)_2 \rightarrow R-NH(CH_3) + CH_2O$
O-dealkylation	$R-O-CH_3 \rightarrow R-OH + CH_2O$
S-dealkylation	$R-S-CH_3 \rightarrow R-SH + CH_2O$
Metaloalkane dealkylation	$Pb(C_2H_5)_4 \rightarrow PbH(C_2H_5)_3$
N-oxidation	$R_3N \rightarrow R_2N=O + H^+$
N-hydroxylation	Ph-NH_2 → Ph-$NHOH$
Sulfoxidation	$R-S-R \rightarrow R-S(=O)-R$
Desulfuration	$R_2C=S \rightarrow R_2C=O$
Dehalogenation	Ph-F → Ph-OH
(b) Nonmicrosomal oxidations	
Monoamine and diamine oxidation	$RCH_2NH_2 \xrightarrow{O_2} RCH=NH \xrightarrow{H_2O} RCHO + NH_3$
Alcohol dehydrogenation	$RCH_2OH + NAD^+ \rightarrow R-CHO + NADH + H^+$
Aldehyde dehydrogenation	$R-CHO + NAD^+ \rightarrow R-COOH + NADH + H^+$

Table 5-1: Metabolic Transformations (continued).

108 ■ Basics of Toxicology

PHASE I (continued)	
II. REDUCTIONS	
(a) Microsomal reductions	
Nitro reduction	$RNO_2 \longrightarrow RNO \longrightarrow RNHOH \longrightarrow RNH_2$
Azo reduction	$RN=NR \quad RNHNRH \longrightarrow RNH_2 + RNH_2$
Reductive dehalogenation	$R-CCl_3 \longrightarrow R-CHCl_2$
(b) Nonmicrosomal reductions	
Aldehyde reduction	$R_2C=O \longrightarrow R_2CHOH$
III. HYDROLYSIS	
Ester hydrolysis	$R-CO-O-R_1 \longrightarrow R-COOH + R_1-OH$
Amide hydrolysis	$R-CO-NH_2 \longrightarrow R-COOH + NH_3$
PHASE II	
I. CONJUGATION	
(a) Amide hydrolysis	
O-glucuronide formation ether type	Ph–OH → Ph–O–$C_6H_9O_6$
O-glucuronide formation ester type	Ph–COOH → Ph–COO–$C_6H_9O_6$
N-glucuronide formation	Ph–NH_2 → Ph–NH–$C_6H_9O_6$
S-glucuronide formation	Ph–SH → Ph–S–$C_6H_9O_6$

Table 5-1: Metabolic Transformations (continued).

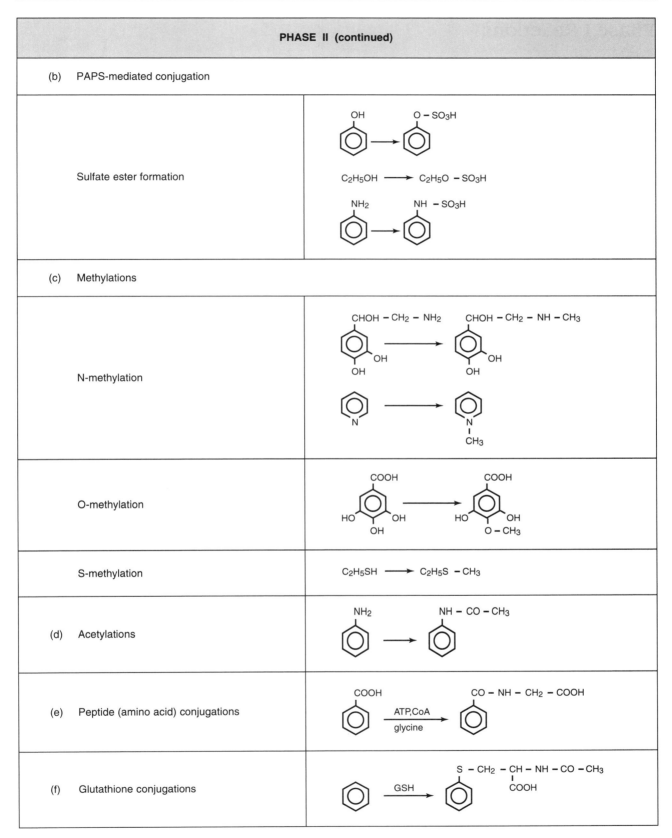

Table 5-1: Metabolic Transformations (continued).

Phase I Reactions

Oxidation

Many toxicants are metabolized by the enzymes cytochrome P-450 reductase and cytochrome P-450 in association with **NADPH (nicotinamide adenine dinucleotide phosphate)**.

> NADPH is a co-enzyme present in most cells; it interacts with various substances during normal cell metabolism.

These two enzymes are found in abundance in the endoplasmic reticulum of liver cells. The liver is the primary site of detoxification of most toxic substances that are absorbed into the bloodstream. When these enzymes interact with the toxic molecule, one atom of oxygen is attached to the toxic molecule and another oxygen atom interacts with hydrogen to form water. The following generalized equation illustrates the chemical reactions that take place:

Toxic Molecule (RH) + O_2 + NADPH + H^+ ⟶ Product (ROH) + H_2O + $NADP^+$

> The letter "R" may be used in chemical formulae to represent other chemical groups, a single atom, or a complex chemical structure.

Figure 5-18 illustrates various types of toxic substances that undergo oxidation reactions. The chemical intermediates may be excreted or undergo further metabolism in Phase II reactions.

Alcohols also undergo oxidation. Ethanol is first converted to acetaldehyde, then to acetic acid, which is further oxidized to carbon dioxide and water. Methanol, however, is converted to the more toxic intermediate, formic acid, which reacts with cell components and coagulates protein.

The organochlorine pesticide DDT is very lipid-soluble and can only be excreted from the body if it is converted to a water-soluble form. DDT undergoes a series of chemical reactions ultimately resulting in the oxidation of the intermediate p,p'-**DDOH** [2,2-bis (p-chlorophenyl) ethanol] to p,p'-DDA [bis (p-chlorophenyl) acetic acid], which has a water-soluble form. The following equation represents the chemical reaction that occurs:

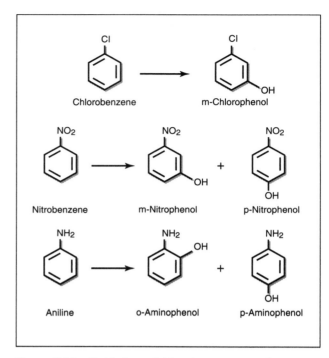

Figure 5-18: Oxidation of chlorobenzene, nitrobenzene, and aniline.

$$R-\underset{\underset{\underset{H}{|}}{\underset{H-C-OH}{|}}}{\overset{H}{\underset{|}{C}}}-R + O_2 \xrightarrow{P-450} R-\underset{\underset{O}{\|}}{\underset{C-OH}{\overset{H}{\underset{|}{C}}}}-R + H_2O$$

P,P'-DDOH   P,P'-DDA

Oxygen was added to the CH_2OH group with the subsequent formation of the COOH group and water. Oxidation reactions involving the pesticides **aldrin** and **heptachlor** result in the formation of **dieldrin** and heptachlor epoxide, respectively. However, these metabolites are still very lipid-soluble and therefore are not readily excreted from the body.

As mentioned previously, some reactions can produce toxic intermediates. Aromatic compounds (compounds that contain the benzene ring symbolized by ⊙ or ◯) that undergo oxidation reactions may produce toxic chemical intermediates referred to as polycyclic aromatic hydrocarbons, such as benzo[a]pyrene. These substances are thought to be the cause of certain types of cancer. Similarly, oxidation of the organophosphate pesticide malathion produces malaoxon, which is responsible for the toxic effect associated with this substance.

Reduction

Reduction reactions occur as a result of 1) an atom or ion accepting one or more electrons, 2) removal of oxygen from a molecule, or 3) addition of hydrogen to a molecule. A simple example of a reduction reaction is:

$$Fe^{3+} + e^- \Leftrightarrow Fe^{2+}$$

All three types of reduction reactions are involved in the metabolism of various toxic substances.

Organochlorine pesticides such as DDT (Figure 5-19) and **toxaphene** undergo reduction to form metabolites that undergo either further reduction or oxidation. Various toxic metals such as pentavalent arsenic (As^{5+}) undergo reduction as well. The final end product is more water soluble and easier to excrete.

Hydrolysis

Hydrolysis is a chemical reaction in which water reacts with another substance to form two or more new substances. This reaction involves the ionization of the water molecule ($H_2O \Leftrightarrow H^+ + OH^-$) and splitting of the substrate molecule reacting with the water molecule. The following equation is representative of hydrolytic reactions:

$$CH_3COOC_2H_5 + H \cdot OH \Leftrightarrow CH_3COOH + C_2H_5OH$$

Examples include conversion of starch to glucose, or conversion of sucrose to fructose and glucose. All these reactions occur in the presence of enzymes that facilitate the splitting of the chemical bonds.

Hydrolysis plays an important role in the detoxification of organophosphate and carbamate pesticides (Figure 5-20). Hydrolysis produces reactive metabolites that interact with Phase II enzymes and substrates to detoxify these pesticides.

> Phase I reactions result in the breakdown of toxic substances that are excreted or involved in Phase II reactions.

Phase II Reactions

During Phase II reactions the toxic substance or its metabolite is bound with a compound that generally renders the newly synthesized molecule less toxic. This type of reaction is referred to as conjugation.

> Conjugation is the joining of two substances to form a single molecule, which can increase water solubility and excretion.

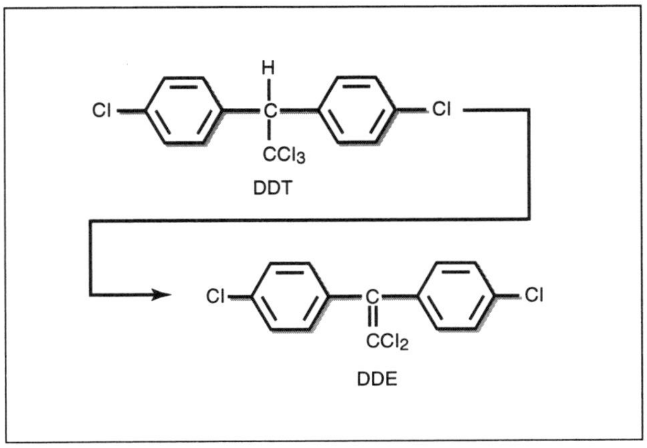

Figure 5-19: Reduction of the organochlorine pesticide DDT. Hydrogen is removed to form DDE, which is more stable.

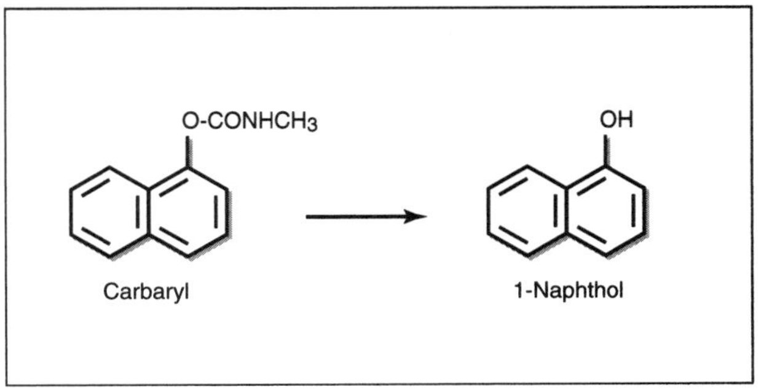

Figure 5-20: Hydrolysis of the insecticide carbaryl to produce the chemical intermediate 1-Naphthol.

The molecule is also more water soluble and readily removed from the body. As mentioned previously, there are several types of Phase II reactions, of which glucuronidation is probably the most important.

Conjugation with Glucuronidation

Glucuronidation occurs as a result of the conjugation of glucuronic acid with either a metabolite from Phase I reactions or with the parent compound (Figure 5-21). There are several chemical compounds that interact with glucuronic acid: 1) alcohols (R-OH); 2) carboxylic acids (R-COOH); 3) sulfhydryl compounds (R-SH); and 4) amines ($R-NH_2$). Glucuronic acid can be conjugated with organophosphate pesticides and organic solvents such as benzene. The conjugated molecules are more water soluble and excreted by the kidneys.

Conjugation with Glutathione

Conjugation with glutathione is an important Phase II reaction that renders highly toxic metabolites harmless. Chemical intermediates (metabolites) of some toxic substances cause cellular damage and genetic **mutations** as a result of binding with **nucleic acids**.

Mutation is a permanent change in the genetic material.

A nucleic acid is a molecule consisting of many nucleotides chemically bound together. Examples are **deoxyribonucleic acid (DNA)** and **ribonucleic acid (RNA)**.

Conjugation of these intermediates with glutathione prevents binding with the nucleic acids, therefore preventing the occurrence of mutations. Metabolites of organic solvents such as benzene, chloroform, and carbon tetrachloride are conjugated with glutathione resulting in decreased toxicity.

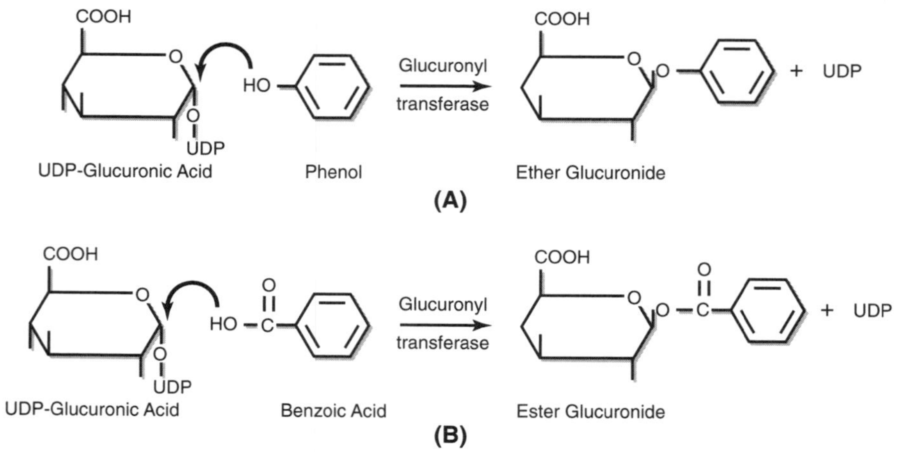

Figure 5-21: Illustration of glucuronidation – a Phase II reaction – involving Phenol (A) in one reaction and benzoic acid (B) in another. Glucuronic acid – in the presence of the enzyme glucuronyl transferase – binds with the toxic substance to produce a chemical intermediate. UDP is uridine diphosphate.

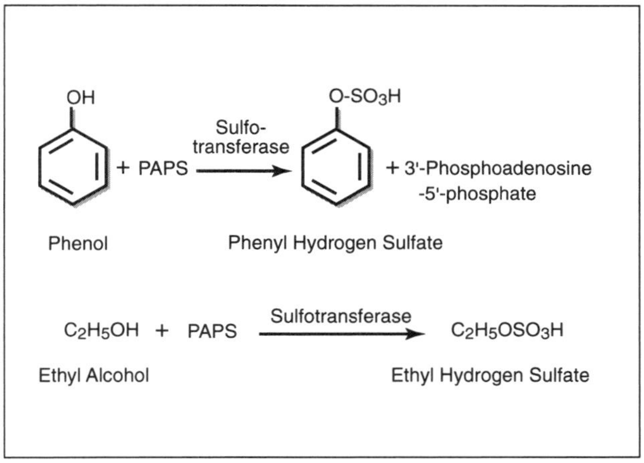

Figure 5-22: Sulfation of phenol and ethyl alcohol. In the presence of the enzyme sulfotransferase phenol may be derived from oxidation (Phase I reaction) of benzene. Ethyl alcohol may not have undergone a Phase I reaction. PAPS is the sulfate donor, 3'-phosphoadenosine-5'phosphosulfate.

Sulfation

Toxic substances that undergo glucuronidation may also undergo **sulfation**, which involves the conjugation of a sulfate group (SO_4^{2-}) with the toxic substance molecule or its metabolite. Phenols and alcohols (ethyl alcohol) undergo sulfation readily and produce a highly water-soluble substance (Figure 5-22). These substances are excreted in the urine.

Substances that do not undergo Phase I reactions can also undergo sulfation. Aromatic amines such as aniline (Figure 5-23) and 2-aminonaphthalene undergo Phase II sulfation to their corresponding sulfate derivatives.

Acetylation and Methylation

Methylation is the addition of one or more methyl groups (CH_3-) to a toxic molecule undergoing metabolism. Methylation – and to some degree **acetylation** – decrease water solubility. Methylation plays only a minor role in the transformation of toxic substances. Toxic substances that undergo these reactions are retained in the body for longer periods of time.

Metals can undergo methylation. Methyl mercury is perhaps the best known methylated metal, which is highly toxic: it is soluble in lipids and concentrates in neural tissue, where it may significantly affect the structure and function of nerve cells. Nicotine (Figure 5-24) arsenic, and selenium can also undergo methylation.

Aromatic amines ($R-NH_2$) are primarily biotransformed by acetylation. Acetylation can either activate or deactivate the aromatic amines. Activated amines can react with sensitive intracellular molecules – such as DNA – causing mutations and disrupting normal cellular function. They may also interact with structural and functional proteins (enzymes).

Phase II reactions result in the synthesis of compounds that are generally less toxic and water soluble.

Acetylation and methylation may increase toxicity by decreasing water solubility and, therefore, excretion.

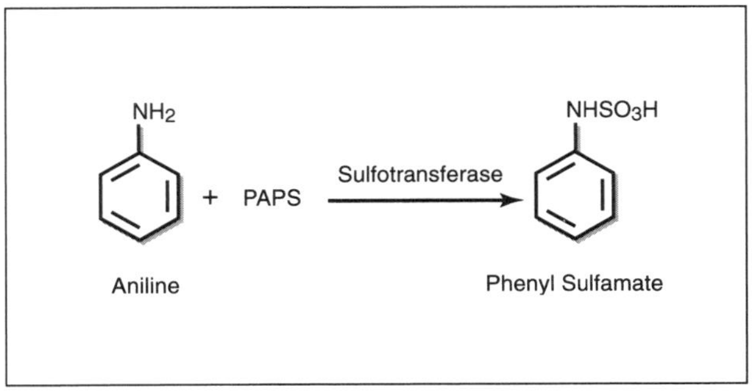

Figure 5-23: Sulfation of aniline. Aniline did not undergo a Phase I reaction. PAPS is the sulfate donor, 3'-phosphoadenosine-5'-phosphosulfate.

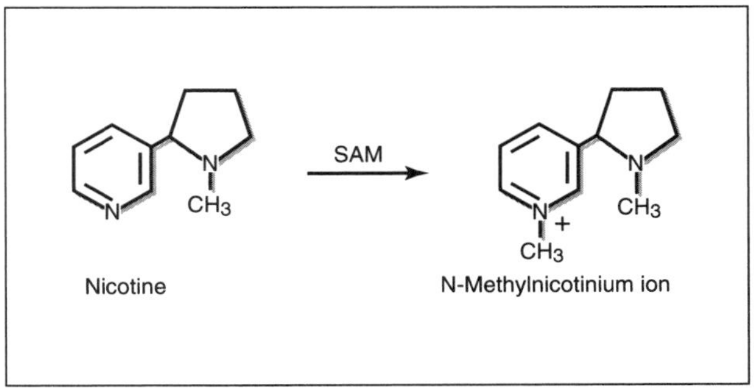

Figure 5-24: Methylation of nicotine. A methyl group is added from SAM (S-adenosyl methione) to nicotine.

Deactivation versus Bioactivation

In general, Phase I and Phase II reactions are designed to deactivate toxic substances by increasing water solubility and excretion of toxic substances. However, these reactions may also bioactivate some toxicants by transforming inactive non-toxic molecules to active toxic forms. In some situations a single toxicant and its metabolites undergo both types of reactions. The rate at which the deactivation and bioactivation reactions occur will in part determine the toxicity of the substance. Figure 5-25 is a generalized illustration of the pathways a toxic chemical may follow during metabolism.

The organophosphate pesticides parathion and malathion are examples of toxicants that undergo both deactivation and bioactivation. Figure 5-26 illustrates the chemical pathways, reactions, and products as a result of biotransformation of parathion.

Benzo[a]pyrene is a polycyclic aromatic hydrocarbon found in cigarette smoke, in exhaust fumes from internal combustion engines, and in smoke from burning various types of organic matter.

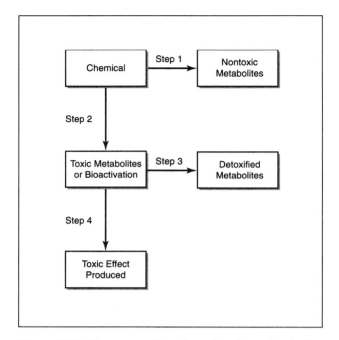

Figure 5-25: Four potential end results of any chemical undergoing metabolism. *Recreated by permission from Williams P. L. and Burson J. L.*, Industrial Toxicology: Safety and Health Applications in the Workplace, *Van Nostrand Reinhold, New York, 1985.*

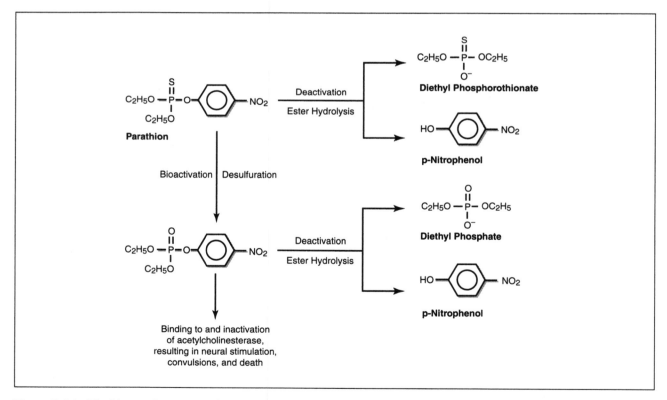

Figure 5-26: The biotransformation of parathion. *Recreated by permission from Williams P. L. and Burson J. L.*, Industrial Toxicology: Safety and Health Applications in the Workplace, *Van Nostrand Reinhold, New York, 1985.*

Benzo[a]pyrene has been studied extensively because it is a potent cancer-causing agent. The parent compound undergoes a series of metabolic reactions producing a variety of chemical intermediates (Figure 5-27). One of the intermediates is 7,8-dihydrodiol,9,10-oxide, which is believed to be the cancer-causing agent. This substance binds to the DNA of the cell; this may lead to cellular dysfunction and development of cancerous growth. The possible mechanisms involved in cancer development are discussed in Chapter 6.

Similarly to benzo[a]pyrene, nitrosamines undergo biotransformation reactions, thereby producing toxic intermediates. These substances cause liver damage and cancer in the liver, lungs, and kidneys. The production of a carbonium ion (Figure 5-28) is thought to be the toxic agent that causes cancer.

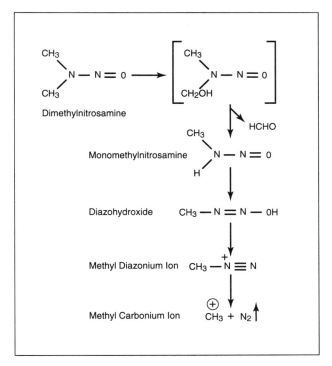

Figure 5-28: Metabolism of dimethylnitrosamine. The carbonium ion produced is thought to be carcinogenic.

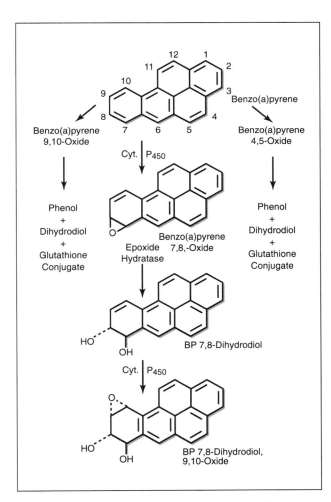

Figure 5-27: Metabolism of benzo(a)pyrene (BP). Several chemical intermediates are produced. The chemical intermediate BP 7,8-dihydrodiol,9,10-oxide is thought to be carcinogenic.

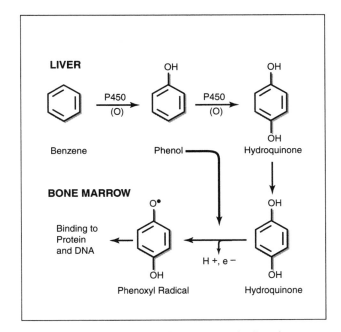

Figure 5-29: Metabolism of benzene in the liver by cytochrome P450 enzymes. Phenol and hydroquinone can be further metabolized to produce the phenoxyl radical (contains a free electron), which binds to the DNA of bone marrow cells. This may be the causative agent of leukemia.

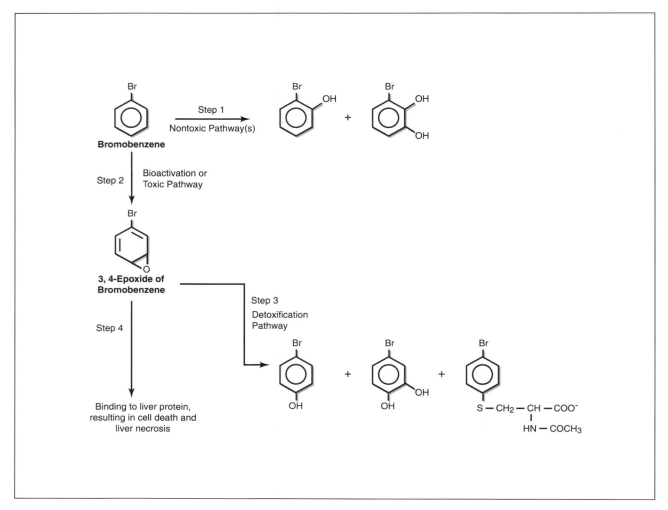

Figure 5-30: Metabolic pathway for the bioactivation of bromobenzene. *Recreated by permission from Williams P. L. and Burson J. L., Industrial Toxicology: Safety and Health Applications in the Workplace, Van Nostrand Reinhold, New York, 1985.*

Benzene is metabolized in the liver to phenol and hydroquinone (Figure 5-29). These two substances are converted to a phenoxyl radical that is capable of interacting with the DNA of cells in the bone marrow. Damage of the DNA by the phenoxyl radical may be the mechanism of action associated with the development of leukemia.

Other chemicals such as bromobenzene and acetaminophen also undergo bioactivation. Figure 5-30 illustrates the biotransformation pathway for bromobenzene. The toxic metabolic product 3,4-epoxide of bromobenzene causes cell death in the liver. Acetaminophen (the active ingredient in Tylenol) will have the same effect as bromobenzene when taken in doses above the recommended daily (Figure 5-31).

> In general, deactivation mechanisms transform toxic substances to non-toxic forms, and bioactivation transforms non-toxic substances to toxic forms.

Table 5-2 lists a variety of substances that are bioactivated, the bioactive agent (if known), and the effects as a result of exposure.

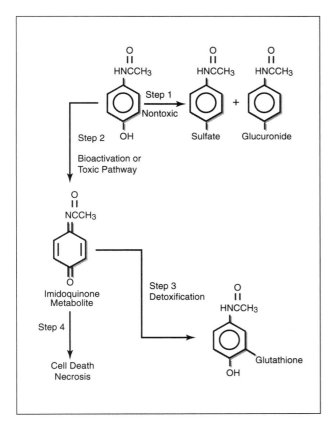

Figure 5-31: Metabolic pathway for the bioactivation of acetaminophen. *Recreated by permission from Williams P. L. and Burson J. L.*, Industrial Toxicology: Safety and Health Applications in the Workplace, *Van Nostrand Reinhold, New York, 1985.*

Factors Affecting Metabolism

The rate at which metabolism of toxic substances occurs is dependent on a variety of factors that can be categorized into two groups: 1) those factors that affect the metabolic processes directly, and 2) those factors that affect the transport of toxic substance to tissues where metabolism occurs.

Factors Affecting the Metabolic Process

In Chapter 4 several factors were described that affect toxicity of a given substance. Changes in toxicity are reflected in different LD_{50} or LC_{50} values for the same chemical in different species or even different categories (age, sex) of the same species. It should not be surprising that the factors affecting toxicity are some of the same factors affecting biotransformation, since the rate of transformation is – in part – directly related to the toxic effects.

Biotransformation is affected by the species of the test animal, age, sex, nutritional status, disease, enzyme induction or inhibition, and genetics. Each of these factors may affect metabolism either on an individual basis or collectively. For example, newborn babies and young infants are more susceptible to a variety of chemicals such as pesticides because the cytochrome P-450 enzymes – important in pesticide detoxification reactions – are not well developed. Therefore, biotransformation and detoxification does not occur as readily.

A balanced diet will provide the necessary protein as well as essential metals and minerals such as copper, zinc, and calcium to assist normal cellular enzymatic activities associated with biotransformation. Protein-deficient diets can result in a decrease in protein synthesis, thus affecting the synthesis of enzymes involved in the metabolic reactions used in detoxification.

Diseases associated with the tissues primarily responsible for detoxification may result in decreased metabolism of the toxicant and, consequently, increased toxicity. Since the liver is the major site of detoxification and storage of toxicants, it is susceptible to cellular damage from exposure to these substances. This problem is somewhat ameliorated by the fact that the liver has a large reserve capacity for performing its normal functions. In other words, some liver damage may occur without affecting the overall rate of transformation of a particular toxic substance. However, prolonged exposure to toxicants or large doses may overwhelm the liver resulting in permanent liver damage and reduced detoxification capacity.

Cirrhosis of the liver is often caused by excessive drinking of alcohol.

> Cirrhosis is a chronic disease of the liver characterized by replacement of normal liver tissue with connective tissue.

During the disease process the liver cells are damaged and replaced by connective tissue. If enough cells are killed, the ability of the liver to metabolize toxic substances is dramatically reduced. At this stage, a chemical substance that could have been

118 ■ Basics of Toxicology

Parent Compound	Toxic Metabolite	Site of Formation	Toxicity
Acetaminophen	N-Hydroxy derivative	Liver	Hepatic necrosis
Aflatoxin B1	Aflatoxin-2,3-epoxide?	Liver	Hepatic cancer
Benzene	Benzene epoxide	Liver	Bone marrow depression
Bromobenzene	Bromobenzene epoxide	Liver, lung	Hepatic, renal bronchiolar necrosis
Carbon tetrachloride	Trichloromethane free radical	Liver	Hepatic necrosis, hepatic cancer
Carcinogenic alkylnitrosamines	Unknown	Liver, many organs	Hepatic cancer
Carcinogenic polycyclic hydrocarbons	Epoxides?	Many organs	Cancers cytotoxicity
Chloroform	Unknown	Liver, kidney	Hepatic, renal necrosis
Cyclamate	Cyclohexylamine	Gut flora	Bladder cancer
Ethanol	Acetaldehyde	Many organs	Varied
Hemolysis-producing aromatic amines	N-Hydroxy metabolites	Liver	Hemolysis
Methanol	Formaldehyde (formic acid)	Liver, retina	Retinal and general toxicity
Methemoglobin-producing aromatic amines and nitro compounds	N-Hydroxy metabolites	Liver, other organs	Methemoglobinemia
Nitrates	Nitrites	Gut flora	Methemoglobin
Nitrites plus secondary or tertiary amines	Nitrosamines	Stomach	Hepatic, pulmonary cancers
Parathion	Paraoxon	Liver	Neuromuscular paralysis
Safrole	1'-Hydroxysafrole	Unknown	Cancers
Thalidomide	Phthalyl-L-glutamic acid?	Nonenzymatic hydrolysis	Teratogenesis
Urethane	N-Hydroxy-urethane	Liver	Cancers, cytotoxicity

Table 5-2: Examples of Bioactivation. *Recreated with permission from Lu F. C., Basic Toxicology: Fundamentals, Target Organs, and Risk Assessment, Taylor and Francis, New York, 1985.*

easily metabolized before the onset of cirrhosis would now be toxic and exert its effects on the target tissues because of the reduced detoxification capacity. Cirrhosis can also be caused by repeated exposure to arsenic or to high levels of vitamin A. Exposure to chemicals such as carbon tetrachloride and vinyl chloride may result in liver cell damage and decreased metabolism of toxic substances. These two substances are also associated with the development of liver cancer.

Similarly, the kidneys – which are the major site of toxic metal metabolism – are damaged by absorption and concentration of heavy metals in the cells of the proximal convoluted tubules of the nephron. Mercury and cadmium bind with intracellular enzymes involved in maintaining normal cellular metabolism. Because of this, the enzymes are not able to perform their normal function, or they do so at a reduced rate resulting in cellular dysfunction or death. If enough nephrons are affected, biotransformation of toxicants is reduced. In addition, excretion of toxicants as well as other molecules and ions may be disrupted. This further enhances the toxicity of the substance.

Factors Affecting Transport and Absorption

Toxic substances must be transported to the appropriate tissues in order for metabolism to occur. Many of the factors discussed earlier in this chapter such as absorption rate, perfusion rate, plasma protein binding, and storage will affect the rate at which a toxic substance is delivered to the tissue where metabolism occurs. In general, organic substances – because of their high lipid solubility – are easily absorbed and undergo metabolism. Lipid-insoluble substances with large molecular weights will not be delivered as rapidly to active metabolic sites such as the liver.

The perfusion rate of a given tissue is important in determining how quickly a toxic substance will be transformed. Organs such as the liver and kidneys have a high perfusion rate relative to other tissue types. These organs have the potential to extract and detoxify larger quantities of toxicants from the blood. Tissues with low perfusion rates – such as fats – generally have low metabolic rates. Any toxic substances absorbed in these types of tissues are metabolized slowly and accumulate over time.

Protein binding can serve a dual purpose relative to metabolism of toxic substances. Toxic substances bound to proteins in the blood do not easily move across cell membranes. In many cases this slows the rate at which the toxicant is metabolized because the substance may not be readily absorbed by the tissue where detoxification occurs. Binding to intracellular protein provides an efficient system for delivery of the toxicant to the enzymes involved in metabolism. The large size of the protein-bound molecules slows their removal from the cell.

Checking Your Understanding

1. List and provide a brief explanation of the types of Phase I reactions.

2. List and provide a brief explanation of how three of the Phase II reactions occur.

3. How do acetylation and methylation differ from other Phase II reactions?

4. What is bioactivation? Give an example. How might bioactivation be related to the development of cancer?

5. Identify the two major factors that affect the metabolism of toxic substances. Give examples of each and how they might affect toxicity.

5-5 Excretion

After undergoing biotransformation many toxic substances are in a form that facilitates excretion. The rate at which excretion of toxic substances occurs is important in determining the toxicity of a substance. The faster a substance is eliminated from the body, the more unlikely a biological effect will be. Excretion of toxic substances occurs by a variety of pathways. The primary organs involved in excretion are the kidneys, liver, and lungs. Minor amounts of substances may leave the body through sweat or milk; they may also be deposited in inactive tissue such as nails, hair, or the outer layer of skin.

Excretion by the Liver

Many toxic substances are stored and detoxified in the liver. Detoxification takes place by one of the many mechanisms discussed previously to produce water-soluble products. The toxic substances are then excreted into the bile. Bile is produced in the liver by the hepatic cells. The bile drains into a system of canals and ducts ultimately converging to form the **common bile duct**, which empties into the duodenum portion of the small intestine (Figure 5-32). This mechanism is important in removing large protein-bound toxicants such as heavy metals.

Excretion of the toxicant from the liver to the intestinal tract will usually result in the substance being removed in the feces. However, the intestinal bacteria are also capable of producing enzymes that cause the detoxified substance to become less water soluble. As a result, the toxic substance may be reabsorbed from the digestive tract. The process of excreting toxic substances from the liver and their subsequent reabsorption from the digestive tract is referred to as **enterohepatic circulation**. Gluconated polycyclic aromatic hydrocarbons and glutathione conjugates of trichloroethylene are reabsorbed by this mechanism and therefore retained.

Because blood flows directly to the liver from the intestines through the hepatic portal veins, a toxic substance absorbed from the digestive tract is often absorbed in the liver and does not enter other parts of the systemic circulation. Once in the liver, the substance may be either stored or detoxified and excreted into the intestinal tract. Detoxification of toxic substances before they reach the other portions of the systemic circulation is referred to as the **"first-pass effect."** This effect can decrease the systemic toxicity of those substances absorbed from the digestive tract.

Excretion by the Kidneys

The kidneys receive 25 percent of the cardiac output. The high perfusion rate not only results in significant exposure to circulating toxic substances, but also facilitates the excretion of the toxicants. The ability of the kidneys to excrete toxic substances is affected by the active and passive transport mechanisms. Toxic substances enter the kidneys as a result of active and passive transport mechanisms present in the glomerulus and the nephron tubules. Toxic substances that are lipophilic – such as carbon tetrachloride, chloroform, and tetrachloroethylene – readily pass through the glomerulus into the nephron. They may also passively diffuse into the kidneys from the extensive capillary system (peritubular capillaries) that surrounds the individual nephrons. In both cases a concentration gradient must exist in order for diffusion of these substances to occur. Because of their lipid solubility, some of these substances are passively reabsorbed in the proximal convoluted tubules. Absorption of these organic toxicants in the tubule cells, where they undergo metabolism, results in the formation of toxic metabolites, which can cause cell death. However, depending on the urine's pH, these organic toxicants may be ionized – instead of reabsorbed – and excreted.

Several heavy metals – such as cadmium, lead, and mercury – are excreted by the kidneys. These metals are bound to plasma proteins. The protein-metal complex has a low molecular weight and is able to pass through the glomerulus to the nephron. However, this complex may be reabsorbed by active transport mechanisms in the proximal convoluted tubules. The metals are then stored in the kidneys for prolonged periods of time, or they may reach a concentration high enough to produce toxic effects causing tubule cell death and, subsequently, kidney failure.

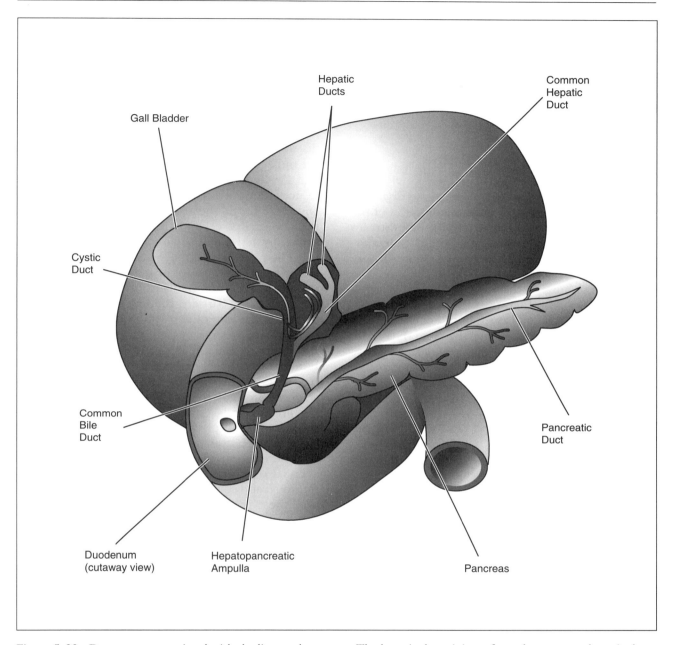

Figure 5-32: Duct system associated with the liver and pancreas. The hepatic ducts join to form the common hepatic duct, which joins with the cystic duct to form the common bile duct. The latter joins with the pancreatic duct and enters the duodenum.

What amount of toxicant is reabsorbed or excreted from the body depends on several factors. If the concentration of the toxicant entering the nephron exceeds the active transport capacity of reabsorption in the tubule cells, a high percentage of the toxicant will be excreted. If the perfusion rate to the kidneys changes, then the amount of toxicant delivered to the kidneys will vary: high perfusion rates will generally result in greater concentrations of the toxicant in the urine. Finally, disease also can affect the amount of toxicant excreted from the body. For example, absorption of certain toxicants – such as cadmium – in the proximal convoluted tubule can result in cell death. The damaged nephron is unable to reabsorb cadmium; therefore, if enough nephrons are damaged, there will be an increased cadmium excretion. Usually, if damage to the kidneys is this severe, the regulation of essential substances such as glucose and amino acids is disrupted and adverse effects begin to appear.

Excretion by the Lungs

The structure of the respiratory system was discussed in detail in Chapter 3. The primary function of the lungs is to supply oxygen-rich air to the bloodstream and to remove gases produced by cellular metabolism – such as carbon dioxide – from the bloodstream. The exchange of gases occurs between the alveoli of the lungs and the extensive capillary system associated with them. Gas exchange is by passive diffusion; it occurs as a result of a concentration gradient for each gas between the air and the blood.

Excretion of volatile toxic gases, such as those associated with organic compounds, occurs in the lungs. The transfer of gases from the blood to the lungs is influenced by concentration gradients and by their solubility in water. Ethylene – which is only slightly soluble in water – will readily diffuse from the blood into the lungs and will therefore be easily removed. Chloroform however, is more water soluble and will not diffuse as easily from the blood into the lungs.

Checking Your Understanding

1. Identify the three major organs involved in excretion. For each organ explain the mechanisms that facilitate excretion.

2. Describe the enterohepatic circulation. What effect might this process have on toxicity?

3. Explain what is meant by the "first-pass effect." What benefit might it provide?

Summary

The amount of a toxic substance that reaches the target tissue is dependent upon the amount absorbed, the distribution, metabolism, and excretion of the substance from the body. There are four major factors that affect absorption and subsequent distribution, metabolism, and excretion: 1) size of the molecule, 2) lipid solubility, 3) electrical charge, and 4) cell membrane carrier molecules. Toxic substances transported by the circulatory system enter and leave the bloodstream by absorption through the endothelial cells of the capillaries. Movement occurs by active and passive transport mechanisms.

There are several mechanisms that oppose distribution of toxicants to the target tissues: the binding to plasma proteins, distribution to storage sites, and specialized barriers. Binding to plasma proteins decreases the rate of absorption by increasing molecular size. Storage of toxicants occurs in various tissue types, primarily in the kidneys, liver, fat, and bone. The blood-brain barrier is a specialized structure that decreases the rate and amount of toxic substances absorbed into the brain tissue.

Phase I and Phase II reactions are involved in the metabolism of toxic substances. Phase I reactions include oxidation, reduction, and hydrolysis. Phase II reactions include glucuronidation, sulfation, acetylation, methylation, conjugation with glutathione, and conjugation with amino acids. In general, these reactions decrease toxicity and increase water solubility and excretion of toxic substances. In some circumstances, such as methylation, the reaction results in bioactivation of the substance by producing a toxic chemical intermediate that is less soluble in water and is not readily excreted by the kidneys.

Excretion of toxic substances occurs primarily from the liver, kidneys, and lungs. Toxic substances excreted by the liver enter the small intestine where they may be eliminated in the feces or reabsorbed. The process of excreting toxic substances from the liver and their subsequent reabsorption from the digestive tract is referred to as enterophepatic circulation. The process of toxic substances being absorbed from the small intestine and detoxified before they reach other portions of the systemic circulation is referred to as the first-pass effect. Both processes can affect toxicity. Diseases that affect the functional capacity of the liver also affect toxicity by decreasing the liver's ability to detoxify the substance.

Toxic substances that enter the kidneys are either reabsorbed in the nephron of the kidneys or excreted in the urine. The amount excreted is affected by the blood perfusion rate, by factors affecting reabsorption, and by diseases affecting either of these two factors.

The lungs primarily excrete toxic gases that have entered the blood. Both the concentration gradient of the toxic gas between the blood and the lungs and the solubility of the substance in the bloodstream will affect the rate of excretion.

Application and Critical Thinking Activities

1. A low protein diet can result in increased blood levels of cadmium and lead. Propose a reason for this effect and describe the mechanism(s) that would be involved in causing this observed response.

2. If the affinity for a given toxic substance is the same for all tissue types, why do some tissue types accumulate larger quantities than others?

3. Choose one of the major storage sites of the body for toxic substances. Identify a substance that is stored in the tissue and the mechanisms that would affect the transport and absorption of the substance to the target tissue site.

4. Identify one toxic substance that is absorbed readily through the blood-brain barrier and one that is not. Provide an explanation that supports each example identified.

5. Explain how deactivation and bioactivation affect the toxicity of a chemical. Why is the understanding of these mechanisms important in establishing toxicity?

6. Explain how cirrhosis of the liver can affect both the transport of a substance to the target tissue and its metabolism. How does this affect toxicity?

7. Given the toxic substances mercury and methylene chloride, identify possible sites of storage, metabolism, and routes of excretion. Explain your choices indicating the mechanisms that might be involved in each of the activities.

6

Target Organ Effects

Chapter Objectives

Upon completing this chapter, the student will be able to:

1. **Identify** the divisions of the nervous system.
2. **Describe** the components of a neuron and their functions.
3. **Identify** and briefly describe the function of neuroglia cells.
4. **Describe** the electro-chemical events associated with response to a stimulus.
5. **Identify** three types of structural changes in neurons associated with neurotoxicity.
6. **Distinguish** the difference between the types of synaptic neurotoxic agents.
7. **List** the cellular components of the blood and their functions.
8. **Define** the terms anemia, granulocytopenia, thrombocytopenia, lymphocytopenia.
9. **Differentiate** between the various types of anemias caused by chemical exposure.
10. **Describe** the difference between chemically induced anemia and hypoxia.
11. **Identify** the components of the immune system.
12. **Differentiate** between nonspecific and specific components of the immune system and the effects of toxicants on each.

Chapter Sections

6-1   Introduction
6-2   Neurotoxicity
6-3   Hematoxicity
6-4   Immunotoxicity
6-5   Cardiotoxicity
6-6   Pulmonary Toxicity
6-7   Hepatotoxicity
6-8   Nephrotoxicity
6-9   Reproductive Toxicity
6-10 Toxicity of the Eye

13. **List** the structures involved in the intrinsic and extrinsic control of the heart.
14. **Identify** some of the major effects toxic substances have on the heart.
15. **List** and explain the primary functions of the respiratory system.
16. **Identify** the components and functions of the specific and nonspecific defense mechanisms of the respiratory tract.
17. **Describe** the effects of various toxic substances on different regions of the respiratory tract.
18. **Identify** and describe the major functions of the liver.
19. **Discuss** how toxic substances affect the liver.
20. **Identify** and describe the major functions of the kidneys.
21. **Explain** the effects of nephrotoxic substances on kidney function.
22. **Discuss** the general anatomy and physiology of the male and female reproductive systems.
23. **Describe** the effects of various toxic substances on the male and female reproductive systems.
24. **Describe** the general anatomy of the eye.
25. **Explain** the effects of various toxic substances on the eye.

6-1 Introduction

Once a toxic substance enters the body it is distributed by the circulatory system, where it may be absorbed by and accumulated in a variety of tissues. The composition of the tissue and the physio-chemical properties of the toxicant will determine where toxic substances will concentrate. There are several different mechanisms by which the toxicant may exert its effects. Usually the observed effects are the result of the cumulative impact of the toxicant on several different aspects of normal cell functioning. For example, heavy metals such as mercury are known to bind to various intracellular enzymes, which are important **catalysts** for chemical reactions that occur in the cell. The chemical reactions may be necessary for maintaining cell membrane integrity or ionic balance, both of which are important in the normal functioning of nerve, muscle, and heart cells. Other toxicants may disrupt the cellular mechanism for producing ATP, which provides the energy necessary to support most body functions. Chemicals such as 2,4-dinitrophenol and pentachlorophenol affect ATP synthesis. Decreased ATP production affects the cell membrane integrity and homeostasis. In the end, the inability of any cell type to maintain intracellular homeostasis results in cell death.

It is important to realize that death of a single cell does not usually result in the appearance of toxic effects. Many tissues such as the liver and kidneys do not begin to demonstrate symptoms associated with toxicity until significant portions of these organs are affected. The liver and kidneys have an excess capacity relative to the functions they perform. Therefore, the loss of some functional areas is compensated for by the previously underutilized segments. Neural tissue, however, is more sensitive to cellular death, and neurotoxic effects are usually observed sooner.

6-2 Neurotoxicity

General Organization of the Nervous System

The nervous system of the body is composed of the **central nervous system (CNS)** and the **peripheral nervous system (PNS)** (Figure 6-1). The central nervous system consists of the brain and the spinal cord. The peripheral nervous system consists of the nerves outside the central nervous system. The peripheral nervous system can be divided into **afferent** (sensory) **nerves** that carry sensory information – such as touch or pain – to the central nervous system; and **efferent** (motor) **nerves** that transmit information from the central nervous system to glands or muscles. The efferent nerves can be divided into the **somatic motor** and the **autonomic nervous systems (ANS)**. The somatic motor nervous system stimulates the skeletal muscles of the body. The ANS can be further divided into the **sympathetic (SNS)** and **parasympathetic (PSNS) nervous systems**. The ANS transmits stimuli from the central nervous system to organs such as the heart, the smooth muscle found in the digestive tract, blood vessels, and various glands such as the salivary glands, pancreas, and adrenal glands.

> The nervous system is comprised of many different components that detect stimuli, which, in turn, elicit a response.

Cells of the Nervous System

Neuron

The basic functional unit of the nervous system is the **neuron**. Neurons function as receptors for internal and external stimuli; they transmit the infor-

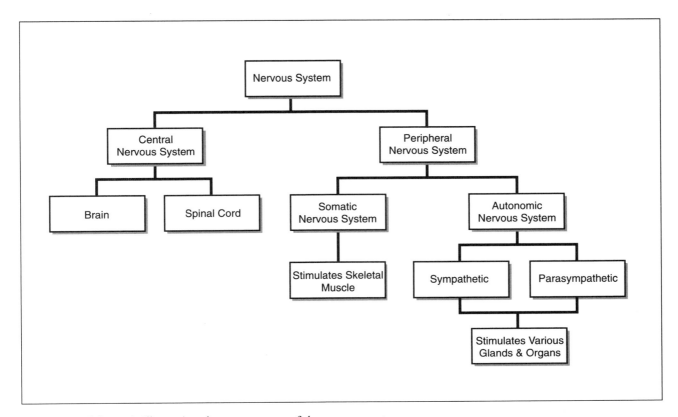

Figure 6-1: Schematic illustrating the components of the nervous system.

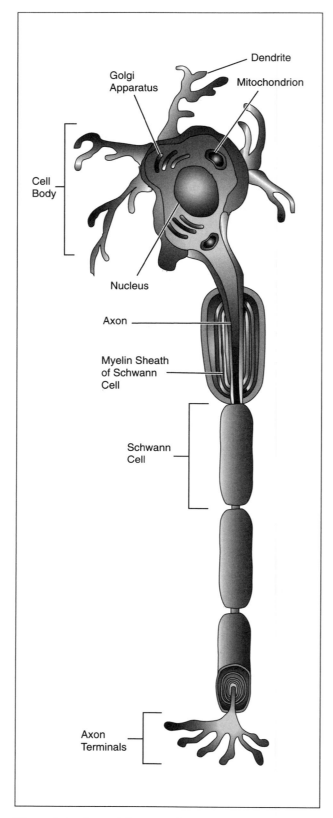

Figure 6-2: General features of a neuron. Nerve cells consist of dendrites, a cell body, and an axon.

mation to the central nervous system and carry impulses from there to muscles and glands.

The neuron is composed of the cell body (**soma**) and the **axon** as well as fiber-like extensions referred to as **dendrites** (Figure 6-2). The soma contains a nucleus, a well-developed endoplasmic reticulum, and a Golgi complex; these elements are involved in protein synthesis. Numerous mitochondria may be found throughout the cell body; they provide the energy needed to maintain and re-establish the **resting membrane potential** associated with neurons. Short, branching cell body **processes** (projections or outgrowths), called dendrites, are primarily responsible for carrying impulses toward the cell body. The dendrites can receive stimuli from a variety of sources such as touch, pain, or smell. They may also receive stimuli from other neurons. The axon is usually a single fiber that carries impulses away from the cell body. The axons of sensory nerves carry information to the central nervous system. The axons of the motor nerves carry information to the muscles and glands (Figure 6-3).

> Neurons are the functional unit of the nervous system; they are important in carrying stimuli to and from the central nervous system.

Neuroglia

Within the nervous system are various types of cells that form the supporting elements of the nervous system. These cells are referred to as the **neuroglia** or **glia** cells. The three types of neuroglia are: astrocytes, **oligodendrocytes**, and **Schwann cells**.

Astrocytes are an important component of the blood-brain barrier. The close association of the astrocyte cell processes with capillary endothelial cells forms an effective barrier that decreases the rate of absorption of many toxic substances into brain tissue. Cell processes of oligodendrocytes in the central nervous system as well as Schwann cells in the peripheral nervous system surround the nerve cell axons. This facilitates the transmission of impulses along the axon (Figure 6-4). If the axon simply lies within an indentation of an oligodendrocyte or Schwann cell, it is referred to as an **unmyelinated nerve** (Figure 6-5). Oligodendrocyte processes or Schwann cells may wrap around the axon forming several concentric layers of cell membrane called a **myelin sheath** (Figure 6-5). These types of axons are referred to as **myelinated nerves**. The myelin

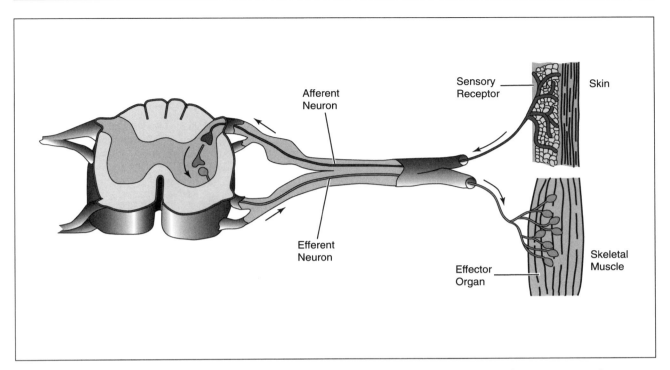

Figure 6-3: Conduction pathway of action potential. Action potentials are carried to the central nervous system by afferent neurons and from the central nervous system to the affected tissue by efferent neurons.

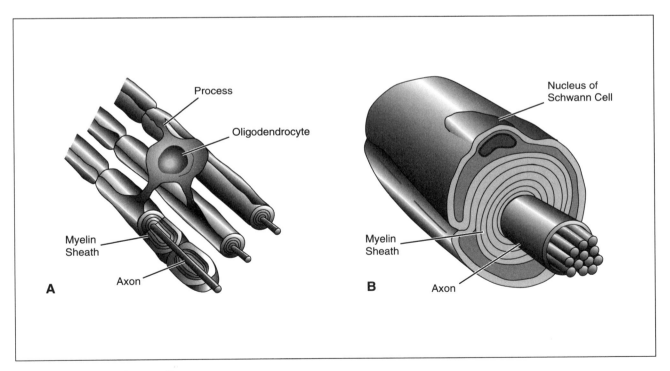

Figure 6-4: Types of neuroglia.
A. Oligodendrocyte processes surrounding axons.
B. Schwann cells wrapped around axon to form myelin sheath.

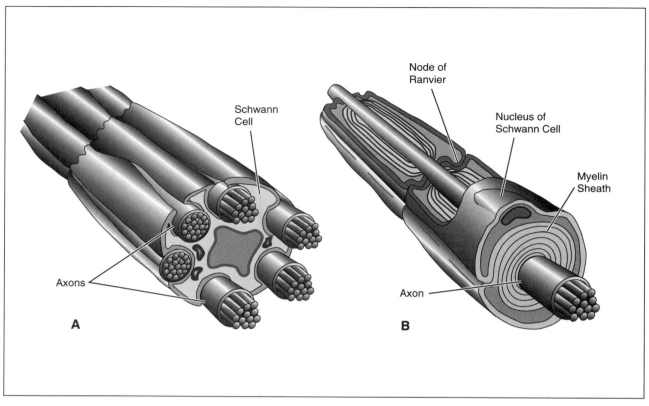

Figure 6-5: Illustration of an unmyelinated and a myelinated axon.
A. Unmyelinated axon. Axons located within indentations of the Schwann cell cytoplasm.
B. Myelinated axon. Schwann cell wraps around the axon forming several concentric layers.

sheath is an efficient insulator and does not allow easy movement of ions, particularly sodium and potassium, into and out of the axon – an important factor in maintaining the membrane potential and in the generation of **action potential**.

> Neuroglial cells are important in regulating ion balance, in the absorption of substances from the bloodstream, and in maintaining normal nerve cell functioning.

General Physiology of the Nervous System

Detection of stimuli and the initiation of responses are the result of electro-chemical events associated with the nervous system. Upon receiving a stimulus, such as pain or the sensation of heat, the cell membrane of the neuron undergoes a change in permeability resulting in the movement of sodium and potassium ions across the cell membrane. This generates an electrical charge called the action potential. The action potential travels along the nerve cell and eventually arrives at the end of the axon, where it causes the release of a chemical substance referred to as a **neurotransmitter**. The neurotransmitter is responsible for initiating a response in another nerve cell, muscle, or gland.

> A neurotransmitter is a chemical substance released from the end of an axon resulting in a stimulatory or inhibitory effect on the target tissue.

Action Potential

The action potential associated with an axon or with a group of axons (referred to as nerves) is responsible for carrying information from the source of

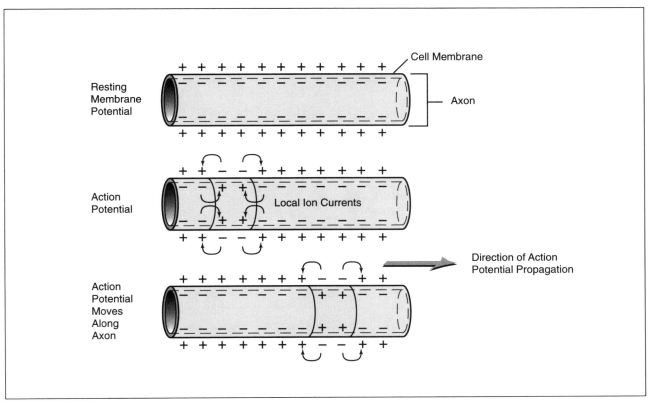

Figure 6-6: Generation of an action potential. A positive charge on the outside of the cell membrane is primarily the result of Na⁺. The negative charge is primarily the result of large negatively charged protein molecules and Cl⁻. An action potential is generated when there is a change in cell membrane permeability to Na⁺ and K⁺. The action potential in one area generates an action potential in the area next to it resulting in propagation.

the stimulus to the central nervous system by the afferent nerves. Nerves originating in the central nervous system transmit an action potential to muscles and glands and initiate a response.

The action potential is generated when the ion distribution across the cell membrane is changed. During the "resting phase" there is a positive charge on the outside and a negative charge on the inside of the nerve cell membrane. The charge difference across the cell membrane at rest – referred to as the resting membrane potential – is the result of the uneven distribution of ions, particularly sodium and potassium, on each side of the cell membrane. There is a higher concentration of sodium ions on the outside of the cell than inside, while the opposite is true for potassium ions. The concentration of negatively charged ions and protein molecules is greater on the inside. The net result of the distribution of these substances is a positive charge on the outside and a negative charge on the inside. It is the uneven distribution of sodium and potassium ions that is primarily responsible for maintaining the resting membrane potential (Figure 6-6).

> An uneven distribution of ions, particularly Na⁺ and K⁺, is responsible for establishing the resting membrane potential.

When a neuron is stimulated the cell membrane of the dendrite becomes more permeable to sodium ions. Because of the concentration gradient, the sodium ions rapidly diffuse into the cell. The inside of the cell becomes positively charged, while the outside of the cell temporarily becomes negatively charged. Potassium ions then diffuse out of the cell in response to the influx of sodium ions. These changes cause the adjacent portion of the cell membrane to depolarize, and a wave of depolarization (action potential) passes down the cell to the end of the axon resulting in the release of the chemical neurotransmitter, which then stimulates the affected tissue (Figure 6-6).

> Stimulation of a neuron results in an increased permeability to Na^+ and K^+, initiating an action potential resulting in a response by the affected tissue.

Chemical Neurotransmitters

There are several chemical neurotransmitters associated with the nervous system. The best known are **acetylcholine** and **norepinephrine**. Neurotransmitters are released from the end of the axon and interact with other neurons, glands, or muscles.

The junction where the neuron interacts with another cell is referred to as a **synapse** (Figure 6-7). The end of the axon is referred to as the **presynaptic terminal**. The cell membrane of the stimulated tissue is referred to as the **postsynaptic membrane**. The space between these two structures is the **synaptic cleft**. The neurotransmitter substance is released from the presynaptic terminal, diffusing across the synaptic cleft, and binding to receptors on the postsynaptic cell membrane. Binding to the postsynaptic cell membrane receptor initiates a response in the affected tissue. In order for the affected tissue not to be constantly stimulated, the neurotransmitter is broken down rapidly by enzymes, or is reabsorbed by the presynaptic terminal.

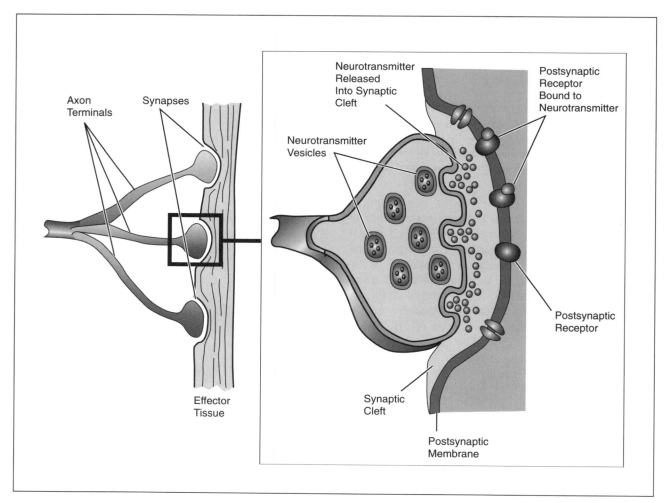

Figure 6-7: Synapse. Neurotransmitter substances are released from the presynaptic membrane, diffusing across the synaptic cleft, and binding to receptors on the postsynaptic membrane.

Neurotoxic Effects

The effects of toxic substances on neurons may be placed into two categories: 1) those that affect the neuron structure, and 2) those that affect the neurotransmission between the presynaptic terminal and the postsynaptic membrane. In each situation normal neuron functioning is disrupted. Damage to the structure of the neuron can be caused by a variety of toxic substances; the kind of damage that occurs can be classified as **neuronopathy, axonopathy,** or **myelinopathy**.

Neuronopathy

Neuronopathy (neuron destruction) is caused by the interaction of toxic substances primarily with the nerve cell body (soma). It is characterized by degeneration of the entire nerve cell including the dendrites and the axon. In general, low concentrations of toxic substances seem to cause neuronopathy in specific regions of the nervous system; higher concentrations tend to cause neuronopathy throughout the whole nervous system.

> Neuronopathy results in total destruction of the nerve cell.

Perhaps one of the best known toxic substances that causes neuronopathy is methyl mercury. Mercury is a byproduct of many different industrial processes such as paper and cosmetics production; it is also used as a fungicide. Mercury-containing waste from these industries is discharged into lakes, rivers, and oceans, where it can be converted to methyl mercury, which, in turn, bioaccumulates in the food chain, concentrating in fish and shellfish that end up on people's dinner table (see Minamata insert in Chapter 1).

High doses of methyl mercury may cause generalized neurotoxicity. Low doses have a localized effect, focusing primarily in specific regions of the brain. In adults, damage occurs to those parts of the brain involved in muscle coordination, resulting in a condition referred to as ataxia, which is loss of muscle coordination and control. Exposure of developing fetuses or young children to methyl mercury can cause a generalized neuronopathy resulting in mental retardation and paralysis.

Methyl mercury appears to cause the observed effects by inhibiting protein synthesis associated with the rough endoplasmic reticulum. It also damages the endothelial cells of the blood-brain barrier by breaking down the components of the capillary endothelial cell membrane. The blood-brain barrier is important in maintaining a "controlled" environment around the brain cells by selectively allowing substances to pass from the blood to the brain. The breakdown of the barrier can result in irregular nerve cell metabolism, which may accentuate the direct toxic effects of methyl mercury on the brain cells.

Much interest has been focused on aluminum and its association with **Alzheimer's disease**. One of the characteristics of Alzheimer's disease is an elevated concentration of aluminum in the brain. How aluminum may be involved in the development and onset of the disease is not clear. It has been speculated that it may interfere with ion uptake, ATP metabolism, or protein synthesis. Aluminum may affect cell membrane integrity, thus affecting ion distribution and normal nerve cell functioning.

Manganese is another toxicant – found in industrial waste – that is discharged into rivers or released in air emissions. Manganese is an essential element for performing a number of enzymatic reactions, particularly those associated with cholesterol and fatty acid synthesis. High concentrations of manganese can result in damage to the central nervous system, particularly the brain. Symptoms are **neuropsychiatric** in nature and are characterized by irritability, difficulty in walking, and speech abnormalities. Continuous, prolonged exposure can cause symptoms similar to **Parkinson's syndrome**, which is characterized by tremors and spastic contractions of skeletal muscle.

> The term "neuropsychiatric" refers to behavioral or functional abnormalities associated with abnormal nerve cell functioning in the brain.

Acute methanol exposure can result in nerve damage leading to blindness or death. Initial exposure results in mild inebriation associated with depression. An asymptomatic period follows for 12-14 hours. **Acidosis** – a decrease in blood pH – may develop during this time. Visual problems will begin to occur including eye pain, blurred vision, and constriction of the visual fields. These symptoms may progress and result in temporary or permanent blindness. If not treated, acidosis will cause nerve cell damage, which can result in cessation of breathing and death.

Exposure to both carbon monoxide and hydrogen cyanide has similar effects. In both cases nerve cells are deprived of the oxygen necessary to maintain normal cell function and integrity. Nerve cell metabolic processes are disrupted, subsequently leading to cell death. Acute exposure to carbon monoxide or hydrogen cyanide leads to death.

Axonopathy

Axonopathies are characterized by damage to the axon. Both the axon and the myelin surrounding it degenerate, while the cell body remains intact. Axons that innervate the extremities, such as hands and feet, are usually affected first, resulting in loss of feeling and motion. If exposure continues, the axons associated with the central nervous system, particularly the spinal cord, will be affected and motor functions will be seriously impaired as a result of progressive weakening of the skeletal muscle. Acrylamide, carbon disulfide, and hexacarbons such as n-hexane, all cause axonopathies. Exposure to these substances is characterized by loss of peripheral sensations and progressive impairment of skeletal muscle functioning.

Acute exposure to organophosphate pesticides affects normal synaptic activities and can lead to delayed effects characteristic of axonopathy. Degeneration of the axon does not begin until seven to ten days after acute exposure. The degenerative process is not related to inhibition of **acetylcholinesterase**, an enzyme that breaks down acetylcholine into acetic acid and choline. Symptoms are similar to those described for other substances that cause axonopathy. If axonopathy progresses to the spinal cord, then symptoms similar to **multiple sclerosis** appear.

> Axonopathy prevents the transmission of the action potential as a result of damage to the axon.

Myelinopathy

The myelin sheath – produced by either Schwann cells or oligodendrocytes – is important in the transmission of the action potential along the axon of myelinated nerves. Loss of the myelin sheath results in impairment or loss of transmission of the nerve impulse.

Hexachlorophene is an antimicrobial substance that was used in fungicides and pesticides and is still found today in germicidal soaps and detergents. Hexachlorophene easily penetrates the skin and once in the body it is absorbed into the myelin sheath surrounding the nerves. This toxic substance appears to affect the ability of the myelin to regulate the intracellular ion-water balance resulting in swelling of the myelin sheath. The swelling of the sheath affects the ion balance across the nerve cell membrane. This impairs the transmission of the impulse resulting in nerve dysfunction. Symptoms include blurred vision, muscular weakness, confusion, and seizures. Coma and death can occur in some instances.

The effects of lead exposure, particularly in young children, have been studied and are well documented. For young children the most widely reported route of exposure has been through consumption of paint chips. However, there are other routes of exposure, such as lead pipes used for drinking water. Other possible sources are inhalation of lead from automobile exhaust in areas where leaded gasoline is still in use, or air contaminated with emissions from smelters.

Acute and chronic exposure of young children to lead result in different symptoms. Acute exposure can result in excess accumulation of fluids in the brain. If this condition persists for a prolonged period of time it can result in mental retardation or, if severe enough, death. Chronic, low-level exposure to lead seems to be correlated with learning disabilities.

Lead appears to produce its effects by more than one mechanism. Demyelination seems to occur in some situations and is associated with a slowing of the rate of transmission of the action potential. In other circumstances lead appears to cause axonopathy. In adults, exposure to lead primarily affects the motor neurons.

Neurotransmission Toxicity

Chemical messengers, such as norepinephrine or acetylcholine, are released from the presynaptic terminal of the neuron in response to the stimulation of the nerve. Neurons release only one type of neurotransmitter substance. Acetylcholine is primarily associated with the neurons that stimulate skeletal muscle. The chemical messenger binds to receptors on the postsynaptic membrane of the effector tissue eliciting a response. Enzymes in the synaptic cleft immediately begin to break down the chemical messenger acetylcholine, preventing continuous stimulation of the affected tissue. Norepinephrine, which stimulates a variety of tissue types such as smooth muscle and glands, is reabsorbed into the presynaptic neuron terminal.

Various types of toxic substances interrupt the normal sequence of events associated with neurotransmission between the presynaptic terminal and the postsynaptic membrane. Based upon their mechanism of action, they may be categorized as: 1) blocking agents, 2) depolarizing agents, 3) stimulants, 4) depressants, or 5) anticholinesterase agents.

1. Blocking Agents – A toxin produced by the bacterium *Clostridium botulinum* is responsible for the illness referred to as botulism. It can be caused by eating contaminated canned foods, smoked fish, or pickled meat. The toxin binds to the presynaptic terminal and prevents the release of acetylcholine. The effects are primarily associated with the parasympathetic nervous system and the neuromuscular junction associated with skeletal muscle. The central nervous system is not affected because the molecule is too large to pass through the blood-brain barrier. Symptoms of exposure include progressively increasing paralysis, weakness, difficulty swallowing, and some visual impairment. When death occurs, it is usually caused by respiratory paralysis.

> Blocking agents prevent the release of neurotransmitters.

2. Depolarizing Agents – As discussed previously, there is a differential charge (resting membrane potential) that exists across the cell membrane during the resting state of nerves; there is a positive charge on the outside of the membrane and it is said to be polarized. The polarization is maintained by an energy-dependent (ATP) **sodium-potassium pump**, which continually transports sodium ions out of the cell and potassium into the cell. Depolarizing agents eliminate the resting membrane potential by altering the permeability of the cell membrane towards sodium and potassium ions. DDT partially depolarizes the presynaptic neuron membrane. As a result, the neuron is in an almost continuous state of excitation, and the neurotransmitter is released in response to weaker stimuli, which under nonexposure conditions would not happen. Depending upon the level of exposure, individuals may develop persistent tremors, irritability, hypersensitivity to external stimuli, and dizziness. Severe convulsions may develop from acute exposures.

These symptoms are the result of the direct effect of DDT on the cellular metabolism of the affected neurons. DDT increases the cell membrane permeability to sodium and decreases it to potassium. It also inhibits the sodium-potassium ATPase enzyme, which is involved in the production of energy for the sodium-potassium pump. Collectively, all of these effects lower the resting membrane potential, making the nerve more sensitive to stimuli.

Pyrethroid insecticides are a relatively new class of insecticides. They are used in agricultural products, household insecticide sprays, flea preparations for pets, and in greenhouses. They are less toxic and less persistent than the organochlorine insecticides such as DDT. Exposure to skin results in a burning, itching, or tingling sensation. Sweating or washing with warm water enhances the symptoms, which usually disappear after 24 hours. Ingestion causes stomach pain, nausea, vomiting, headaches, dizziness, fatigue, tightness of chest, and blurred vision. In cases of acute exposure convulsions may occur that last from 30-120 seconds.

The mechanism of action is similar to that of DDT. Sodium and potassium permeability is affected, resulting in an increased sensitivity of the nerve to stimuli. There is a continuous release of neurotransmitter, which causes continuous stimulation of the affected tissue.

> Depolarizing agents cause an increased sensitivity of nerves to external stimuli, which results in a constant stimulation of affected tissues.

3. Stimulants – Toxic substances that are classified as stimulants increase the excitability of neurons. The increased excitability may be caused by several different cellular mechanisms. Nicotine has received widespread attention as the addictive component of cigarettes. It exerts its effects by binding to and stimulating a subset of receptors that normally bind acetylchloline, called **nicotinic receptors**. These receptors are found in the central nervous system and the neuromuscular junction. It is in the central nervous system that nicotine exerts its addictive properties. In addition, nicotine stimulates

> **Box 6-1 ■ Cocaine Poisoning**
>
> Cocaine is obtained from the shrub *Erythroxylum coca*, which is native to Bolivia and Peru. Street names for cocaine include: snow, coke, lady, flake, gold dust, green gold, blow, toot, and crack. Crack is a smokable form and produces a 3-5 minute "high" within 10 seconds of being inhaled.
>
> Acute and chronic exposures have long-term effects. Acute exposure results in stimulation of the nervous system causing excitement, incoherent talking, and hallucinations. These symptoms may be followed by increased heart rate, dilated pupils, chills or fever, abdominal pain and vomiting, and muscle spasms. In severe cases, irregular breathing may develop, followed by convulsions, coma, and death from respiratory or cardiac arrest.
>
> Chronic exposure results in some of the same symptoms such as heart irregularities; abdominal pain and vomiting; irregular breathing and, in some cases, death. Perhaps the most tragic chronic effect of cocaine exposure is the birth of cocaine babies. Cocaine crosses the placenta and enters the blood of the fetus. As a result physical and psychological abnormalities may be present at birth.

neurons in the sympathetic nervous system, which results in an increased heart rate and elevated blood pressure. Acute exposures to nicotine have occurred as a result of children ingesting tobacco products, tobacco workers exposed to wet tobacco leaves, or exposure to nicotine-containing pesticides. As a result, heart and respiratory rates increase and nausea develops. These symptoms are followed by a slowing of the heart rate and a drop in blood pressure. Death occurs due to respiratory paralysis.

Xanthines, such as caffeine (1,2,7-trimethylxanthine), also excite neurons. Caffeine is present in coffee, chocolate, tea, and various types of over-the-counter medicines. The amount of caffeine found in these beverages varies. In 6 oz (180 ml) of coffee the caffeine content ranges from 110-150 mg; decaffeinated coffee contains from 2-5 mg of caffeine. Teas (150 ml or 5 oz) may contain 20-46 mg caffeine, and 12 oz (360 ml) of a cola drink has from 30-60 mg of caffeine. Caffeine is a stimulant of the central nervous system. It appears to produce this effect by affecting the active transport system that maintains the sodium/potassium concentration gradient across the cell membrane. Caffeine intoxication occurs as a result of acute exposure, which is ingestion of 250 mg. Symptoms that may occur include nervousness, insomnia, flushed face, gastrointestinal disturbance, muscle twitching, rambling thoughts and speech, increased heart rate, and periods of inexhaustibility. Chronic long-term exposure may produce some of the same symptoms as well as elevated blood pressure.

Cocaine easily crosses the blood-brain barrier where it is able to produce its euphoric and addictive effects. It inhibits the reabsorption of a group of neurotransmitters, called **catecholamines**, into the presynaptic neuron. **Epinephrine**, **norepinephrine**, and **dopamine** are catecholamines. Normally, these neurotransmitters are reabsorbed and broken down in the presynaptic neuron. Inhibition of the reabsorption allows the neurotransmitters to remain in the synaptic cleft region. This results in continuous stimulation of the postsynaptic membrane of the affected tissue. In cases of cocaine overdose there is constant stimulation of the heart resulting in an increased heart rate, which can ultimately lead to heart failure and death.

> Stimulants may act by increasing the sensitivity of neurons to stimuli, or by inhibiting neurotransmitter reabsorption.

4. *Depressants* – Some volatile organic toxicants decrease the excitability of neurons. Examples of these types of toxicants include methylene chloride, carbon tetrachloride, and chloroform. These substances are highly lipophilic and are easily absorbed by the myelin sheath and cell membrane of neurons. Accumulation in the neuron interferes with the normal excitability and impairs conduction of nerve impulses. The decreased sensitivity may be related to the ability of these substances to prevent normal ion exchange (Na^+, K^+, and Ca^{2+}) across the cell membrane, which affects the resting membrane and action potentials. The overall reaction is depression of the central nervous system activities result-

ing in a general anesthetic effect. Initially, exposure results in a feeling of being light-headed, dizziness, and loss of coordinated movement. Continued exposure may result in unconsciousness; further fatal symptoms are respiratory or cardiac arrest.

Aromatic organic solvents that consist of one or more benzene rings have a depressant effect on the central nervous system. This class of compounds includes chemicals such as benzene, toluene, and xylene. In addition to its effects on the central nervous system, benzene has other toxic properties related to blood cell production, which will be discussed later in this chapter. Toluene impairs the speed of perception and reaction time. Inhalation of toluene or xylene can result in fatigue, mental confusion, headaches, nausea, and dizziness.

Alcohols are another group of organic solvents that impair normal nerve functioning and are powerful central nervous system depressants. Perhaps the best known alcohols are methanol and ethanol. Methanol can cause temporary or permanent blindness with as little as a 15 ml dose (70-100 ml may be fatal). Other symptoms include headache, vomiting, abdominal pain, labored breathing, restlessness, and delirium. Ethanol primarily affects the central nervous system resulting in **bradycardia** (slow heart rate) and dilation of the blood vessels, which in large doses may lead to cardiac arrest. These effects are the result of ethanol interacting with the brain center, which regulates these structures. Symptoms based on progressive consumption include excitation, intoxication, stupor, hypoglycemia, and coma. In addition to these symptoms, there is a decrease in visual acuity and muscular coordination, before loss of consciousness.

> Depressants cause their effects primarily by interfering with the maintenance of the resting membrane potential and the generation of action potential.

5. *Anticholinesterase Agents* – Toxicants that are classified as anticholinesterase agents affect only those nerves that release and those that are stimulated by the neurotransmitter acetylcholine. Anticholinesterase toxicants inhibit the enzyme (acetylcholinesterase) that is responsible for breaking down acetylcholine. This results in an increased and prolonged stimulation of the postsynaptic membrane. Organophosphate pesticides (parathion, malathion, diazinon, etc.) bind directly to the enzyme acetylcholinesterase, which prevents the breakdown of acetylcholine. The binding of the pesticide to the enzyme is irreversible. As a result the effects of exposure to organophosphate pesticides are prolonged until new enzymes can be synthesized.

Acute organophosphate intoxication can result in muscular twitching, extreme weakness, and paralysis. The paralysis of the **diaphragm** and chest muscles can cause respiratory arrest. Other symptoms include tremors, disequilibrium, constriction of the respiratory airways, and an asthmatic response.

> The diaphragm is a muscle that separates the chest and abdominal cavity.

As discussed previously, chronic effects may cause nerve cell damage. Individuals develop paralysis mainly in the lower legs, but also loss of strength, and difficulty with movement. Central nervous system effects include damage to the spinal cord, which can lead to symptoms similar to **amyotrophic lateral sclerosis (ALS)**.

> Amyotrophic lateral sclerosis is a disease characterized by muscular weakness and atrophy, caused by degeneration of motor neurons in the spinal cord and various regions of the brain.

In addition, psychiatric changes have been observed as a result of exposure to organophosphate pesticides. Symptoms include depression, impaired memory, decreased concentration, nightmares, sleepwalking, and emotional instability.

Carbamate insecticides (carbaryl or Sevin™, aldicarb, etc.) have similar effects as organophosphate pesticides. They bind to the acetylcholinesterase enzyme and inhibit the breakdown of acetylcholine. However, the pesticide-enzyme complex is not stable and easily disassociates (breaks apart). As a result, inhibition is of shorter duration and the subsequent effects are short-lived. Acute symptoms include lightheadedness, nausea, vomiting, blurred vision, muscular weakness and, in severe cases, convulsions.

> Anticholinesterase toxicants cause continuous stimulation of the affected tissue.

Checking Your Understanding

1. Identify the divisions of the nervous system and their functions.
2. What is the functional unit of the nervous system?
3. What is the functional difference between the nerve cell dendrite and the axon?
4. List the support cells of the nervous system and their functions.
5. What is the difference between a myelinated and an unmyelinated nerve?
6. How is an action potential generated and what is its role in nerve cell functioning?
7. Identify components of the synapse and explain how they interact with each other.
8. Distinguish between neuronopathy, axonopathy, or myelinopathy. Give an example of a toxic substance that causes each. How does each affect the action potential?
9. List the general categories of toxic substances that may affect neurotransmission and explain their mechanism of action. Give an example of each.

6-3 Hematoxicity

Blood is composed of cellular material in a noncellular liquid called plasma. The cellular material in an adult is derived primarily from nondifferentiated **stem cells** located in the bone marrow. The stem cells give rise to: 1) the red blood cells (RBCs) or **erythrocytes**, 2) the white blood cells (WBCs) or **leukocytes,** and 3) the **platelets** or **thrombocytes** (Figure 6-8).

The white blood cells can be divided into two types: **granulocytes** and **agranulocytes**. There are three kinds of granulocytes: **neutrophils, eosinophils,** and **basophils** (Figure 6-8). Neutrophils represent the largest portion of leukocytes (60-70 percent) and can be distinguished by a nucleus that has two to four lobes connected by a thin filament and light-pink staining granules found in the cyto-

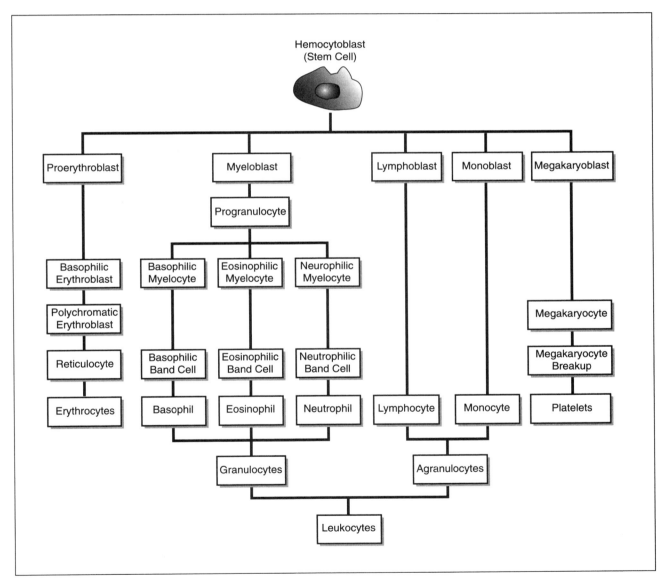

Figure 6-8: Blood cell hematopoiesis and cell types. Stem cells differentiate to form erythrocytes, leukocytes, or cells that produce platelets.

plasm. Neutrophils are primarily involved in the phagocytosis of bacteria and foreign matter circulating throughout the body. Eosinophils comprise 1-4 percent of the leukocytes; they have a bi-lobed nucleus, and the granules stain an orange-red or bright red. Eosinophils are involved in the inflammatory response. They release chemicals that decrease the degree of **inflammation** associated with injury or infection. Basophils comprise 0.5-1 percent of the leukocytes. The nucleus has two indistinct lobes, and the granules stain a dark purple. These cells release **histamine**, which promotes inflammation, and **heparin**, which prevents blood clot formation.

> Granulocytes represent approximately 65-75 percent of the white blood cells and perform a variety of functions.

The agranulocytes include **lymphocytes** and **monocytes**. The lymphocytes are connected with the immune response of the body and make up 20-30 percent of the leukocytes. They are involved in the production of antibodies, which are important in several processes: the killing of microorganisms entering the body, allergic response reactions, rejection of tissue grafts, and controlling of cancerous tumors. Lymphocytes are characterized by a large nucleus surrounded by a thin ring of cytoplasm. The monocytes are the largest of all leukocytes and can be distinguished by a kidney- or horseshoe-shaped nucleus. These cells make up 2-8 percent of all leukocytes. Monocytes migrate out of the blood, enlarge, and become **macrophages**. Macrophages are involved in the phagocytosis of bacteria, cell fragments, or other debris found within the body tissue. They also interact with lymphocytes resulting in the production of antibodies.

> Agranulocytes comprise 25-35 percent of the white blood cells and are the primary cellular components of the immune system.

Red blood cells are by far the most common and numerous cellular component of the blood. There are 4.2-5.8 million/mm^3 of blood. One milliliter (ml) or one cubic centimeter (cm^3) is equal to 1,000 mm^3. The primary function of red blood cells is to carry oxygen from the lungs to all the cells found throughout the body. Red blood cells are also involved in the transport of the carbon dioxide derived from cellular metabolism to the lungs, where it is exhaled. Hemoglobin found in the red blood cells facilitates the absorption and transport of oxygen from the lungs to the cells. The hemoglobin molecule contains one atom of iron, which binds with oxygen. Ninety-seven percent of all oxygen transported in the blood is bound to the hemoglobin molecule, and the remaining three percent is dissolved in the plasma. Only about 20 percent of the carbon dioxide is carried by the hemoglobin molecule. A majority of the carbon dioxide is converted to a bicarbonate ion by an enzyme in the red blood cells and transported in this form; the other 8 percent is dissolved in the plasma.

> Red blood cells are the most common cellular component of the blood and are primarily responsible for transporting oxygen to tissues of the body and CO_2 from tissues to lungs.

Platelets or thrombocytes are small cell fragments. They consist of a small amount of cytoplasm surrounded by a cell membrane. The platelets are fragments derived from **megakaryocytes** in the red bone marrow. There are 150,000 to 400,000 platelets per mm^3 of blood. The fragments contain granules that release chemicals important in the clotting of blood.

> Platelets are an important cellular component of the bloodclotting mechanism.

The noncellular portion of the blood is the plasma. The plasma consists of approximately 92 percent water and eight percent other substances such as proteins, ions, nutrients, gases, and waste products. The proteins include albumin, globulin, and **fibrinogen**, the latter being important in the clotting of blood.

Blood Cell Hematopoiesis

Hematopoiesis is the process of blood cell formation. In adults, blood cell production occurs primarily in the bone marrow. Marrow is a soft tissue in the medullary (middle) cavity of the bone. Bloodforming cells are found in the red marrow of the bone. The blood cells are derived from a single, multipotential cell referred to as stem cell or **hemocytoblast** (Figure 6-8). The stem cells differentiate into specific cell types. Blood cell production is affected by growth factors, which will determine the type and number of cells that are formed.

> Hematopoiesis occurs in red bone marrow and is the result of stem cell differentiation.

Cells undergoing cellular division are more susceptible to toxicants than nondividing cells. Therefore, stem cells are more susceptible to toxicants than mature blood cells. However, the various cell functions of the mature cells may also be affected by toxicants.

Hematoxic Effects

Toxic substances may have a direct effect on mature blood cells; further, they may affect the bone marrow, where the blood cells are produced. Exposure of the bone marrow to toxic substances may result in a variety of effects and is dependent on the type of substance, length of exposure, and concentration. There may be a transient decrease in the production of one or more cell types with subsequent recovery after cessation of exposure; or exposure may result in a persistent, long-term decrease in the number of cells. Changes in the number and type of circulating blood cells can be classified as **anemia**, **granulocytopenia**, **thrombocytopenia**, **lymphocytopenia**, or **pancytopenia**.

Anemia

Anemia is the condition resulting from a decrease in the number of red blood cells, a decrease in the size of the red blood cell referred to as **microcytic anemia**, a decrease in red blood cell hemoglobin content, or a combination of any of these factors. Toxic and nontoxic factors such as excessive blood loss and nutritional deficiencies may cause anemia.

Toxicants that cause anemia do so by interacting with mature red blood cells or by affecting red blood cell synthesis. Some insecticides cause destruction of circulating red blood cells by binding to the surface of the cell. The red blood cell/insecticide complex is recognized as a foreign substance by the body's immune system. As a result a complex series of cellular-chemical interactions occurs by which the red blood cell-insecticide complex is destroyed. Other toxic substances such as benzene, lead, methylene chloride, nitrobenzene and naphthalene are all capable of red blood cell destruction. These substances interfere with the metabolic activities associated with cell membrane integrity and ion balance. The membrane of the red blood cell is not able to maintain normal intracellular homeostasis and cellular dysfunction and death result.

Oxygen is transported throughout the body bound to the hemoglobin molecules located in red blood cells. Therefore toxic substances that decrease the synthesis of hemoglobin may cause anemia. Lead decreases hemoglobin synthesis by inhibiting enzymes involved in the transport of amino acids used to synthesize the protein portion (globin) of the molecule. Lead also inhibits the incorporation of iron into the heme portion of the molecule by inhibiting enzymes associated with this cell activity. Chronic alcohol (ethanol) use has a similar effect on hemoglobin synthesis.

Whether the decrease occurs in the number of red blood cells or in the amount of red blood cell hemoglobin, the end result is a decrease in the oxygen-carrying capacity of the blood. Oxygen-sensitive tissues such as the brain and the heart are the first to demonstrate symptoms associated with a decrease in oxygen availability. Neural tissue has a high demand for oxygen to maintain normal cell function and initiate action potentials. The same is true for the cells that comprise the conduction pathway of the heart. Control of the heart rate is dependent on the normal functioning of the conduction pathway starting in the **sinoatrial (SA) node** and ending with the **Purkinje fibers**.

> The sinoatrial node is a specialized heart muscle fiber located in the right atrium that initiates contraction of the heart.

In addition, the heart is regulated by the **cardioregulatory center** in the brain.

> The cardioregulatory center is a specialized area of the brain that helps maintain a normal heart rate.

If a decrease in oxygen occurs, then normal nerve and heart cell function may be disrupted. The cellular malfunction can cause **arrhythmias**.

> Arrhythmia is any irregular heartbeat.

Other symptoms include shortness of breath, pale skin, and fatigue. In the case of anemia, the effect on the nerves and heart are indirect. These responses are the result of a decrease in oxygen supplied to the tissue because of the direct effect of toxic substances on red blood cells.

> A decrease in the oxygen-carrying capacity of the blood directly affects cellular metabolism of oxygen-sensitive tissues, which may result in adverse effects.

Aplastic Anemia

Toxic chemical substances such as carbon tetrachloride, chlordane (pesticide), benzene, and physical agents like ionizing radiation, can affect the bone marrow directly, suppressing the production of red blood cells, which leads to a condition known as aplastic anemia. Benzene produces its effects primarily on the stem cells in the bone marrow. The stem cells are unable to differentiate and produce enough red blood cells to sustain normal body functions. Ionizing radiation affects the stem cells directly by damaging the DNA. If the damage is severe enough, cellular replication is inhibited.

> Aplastic anemia is caused by a decrease in circulating red blood cell production.

Nephrotoxicity-Induced Anemia

The kidney is the source of **erythropoietin**, which it releases in response to low oxygen levels in the blood and which stimulates bone marrow to produce more red blood cells. Toxic substances – such as mercury and cadmium – that affect the normal functioning of the kidneys may induce anemia by decreasing the secretion of erythropoietin.

Chemically Induced Hypoxia

Some chemicals have the ability to decrease the amount of oxygen reaching body tissues by affecting oxygen transport of red blood cells. **Hypoxia** is any condition in which inadequate amounts of oxygen are delivered to the tissues. Neither a decrease in the number of red blood cells nor changes in red blood cell hemoglobin have to be present to cause hypoxia.

Carbon Monoxide-Induced Hypoxia

During normal respiration oxygen enters the bloodstream from the lungs and binds to the hemoglobin molecule in the red blood cell. Carbon monoxide – when present in the inhaled air – also enters the bloodstream and binds with hemoglobin forming **carboxyhemoglobin**. However, the affinity of hemoglobin for carbon monoxide is approximately 225 times greater than that for oxygen. Therefore, carbon monoxide displaces oxygen in the red blood cells.

> Carbon monoxide actively competes with oxygen for the hemoglobin-binding site resulting in decreased oxygen availability for tissues of the body.

The decreased oxygen-carrying capacity of the blood has several effects. Smooth muscle in the walls of the blood vessels requires large amounts of oxygen to maintain sustained contraction. As oxygen in the blood decreases, the smooth muscle cannot sustain contraction. It relaxes, and blood pressure begins to drop resulting in dizziness, fainting, and an increased heart rate. Other symptoms may include headaches, muscular weakness, and nausea. Death occurs as a result of prolonged exposure.

Methemoglobinemia

The hemoglobin molecule contains an iron atom in the ferrous state (Fe^{2+}). Exposure to certain toxic chemicals – such as sodium nitrite, hydroxylamine hydrochloride, or nitrobenzene – converts the ferrous atom to the ferric state (Fe^{3+}) resulting in the conversion of hemoglobin to methemoglobin (MetHb). MetHb does not bind to oxygen and therefore reduces the supply of oxygen to body tissues. The observed symptoms are similar to those of individuals exposed to carbon monoxide. Vasodilation (opening) of peripheral blood vessels occurs, resulting in decreased blood pressure, increased heart rate, and dizziness. If exposure is prolonged death may occur.

Cytotoxic (Histotoxic) Hypoxia

Certain toxic chemicals produce symptoms that mimic chemically-induced hypoxia without changing the oxygen-carrying capacity of the blood. Exposure to hydrogen cyanide or hydrogen sulfide interferes with the ability of cells to use oxygen. Thus, the body reacts by increasing heart rate and respiratory rate as a result of a perceived decrease in blood-borne oxygen. In reality, these chemicals have a stimulatory effect upon specialized **chemoreceptor** cells found in the aorta, and it is this effect that produces the increased heart and respiratory rate.

> A chemoreceptor is a specialized nerve cell ending that is stimulated by certain chemicals.

These symptoms are very similar to those associated with chemically induced hypoxia. Death may occur upon exposure to high concentrations of hydrogen cyanide or hydrogen sulfide due to respiratory failure.

Granulocytopenia, Lymphocytopenia, and Thrombocytopenia

Granulocytopenia, lymphocytopenia, and thrombocytopenia are the result of a decrease in granulocytes, lymphocytes, and thrombocytes (blood platelets), respectively. These conditions occur primarily as a result of suppression of stem cell proliferation and differentiation. Benzene, carbon tetrachloride, and trinitrotoluene are capable of suppressing each of these cell types. Pesticides such as DDT and dinitrophenol seem to be more selective and affect only the production of white blood cells.

Thrombocytes are an important component of the clotting mechanism in the bloodstream. Thrombocytopenia (a decreased platelet count) results in slower clotting of the blood in response to injury. Individuals with lymphocytopenia and granulocytopenia have a decreased number of lymphocytes, macrophages, and granulocytes in the body. These cells are important in the defense of the body against infection; therefore, individuals with these symptoms are more susceptible to infection and disease.

Leukemia

Leukemia is characterized by a higher than normal number of leukocytes in the bloodstream. Benzene is probably the best known toxicant responsible for causing leukemia. Exposure to benzene may initially cause a total suppression of all stem cell differentiation (pancytopenia), which will be manifested as anemia, leukocytopenia, and thrombocytopenia. This is followed by an explosive growth and development of white blood cells in quantities above and beyond the normal number. If the individual does not respond to treatment – usually chemotherapy – the disease results in death.

Checking Your Understanding

1. Identify the major components of blood and their functions.

2. What is hematopoiesis and where does it occur?

3. Differentiate between anemia, granulocytopenia, thrombocytopenia, and lymphocytopenia.

4. Define anemia. Give two examples of toxic substances that cause anemia and explain their mechanism of action.

5. How does hypoxia differ from anemia? How are they similar?

6. Distinguish between carboxyhemoglobin and methemoglobin and explain their effects on the heart.

6-4 Immunotoxicity

Structure and Function of the Immune System

The immune system functions to protect the body against illness and disease caused by microorganisms and other environmental factors. This function is performed by various organs, tissues, and cellular components of the body. The major components of the immune system are the cells of the lymphatic system, the cellular components of the blood, and the blood-borne proteins called antibodies. These elements work together in an integrated manner to remove or destroy agents that may cause disease or illness.

Lymphatic System

One of the major functions of the lymphatic system (Figure 6-9) is to remove microorganisms and other harmful substances from the **lymph**. Lymph is a fluid derived from the plasma of the blood (Figure 6-10). Liquid from the blood is filtered through the capillary walls and into the space located between cells (interstitial space). Most of this interstitial liquid is reabsorbed into the blood capillaries except for a small amount that remains there and moves into lymphatic vessels. Entry into the lymph vessels occurs in the **lymphatic capillaries**. Anatomically they are similar to the blood vessel capillaries, as the walls are a single cell thick. The lymph is eventually returned to the bloodstream near the heart by either of the two large lymph vessels: the right lymphatic duct or the thoracic duct.

Before the lymph is returned to the blood, it passes through structures referred to as lymph nodes (Figure 6-11). Lymph nodes vary in size and are found at various locations along the lymph vessels throughout the body. Within the lymph nodes are dense aggregations of lymphocytes and other cell types that form **lymph nodules**. Located between the lymph nodules are sinuses that contain macrophages associated with a network of connective tissue. When lymph enters the nodes through afferent lymph vessels, it passes through the lymphocytes and sinuses and exits through efferent vessels. The lymphocytes and macrophages serve as an effective filter, capturing microorganisms and cellular debris present in the lymph. The macrophages are able to

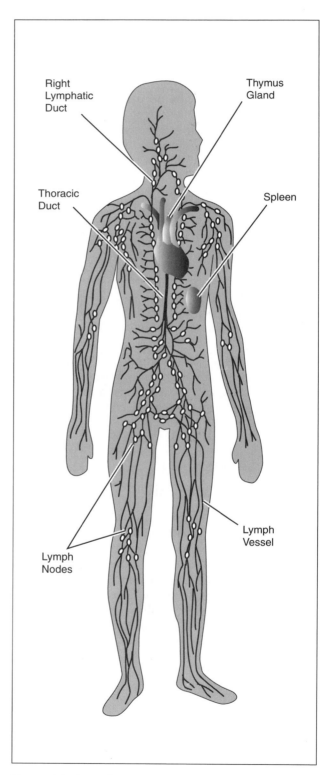

Figure 6-9: Lymphatic system.

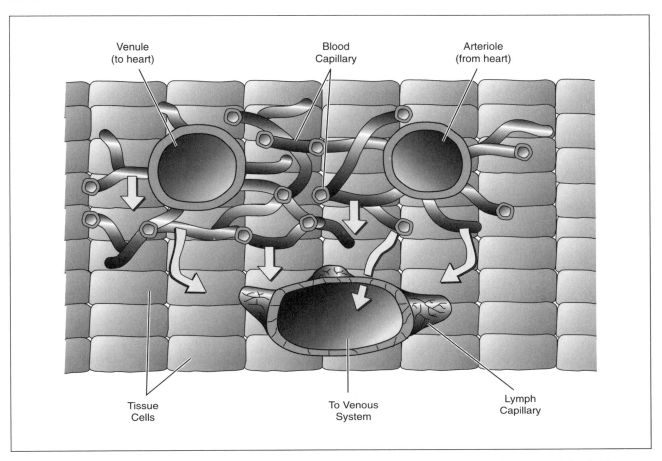

Figure 6-10: Lymph formation. Fluid diffuses out of the capillaries into the spaces between cells. Some of the fluid diffuses into the lymph capillaries to form lymph.

destroy microorganisms and also assist in lymphocyte proliferation, which occurs in response to the presence of microorganisms or other foreign substances. The lymphocytes then enter the lymph and eventually are distributed to other lymph nodes throughout the body via the bloodstream.

> Cellular components, such as lymphocytes and macrophages, in the lymph nodes are important in protecting the body against disease.

Two other organs that may be considered part of the lymphatic system are the spleen and the **thymus**. Lymph tissue in the spleen can be divided into white pulp and red pulp (Figure 6-12). White pulp surrounds the arteries and red pulp is associated with the veins. The white pulp is composed primarily of lymphocytes, which respond to microorganisms in the blood in the same manner as the lymphocytes in the lymph nodes. Macrophages are associated with the red pulp and remove and destroy microorganisms as the blood passes through the spleen.

The thymus is a triangular organ located in the chest behind the upper end of the sternum. Lymphocytes are produced and mature in the thymus. Once they mature, the lymphocytes enter the bloodstream, and eventually other lymphatic tissue such as the spleen or lymph nodes, where they respond to microorganisms to protect the body.

Cellular Components of Immune System

The cellular components of the immune system are derived primarily from the bone marrow in adults. The cell types include lymphocytes, macrophages,

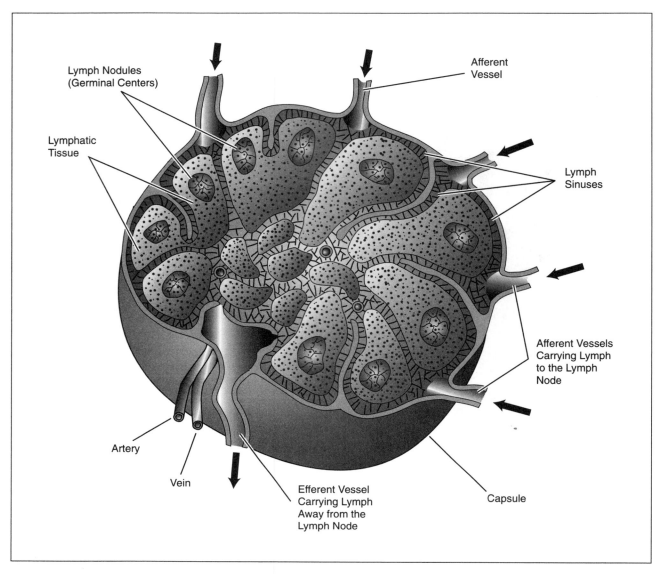

Figure 6-11: A lymph node. Lymph is filtered as it passes through the lymph node. Macrophages phagocytize foreign materials.

and neutrophils. Lymphocytes can be further differentiated into T cell lymphocytes and B cell lymphocytes. Natural killer (NK) cells – which can identify and destroy virus-infected cells or cancer-like cells – are thought to be a specialized type of T cell. The neutrophils and macrophages are the primary lines of defense against infectious agents and are able to phagocytize and destroy bacteria.

Phagocytosis by macrophages and neutrophils is a rather nonspecific form of response to foreign substances that enter the body. A more specific type of response is carried out by the T and B lymphocytes in the form of antibody production.

Antibody-Mediated Immunity

The production of antibodies is the result of a complex interaction between T cells, B cells, and macrophages. Following is a generalized description of the events involved in the production of antibodies. When a foreign substance or an antigen enters the body it binds to antigen-binding receptors found on the surface of B lymphocytes and macrophages. The antigen-receptor complex is processed by these cells, with the processed antigen being returned to the surface of the B cells and macrophages. The altered antigen is presented to T cells by the mac-

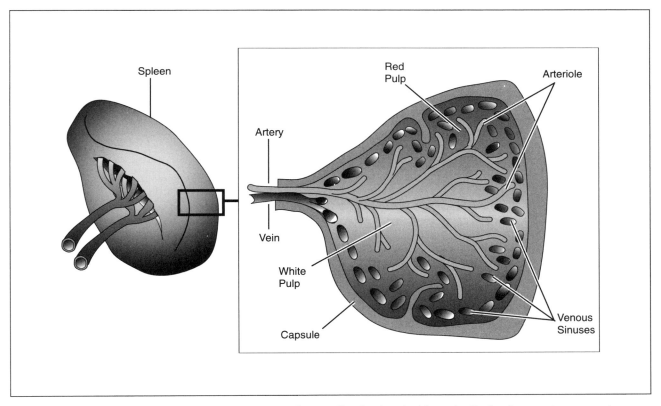

Figure 6-12: Section of the spleen. Spleen is comprised primarily of red and white pulp. Red pulp contains macrophages, white pulp contains lymphocytes.

rophage, which stimulates division of the T cells. The T cells interact with the B cells that have processed the antigen. The B cells divide forming plasma cells, which produce the antibody specific for the antigen.

Symptoms of Immunotoxicity

Immunotoxins affect the cellular component of the immune system by inhibiting the production of leukocytes in the bone marrow, or by inhibiting their proliferation in response to an antigen. Some toxicants act specifically on the cellular mechanisms associated with producing antibodies. In either case, the immune response may be suppressed; individuals then become more susceptible to illnesses and disease.

Halogenated Aromatic Hydrocarbons

Halogenated aromatic hydrocarbons (HAHs) include polychlorinated biphenyls (PCBs), polybrominated biphenyls (PBBs), polychlorinated dibenzo-p-dioxins [2,3,7,8-tetrachlorodibenzo-p-dioxin (TCDD)], and dibenzofurans. These substances are produced from a variety of sources. In the past PCBs were used in the fluids of capacitors and transformers as well as in adhesives and plasticizers. Polybrominated biphenyls are excellent flame retardants. The dibenzofurans are formed as a byproduct during the production of other chemicals such as chlorophenoxy acids, pentachlorophenol, and PCB mixtures. Probably the most widely studied toxic chemicals of the group are the polychlorinated dibenzo-p-dioxins, commonly referred to as dioxins. These substances are formed as a byproduct of chlorinated phenol production or as the result of the combustion of chlorinated compounds. TCDD is a contaminant of the herbicide 2,4,5-trichlorophenoxyacetic acid (2,4,5,-T) and the defoliant Agent Orange, which is a combination of 2,4-dichlorophenoxyacetic acid (2,4-D) and 2,4,5-T.

The effects HAHs have on the immune system are difficult to characterize. Most information is derived from animal studies. In general, this group of compounds has been shown to cause atrophy of the thymus, decrease in the number of all blood cell types, decrease in antibody production, and tumor promotion. PCBs may cause atrophy of the various lymphoid organs such as the spleen, and reduce the number of circulating lymphocytes as well as antibody production. The effect of PCBs on cell-mediated immunity is complicated. In some situations natural killer cell activity is suppressed, and in other studies it has been enhanced. PBBs have been shown to have similar effects as PCBs.

Of the dioxin group, exposure to TCDD causes atrophy of lymphoid organs. In addition, TCDD affects lymphocyte development and maturation in the bone and thymus. In the bone marrow, TCDD affects stem cell division and subsequent maturation. The effects on antibody production are variable. TCDD appears to affect development and maturation of B cells following antigen stimulation, thereby decreasing antibody production. This effect is not seen in all cases. There may be a limited effect on the phagocytic capabilities of neutrophils. However, there appears to be no significant effect on macrophages or natural killer cell functions.

The dibenzofurans have similar effects on the immune system as TCDD. Individuals in Japan in 1968 and Taiwan in 1979 were accidentally exposed to dibenzofurans in contaminated commercial rice oil. These individuals had a decreased number of circulating T cells and developed recurring respiratory infections.

Polycyclic Aromatic Hydrocarbons

Polycyclic aromatic hydrocarbons (PAH) are produced primarily from the burning of fossil fuels. This class of compounds includes anthracene, benzanthracene, benzo[a]pyrene, dimethylbenz[a]anthracene, and methylcholanthrene. These compounds are best known for their carcinogenic properties, which may in turn be due to their immunosuppressive characteristics. The most widely studied PAHs are benzo[a]pyrene (BaP) and 7,12-dimethylbenz[a]anthracene (DMBA).

BaP suppresses the division of B cells and their subsequent differentiation into plasma cells. Since plasma cells produce antibodies, antibody production is suppressed. How BaP causes the observed effects is not well understood. It may interfere with normal macrophage function; or it may act directly on the B cells.

DMBA is a more potent inhibitor of the immune system. Not only does it suppress antibody production, but it also suppresses natural killer cell activity and tumor resistance. The suppression of the immune system is believed to be the reason for the carcinogenic effect.

Heavy Metals

The effects of heavy metals on the immune system are related to the concentration and type of exposure. Lead, arsenic, mercury, and cadmium all have some effect on the immune system. In general, high concentrations result in immunosuppression, but in several instances low concentrations appear to enhance the immune system.

Exposure to lead causes an increased susceptibility to infections. Children and workers exposed to lead have more colds and influenza. Decreased production of antibodies to the infectious agents appears to be the cause.

Arsenic seems to have the same effect on the immune system as lead. At low concentrations there is enhancement of the immune response, at high concentrations there is suppression. The suppression appears to be related to inhibition of circulating lymphocyte proliferation, i.e., decreased formation of plasma cells. A decrease in the number of plasma cells would result in a decrease in antibody production. Macrophage and T cell functioning, both important in the interaction with B cells, may also be affected.

Mercury is known to decrease antibody response by binding with enzymes involved in antibody synthesis in the B cells. Exposure to mercury can also result in the development of an **autoimmune response**. An autoimmune response develops when the immune system attacks normal tissue of the body. The result is destruction of the tissue. As discussed in a previous chapter, one of the major storage areas for mercury is in the kidneys. Mercury binds to structural proteins in the various cells that compose the kidneys. The protein-mercury complex is identified as a foreign substance, and the immune system begins to attack the tissue. This autoimmune disease of the kidneys is referred to as **glomerular nephritis**. This may result in kidney malfunction and possible kidney failure.

Cadmium exposure causes an increased susceptibility to infections. The effects are the result of decreased lymphocyte proliferation. Both macrophage and antibody production are also suppressed when exposed to cadmium.

Pesticides

The response of the immune system to pesticides is as varied as the response to metals. Organophosphate pesticides like malathion seem to enhance the immune response in some circumstances. High doses of the insecticide seem to decrease antibody production. The decreased antibody production is believed to be the result of increased stress due to the direct effect of malathion on the nervous system.

Parathion suppresses antibody production as well as natural killer cell and macrophage activity. In addition, it causes the thymus to atrophy and decreases lymphocyte proliferation in response to interaction with an antigen.

Organochlorine pesticides such as DDT are known to decrease antibody production. DDT and dieldrin also affect the phagocytic activity of macrophages, causing decreased production of antibodies and greater susceptibility to infection.

Organic Solvents

In addition to the already discussed benzene, several organic solvents have been studied. Toluene, which is structurally very similar to benzene, has little effect on the immune system. Toluene reduces the effect of benzene, probably by competing with it for the metabolic enzymes that are responsible for its bioactivation (see Chapter 5). Nitrobenzene, 2,4-dinitrotoluene and carbon tetrachloride all seem to have varying effects on the immune system. In general, they tend to suppress antibody production but have differing effects on cellular activity. 2,4-dinitrotoluene suppresses natural killer cell and macrophage activity. Carbon tetrachloride has little or no effect.

> Effects of toxicants on the immune system are varied and depend on the type and concentrations of the toxicant.

Checking Your Understanding

1. Identify the components of the lymphatic system and discuss their functions.

2. What are the cellular components of the nonspecific and specific defense mechanisms of the immune system?

3. Describe, in general, how antibodies are produced.

4. List three possible effects of toxic substances on the components of the immune system.

6-5 Cardiotoxicity

Structure and Function of the Heart

The heart is responsible for moving blood through the extensive vascular network found throughout the body. It pumps oxygenated blood through the arteries and arterioles to the capillaries. The capillaries serve as the point of exchange between blood-borne substances, such as oxygen and nutrients, and the various tissues that comprise the body. Metabolic products and waste from cellular metabolism such as carbon dioxide enter the capillaries and are removed from the body by various organs such as the lungs, kidneys, and liver. Toxic substances follow a distribution pattern similar to oxygen and carbon dioxide: either they enter various tissues from the capillaries, or they enter the capillaries from the tissues.

Intrinsic Control of the Heart

The heart performs its function by continuous rhythmic contractions of the cardiac muscles, which form the walls of the heart. Contraction is initiated in the right atrium of the heart by specialized cardiac muscle cells called the sinoatrial (SA) node (Figure 6-13). The SA node serves as the pacemaker for the heart. Action potentials originate in the SA node and are carried to the muscle fibers of the right and left atria resulting in contraction. In the lower portion of the right atrium is the **atrioventricular (AV) node**. The action potential passes through the AV node to specialized cardiac muscle cells called the **atrioventricular bundle** or **bundle of His**. The bundle of His enters the tissue separating the ventricles and divides to form the left and right **bundle branches**. The bundle branches subsequently divide into many small conducting Purkinje fibers. These fibers carry the action potential to the muscle cells of the ventricle. Contraction of the ventricles begins at the apex of the heart, pushing the blood through the pulmonary artery to the pulmonary circulation and through the aorta, which carries blood to the systemic circulation. Both arteries are attached to the upper walls of the right and left ventricles, respectively.

> Contraction of the heart is the result of an action potential generated in the SA node and conducted along specialized cells to the ventricular muscle.

Electrical activity in the heart is a result of the differential distribution of sodium (Na^+), potassium (K^+), and calcium (Ca^{2+}) ions across the cell membrane. In general, the concentration of Na^+ and Ca^{2+} is greater on the outside of the cell, while the concentration of K^+ is greater on the inside of it. As a result, there is a positive charge on the outside of the cell membrane and a negative charge on the inside. The Na^+ slowly leaks into the cells of the SA node resulting in an increase in positive charges inside the cells. Potassium begins to leave the cells to reestablish the negative charge inside the cells The inflow of Na^+ plus Ca^{2+} and the outflow of K^+ are responsible for generating the action potential, which is conducted along the specialized fibers of the heart to the muscle, initiating contraction. Intracellular calcium is also necessary for the contractile mechanism of the individual muscle cells.

> Initiation of the action potential occurs in the SA node as a result of changes in cell membrane permeability to Na^+, K^+, and Ca^{2+}.

Extrinsic Control of the Heart

The heart rate is affected by the nerves of the autonomic nervous system (ANS). The parasympathetic nervous system (PNS) or **vagus nerve** innervates the SA node, atrial muscle fibers, AV node, and some of the muscle fibers of the ventricle where acetylcholine is released. Stimulation of the PNS results in a decrease in the heart rate and the strength of atrial and ventricular muscle contraction. The sympathetic nervous system (SNS) also innervates the SA node and the ventricles, releasing norepinephrine. Stimulation of the sympathetic nervous system results in an increase in the rate of action potentials generated by the SA node, increasing the heart rate. The strength of contraction is also increased.

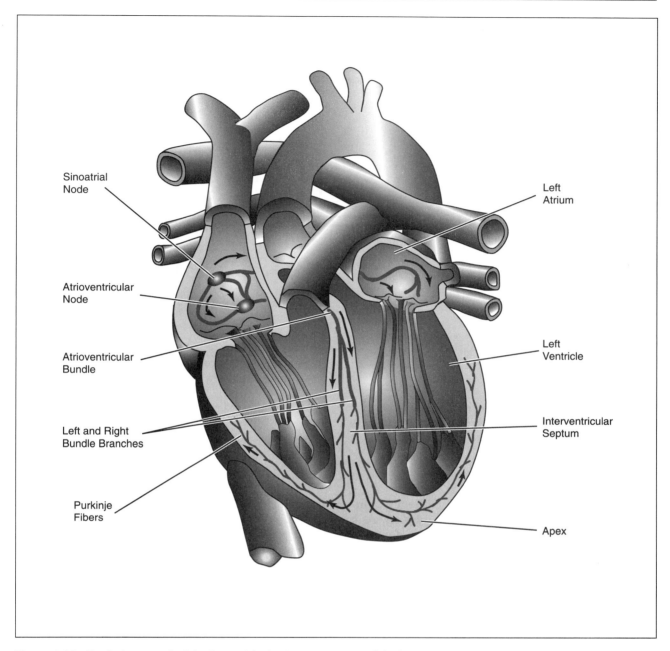

Figure 6-13: Intrinsic control of the heart. Rhythmic contractions of the heart are controlled by the SA node. An action potential is generated and conducted along specialized cells in the heart, causing contraction of the cardiac muscle.

The sympathetic and parasympathetic nerves are components of the **baroreceptor reflex** (Figure 6-14), which plays an important role in regulating the heart rate. Located in the aorta are stretch receptors (**baroreceptors**), which are modified nerve fibers. The stretch receptors are sensitive to changes in aorta blood pressure. Nerve fibers from the baroreceptors carry impulses from the baroreceptors to the cardioregulatory center located in the **medulla oblongata** of the brain. When the blood pressure in the aorta increases, action potentials are carried to the cardioregulatory center. The cardio-

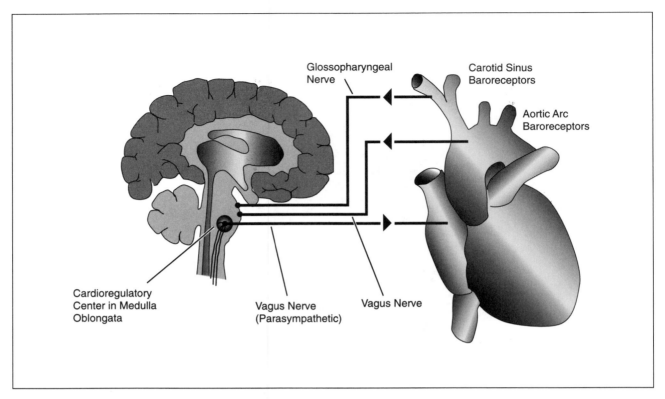

Figure 6-14: The baroreceptor reflex. Increased blood pressure in the carotids and aorta generate action potentials transmitted to the cardioregulatory center. Stimulation causes increased activity of the parasympathetic nerves innervating the heart, decreasing heart rate.

regulatory center increases parasympathetic stimulation and decreases sympathetic stimulation. As a result, the heart rate decreases, which subsequently decreases blood pressure. When the blood pressure decreases in the aorta, sympathetic stimulation is increased and parasympathetic stimulation is decreased resulting in an increased heart rate and blood pressure.

> The heart rate is controlled by components of the nervous system, which include nerves of the parasympathetic and sympathetic nervous system.

Neurotransmitters

Innervation of the cardiac tissue is mediated by chemical neurotransmitters. The parasympathetic nerves release acetylcholine, the sympathetic nerves release norepinephrine. Acetylcholine makes it more difficult for cells of the SA and AV node to generate an action potential, thus slowing the heart rate. Action potentials are generated more readily in response to norepinephrine, resulting in an increased heart rate.

Cardiotoxic Effects

Toxic substances affect the heart in several ways. The toxicant may have a direct effect on the cardiac tissues by affecting the cell membrane integrity or cellular metabolism, such as enzyme synthesis, or ATP production. Toxicants may also produce their effects indirectly by altering the functions of other systems in the body, such as the nervous system.

The specific cellular mechanism affected by a toxicant varies and depends on the type of toxicant exposure. However, the symptoms observed are a change in the strength of contraction, in the rhythm of contraction (arrhythmia), or both.

> Toxic substances affect the heart by changing the metabolism of cardiac tissues, or by affecting the neural control of the heart by altering nerve cell functioning.

Symptoms of Cardiotoxicity

Arrhythmias are any variation in the normal rhythmic beat of the heart. The variations may occur as a result of exposure to different types of toxic substances; and can be caused by changes in the rate at which action potentials are generated or the rate at which conduction of the action potential occurs. Changes in the heart rate are categorized as either **tachycardia** or bradycardia. Tachycardia is an increase in the heart rate, while slowing of the heart rate is bradycardia.

Normally all of the muscle fibers of the atria contract at about the same time followed by the coordinated contraction of the ventricle. **Atrial fibrillation** is characterized by irregular unsynchronized contractions of the atrium. This produces irregular filling of the ventricles with blood, leading to decreased transport of blood and oxygen to the body. **Ventricular fibrillation** is more serious and is characterized by repetitive, rapid, and randomized excitation of the ventricles. This produces a life threatening, uncoordinated, and inefficient contraction of the ventricles.

> Changes in the rate of contraction or uncoordinated contraction may occur in the atria or the ventricles.

Certain metals affect the strength of contraction and the conduction of the electrical signal. Barium has been shown to cause tachycardia by interfering with calcium diffusion. Manganese and nickel have a similar effect on calcium, affecting the generation of the action potential. Lead results in arrhythmias and in decreased strength of contraction. These symptoms may be attributed to lead interference with energy metabolism and ATP synthesis in the heart. Lead displaces calcium, which is necessary for the contractile mechanism of the heart. Lead, cadmium, and cobalt affect the cardiac cells directly by interfering with cellular metabolism; this may cause cell death. Both the displacing of calcium and the death of heart cells – if large in number – decrease the strength of contraction.

Toxic substances such as organic solvents may have both direct and indirect effects on the heart. Organic solvents are lipid soluble and accumulate in the lipid membrane of nerve cells and cardiac muscle cells. This interferes with the generation and conduction of the action potential resulting in altered heart rate and strength of contraction. Changes in the generation and conduction of the action potential may be related to changes in cellular ATP production, which may affect cell membrane integrity and ion permeability.

One of the primary targets of solvents is represented by nerve cells comprising the central nervous system. Solvents accumulate in the nerve cell membrane and disrupt the generation of action potentials. In general, this may result in depressing the activity of the central nervous system. This in turn affects the cardioregulatory center in the brain, affecting sympathetic and parasympathetic innervation of the heart. Arrhythmias and decreased contractility of the heart may occur. Toluene – found in many paint-related products – and ketones (i.e., acetone) can cause arrhythmias as a result of depressing parasympathetic activity. Halogenated hydrocarbons like methylene chloride can also cause arrhythmias as well as decreased contractility by depressing the parasympathetic activity of the central nervous system.

Carbon monoxide (CO) may have both a direct and an indirect effect on the heart. Acute exposure to carbon monoxide will result in a decreased oxygen supply either directly to cardiac tissue by binding to hemoglobin (see explanation above), or indirectly to the nerve tissue of the central nervous system. The elevated CO levels and decreased O_2 levels result in tachycardia. The central nervous system is extremely sensitive to O_2 concentrations in the blood. Low O_2 levels in the blood result in malfunctioning of the nervous tissue that comprise the cardioregulatory center. As a result the sympathetic and parasympathetic control of the heart rate can be disrupted producing arrhythmias.

Whether the arrhythmias or changes in contractility are due to exposure to metals, or solvents, or any other cardiotoxic substance, the results are the same. There is a change in the amount of O_2 and in the amount of nutrients being delivered to important tissues of the body such as the brain; also, removal of toxic cellular metabolites may be slowed. This may produce symptoms such as muscular weakness and fatigue, or changes in cardiac function, which – in extreme cases – may result in death.

> Several substances affect the heart directly or indirectly by a variety of mechanisms. The end result is a change in heart rate, coordination, strength of contraction, or all three.

Checking Your Understanding

1. Differentiate between the intrinsic and extrinsic control of the heart and identify the associated structures.
2. What is the role of Na^+, K^+, and Ca^{2+} in contraction of the heart?
3. What is the baroreceptor reflex and how does it work?
4. Define arrhythmia.
5. Explain the difference between bradycardia and tachycardia.
6. Differentiate between atrial and ventricular fibrillation. Which is more life threatening?
7. List three types of toxic substances that affect the heart and briefly explain their mechanisms of action and symptoms.

6-6 Pulmonary Toxicity

Functions of the Respiratory System

The primary function of the respiratory system is the exchange of oxygen and carbon dioxide between the lungs and the bloodstream; however, other functions are also performed by it. These functions include protection of the respiratory tract from particulates, chemical irritants, and infectious agents by nonspecific and specific defense mechanisms as well as the biotransformation of toxic substances.

Gas Exchange

Movement of air into the lungs is dependent upon the rhythmic expansion and contraction of the rib cage and the diaphragm. During inhalation the contraction of muscles associated with the rib cage causes it to expand upward and outward. The diaphragm contracts at the same time and moves downward. This causes a negative pressure in the lungs compared to the atmosphere, and air moves into the lungs. When the rib cage muscles and the diaphragm relax, the chest cavity returns to its normal position and shape. This increases the pressure in the lungs and forces air out through the mouth and nose.

Portions of the inhaled air reach the alveoli of the lungs. The alveoli are the small saclike structures surrounded by an extensive capillary network. Gas exchange – as well as the exchange of some toxic substances – occurs in the alveoli. Movement of these substances is primarily by passive diffusion based on concentration gradients. Oxygen moves from the lungs to the blood and carbon dioxide moves from the blood into the lungs.

> Inhaled toxic substances enter the bloodstream primarily through the alveoli.

Damage to the muscles involved in the expansion of the rib cage will affect the efficiency of removal of substances that diffuse from the blood into the lungs. It will also affect the absorption of oxygen into the blood and it will subsequently affect those tissues that are sensitive to oxygen changes in the blood, such as nerve tissue. Also, changes in the structure of the alveoli, such as thickening of the alveoli wall, will impede diffusion of oxygen and toxic substances into and out of the blood.

Defense Mechanisms

The respiratory tract is equipped with several different types of defense mechanisms to protect it from damage by inhaled substances and microorganisms. The defense mechanisms may be classified as nonspecific or specific.

Nonspecific Defense Mechanisms – Hairs, cilia, and mucus-secreting cells are found throughout the respiratory tract. They are important in trapping substances, especially particulates, before they reach the delicate alveoli. Large particulates – 5-30 μm in diameter – are trapped in the nasopharyngeal region. Particles 1-5 μm are trapped primarily in the region of the respiratory tract between the trachea and the bronchioles. Many of the particles that are 1 μm or less in diameter reach the alveoli.

Once the smaller particulates reach the alveoli, macrophages may engulf the particles and destroy them (phagocytosis). After this process the macrophages may be carried to the upper regions of the respiratory tract by the mucociliary escalator to the nasopharyngeal region to be swallowed or expectorated. Neutrophils also play an important role in the phagocytosis of particulates in the respiratory tract.

> Nonspecific mechanisms – primarily responsible for removing particulates – are composed of cilia, hairs, mucus-secreting cells, as well as phagocytic neutrophils and macrophages.

Specific Defense Mechanisms – Lymphocytes are found throughout the respiratory tract or in close association with it. Foreign substances such as bacteria entering the respiratory tract interact with antibodies produced by lymphocytes. The antibody binds specifically to the surface of the foreign substance; this initiates a series of chemical reactions, ultimately resulting in the destruction of the substance. Macrophages are also an important part of the specific defense mechanism. They are responsible for phagocytizing the foreign substance and presenting it to the lymphocytes. This is an important step in initiating the immune response and producing antibodies.

Cellular immunity may also play an important role in the defense mechanisms of the lungs. Natural killer cells, which are modified lymphocytes, may be present in large numbers in the lungs. They interact directly with the foreign substance and destroy it.

> Antibody-producing lymphocytes are the primary components of the specific defense mechanism of the respiratory tract.

Biotransformation of Toxic Substances

There are several metabolically active enzymes present in the respiratory tract similar to those found in the liver. These enzymes either detoxify or activate various toxic substances including carbon tetrachloride and polycyclic aromatic hydrocarbons such as benzo[a]pyrene. The enzymes occur throughout the respiratory tract; the specific types and concentrations vary from region to region.

Pulmonary Toxic Effects

Toxic substances that are inhaled can be divided into two categories: particulates and nonparticulates. As described previously, many particulates are removed by nonspecific defense mechanisms associated with various regions of the respiratory tract. Those that reach the alveoli have the potential to exert toxic effects. Gases and vapors are nonparticulate substances and may be absorbed along the entire length of the respiratory tract. Gases and vapors with high water solubility such as sulfur dioxide (SO_2) are absorbed primarily in the upper respiratory tract. Gases with low water solubility will reach the alveoli.

Exposure to particulates such as silica and asbestos occurs in various occupations. Emissions from industrial processes contain potentially toxic gases such as nitrogen oxides, carbon dioxide, sulfur dioxide, and ozone-forming substances. Carbon monoxide and particulates are produced from vehicle exhaust. All of these substances have the potential to cause injury. Some kinds of injury are reversible while others are permanent.

The toxic effects may be the result of: direct interaction of the toxicant with cellular components of the respiratory tract; bioactivation of the toxicant, which then may bind directly with intracellular targets resulting in cellular dysfunction; or by immuno-mediated mechanisms, which affect various aspects of the respiratory tract. Tissue damage resulting from chemical exposure is characterized by cell degeneration and, if severe enough, cell death (necrosis), followed by inflammation and, subsequently, repair of the affected area accompanied by cellular proliferation. The cellular proliferation in response to tissue damage can be associated with the development of cancer in the various regions of the respiratory tract.

The following description identifies the effects of various toxic substances on different regions of the respiratory tract. These regions are the nasopharyngeal region, the tracheobranchiole region, and the alveoli.

Nasopharyngeal Region

The nasopharyngeal region includes that portion of the respiratory tract from the nasal cavity to, and including, the pharynx. It is characterized by an epithelium containing ciliated and mucus-secreting cells. In addition, the nasal cavity contains hairs. All of these structures are designed to trap and filter the inhaled air. Since this portion of the respiratory tract is the first section to contact the air, it is also the first segment to be exposed to toxic substances in the form of particulates, gases, vapors, or toxicants absorbed to particulates. In all cases the toxicant may affect the functioning of the cells that line this portion of the respiratory tract.

Exposure of the nasopharyngeal epithelium to toxicants may affect the normal functioning on the mucus cells and cilia, it may irritate or damage the nasal mucosa, and it may also damage the nerve cells involved in the detection of smell. Chemicals such as ammonia, hydrochloric acid, formaldehyde, hydrofluoric acid, chlorine, and chromium (VI) are irritants to the nasopharyngeal epithelium. These substances stimulate nerve endings in the nose that usually cause a reflex holding of the breath in exposed individuals.

Prolonged exposure to formaldehyde, various acids such as hydrochloric or hydrofluoric, or other irritants such as cigarette smoke, may result in changes in the nasopharyngeal epithelium. Some of these changes are reversible. The cells may increase in number (**hyperplasia**) or undergo transformation to a different cell type (**metaplasia**). Mucous cell hyperplasia is associated with changes in the chemical characteristics of mucus. Hyperplasia causes mucus viscosity to increase. Metaplasia can cause the epithelial cells to become keratinized. Hyperplasia and metaplasia of the epithelial cells can also cause constriction of airflow in this part of the respiratory tract. In addition to changes in cellular structure and mucus secretion, ciliary activity is decreased; this – along with the change in mucus secretion – decreases the effectiveness of the mucociliary clearance mechanism.

> Changes in the composition and type of mucus along with decreased ciliary activity impair the removal of particulates from the respiratory tract.

Following injury to the nasopharyngeal epithelium, cell proliferation occurs to repair the damage. The cellular proliferation may result in uncontrolled growth of the epithelial cells with the potential for development of cancer. Occupations with a higher potential for the development of nasal cancer include: chromate workers, nickel refiners and workers, mustard gas makers, isopropyl alcohol workers, wooden furniture makers, and workers making boots and shoes. Welders and solderers may be exposed to either hexavalent chromium compounds or nickel compounds, which may result in nasal cancer. Prolonged exposure to toxicants such as formaldehyde, nitrosamines, acrolein, acetaldehyde, or vinyl chloride may also cause nasal cancer.

Tracheobronchiole Region

The epithelium that lines the tracheobronchiole region is composed of a variety of cell types including mucous cells and ciliated epithelial cells. The ciliated cells are the type of cell most numerous in this region of the respiratory tract. The diameter of the airway progressively decreases from the trachea to the alveoli. As the diameter decreases, so does the number of ciliated cells.

The cilia continuously beat moving mucus toward the mouth where it is either swallowed or expectorated. This process removes embedded particulates, which – if left in the respiratory tract – may cause damage to the alveoli. The cilia are easily damaged by inhaled toxicants such as nitrogen dioxide (NO_2), sulfur dioxide (SO_2), ozone (O_3), and cigarette smoke. The cilia may stop beating (**ciliostasis**), or decrease in number. A decrease in number occurs either as a result of resorption of the cilia by the cell, or by detachment of the cilia, or also by cell death. The effects may be reversible if the exposure to the toxic substance stops.

Mucous cells are also found in the same regions as the ciliated cells. Mucus secretion increases in response to ammonia (NH_3), SO_2, O_3, and NO_2. This increased secretion irritates nerve endings in the tracheobronchiole airway, which initiates the cough reflex. Irritation of the autonomic sensory nerve fibers located in the epithelium of the bronchioles results in the reflex constriction of the bronchiole smooth muscle and restriction of air flow. Chronic exposure may result in continued hypersecretion of mucus, constriction of the bronchioles with epithelial cell hyperplasia and metaplasia. If these conditions persist, hypoxia may occur causing a decrease in the oxygen supply to the alveoli. Death of the alveolar cells may occur, which leads to the development of **emphysema**. Proliferation of cells following damage may lead to the development of cancer.

Respiratory diseases like **bronchitis** may also be caused by exposure to inhaled toxicants. Bronchitis is characterized by an increased number of hypersecreting mucous cells. As discussed previously, the increased mucus secretion results in vasoconstriction of the bronchioles producing increased difficulty in breathing. Cigarette smoke, along with chlorine, chromium (VI), ammonia, sulfur dioxide, and cotton dust may cause bronchitis.

Alveoli

If the toxic substance is not trapped and removed by the nonspecific defense mechanisms in the respiratory tract, then the alveoli may be damaged. If the damage is severe enough, the exchange of oxygen between the lungs and the blood may be impaired. The decreased oxygen supply may affect the functioning of other tissues of the body, but it also affects the lung tissues; this may exacerbate the damage already sustained by the lungs from exposure to the toxicant. Several types of respiratory diseases are associated with damages to the alveoli.

Edema – **Edema** is the excessive accumulation of fluid. **Pulmonary edema** is therefore the excessive accumulation of fluid in the lungs. Fluid accumulation in the alveoli results in decreased diffusion of oxygen to the bloodstream and carbon dioxide into the lungs. The decreased oxygen concentration affects other tissues particularly those that are oxygen sensitive such as nerve and cardiac tissue.

Excess fluid accumulates when there is damage to the alveolar cells. Healthy, alveolar cells are closely associated with each other, therefore very little liquid moves into the alveolar lumen through the intercellular space. When damage to the alveolar epithelial cells occurs, liquid and some larger molecules migrate into the alveolar lumen. If the damage is not fixed and fluid continues to accumulate in the lungs, then pneumonia may develop. Ammonia, chlorine, hydrogen fluoride, nickel, oxides of nitrogen, ozone, and perchloroethylene are all capable of causing edema.

Fibrosis – **Fibrosis** is a disease of the lungs characterized by an increased amount of connective tissue associated with the alveoli. The connective tissue may be generalized or restricted to specific locations within the lungs. Replacement of alveolar tissue by connective tissue results in decreased lung elasticity. This impairs the movement of air into and out of the lungs and, more importantly, affects the diffusion of gases between the lungs and the bloodstream.

Fibrosis may be caused by a variety of etiological agents including silica, coal dust, asbestos, beryllium, cadmium, or ingestion of the pesticide paraquat. Exposure to all of these substances may result in damage to the alveolar tissue. The specific mechanism of action for each of these toxicants varies; however, the end result is the replacement of the damaged tissue by connective tissue.

Asbestosis – **Asbestosis** occurs as the result of exposure to asbestos. Asbestos mining, various types of construction activities, and working in shipyards are all possible sources of exposure. Because of its fire- and heat-resistant properties, asbestos was used in pipe insulation, ceiling, and floor tile.

It is important to recognize that the presence of asbestos in a material does not necessarily mean that an individual will be exposed. The asbestos must be in a form that allows it to become airborne. In this state, inhalation is more likely to occur. As with other toxic substances, inhalation of asbestos does not necessarily mean that lung damage will occur. However, asbestos exposure is associated with three illnesses: asbestosis, lung cancer, and **malignant mesothelioma**. The onset of these illnesses seems to be correlated with length and diameter of inhaled asbestos fibers. Fibers 2 µm in length cause asbestosis. Mesothelioma is associated with fibers around 5 µm in length. Fibers larger than 10 µm in length produce lung cancer. Fiber diameter greater than 3 µm is more likely to cause asbestosis or lung cancer, while fibers 3 µm or less are associated more readily with mesothelioma.

Asbestos fibers that enter the lungs are phagocytized by macrophages. Shorter fibers are totally digested by macrophages. Longer fibers are not totally digested. The incomplete digestion is associated with the inability of the asbestos-containing macrophage to be removed from the alveoli. The macrophage remains in the alveoli and releases a variety of chemicals; this leads to death of alveolar cells. The dead alveolar tissue is replaced by connective tissue, which is characteristic of asbestosis, resulting in breathing difficulties. Conversely, the chemicals released by the macrophage may result in cellular proliferation. If the cellular proliferation is uncontrolled, lung cancer or mesothelioma may develop.

Silicosis – Like asbestosis, **silicosis** is characterized by increased deposition of connective tissue in the lungs, specifically the alveoli. The disease is caused by crystalline silica. Silica-containing dust is generated from a variety of occupations such as rock mining, sandblasting, stone cutting, and foundry work. Fine silica-containing particles are generated from these activities and inhaled. The nonspecific defense mechanism entraps and removes some of the particles. However, small particles from 0.5-3 µm in diameter may reach the alveoli where they may cause lung damage.

Macrophages engulf the silica particles that reach the alveoli. The phagocytized particles can damage the macrophage, causing it to release enzymes that damage the alveoli. In response, there is an increased deposition of connective tissue in the damaged area.

Acute silicosis is almost always fatal. This condition develops from exposure to very high levels of silica particles, 5 µm or less in diameter, over a few months to a year. Prior to death, individuals will develop a cough and labored breathing, fever, and weight loss. Chronic silicosis has symptoms similar to acute silicosis; however they both develop after a prolonged period of time. Individuals may be asymptomatic even though **x-rays** may show localized nodules of connective tissue indicative of lung damage.

Emphysema – Emphysema is characterized by permanent enlargement of the alveoli. Unlike fibrosis, emphysema is not characterized by an increase in connective tissue deposition. In contrast the elastic connective tissue – which is one of the components of the alveolar cell wall – is destroyed. As a result, there is a fusion, or joining of the individual alveoli, into a single larger alveolus. The lungs lose some of their ability to expand and contract as destruction of the elastic connective tissue progresses. In turn, gas exchange in the lungs is impaired resulting in labored breathing and cough. Emphysema may also be associated with chronic bronchitis. Cigarette smoke, ozone, NO_2, and exposure to some cadmium compounds are all capable of causing emphysema.

Lung Cancer–Exposure to certain toxicants may also result in the development of lung cancer. Cigarette smoke is believed to be a cancer-causing substance. Asbestos may cause either lung cancer or mesothelioma, which is a form of lung cancer associated with the tissues that line the external surface of the lung and the wall of the rib cage. Exposure to cigarette smoke and asbestos has a synergistic effect. Smokers exposed to asbestos have a significantly increased chance of developing lung cancer. Other substances that may cause lung cancer are the polycyclic aromatic hydrocarbons such as benzo[a]pyrene, nickel and chromium compounds, and the elements arsenic and beryllium. Exposure to radiation may also cause cancer.

Checking Your Understanding

1. Identify the components of the nonspecific defense mechanisms in the respiratory tract and briefly explain how they function to protect the respiratory system.

2. List the components of the specific defense mechanisms of the respiratory tract and their functions.

3. Give two reasons why exposure to a toxic substance that causes granulocytopenia results in increased respiratory infections.

4. What are the effects of chemical exposure on the nasopharyngeal region, tracheobronchiole region, and alveoli? Give examples of toxic substances that affect these areas.

5. Define the terms metaplasia, hyperplasia, and ciliostasis. How are they related to toxic chemical exposure?

6. Explain how edema is caused by exposure to toxic substances. Give two examples. Why may edema be life threatening?

7. How does fibrosis occur? Give examples.

6-7 Hepatotoxicity

Functions of the Liver

The liver plays a major role in metabolism and the body's defense mechanisms. There are six major metabolic functions associated with the liver: 1) interconversion of carbohydrates, fats, and proteins, 2) removal of excess nitrogen from the body, 3) storage, 4) destruction of old red blood cells, 5) synthesis of substances needed for metabolic activities, and 6) detoxification.

The liver is able to meet the changing nutritional needs of the body by interconverting carbohydrates, fats, and proteins. For example, glucose can be derived from amino acids when blood glucose levels are lower than normal. If blood glucose levels are normal then excess protein and carbohydrates can be converted to fat.

Excess amino acids are metabolized by the liver cells and ammonia is produced as a byproduct. The ammonia (NH_3) is removed and converted to urea in the liver. Urea is subsequently carried by the bloodstream to the kidneys, where it is excreted. The remainder of the amino acid molecule is used in other biochemical reactions.

Beneficial substances such as carbohydrates, fats, proteins, and fat-soluble vitamins A, D, E, and K are stored in the liver. Metabolically important metals are also found in the liver. However, the same mechanisms that allow the liver to store beneficial substances such as vitamins and metals also contribute to the concentration and storage of toxic substances. For instance, lipid-soluble toxic substances, such as organochlorine pesticides (DDT) or organic solvents like trichloroethylene, can concentrate in the liver and disrupt many of the normal metabolic processes. Similarly, toxic metals such as cadmium can be concentrated in the liver and have similar effects.

Red blood cells have a life span of about 120 days. As red blood cells begin to degenerate, they are engulfed by macrophages throughout the body. Macrophages in the liver are referred to as **Kupffer cells** and are actively involved in this process. One of the byproducts of red blood cell destruction is a protein called **bilirubin**. The liver metabolizes bilirubin and the metabolites are either excreted in the bile to the small intestine or removed by the kidneys.

The liver synthesizes a variety of substances. Energy in the form of **glycogen** is produced and stored for future use when blood glucose levels decrease below normal. Albumin, a **glycoprotein**, is produced by the liver and released into the blood where it is important in maintaining proper osmotic balance. It also helps to maintain blood pH; it binds to various substances and carries them in the blood. Alpha and beta globulins, which are also glycoproteins, are produced in the liver. These proteins perform a variety of functions in the body. For example, the beta globulin transferrin carries iron to various tissues of the body such as bone marrow where it is needed for incorporation into the hemoglobin of developing red blood cells. Proteins important in blood clotting are also synthesized in the liver. **Prothrombin** and fibrinogen are two proteins secreted into the bloodstream where they are available to participate in the blood clotting mechanism when the need arises. **Heme** is an organic compound containing iron and nitrogen; it is produced in the liver cells, secreted into the blood, and then transported to the red bone marrow. There, heme is absorbed from the bloodstream into developing red blood cells called **reticulocytes**. In the reticulocytes, it is combined with the protein **globin** to form hemoglobin. Bile is also synthesized in the liver cells. It is stored in the **gall bladder** and is released into the small intestine where it is involved in the digestion and absorption of fats.

One of the most important functions of the liver is the detoxification of substances generated from normal metabolic processes, as well as of toxic substances that enter the body through ingestion, inhalation, or dermal absorption. As discussed in Chapter 5, there are several oxidation-reduction metabolic processes that are involved in detoxifying foreign substances. In general, these metabolic processes convert toxins to more water-soluble and less toxic forms. The increased water solubility facilitates the excretion of the substance by the kidneys. In some circumstances, however, the conversion of toxic substances to an intermediate compound results in increased toxicity, resulting in bioactivation rather than detoxification. For example, pyridine, which is somewhat toxic, is converted in the liver to methylpyridine, which is more toxic.

The liver macrophages (Kupffer cells) are an important component of the body's defense mechanism. They are able to engulf and destroy foreign substances that enter the body. Phagocytosis also serves as one of the mechanisms for removing from the bloodstream toxic substances that are either too large to pass through cell membrane pores or that are not lipid soluble.

Hepatotoxic Effects

The liver is one of the primary organs of the body involved in the detoxification of harmful substances that enter the bloodstream. But, as with pyridine, some metabolites are more toxic than the original substance. Chromium (VI) is easily aborbed into the liver cells where it is converted to chromium (III), which cannot pass through the cell membrane, so it is retained in the liver cells. Arsenic undergoes reductio to an ionic form that cannot penetrate the cell membrane. When these toxic substances are retained within the liver cells they cause intracellular damage. They interfere with normal protein synthesis and enzyme functions. This interference, as well as damage to structural components of the cell, results in death of the liver cells. If significant cell death occurs in the liver, then many of the metabolic processes and defense mechanisms discussed previously are affected.

Cirrhosis of the liver – caracterized by damage to and subsequent death of liver cells – is caused by a variety of toxic substances. The dead cells are replaced by connective tissue. Excessive exposure to alcohol or to the organic substance carbon tetrachloride reults in liver cell death and replacement of the cells with connective tissue. One of the functions of the liver is to remove bilirubin from the blood by excreting it with the bile, or by converting it to a water-soluble product to be excreted by the kidneys. Cirrhosis can produce an accumulation of bilirubin in the bloodstream because the liver's ability to absorb the substance from the bloodstream is impaired. As a result, the bilirubin – along with another byproduct of red blood cell metabolism, **biliverdin** – begins to accumulate in the bloodstream. This accumulation produces a yellowish coloring in the skin, the whites of the eyes, and body fluids. This condition is referred to as **jaundice**.

Hepatitis is an inflammatin of the liver that can be caused by exposure to chemicals or viruses. In both cases, cell death occurs leading to liver dysfunction and death. When cell death occurs, macrophages and neutrophils migrate to the damaged area to remove the cell debris. In the process, various enzymes are released causing inflammation of the surrounding tissues. If enough liver cells are affected, the liver will enlarge and produce symptoms like jaundice and tenderness to the touch. Further, the reduced ability to produce albumin and blood-clotting proteins will result in edema and uncontrolled bleeding.

Checking Your Understanding

1. What are the major functions of the liver?
2. Where are Kupffer cells located? Give at least two functions for this type of cell.
3. List the various types of proteins synthesized in the liver and their function in the body.
4. Explain how cirrhosis and hepatitis develop as a result of exposure to toxic substances. What types of toxic substances may cause these illnesses?
5. Define jaundice and how is it caused by exposure to toxic substances.

6-8 Nephrotoxicity

Functions of the Kidneys

The kidneys perform several important functions. They play a major role in excretion of metabolic waste such as ammonia in the form of urea, which is derived from the metabolism of amino acids. They are also involved in maintaining blood pH, ion and water balance, and the secretion of hormones.

Excretion of excess ions and water is directly related to the need to maintain homeostasis in the body. Water and ions, such as sodium and potassium, are selectively absorbed along various sections of the nephron, particularly the proximal and distal convoluted tubules, and the Loop of Henle. The processes involved in absorption are both active and passive. In general, if there is an excess amount of ions in the bloodstream, ions will be excreted in high concentrations in the urine. If there is an excess amount of water, diluted urine will be excreted, the ions having been reabsorbed as they pass through the nephrons. The transport mechanisms involved in the various segments of the nephron were described in Chapter 5. A proper balance of ions such as sodium and potassium is important in maintaining normal cell membrane integrity and function. In addition nerve cells and the specialized cells of the SA and AV nodes, require proper sodium/potassium balance in order to initiate and conduct electrical signals.

It is important to maintain the blood pH between 7.2 and 7.4 in order to maintain normal body functions. If the blood pH decreases, a condition referred to as acidosis can occur. This may cause an increased heart and respiratory rate. The kidneys play an important role in maintaining the blood pH by removing excess hydrogen ions that are produced by various cellular metabolic activities. If excess hydrogen ions were allowed to accumulate, the pH of the blood would decrease. Instead, the hydrogen ion is exchanged with sodium in the kidney tubules and excreted from the body.

Finally, the kidneys secrete a hormone call **renin**, which is involved in regulating blood pressure. It is secreted into the blood where it interacts with another protein, **angiotensin**, which stimulates contraction of the smooth muscle of the arterioles. This contraction causes an increase in blood pressure.

Nephrotoxic Effects

Like the liver, the kidneys are a metabolic site for the deactivation of some toxic substances. Many of the mechanisms that result in the accumulation and detoxification of toxic substances in the liver are similar to those found in the nephron cells of the kidneys. In addition, the kidneys have a higher amount of blood flowing through them than most other tissues in the body. This facilitates the accumulation of toxic substances. Many of the active and passive transport mechanisms involved in the excretion and reabsorption of ions, glucose, amino acids, and metabolic waste also enhance accumulation of toxic substances such as cadmium and lead, which compete for the same carrier molecules.

Heavy Metals

Heavy metals such as lead, mercury, and cadmium typically accumulate in the kidneys and generally affect the proximal convoluted tubules of the nephrons. Cadmium binds with proteins circulating in the blood and is able to freely pass into the glomerular filtrate of the kidneys. The cadmium-protein complex is absorbed into the cells of the proximal convoluted tubule where noncomplexed free cadmium is formed as a result of cellular metabolic processes. The cadmium may bind with intracellular metallothionein or with other intracellular proteins such as enzymes and components of the cell membrane. Binding of cadmium to proteins other than metallothionein can disrupt normal cellular metabolism and result in cell death.

Chronic and acute exposure to mercury may also result in death of proximal convoluted tubule cells. Mercury inhibits enzymes in the mitochondria, which are associated with the production of the energy-containing molecule, ATP. ATP is necessary to maintain many normal cell functions as well as cell membrane integrity. Without ATP, functions such as protein and carbohydrate synthesis – which are important in intracellular metabolism – do not occur. Consequently, the cell eventually dies.

Lead is rapidly accumulated in the kidneys after exposure. It is transported into the cells of the proximal tubules, where it is stored in the cell cytoplasm or nucleus. Storage in the nucleus and subsequent interaction with nuclear DNA may explain the carcinogenic effect of lead seen in experiments with rats. Like mercury, lead binds with enzymes associated with mitochondria and ATP production. The results are similar: impaired cellular function and death.

Toxic Organic Substances

Organic substances may cause cellular dysfunction and death; they also have the potential to cause cancer. Chloroform has the potential to do both. The toxic effects are the result of formation of phosgene derived from intracellular metabolism of chloroform. Nephrotoxicity results in degradation of the cell membrane structure and subsequent death. Binding of chloroform with DNA may be associated with the cancer-causing potential of chloroform in the kidneys.

Ethylene glycol may cause renal failure because oxalic acid is formed from the breakdown of ethylene glycol. Oxalic acid reacts with calcium in the lumen of the nephron to form an insoluble calcium oxalate precipitate, which obstructs the normal flow of liquid through the nephron, subsequently leading to renal failure.

Cell Death in the Kidneys

Toxic substances result in the death of cells by one of two mechanisms: necrosis or apostosis. Apostosis is characterized by a decrease in cell volume. Cell components such as the mitochondria remain intact until eventually the cell breaks into small fragments, which are phagocytized by macrophages. Cell death is limited to individual cells scattered throughout the kidneys. **Necrosis** affects many contiguous cells and is characterized by the swelling of cell organelles and increased cell volume. The cells eventually rupture resulting in the release of the cell contents, which in turn causes inflammation.

A toxic substance can initiate cell death by several different mechanisms. The mechanisms may involve the original toxic substance or a metabolic intermediate that is more toxic. Either of these substances may bind with macromolecules within the cells causing cell injury and/or death. Interaction of toxicants such as mercury or cadmium with macromolecules of the cell membrane can affect transport mechanisms associated with maintaining cell volume and ion balance. Mercuric chloride is thought to be nephrotoxic as a result of altering cell membrane activity, cellular enzyme activity, membrane permeability and transport activities, as well as inducing breaks in DNA and **chromosomes**.

Checking Your Understanding

1. List and explain the major functions of the kidneys.
2. What is the importance of the kidneys in maintaining the normal function of the heart?
3. How might a nephrotoxic substance such as lead affect blood pressure?
4. Differentiate between necrosis and apostosis.
5. What portion of the nephron is affected the most as a result of exposure to toxic substances? Give two examples of the observed effects.
6. List and briefly explain the possible causes of cell death in the kidneys as a result of exposure to toxic substances.

6-9 Reproductive Toxicity

Continuous existence of any species of organism relies on the ability of individuals within the population to successfully mate and reproduce. There are many physiological and psychological factors that under normal circumstances may impact the successful generation of offspring. The effect of toxic chemicals on reproductive tissues is poorly understood compared to their effects on other tissues. The number of men with sperm counts below what is considered a fertility threshold has increased over the past 20 years. The number of women who are infertile has increased in some age groups. This leads people to suspect environmental factors such as chemical exposure to be the cause.

The successful development of the egg or sperm is the result of the interaction of the nervous and endocrine systems with the reproductive structures. Toxic substances may affect the reproductive systems either directly or indirectly by affecting the endocrine or nervous systems. This section will focus on how toxic substances affect the development of egg and sperm. Effects of toxic substances on the development of the fetus will be discussed in Chapter 9, which addresses the topic of teratogens and fetal development.

The male and female reproductive systems are comprised of a variety of structures and cell types, which are influenced by various hormones. The following provides a general description of the anatomy and physiology of the male and female reproductive systems as well as of the effects of various toxic substances on these systems. Since most of the toxicological studies have been performed on rats, the effects described are primarily based on these studies.

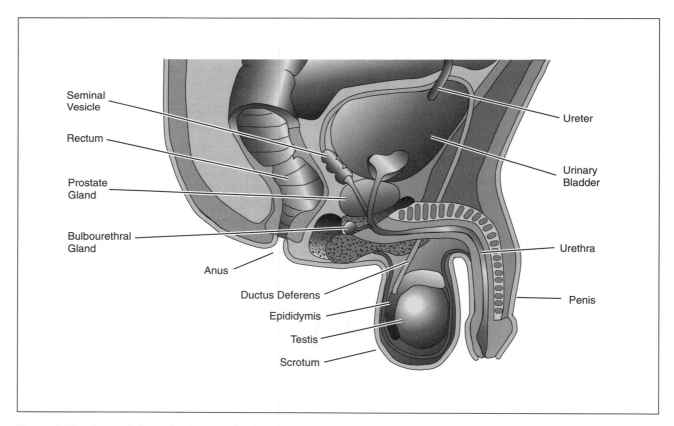

Figure 6-15 : Sagittal view of male reproductive structures.

Anatomy and Physiology of the Male Reproductive System

Anatomy

The male reproductive tract consist of the **testes, epididymis, ductus deferens, seminal vesicles, urethra, prostate gland, bulbourethral gland**, and penis (Figure 6-15).

The testes are oval structures that are divided into sections called lobules. Within each lobule are structures referred to as **seminiferous tubules**, where the sperm cells develop. The seminiferous tubules are surrounded by connective tissue, which contains interstitial cells (**cells of Leydig**). Interstitial cells secrete the hormone **testosterone**. The seminiferous tubules join with a series of tubules referred to as the **efferent ductules**, which join the epididymis (Figure 6-16).

Spermatogenesis in the seminiferous tubules is the result of a series of mitotic and meiotic cell divisions. **Spermatogenia** (germ cells) undergo mitotic division to produce other spermatogenia and **primary spermatocytes**. The primary spermatocytes undergo the first meiotic division and produce two **secondary spermatocytes**. Each of the two secondary spermatocytes undergoes the second meiotic division resulting in the production of four smaller sex cells called **spermatids**. Structural changes occur in the spermatids, in which the cells become smaller and develop flagellum. Once these changes have occurred, the spermatids become sperm cells (**spermatozoan**).

Sertoli cells are also found within the seminiferous tubules. These cells serve a variety of functions. They are involved in regulating spermatogenesis, in providing structural and metabolic support for the germ cells, and in secreting tubular fluid for sperm

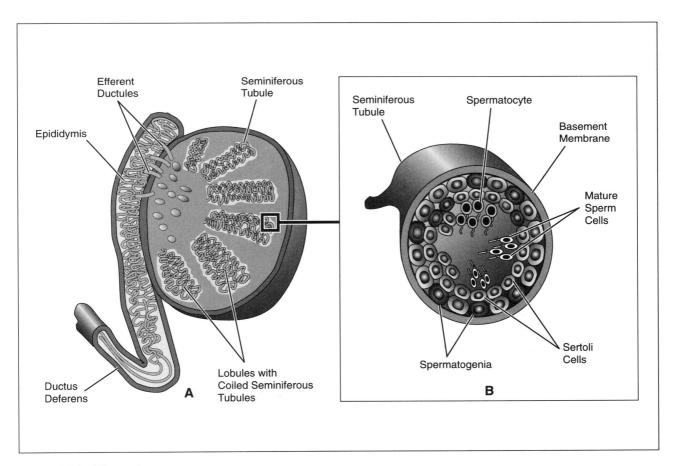

Figure 6-16 : The testis.
A. Gross anatomy of the testis.
B. Cross section of seminiferous tubule illustrating stages of sperm development. Mature sperm cells located toward the center of the tubule lumen.

transport; they also comprise part of the blood-tubule barrier in the testis, which is analogous to the blood-brain barrier. Through this barrier the cells of the tubules are protected from some of the blood borne toxicants.

Transport of the sperm cell from the seminiferous tubules to the exterior of the body occurs by a series of ducts. The seminiferous tubules join with the epididymis, where final maturation of the sperm cell occurs. The epididymis empties into the ductus deferens. The walls of the ductus deferens contain smooth muscle that moves sperm cells from the epididymis to the ductus deferens. The two ductus deferens empty into the urethra, and the sperm cells move from the body through the urethra.

There are several glands located along this path that synthesize and secrete materials into the ducts. These materials are important in sperm mobility and survival. Bulbourethral glands and mucous glands located in the urethra secrete mucus, which assists sperm mobility. The seminal vesicle glands secrete a thick, mucus-like substance that is high in fructose and other nutrients. These substances provide nourishment to the sperm cells. The prostate gland secretions are important in regulating the pH within the urethra to maintain normal sperm cell functioning.

Physiology

The nervous system and various hormones are involved in regulating the development of sperm and the function of the male reproductive system. Of specific interest are the hormones released from the **anterior pituitary** in response to secretions produced by neurons in the **hypothalamus** of the brain. Neurons in the hypothalamus of the brain release a hormone referred to as **gonadotropin-releasing hormone (GnRH)**. GnRH causes the release of **luteinizing hormone (LH)** and **follicle-stimulating hormone (FSH)** from the anterior pituitary. LH binds to the cells of Leydig and causes testosterone to be secreted. Testosterone is important in developing the secondary sexual characteristics and spermatogenesis. FSH binds to the Sertoli cells resulting in spermatogenesis.

Toxic Effects in the Male Reproductive System

Like other target tissues and organs of the body, the components of the reproductive system may be affected directly or indirectly by toxic substances. Cells undergoing division during spermatogenesis are more sensitive to toxic substances than nondividing cells. Therefore, of the limited studies performed on the effects of toxicants on the reproductive system, most of the information addresses the toxic effects on spermatogenesis. The effects of toxicants on the reproductive system range from a decrease in spermatogenesis to cellular necrosis.

Workers exposed to the fungicide 1,2-dibromo-3-chloropropane (**DBCP**) were shown to have a decreased sperm count in the presence of elevated FSH and LH levels in the blood. In addition, sperm motility was decreased and the number of abnormal sperm increased. In most cases this was associated with damage to the seminiferous tubules.

Soil, grain, fruit, and timber are sprayed with the fumigant ethylene dibromide (**EDB**). Workers from a Hawaii fruit processing plant who were chronically exposed to EDB demonstrated a decreased sperm count and an increased number of abnormal sperm.

Glycol ethers are used throughout the chemical industry. Two types of glycol ethers are in paints: 2-ethoxyethanol and 2-methoxyethanol. All stages of spermatogenesis seem to be affected. Painters exposed to these chemicals demonstrated a decreased number of sperm and an increased number of dead sperm cells.

The effects of lead on the reproductive system of men have been widely studied. As discussed previously, lead can affect the nervous system. It is possible that lead may either affect the release of GnRH from the hypothalamus, or act on the pituitary directly. In either case there may be a decrease in FSH and LH, which could explain the decreased spermatogenesis observed in men exposed to lead.

Other reproductive structures such as the Sertoli cells, Leydig cells, and the epididymis may be affected by toxicants. Ethanol, acetaldehyde, and ethylene dimethane sulfonate (**EDS**) all interact with the Leydig cells and decrease testosterone levels, which reduces spermatogenesis.

Sertoli cells are important in regulating the normal progression of events associated with spermatogenesis and subsequent formation of sperm. The germ cells are held in contact with the Sertoli cells. Exposure to phthalate esters such as di-(2-ethylhexyl)phthalate (**DEHP**) and 1,3-dinitrobenzene cause the germ cells to detach from the Sertoli cells prematurely. The end result is decreased spermatogenesis.

Several substances affect the epididymis, where final maturation of sperm cells occurs. Inhalation of methylene chloride causes damage to the epididymis epithelium. Formation of **granulomas** (tumor or growth) occurs, which inhibits the movement of sperm from the epididymis, resulting in infertility. Dibromochloropropane (DBCP) – a soil fumigant – and cadmium cause damage to the blood vessels supplying the epididymis. The net result is decreased viability of sperm cells.

A great deal of attention has been focused lately on toxic substances that mimic the effects of estrogen on the male reproductive tract. The pesticides DDT, DDE, methoxychlor, toxaphene, chlordecone, and endosulfan demonstrate some level of estrogenic activity in laboratory studies. Exposure to these chemicals produces testicular atrophy caused by a decrease in the blood levels of GSH and testosterone, decreased sperm count, prostate and testicular cancer.

> Toxic substances impair normal spermatogenesis as a result of interaction with the various reproductive structures and hormonal controls associated with sperm cell formation.

Anatomy and Physiology of the Female Reproductive System

Anatomy

The female system includes the **ovaries**, the uterine or fallopian tubes, the **uterus** and the vagina (Figure 6-17). Although in many texts the vagina is dis-

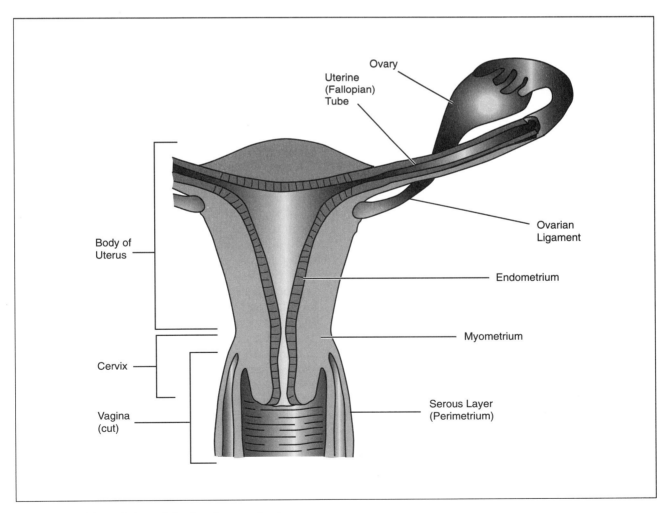

Figure 6-17: Frontal view of the female reproductive organs.

cussed as part of the reproductive tract, in the context of this chapter it will not be discussed in detail. The ovary is responsible for development and release of the egg (ovum) to the fallopian tube where fertilization of the egg may occur. If fertilization occurs, then implantation in the uterus may take place.

Located in the outer portion of the ovary are the **ovarian follicles**. Within the follicles are **oocytes**, which are analogous to the spermatogenia in the seminiferous tubules. The oocyte in the ovary of a newborn girl has already begun the first meiotic division, but it is stopped at an early stage. These oocytes are referred to as **primary oocytes**. Surrounding the primary oocyte is a single layer of cells referred to as **granulosa cells**. The primary oocytes together with the granulosa cells are referred to as the **primordial follicles**, of which there are 300,000 to 400,000 at birth. After the onset of puberty, the specific primordial follicles undergo a series of changes in response to hormones secreted during the monthly cycle. Each month a follicle enlarges to form a mature or **Graafian follicle**. The first meiotic division is completed and the follicle ruptures (**ovulation**) releasing the egg. The second meiotic division is initiated but not completed. The second division is completed only if fertilization occurs.

Once the ovum is released from the follicle it enters the uterine tube. The uterine (fallopian) tube extends from the surface of the ovaries and opens into the uterus. If fertilization occurs it will usually happen in the upper third of the uterine tube. The uterus is a pear-shaped organ. It is composed of three layers: an external serous layer; the middle layer or **myometrium**, containing primarily smooth muscle; and the innermost layer called the **endometrium**. It is the endometrium that undergoes significant changes during each **menstrual cycle** and during pregnancy.

Physiology

Ovulation and changes in the endometrium of the uterus are under the control of hormones secreted from both the pituitary gland and specialized cells in the ovary. The levels of these hormones are cyclic in nature. Changes in these hormones are responsible for ovulation, which occurs on average every 28 days, although there is considerable variation among women.

FSH levels increase in the bloodstream and cause the follicles to begin growing. Generally only one follicle grows large enough, eventually rupturing and releasing its egg. The developing follicle secretes another hormone, **estrogen**, which initiates changes in the endometrium. Actual ovulation occurs as a result of increased levels of LH. Cells of the ruptured follicle, under the influence of LH, form a new structure referred to as the **corpus luteum**. The corpus luteum secretes another hormone, **progesterone**, along with estrogen; they both continue to cause changes in the endometrium, preparing it for the potential implantation of a fertilized egg. If pregnancy does not occur the corpus luteum degenerates and the cycle starts again.

Toxic Effects in the Female Reproductive System

The anatomy and physiology of the female reproductive system involves complex interactions between the endocrine system, the brain, and the reproductive tract, making it sometimes difficult to recognize the toxic effects of chemicals. Toxic effects may be expressed as a decrease in GnSH, or impairment of follicle development, or as changes in the manufacture of estrogen, progesterone, and other hormones.

Exposure to several different types of toxicants has been shown to affect GnSH secretion. Heavy metals such as inorganic mercury, lead, manganese, tin, and cadmium can affect GnSH secretion. Polystyrene and styrene, which are byproducts of gasoline, may alter LH and FSH secretion. A decrease in these hormones will result in decreased estrogen and progesterone secretion. As a result, the uterus will not undergo the necessary anatomical changes required for successful implantation of a fertilized egg. Drugs such as cocaine and marijuana also affect secretion of LH and FSH. Low doses of cocaine stimulate LH secretion, while high doses inhibit it. Marijuana depresses LH and FSH secretion.

Impairment of follicle development will result in the loss of control of FSH and LH secretion. As levels of circulating estrogen and progesterone increase, the levels of FSH and LH decrease. This is referred to as negative feedback. Toxicants such as cadmium, dibromochloropropane, and polycyclic aromatic hydrocarbons inhibit follicle development. This results in decreased fertility. Also, if the damage to the follicles occurs early in life, cancer of the ovaries may occur.

Synthesis of the steroids estrogen and progesterone requires the presence of enzymes. Heavy metals like mercury and cadmium are known to inhibit enzyme function. These toxicants can affect steroid synthesis by inactivating the enzymes necessary for synthesis.

Smoking cigarettes also appears to affect the female reproductive tract: in rats ovulation is inhibited, while in humans estrogen production is decreased. It has been shown that women who smoke tend to enter menopause 1-4 years earlier than non-smoking women. The mechanism of action for each of these effects is not clearly known; however, polycyclic aromatic hydrocarbons found in cigarette smoke may be the cause, since they destroy oocytes.

The estrogenic effect of pesticides discussed previously also affects the female reproductive system. Epidemiological studies suggest that there is a correlation between blood levels of DDE and breast cancer in women.

Checking Your Understanding

1. Identify the steps involved in the production of sperm cells and list the structures sperm must pass through to leave the body.

2. Where are the Sertoli cells and cells of Leydig located and what is their function?

3. List the hormones involved in regulating the development of sperm, their source, and their function.

4. How do toxic substances affect hormonal control of spermatogenesis? What are the symptoms?

5. Explain how the effects of ethanol on the Leydig cells result in decreased spermatogenesis.

6. List the structures associated with the development of ovum and briefly explain what events occur during each step of development.

7. Identify the hormones and their functions involved in regulating development of the ovum.

8. Explain why the weakly estrogenic effects of organochlorine pesticides such as DDT might cause cancer in the ovaries.

9. How does exposure to a toxic substance affect production of a mature egg?

6-10 Toxicity of the Eye

The function of the eye can be damaged as a result of internal and external exposure to toxic substances. The eyelid and the blink reflex are the primary physical and functional defense mechanisms against external toxic exposure. Protection by these mechanisms is not always successful. In fact, the most common type of exposure occurs as a result of accidental splashing of the eye with a toxic substance. If the substance is not removed immediately, permanent damage may occur. To understand the effects of toxic exposure to the eye, an understanding of its general anatomy is necessary. The following is an overview of the structure of the eye.

When the eye is exposed to a toxic substance it is the cornea (Figure 6-18) that is the primary site of contact. It contains no blood vessels or pigmented cells and is therefore transparent. This transparency is important in allowing light to pass through the outer structures of the eye to the retina. The outer layer of the cornea, the corneal epithelium, is five to

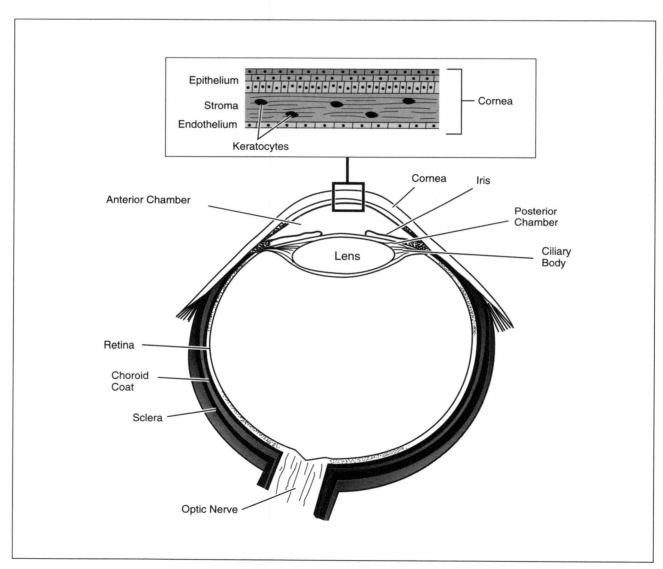

Figure 6-18: The cornea.

six cell layers thick. The basal cells comprise the deepest layer. The basal cells rapidly divide, replacing the cells in the outer layers of the epithelium as they die and are sloughed off. Significant damage to the basal cells can produce permanent damage to the cornea in the form of scar tissue and reduced vision.

Underneath the corneal epithelium is the **corneal stroma**. The stroma makes up about 90 percent of the corneal thickness. It consists primarily of connective tissue and it contains a small number of corneal cells called **keratocytes**. Lying under the stroma is the corneal endothelium, which is composed of a single layer of cells. Unlike the epithelium, the endothelium does not regenerate itself when damaged. Therefore damage of the corneal endothelium from toxic chemical exposure can cause permanent damage to the eye.

Under the cornea and in front of the lens is the anterior chamber of the eye. It is filled with a liquid substance called the aqueous humor. The aqueous humor is important in supplying oxygen and nutrients to the cornea and the lens. The lens is located between the anterior and posterior chambers of the eye. It is transparent and – similarly to the cornea – does not contain blood vessels. The lens obtains oxygen and nutrients from the surrounding fluids, the aqueous humor (anterior chamber) and the vitreous humor (posterior chamber). Once light passes through the cornea, the lens focuses it on the retina.

The retina is situated on the posterior wall of the eyeball. It is composed of multiple layers of nerve cells. The axons of the retinal nerve cells converge at a central location in the posterior wall of the eyeball to form the **optic nerve**. The optic nerve carries the electrical signals from the retinal nerve cells to the brain where images and colors are formed.

Toxic Effects to the Eye

The eye is very sensitive to toxic chemical exposure and is easily damaged. External exposure may result in symptoms ranging from mild irritation to blindness, depending on the type of toxic chemical and the duration of exposure. The inadvertent exposure to acids – such as sulfuric acid found in car batteries – as well as bases (sodium hydroxide) in drain cleaners are some of the most common sources of exposure.

Injuries caused by acid or alkali exposure are primarily the result of rapid pH changes within the tissues. Acids cause the proteins in cells of the corneal epithelium to break down and coagulate. This kills the cells of the corneal epithelium, which are then sloughed off. Initially, there may be some cloudiness associated with the cornea. However, if the basal cells of the corneal epithelium have not been damaged, the cloudiness will disappear as new epithelial cells are generated. If the duration of exposure is prolonged and/or the concentration of the acid is high, the basal cells and the underlying stroma and endothelium can be permanently damaged. This will result in scar tissue formation and permanent cloudiness of the cornea. The reduced transparency associated with the corneal damage will result in a reduced visual field.

Although exposure to acids causes cell proteins to coagulate, the coagulated proteins in the outer layers of the corneal epithelium protect the underlying tissue by inhibiting further penetration of most acids. Hydrofluoric acid, however, is capable of penetrating the outer layers of the corneal epithelium. It can penetrate the stroma and damage the corneal endothelial cells, potentially leading to glaucoma.

Similar toxic effects to the eye can be observed as a result of exposure to alkaline toxic substances such as ammonia or sodium hydroxide. However, the mechanisms of action for alkaline substances are different from the ones in acid exposure. Alkaline substances alter cell protein structure and function, also causing the hydrolysis of lipids in the cell membrane. This causes the cells to disintegrate. Temporary cloudiness of the cornea, with some reduced vision, may occur if the exposure is of short duration or if the toxic substance is a weak base. The corneal epithelial cells will be regenerated and normal vision will return.

More extensive effects are associated with exposure to alkalis. Unlike the diffusion of acids – which is hindered by coagulated protein – alkaline substances are fat soluble, and therefore diffuse more readily into the underlying tissue. This characteristic contributes to the delayed toxic effects observed when exposed to these substances. If the alkaline substance is not washed from the eye, it can cause severe damage to the deeper tissue such as the stroma, the endothelium, and the lens. This damage may cause scar tissue to form in the cornea and cloudiness in the lens. Collectively, these effects may cause reduced vision or blindness.

Lime is a component of Portland cement and many types of wall plaster. Lime interacts with the moisture in the eye to form calcium hydroxide (a base) and the generation of heat. The calcium hydroxide interacts with proteins to form small particles that remain lodged in the eye. The particles are not easily removed by flushing the eye and will continually release calcium hydroxide over a prolonged period of time. Complexing agents – such as the disodium salt of EDTA (ethylenediaminetetraacetic acid) – may be used to remove the residual material.

Toxic metals and solvents can also affect the functioning of the eye. In the past, compounds containing arsenic (arsenicals) and/or silver were used for medicinal purposes. A silver-containing solution was administered to the eyes of newborn babies to prevent a disease called **gonococcal ophthalmitis**, which is an inflammatory disease of the eye caused by the bacterium *Neisseria gonorrhoeae*. Application of the silver-containing solution resulted in the formation of a bluish-gray pigment in the cornea and conjunctiva (lining of the eyelid). The pigmentation is associated with the complexing of silver with the protein albumin. Similar effects have been observed in workers involved in the manufacturing of various silver-containing chemicals such as silver nitrate, silver oxide, and silver solder.

Pentavalent arsenicals (As^{5+}) were used to treat individuals suffering from **sleeping sickness** or **neurosyphilis**.

> Sleeping sickness is characterized by lethargy, drowsiness, muscular weakness, and tremors.

The primary toxic effect associated with arsenical exposure was reduction in peripheral vision, which in turn appeared to be associated with the arsenic's effect on the nerve cells in the retina.

> Neurosyphilis is an infection of the central nervous system by the bacterium *Treponema pallium*.

Exposure to organic mercury (e.g., methyl mercury) affects the eye indirectly. The mercury penetrates the blood-brain barrier and affects the visual cortex of the brain. The visual cortex undergoes neuropathy, which affects the transmission of nerve impulses to and from the cortex. This in turn affects all aspects of vision.

Unlike mercury, lead can affect the eye directly by acting on the specialized nerve cells of the retina called rods. The rods are light-sensitive cells associated with visual acuity. The lead alters cell metabolism, which produces cell degeneration and death.

Exposure to organic solvents (acetone, benzene, toluene) produce a variety of effects. In general, external exposure can cause irritation of the eye and slight cloudiness of the cornea. The irritation is caused by stimulation of nerves associated with the cornea and the surrounding tissues. Solvents break down the corneal epithelial cells, which causes the cloudy appearance. The damaged cells are then replaced by newly generated epithelial cells, recreating a clear cornea.

Other organic toxic chemicals have more pronounced effects. Carbon disulfide causes damage to the blood vessels that deliver blood to the neural components of the retina. The end result is decreased blood flow to these tissues, which affects visual acuity, reduces the field of vision, and may cause blindness.

DNP (2,4-dinitrophenol) has its primary effect on the lens of the eye. DNP appears to decrease the production of ATP. This affects the normal functioning of the lens tissue and is characterized by altered proteins, edema, necrosis, and disruption of the lens fibers. The end result is increased opacity of the lens and cataract formation. Naphthalene, dimethyl sulfoxide, 2,4,6-trinitrotoluene, diquat, and heptachlor also cause cataracts.

Chloroquinone and hydroxyquinone are drugs used to treat malaria and inflammatory diseases such as rheumatoid arthritis. These drugs may have toxic effects on the retina. Symptoms include decreased visual acuity, night blindness, and sensitivity to light. These symptoms are associated with the loss of rods and cones (nerve cells) in the retina.

Ingestion of methanol as a result of accidental poisoning, or as a component (contaminant) in ethyl alcohol, can have long-term effects on the eye. Metabolism of methyl alcohol results in the formation of formaldehyde and formic acid, both of which are toxic. Decreased visual acuity, inflammation of the optic nerve (which may cause blindness), and the appearance of blind spots can appear within 18-48 hours after exposure. Within 30-60 days degeneration of the optic nerve is observed. Eventually this may lead to complete blindness.

Smog, dust (particulate matter), and allergens are other environmental agents that may cause irritation of the eye. Irritation can be associated with tearing, swelling of the tissues of the eyelid, and increased mucus secretion. In addition, particulates may scratch the surface of the cornea, which may temporarily impair vision. Excessive exposure to ultraviolet light may increase the chance of cataract formation.

Checking Your Understanding

1. Draw the general structure of the eye (additional information can be found in Chapter 3).
2. Using the illustration from question 1, identify the structures that are affected by acids and solvents.
3. What is the difference in symptoms associated with exposure to acids and alkaline substances?
4. What symptoms are associated with exposure to toxic organic chemicals? How are the symptoms produced?
5. Lead and mercury compounds can affect the functioning of the eye. What are the symptoms associated with exposure? What is the mechanism of action for each?
6. What is the effect of methyl alcohol on the eye?

Summary

Toxic substances can have a variety of effects, depending on the type of toxic substance, concentration, and the target tissue affected. The observed effects are the result of the cumulative impacts on cell and organ structure functioning.

Neurotoxicity

The two major divisions of the nervous system are the peripheral nervous system and the central nervous system. They are responsible for detecting stimuli and initiating a response. The functional unit of the nervous system is the neuron. The neuron transmits information to and from the central nervous system along cellular processes referred to as dendrites and axons. The signal is carried along these processes by an action potential to the synapse where neurotransmitters such as acetylcholine and norepinephrine are released.

Neurotoxic effects occur as a result of direct interaction on the neuron structure, or by affecting neurotransmission between the presynaptic terminal of the neuron and the postsynaptic membrane. Direct effects on the neuron structure result in neuronopathy, axonopathy, and myelinopathy, all of which affect nerve transmission. Various toxic substances affect the normal sequence of events associated with neurotransmission between the presynaptic terminal and postsynaptic membrane. These types of toxic substances may either inhibit stimulation or result in constant stimulation of the affected tissue. Symptoms associated with neurotoxicity cover a wide spectrum. Normal skeletal muscle functioning is affected resulting in ataxia, muscular weakness, or paralysis, depending on the toxicant. Other symptoms include loss of sensation, blurred vision, mental disorders, changes in breathing patterns, heart rate, and possibly death.

Hematoxicity

The primary elements of the blood system are the noncellular plasma and the cellular components: erythrocytes, leukocytes, and thrombocytes, which are produced from stem cells in the bone marrow. The red blood cells are important in the exchange of gases in the lungs; the leukocytes are important in the defense of the body against foreign substances; and the thrombocytes play a role in blood clotting. The leukocytes include eosinophils, basophils, neutrophils, lymphocytes, and monocytes (macrophages).

Hematoxic effects are observed either by direct action of toxicants on mature circulating cells or by interference of toxicants with blood cell formation. Symptoms include the development of anemia, granulocytopenia, thrombocytopenia, lymphocytopenia, or pancytopenia. Anemia results in decreased oxygen to sensitive neural and cardiac tissues, which produces abnormal functioning of these tissues. This is most often manifested as an increase in heart rate and breathing. Granulocytopenia, lymphocytopenia, and pancytopenia are characterized by a decrease in the number of leukocytes and increased susceptibility to infections and disease.

Immunotoxicity

The immune system consists primarily of the lymphatic system, the cellular components, and the antibodies. The lymph nodes, the spleen, and thymus contain lymphocytes and macrophages, which are involved in responding to foreign substances that enter the body. The T cells, B cells, and macrophages make up the principal cellular components of the immune system. It is the interaction between these three cell types that results in the production of antibodies involved in the destruction of foreign substances.

Immunotoxic effects from exposure to various chemicals create a variety of responses. In general, exposure to low doses tends to have a stimulatory effect on different components of the immune system, while exposure to high doses has an inhibitory effect.

Cardiotoxicity

The heart is responsible for moving blood throughout the body in order to provide a consistent supply of nutrients and oxygen to the various tissues. The rhythmic contractions of the heart are controlled by intrinsic and extrinsic systems. The intrinsic control is maintained by the SA node and by the associated specialized cells of the conduction pathway. Extrinsic control is maintained by both the parasympathetic and sympathetic nervous systems.

Toxic substances may act directly on the heart by affecting the intrinsic control mechanism or muscle cell functioning. Or they may also affect the extrinsic control of the heart by acting on various aspects of the nervous system. In either case, the net result is a change in the heart rate, a change in the strength of contraction, or both.

Pulmonary Toxicity

The respiratory system performs several different functions: 1) exchange of gases between the lungs and the bloodstream; 2) protection against foreign substances inhaled with air; and 3) biotransformation of toxic substances. Protection is provided by nonspecific and specific defense mechanisms. Nonspecific mechanisms include mucus cells, cilia, and macrophages. Specific defense mechanisms are associated with lymphocytes and antibody production.

Toxic substances affect both types of defense mechanisms. Effects on the nonspecific defense mechanisms include increased mucous secretion, decreased ciliary activity, and impaired clearance of foreign substances by the mucociliary escalator. Consequently, it is easier for particulates to reach the alveoli, where they can cause cellular damage resulting in the development of fibrosis, emphysema, or edema. High doses of toxic substances cause a decrease in the production of antibodies creating increased susceptibility to infections. Lung cancer is associated with exposure to asbestos, chemicals in cigarette smoke, chromium, and nickel compounds.

Hepatotoxicity

There are six major metabolic functions performed by the liver: 1) interconversion of carbohydrates, fats, and proteins, 2) removal of excess nitrogen from the body, 3) storage, 4) destruction of old red blood cells, 5) synthesis, and 6) detoxification. The Kupffer cells are an important element of the nonspecific defense mechanism of the body. They remove from the blood both foreign debris and toxic substances that are lipid insoluble or too large to pass through the cell membrane.

The liver is the site of biotransformation of toxic substances resulting in either deactivation or bioactivation of the substance. Biotransformation may result in retention of toxic substances within the cells where they can cause cellular dysfunction and death. Cirrhosis and hepatitis are two diseases characterized by injury or death of liver cells.

Nephrotoxicity

The kidneys play an important role in maintaining homeostasis by regulating the ion and water balance in the body fluids. They also secrete hormones involved in regulating blood pressure.

The proximal convoluted tubule of the nephron is the primary target for nephrotoxic substances. The toxic substances disrupt tubule cell functions, particularly those associated with the active and passive transport mechanisms involved in the reabsorption of ions and molecules from the tubular fluid. Other effects include cell death, apostosis or necrosis, or the development of cancerous growths.

Reproductive Toxicity

The primary effect of toxic substances on the male and female reproductive system is on spermatogenesis and development of a mature egg. The effects may be mediated through interaction with the nervous system, through hormones involved in the formation of the sperm and egg, or by directly affecting the reproductive structures.

Toxic effects associated with the male reproductive system include decreased spermatogenesis, sperm viability, or both. Similar effects are observed in the female system, which is characterized by impairment of follicular development and production of mature eggs. In addition, some toxicants have estrogenic effects that may cause cancer of the ovaries.

Eye Toxicity

The eye can be damaged as a result of the external and internal exposure to toxic substances. The cornea is the primary site of contact for external exposure. Damage of the cornea can be repaired if the basal cells of the corneal epithelium are not destroyed. Penetration of the corneal epithelium to the underlying tissue may result in permanent eye damage.

The symptoms associated with exposure are dependent on the type of toxic substance. Acids, alkaline substances, and solvents can cause cloudiness of the cornea. Toxic metals can cause increased pigmentation of the cornea and eyelids, decreased visual acuity, and decrease in peripheral vision.

Organic toxic substances can produce symptoms similar to toxic metals. In addition, some organic substances can affect the lens of the eye producing cataracts. Methanol can cause blindness by exerting its effect on the optic nerve.

Application and Critical Thinking Questions

1. What are the symptoms associated with Alzheimer's disease, multiple sclerosis, and Parkinson's syndrome? Explain the possible mechanisms involved in initiating the symptoms observed.

2. Mercury is accumulated in the kidneys. Propose a possible mechanism by which cadmium accumulation in the kidneys could lead to the development of arrhythmias.

3. Briefly explain what causes glomerular nephritis. Hypothesize how this disease may cause anemia and the appearance of symptoms related to nerve cell and cardiac tissue dysfunction.

4. Explain how exposure to benzene might have direct and indirect effects on the intrinsic and extrinsic control of the heart.

5. Hypothesize why individuals who smoke and are exposed to asbestos have a significantly higher rate of asbestosis than individuals who do not smoke and are exposed to asbestos.

6. Cadmium is bioaccumulated in various cells of the kidney. Identify the portion of the kidneys where cadmium accumulates and explain how the nephrotoxic effects may affect at least two other tissue types discussed in this chapter.

7. How might a neurotoxic substance that affects the central nervous system result in decreased spermatogenesis or follicular development?

7

Survey of Toxic Substances

Chapter Objectives

Upon completing this chapter, the student will be able to:

1. **Identify** various types of indoor and outdoor pollutants and their toxic effects.
2. **Describe** the causes and symptoms of building-associated illnesses.
3. **List** various types of petroleum distillates and their toxic effects.
4. **Identify** the toxic effects of various types of metals, pesticides, and other important chemicals.

Chapter Sections

7–1 Introduction

7–2 Outdoor and Indoor Air Pollution

7–3 Petroleum Distillates

7–4 Metals and Metalloids

7–5 Other Important Chemicals

7–6 Pesticides

7-1 Introduction

The quantity of chemical, physical, and biological agents with the potential to cause toxic effects seems almost unlimited. Hundreds to thousands of new chemicals and chemical-containing products are produced every year. Usually they are produced with the intent of improving our physical wellbeing at home, in the workplace, or in our recreational activities. Science and medicine continue to make great strides in dealing with illnesses and diseases by improving the general health of the world population. This is primarily the result of better hygiene practices and improved medicines such as antibiotics.

Along with the benefits associated with the increased development and use of chemical agents has come an increased risk of being exposed to these substances, and the development of illnesses and diseases due to either acute or chronic exposure. Many respiratory problems have been attributed to elevated levels of sulfur and nitrogen oxides found in the emissions of vehicles and industrial smokestacks. These same sources have been responsible for emitting various types of toxic metals such as lead and cadmium into the environment. Even individuals not located near the source are at risk of exposure when these substances are transported in the food chain, air, or water.

The development of pesticides has resulted in the availability of an abundant food supply due to improved control of nuisance plants and insects. Larger pests such as mice, rats, and insects which spread disease and damage stored food crops, were initially controlled by using pesticides. Unfortunately, many of these substances have had undesirable effects on humans as well as on plants and animals. In addition, the mice and rats have developed resistance to many of the pesticides, which now have to be stronger to control the pests.

This chapter represents a survey of various toxic agents. It is not designed to be a comprehensive discussion on the effects of each class of toxic agent mentioned. Indeed, there are entire texts on each of these topics. Enough information is provided to identify some of the physical and chemical characteristics of the toxic agents discussed as well as some of their toxic effects. Where possible, information is summarized in tables.

A number of organizations have classified chemical substances based upon various properties and effects. The **National Fire Protection Association (NFPA)** and the American Conference of Governmental Industrial Hygienists (ACGIH) are two such organizations. The NFPA has developed a system for identifying hazards associated with some, but not all, chemicals. Usually this information is found on a diamond-shaped and color-coded label of the product, which provides information relative to the chemical's flammability, reactivity, and health hazards. The health hazard (code H), flammability hazard (code F), and reactivity hazard (code R) are all ranked using numbers from 0 to 4, the latter being the most severe. Where applicable, this information is presented for various chemicals in this chapter. Table 7-1 provides information on how to understand the coding system used.

The ACGIH publishes a booklet – updated every year – concerning exposure limits in the workplace and carcinogenic potential for a variety of chemicals. The exposure levels are referred to as Threshold Limit Values (TLVs). The TLVs are used as guidelines in establishing the acceptable risk of exposure to a toxic substance. The TLVs are based on information obtained from human and animal studies as well as on experiences from various industries. "The basis on which the values are established may differ from substance to substance; protection against impairment of health may be a guiding factor for some, whereas reasonable freedom from irritation, narcosis, nuisance, or other forms of stress may form the basis for others" (from 1996 TLVs and BEIs, ACGIH, see below). They are not intended to determine safe exposure levels.

TLVs are provided for chemicals evaluated by the ACGIH. Some of the TLVs will be followed by the letter "S." This indicates the TLV is based upon skin exposure. Additional information concerning TLVs can be obtained from the booklet published annually entitled, "1996 TLVs and BEIs, Threshold Limit Values for Chemical Substances and Physical Agents Biological Exposure Indices."

Several organizations are involved in evaluating and classifying chemicals as carcinogens. In addition to the ACGIH, carcinogenicity is evaluated by the **National Toxicology Program (NTP)**, the **International Agency for Research on Cancer (IARC)**, and the National Institute for Occupational Safety and Health (NIOSH). Because the criteria used by these organizations are slightly different from each other, the list of carcinogens identified may vary. Table 7-1 lists carcinogenicity by organization. Information concerning TLVs and NFPA hazard codes is also provided as well as information concerning the concentration considered to be **Immediately Dangerous to Life or Health (IDLH)**. The IDLH represents "a maximum concentration from which one could escape within 30 minutes without any escape-impairing symptoms or any irreversible health effects" (Olson, K.R., *Poisoning and Drug Overdose*, Appleton & Lange, 1994).

Chemical Name	TLV [1]	IDLH [2]	Carcinogenicity				NFPA [7]		
			IARC [3]	NTP [4]	ACGIH [5]	NIOSH [6]	H	F	R
1,1-Dichloroethane – $C_2H_4Cl_2$	100 ppm	4,000 ppm					2	3	0
1,2-Dichloroethane (Ethylene dichloride) – $C_2H_4Cl_2$	10 ppm	1,000 ppm	2B	B			2	3	0
1,1,1,2-Tetrachloroethane – $C_2H_2Cl_4$									
1,1,2,2-Tetrachloroethane – $C_2H_2Cl_4$	1 ppm(S)[8]	150 ppm	3			carc			
1,1,1-Trichloroethane (Methyl Chloroform) – $C_2H_3Cl_3$	350 ppm	1,000 ppm					3	1	1
1,1,2-Trichloroethane – $C_2H_3Cl_3$	10 ppm(S)	500 ppm	3			carc	3	1	0
2,3,7,8-Tetrachlorodibenzo-p-dioxin (TCDD) – $C_{12}H_4Cl_4O$			2B	B					
2,4-D-dichlorophenoxyacetic acid – $C_8H_6Cl_2O_3$	10 mg/m³	500 mg/m³	2B						
2,4,5-T-trichlorophenoxyacetic acid – $C_8H_5Cl_3O_3$	10 mg/m³	5,000 mg/m³	2B						
Aldrin (dry) – $C_{12}H_8Cl_6$	0.25 mg/m³(S)	100 mg/m³	3			carc	2	0	0
Aldrin (solution) – $C_{12}H_8Cl_6$	0.25 mg/m³(S)	100 mg/m³	3			carc	3	1	0
Ammonia – NH_3 (gas)	25 ppm	500 ppm					2	1	0
Arsenic (As) Compounds:	0.1 mg/m³		1	A	A1	carc	3	2	0
Arsenic trioxide – As_2O_3							3	0	0
Arsenic trifluoride – AsF_3							3	1	0
Arsenic acid – H_3AsO_4							3	0	0
Arsenic disulfide – As_2S_2							3	1	1
Arsine – AsH_3	0.05 ppm	6 ppm				carc			
Asbestos:			1	A	A1	carc			
Amosite	0.5 fibers/cc				A1				
Chrysotile	2 fibers/cc				A1				
Crocidolite	0.2 fibers/cc				A1				
Other forms	2 fibers/cc				A1				
Benzene (C_6H_6)	10 ppm	3,000 ppm	1	A	A1	carc	2	3	0
Benzene hexachloride – $C_6H_6Cl_6$ (Lindane) (solid)	0.5 mg/m³(S)	1,000 mg/m³		B			2	0	0
Benzene hexachloride – $C_6H_6Cl_6$ (Lindane) (solution)	0.5 mg/m³(S)	1,000 mg/m³		B			2	1	0

Table 7-1: Exposure levels and chemical characteristics associated with selected chemical and physical agents.

Chemical Name	TLV [1]	IDLH [2]	Carcinogenicity				NFPA [7]		
			IARC [3]	NTP [4]	ACGIH [5]	NIOSH [6]	H	F	R
Cadmium (Cd) Compounds:			2A	B	A2	carc			
Cadmium dust	0.01 mg/m^3	50 mg/m^3							
Cadmium fume	0.01 mg/m^3	9 mg/m^3							
Cadmium chloride – $CdCl_2$							3	0	0
Cadmium oxide – CdO							3	0	0
Carbaryl – $C_{12}H_{11}NO_2$	5 mg/m^3	600 mg/m^3							
Carbon dioxide – CO_2	5,000 ppm								
Carbon monoxide – CO	25 ppm	1,500 ppm					2	4	0
Carbon tetrachloride – CCl_4	5 ppm(S)	300 ppm	2B		A2	carc	3	0	0
Chlordane – $C_{10}H_6Cl_8$	0.5 mg/m^3(S)	500 mg/m^3	2B			carc			
Chlorine – Cl_2	0.5 ppm	30 ppm					3	0	0
Chromium (Cr) Compounds:									
Metal and Cr(III) compounds	0.5 mg/m^3				A3				
Water-soluble Cr(VI) compounds	0.05 mg/m^3		1	A	A1	carc			
Insoluble Cr(VI) compounds	0.01 mg/m^3		1	A	A1	carc			
Chromium oxychloride – CrO_2Cl_2						carc	3	0	1
Chromic acid – CrO_3	0.05 mg/m^3						3	0	1
Creosote			2A				2	2	0
Cyanide (CN^-) Compounds:		50 ppm							
Calcium cyanide – $Ca(CN)_2$	5 ppm(S)								
Hydrogen cyanide – HCN	4.7 ppm(S)						4	4	2
Potassium cyanine – KCN	5 ppm(S)								
Sodium cyanide – NaCN	5 ppm(S)						3	0	0
DDT – $C_{14}H_9Cl_5$	1 mg/m^3		2B	B		carc			
Dichloromethane (Methylene Chloride) – CH_2Cl_2	50 ppm	5,000 ppm	2B	B	A2	carc	2	1	0
Dichlorvos – $(CH_3O)_2POOCH=CCl_2$	0.1 ppm(S)	200 mg/m^3	2B						
Dieldrin – $C_{12}H_8Cl_6O$	0.25 mg/m^3(S)	450 mg/m^3	3			carc			
Diquat – $C_{12}H_{12}Br_2N_2$	0.5 mg/m^3(S)								
Endrin – $C_{12}H_8Cl_6O$	0.1 mg/m^3(S)	200 mg/m^3							
Ethyl chloride – C_2H_5Cl	1,000 ppm	20,000 ppm			A3		2	4	0
Formaldehyde – HCHO	0.3 ppm	30 ppm	2A	B	A2	carc	2	4	0
Hexachlorobenzene – C_6Cl_6	0.025 mg/m^3(S)		2B	B	A3				
Hexachloroethane – C_2Cl_6	1 ppm(S)	300 ppm	3		A2	carc			

Table 7-1: Exposure levels and chemical characteristics associated with selected chemical and physical agents (continued).

Chemical Name	TLV [1]	IDLH [2]	Carcinogenicity				NFPA [7]		
			IARC [3]	NTP [4]	ACGIH [5]	NIOSH [6]	H	F	R
Hydrofluoric acid – HF	3 ppm	30 ppm					4	0	0
Hydrogen fluoride (HF) Gas	3 ppm	30 ppm					4	0	0
Hydrogen sulfide – H_2S	10 ppm	300 ppm					3	4	0
Kerosene									
Lead (Pb) Compounds: Elemental lead and inorganic lead compounds	0.05 mg/m^3	700 mg/m^3	2B		A3				
Lead arsenate – $Pb_3(AsO_4)_2$	0.15 mg/m^3	300 mg/m^3					2	0	0
Lead arsenite – $Pb(AsO_2)_2$							2	0	0
Lead chromate – $PbCrO_4$: Lead Chromate	0.05 mg/m^3 0.012 mg/m^3				A2	carc			
Lead fluoroborate – PbB_2F_8							0	1	1
Lead nitrate – $Pb(NO_3)_2$							1	0	1
Lead thiocyanate – $Pb(SCN)_2$							1	1	1
Tetramethyl lead – $(CH_3)_4Pb$	0.15 mg/m^3(S)	40 mg/m^3					3	3	3
Malathion – $C_{10}H_{19}O_6PS_2$	10 mg/m^3(S)	5,000 mg/m^3							
Mercury (Hg) Compounds:									
Mercury compound, liquid							3	0	0
Mercury compound, solid							3	0	0
Organomercurial Compounds					A4				
Alkyl mercury compounds	0.01 mg/m^3(S)	10 mg/m^3							
Aryl mercury compounds	0.1 mg/m^3(S)	10 mg/m^3							
(Acetate) phenyl mercury	0.01 mg/m^3(S)	10 mg/m^3					3	1	0
Inorganic compounds including metallic mercury	0.025 mg/m^3	28 mg/m^3 (metallic mercury)							
Mercuric chloride – $HgCl_2$	0.1 mg/m^3	10 mg/m^3					3	0	0
Methyl bromide – CH_3Br	5 ppm(S)	2,000 ppm	3			carc	3	1	0
Monochloroethane – C_2H_5Cl		20,000 ppm					2	4	0
Nickel (Ni) Compounds:			1	B	A1	carc			
Nickel metal and insoluble compounds	1 mg/m^3								
Soluble metal compounds	0.1 mg/m^3								
Nickel carbonyl – $Ni(CO)_4$	0.12 mg/m^3	7 ppm					4	3	3

Table 7-1: Exposure levels and chemical characteristics associated with selected chemical and physical agents (continued).

Chemical Name	TLV [1]	IDLH [2]	Carcinogenicity				NFPA [7]		
			IARC [3]	NTP [4]	ACGIH [5]	NIOSH [6]	H	F	R
Nickel nitrate – $Ni(NO_3)_2$	$0.1\ mg/m^3$						1	0	1
Nickel sulfate – $NiSO_4$	$0.1\ mg/m^3$						1	0	0
Nitric Acid – HNO_3	2 ppm	100 ppm					3	0	0
Nitric oxide – NO	25 ppm	100 ppm					3	0	3
Nitrogen dioxide – NO_2	3 ppm	50 ppm					3	0	1
Nitrous oxide (N_2O)	50 ppm								
Ozone – O_3	0.1 ppm	10 ppm							
Paraquat (respirable fraction) – $C_{12}H_{14}N_2$	$0.1\ mg/m^3$	$1.5\ mg/m^3$							
Paraquat (total dust) – $C_{12}H_{14}N_2$	$0.5\ mg/m^3$	$1.5\ mg/m^3$							
Parathion – $C_{10}H_{14}NO_5PS$	$0.1\ mg/m^3$(S)	$20\ mg/m^3$					4	1	2
Pentachlorophenol (dry) – C_6Cl_5OH	$0.5\ mg/m^3$(S)	$150\ mg/m^3$	2B				3	0	0
Pentachlorophenol (solution) – C_6Cl_5OH	$0.5\ mg/m^3$(S)	$150\ mg/m^3$	2B				3	2	0
Petroleum ether	400 ppm	10,000 ppm					1	4	0
Phosgene – $COCl_2$	0.1 ppm	0.2 ppm					4	0	0
Phosphine – PH_3	0.3 ppm	200 ppm					3	4	1
Sodium fluoracetate – $CH_2FCOONa$	$0.05\ mg/m^3$(S)	$5\ mg/m^3$							
Sodium hydroxide – NaOH	$2\ mg/m^3$	$250\ mg/m^3$					3	0	1
Stoddard Solvent and Mineral Spirits	100 ppm	5,000 ppm					1	2	0
Strychnine – $C_{21}H_{22}N_2O_2$	$0.15\ mg/m^3$	$3\ mg/m^3$							
Sulfur dioxide – SO_2	2 ppm	100 ppm					3	0	0
Sulfuric acid – H_2SO_4	$1\ mg/m^3$	$80\ mg/m^3$					3	0	2
Tetrachlorethylene (perchlorethylene) – C_2Cl_4	25 ppm	500 ppm			A3	carc	2	0	0
Tetraethyl pyrophosphate – $(C_2H_5)_4P_2O_7$	0.004 ppm(S)	$10\ mg/m^3$							
Toluene – C_7H_8	50 ppm(S)	2,000 ppm					2	3	0
Trichloroethylene – C_2HCl_3	50 ppm	1,000 ppm				carc	2	1	0
Warfarin – $C_{19}H_{16}O_4$	$0.1\ mg/m^3$	$200\ mg/m^3$							
Xylene – C_8H_{10}	100 ppm	1,000 ppm					2	3	0
Zinc (Zn) Compounds:									
Zinc chlorate – $Zn(ClO_3)_2$							2	0	2
Zinc chloride – $ZnCl_2$							2	0	2
Zinc chloride fumes	$1\ mg/m^3$								

Table 7-1: Exposure levels and chemical characteristics associated with selected chemical and physical agents (continued).

[1] TLV = Threshold Limit Value obtained from "1996 TLVs and BEIs Threshold Limit Values for Chemical Substances and Physical Agents, Biological Exposure Indices." 1996 ACGIH. Cincinnati, OH, pp. 138

[2] IDLH = Immediately Dangerous to Life or Health. The IDLH represents "a maximum concentration from which one could escape within 30 minutes without any escape – impairing symptoms or any irreversible health effects" (Olson, K.R., 1994)

[3] IARC = International Agency for Research on Cancer
1 = Agent is carcinogenic
2A = Agent is probably carcinogenic
2B = Agent is possibly carcinogenic
3 = Agent is not classifiable as to carcinogenicity
4 = Agent is probably not carcinogenic

[4] NTP = National Toxicology Program
A = Known human carcinogen
B = Suspected or reasonably anticipated to be human carcinogen

[5] ACGIH = American Conference of Governmental Industrial Hygienists
A1 = Confirmed human carcinogen
A2 = Suspected human carcinogen
A3 = Animal carcinogen
A4 = Not classifiable as a human carcinogen
A5 = Not suspected as a human carcinogen

[6] NIOSH = National Institute for Occupational Safety and Health
carc = carcinogenic

[7] NFPA = National Fire Protection Association Identification of the Hazards of Materials (see below)

[8] (S) = Indicates measurement by skin exposure

[7] National Fire Protection Association (NFPA) Identification of the Hazards of Materials					
Identification of Health Hazard Color Code: BLUE		Identification of Flammability Color Code: RED		Identification of Reactivity (Stability) Color Code: YELLOW	
Signal	Type of Possible Injury	Signal	Susceptibility of Materials to Burning	Signal	Susceptibility of Release of Energy
4	Materials that on very short exposure could cause death or major residual injury even though prompt medical treatment was given	4	Materials that will rapidly or completely vaporize at atmospheric pressure and normal ambient temperature, or that are readily dispersed in air and will burn readily.	4	Materials that in themselves are readily capable of detonation or of explosive decomposition or reaction at normal temperatures and pressures.
3	Materials that on short exposure could cause serious temporary or residual injury even though prompt medical treatment was given.	3	Liquids and solids that can be ignited under almost all ambient temperature conditions.	3	Materials that in themselves are capable of detonation or explosive reaction but require a strong initiating source, or that must be heated under confinement before initiation, or that react explosively with water.
2	Materials that on intense or continued exposure could cause temporary incapacitation or possible residual injury unless prompt medical treatment is given.	2	Materials that must be moderately heated or exposed to relatively high ambient temperatures before ignition can occur.	2	Materials that in themselves are normally unstable and readily undergo violent chemical change but do not detonate. Also materials that may react violently with water or that may form potentially explosive mixtures with water.
1	Materials that on exposure would cause irritation but only minor residual injury even if no treatment is given.	1	Materials that must be preheated before ignition can occur.	1	Materials that in themselves are normally stable, but can become unstable at elevated temperatures and pressures, or that may react with water with some release of energy but not violently.
0	Materials that on exposure under fire conditions would offer no hazard beyond that of ordinary combustible material.	0	Materials that will not burn.	0	Materials that in themselves are normally stable, even under fire exposure conditions, and are not reactive with water.

Table 7-1: Exposure levels and chemical characteristics associated with selected chemical and physical agents (continued).

7-2 Outdoor and Indoor Air Pollution

The industrial revolution brought many benefits to society. In general, it provided an opportunity to improve the quality of life by creating well-paid occupations and products that made life easier. Oil and coal were essential elements that fueled production during that period, but that also brought air pollution. It was not uncommon during the latter part of the 19th century and the first two thirds of the 20th century to see major industrialized areas of the United States enveloped in thick, dense clouds of smoke and/or fog, which significantly reduced visibility. Well-paid workers found they could afford more material goods including the automobile. The increase in the number and use of automobiles, trucks, and other types of vehicles created new sources and types of air pollutants in escalating quantities.

Efforts to address the issue began in the late 1940s when California passed the Air Pollution Control Act. The federal government passed its version of the Air Pollution Control Act in 1955, but it wasn't until 1963 that the first major legislation was passed. It was only with the establishment of the Environmental Protection Agency in 1970 that an agency had the specific task of determining which air pollutants were potentially hazardous and of establishing safe limits for exposure. The initial efforts by EPA resulted in standards being established for only eight hazardous air pollutants. The 1990 amendment of the Clean Air Act expanded the list to 189 hazard pollutants for which standards must be established by the end of the decade.

Industry has invested large amounts of money to install and operate pollution control equipment for smokestack emissions. Large quantities of particulates, sulfur, and nitrogen oxides are removed from the emissions prior to their discharge. All new automobiles must now use unleaded gas, which has helped reduce the amount of airborne lead released into the environment. Alternate fuels – such as natural gas, a combination of methanol/gasoline, and electrical power – are being investigated to help further reduce harmful emissions. Nuclear power had been touted as the energy resource of the future, largely because no significant air emissions are produced under normal operating conditions. It was thought that electricity generated by fossil fuels would be replaced by nuclear energy sources. However, many issues have slowed the building of nuclear power plants including considerable debate about the harm of radiological emissions from these facilities. An even more serious concern involves the handling, storage, and subsequent disposal of waste generated, particularly the spent fuel rods from nuclear reactor operations. The solid waste is extremely hazardous and will remain so for a very long time.

Although great strides have been made in the United States, there are still major cities, such as Los Angeles and New York City, that have air pollution problems even with the passage of new federal and state laws regulating air emissions. Additional efforts will have to be made to reduce air pollution in these areas.

Indoor air pollution – in various forms – has always existed, but has only been recognized within the last 20-30 years. Changes in lifestyle as well as in home and workplace construction have raised the visibility of – and the interest in – the types and amounts of indoor air pollutants and their health effects.

The oil crisis of the 1970s is often associated with the apparent increase in the number of health problems due to indoor air pollution. During this period homes, factories, office buildings, and other types of occupied structures were constructed or remodeled according to energy-efficient principles. This reduced oil consumption, but resulted in less air being exchanged between the inside and the outside of the dwelling, making it more difficult for indoor air pollutants to be removed. In addition, outdoor pollutants transferred indoors through ventilation systems affected the indoor air quality as well.

Outdoor Air Pollution

Any of the thousands of chemicals used daily can be a component of outdoor air pollution. However, EPA has focused on a limited number of pollutants, which current evidence suggests may pose a health risk. The main outdoor air pollutants are: carbon monoxide (52 percent), sulfur oxides (14 percent), nitrogen oxides (14 percent), volatile organic compounds (14 percent), and particulates (4 percent).

Focusing on certain chemicals does not imply that other chemicals are not potentially toxic. Generally, the first chemicals to be investigated are those produced in the largest quantity and/or those that have the greatest potential to produce a toxic effect. The outdoor air pollutants discussed in this section will include carbon oxides, nitrogen oxides, sulfur oxides, ozone, and particulates.

Both acute and chronic effects have been associated with exposure to air pollutants. Long-term exposure of individuals in large, urban, and industrialized areas such as the Los Angeles basin has been shown to have a negative affect on the respiratory system as measured by pulmonary function tests. In addition, increased incidence of bronchitis in children has been documented. These effects have been correlated primarily with elevated levels of particulates, sulfates, and sulfur dioxide.

Levels of air pollutants vary during the day, usually reaching their peak beginning around midday and declining during the evening. Levels may also increase as a result of accidental release. In either case, acute exposures may have some effect on individuals, particularly individuals with pre-existing cardiopulmonary diseases. Studies focusing on changes in particulate matter levels in 16 urban areas were able to show an association between increases in particulate matter and an increase in the number of deaths. For every 100 mg/m^3 increase in particulate matter, there appears to be an increase in mortality of about six percent. Other studies have shown an increased number of respiratory problems as a result of elevated ozone and sulfates, particularly in asthmatic children.

Carbon Monoxide – CO
—TLV = 25 ppm
—NFPA: 2H, 4F, 0R

Carbon monoxide (CO) is a colorless, odorless, and tasteless gas produced by the incomplete combustion of carbon or carbonaceous materials. It is lighter than air and slightly soluble in water. Common sources of carbon monoxide include automobile exhaust and faulty or poorly ventilated charcoal, kerosene, or gas stoves. Incomplete combustion in foundries, coke ovens, and refineries are other sources of carbon monoxide. Tobacco products are also a source due to the same principle of incomplete combustion. Worldwide annual emission of carbon monoxide is 232 million tons, of which 88 million tons are produced in the United States alone.

Emissions from the incomplete combustion of natural gas or petroleum fuels may contain as much as five percent carbon monoxide. A natural gas heater can emit up to one cubic foot per minute of carbon monoxide, if it is not properly vented. In the past automobile emissions contained 3-7 percent carbon monoxide, which would equal 2.7 pounds of carbon monoxide per gallon of fuel used by an automobile with no emission control device. New car standards limit carbon monoxide emissions to 0.5 percent. Diesel exhaust can emit 0.074 pounds of carbon monoxide per gallon of fuel. Cigarette, pipe, and cigar smoke can contain up to four percent carbon monoxide.

Exposure to carbon monoxide at 100 ppm for eight hours generally does not produce any symptoms. Depending on the level of activity, exposure to 500 ppm for one hour may produce a slight headache and shortness of breath. Longer exposures cause symptoms of headache, nausea, irritability, increased respiration, chest pain, impaired judgement, cyanosis, and fainting. Unconsciousness, respiratory failure, and death occur as a result of exposure to 1,000 ppm for one hour. Chronic exposure to the anoxic conditions due to carbon monoxide may result in damage to the central nervous system, loss of peripheral sensations, poor memory, and general mental deterioration. A concentration of 1,500 ppm is considered to be IDLH (Immediately Dangerous to Life or Health).

> The term anoxic pertains to a general lack of oxygen.

The effects associated with carbon monoxide exposure are the result of the binding of carbon monoxide with hemoglobin to form carboxyhemoglobin. Hemoglobin has an affinity for carbon monoxide that is 225 times greater than for oxygen. Since oxygen cannot bind to carboxyhemoglobin, tissue anoxia results. In addition to this, the presence of carbon monoxide causes a decrease in the rate of oxygen disassociation from hemoglobin, which means that the release of oxygen to oxygen-deprived tissues is reduced. For example, smoking one cigarette reduces the amount of oxygen available to the tissues by about eight percent. This change in available oxygen is the same as the change that takes place when moving from sea level to an altitude of 4,000 feet. The decreased oxygen availability to metabolically active tissues – such as neural tissue – results in disruption of cellular functions, producing the symptoms discussed above. Table 7-2 lists symptoms associated with exposure to various concentrations of carbon monoxide as well as the percentage of carboxyhemoglobin formed.

> Carbon monoxide decreases the amount of oxygen available to cells resulting in cellular dysfunction.

Carbon Dioxide – CO_2

—TLV = 5,000 ppm

Carbon dioxide is odorless and colorless with low water solubility. It is estimated that worldwide five billion tons are produced annually from a variety of sources including fuel combustion, putrefaction and fermentation processes, sugar refineries, and chemical manufacturing.

Carbon dioxide acts as a **simple asphyxiant**, which means that it produces its effects primarily by displacing oxygen. Air normally contains approximately 20 percent oxygen, 0.03 percent carbon dioxide, while the rest is primarily composed of nitrogen. Carbon dioxide concentrations of 2.5-5 percent in the air cause headaches, irritation of the upper respiratory tract, an increase in the heart and breathing rate along with an increase in blood pressure. At ten percent concentration there are visual disturbances, tremors, and loss of consciousness in about one minute. Respiratory arrest occurs as a result of breathing air containing 20 percent or more carbon dioxide.

Caution should be exercised when entering any confined space, particularly if an asphyxiant like carbon dioxide may be present. The lack of smell or taste creates a particularly treacherous situation, since there is no forewarning. Cases of carbon dioxide intoxication have been associated with the cleaning of fermentation tanks, sewer pipes, and wine casks. Fatalities have occurred as a result of entering a cellar containing rotten potatoes, manholes, and cargo holds of fishing vessels, where decaying fish release toxic gases such as hydrogen sulfide, ammonia, and carbon dioxide. Reports from 1966 to 1971 indicated that there were three deaths and four cases of unconsciousness aboard Danish commercial fishing vessels, all due to **asphyxia**.

Estimated Carbon Monoxide Concentration	Carboxy-hemoglobin (%)	Symptoms[a]
Less than 35 ppm (cigarette smoking)	5	None, or mild headache
0.005% (50 ppm)	10	Slight headache, dyspnea on vigorous exertion
0.01% (100 ppm)	20	Throbbing headache, dyspnea with moderate exertion
0.02% (200 ppm)	30	Severe headache, irritability, fatigue, dimness of vision
0.03-0.05% (300-500 ppm)	40-50	Headache, tachycardia, confusion, lethargy, collapse
0.08-0.12% (800-1,200 ppm)	60-70	Coma, convulsions
0.19% (1,900 ppm)	80	Rapidly fatal

[a]Symptoms and signs of toxicity do not always correlate with measured carboxyhemoglobin level.

Table 7-2: Symptoms associated with exposure to different concentrations of carbon monoxide. *Recreated with permission from Ho M.T. and Saunders C.E. editors, Current Emergency and Diagnosis Treatment, 3rd edition, Appleton & Lange, Norwalk CT, 1990.*

> Asphyxia is an extreme condition characterized by lack of oxygen to the body.

Measurement of the atmosphere in the holds of the ships indicated that the average carbon dioxide concentration was 20 percent, reducing the oxygen concentration to only seven percent (from Zenz, C. et al., *Occupational Medicine*, 3rd edition, Mosby-Year Book, Inc., St. Louis, 1994).

Nitrogen Oxides – NO_x

Nitrogen dioxide – NO_2
—TLV = 3 ppm
—NFPA: 3H, 0F, 1R

Nitric oxide – NO
—TLV = 25 ppm
—NFPA: 3H, 0F, 3R

Nitrous oxide – N_2O
—TLV = 50 ppm

It is estimated that 300,000 tons of nitrogen oxides are produced annually from industrial processes; fuel combustion adds 10 million tons to that. The collective oxides of nitrogen (NO_x) include: nitric oxide (NO), which is a colorless gas; nitrogen dioxide (NO_2), which is a reddish-brown or dark orange gas; nitrogen trioxide (N_2O_3), which is a colorless gas; nitrogen tetroxide (N_2O_4), which is colorless; nitrogen pentoxide (N_2O_5); and nitrous oxide (N_2O), which is colorless and often referred to as laughing gas.

Nitrogen oxides are produced from a variety of sources. The combustion of any nitrogen-containing substance will result in the production of one or more oxides of nitrogen. Specific sources include combustion of fuels in furnaces and internal combustion engines, detonation of explosives, welding, and tobacco smoke. Diesel exhaust may contain up to 1,100 ppm of nitric oxide, which is equivalent to 0.13 pounds per gallon of fuel or four grams per mile, if fuel consumption is 15 miles/gallon. New automobiles are required by federal regulations to emit no more than 0.31 grams per mile. Smoke from a single cigarette contains 200-650 ppm nitrogen oxides, while smoke from a pipe contains up to 1,100 ppm. The decomposition of plant material will also produce nitrogen oxides. For example, nitrogen dioxide levels up to 1,500 ppm have been produced by the decomposition of plant material in enclosed silos.

The most prevalent forms of nitrogen oxides in the air are nitric oxide and nitrogen dioxide. Nitric oxide is a relatively reactive substance that rapidly oxidizes to nitrogen dioxide when released into the atmosphere. The reaction occurs more slowly if the nitric oxide concentration is less than one ppm. Because of these factors most health effects are expressed in terms of exposure to nitrogen dioxide.

Nitrogen dioxide – which is what gives smog its brown color – can be detected by its odor or taste at a concentration as low as one ppm. General symptoms of exposure include coughing, labored breathing, chest pain, pulmonary edema, cyanosis, **tachypnea**, tachycardia, and eye irritation.

> Tachypnea means rapid breathing; tachycardia means rapid heart rate.

The specific symptoms observed are dependent on the level of exposure. Chest discomfort can occur at concentrations as low as 15 ppm; it increases and may begin to become painful at 25 ppm. Eye irritation may also occur at this level. Pulmonary edema, with possible subacute or chronic lesions in the lungs as well as irritation of the eyes and throat occur following exposure to 50 ppm. One hour of exposure to 100 ppm nitrogen dioxide or a brief exposure to 250 ppm both result in death due to pulmonary edema or **bronchopneumonia**.

> Bronchopneumonia is a swelling of the lungs and bronchioles characterized by chills, fever, and chest pain.

Concentrations of 100 ppm nitric oxide and 50 ppm nitrogen dioxide are considered to be IDLH.

Symptoms of exposure to higher concentrations (250 ppm) may occur in two stages. During the initial stage individuals may develop pulmonary edema, tachypnea, cyanosis, and breathing characterized by crackling and wheezing sounds. In some individuals only labored breathing and coughing may develop initially, and then the symptoms begin to subside. Several weeks later these individuals may enter a second stage characterized by fever and chills, increased dyspnea, cyanosis, and pulmonary edema. Death may occur at the end of either stage. Pathological changes in the lung are associated with both stages. Lesions are associated with extensive mucosal edema and inflammatory **exudate**.

> Exudate is a fluid that has moved out of the tissue or its capillaries due to injury or inflammation.

The inflammatory exudate in the small bronchi and bronchioles undergoes a fibrinous ("thickening") organization, which finally blocks or obliterates the lumen. The result is decreased diffusion capacity and lung volume, which contributes to the labored breathing.

Sulfur Oxides – SO_x

Sulfur dioxide
—TLV = 2 ppm
—NFPA: 3H, 0F, 0R

Sulfur oxides are a major source of air pollutants. They are produced from automobile exhaust, smelters, combustion of coal and oil high in sulfur content, petroleum refineries, paper manufacturing, and chemical industries. Like nitrogen oxides there are two types of sulfur oxides: sulfur dioxide (SO_2) and sulfur trioxide (SO_3).

Sulfur dioxide is the primary air pollutant of concern. It has a sharp, pungent odor and reacts with water to form sulfurous acid. Sulfur dioxide can also be oxidized forming sulfur trioxide gas, which forms sulfuric acid when dissolved in water. Both sulfurous and sulfuric acids are components of acid rain. Sulfur dioxide is also involved in reactions between organic compounds and nitrogen oxides resulting in the formation of particulates, which – along with the droplets of acid formed by sulfur dioxide and trioxide – is responsible for reduced visibility.

The high water solubility of sulfur dioxide and the subsequent formation of sulfurous and primarily sulfuric acid explains the irritating effect these substances have on the eye, the upper respiratory tract, and the skin. Approximately 90 percent of inhaled sulfur dioxide is absorbed in the upper respiratory tract, where most of the toxicological effects occur. Exposure to high concentrations may result in the lower air passageways being affected, causing chemical **pneumonitis** and pulmonary edema. Concentrations up to one ppm have little or no effect on the respiratory tract of most individuals. Usually only a slight increase in the rate of breathing may begin to be observed as the concentration increases above one ppm. A sharp odor or taste can be noticed when the concentration approaches 3-5 ppm. At 8-12 ppm eye and throat irritation begin to be noticed. Eye, nose, and throat irritation, choking, cough, and broncho-constriction have been shown to occur in individuals exposed for 5-15 minutes to 10-50 ppm sulfur dioxide. Extremely high concentrations of sulfur dioxide can cause pulmonary edema and death within hours of exposure. Lungs from exposed individuals show extensive damage to the mucosa lining the air passages as well as edema and bleeding.

Individuals who recover from the initial exposure to sulfur dioxide may develop additional symptoms two to three weeks later, in the same way as individuals exposed to nitrogen oxides. Symptoms like difficulty in breathing and chest congestion occur, resulting in an association with the development of pulmonary fibrosis and the destruction of the bronchial epithelium. Edema develops as a result of tissue damage, which can be fatal at this stage. Concentrations of 100 ppm sulfur dioxide are considered to be IDLH.

Ozone – O_3

—TLV = 0.1 ppm

Ozone is a colorless gas and its odor can be detected at a concentration as low as 0.01 ppm. It is formed as a result of the interaction between organic compounds – such as ketones, aldehydes, and unsaturated hydrocarbons – and nitrogen oxides in the presence of sunlight. It is also formed by any other high-energy source – such as lightning, arc welding, high voltage electrical equipment (600 volts), and air- and water-purifying devices.

Ozone is one of the major air pollutants in heavily industrialized areas and in cities with heavy vehicular traffic. More than half of all ozone precursors can be attributed to vehicle traffic. Formation of ozone from organic compounds and nitrogen oxides in conjunction with sunlight peaks during early midday and begins to decrease during late afternoon and evening.

Typical symptoms associated with exposure to low concentrations of ozone (0.05-0.1 ppm) are irritation of eyes, nose, throat, and lungs. This occurs after only 10-30 minutes of exposure. At progressively higher concentrations, coughing and impairment of lung function occurs. Concentrations of 0.8 ppm or greater result in a marked increase in respiratory tract irritation, chest pain, and dyspnea. Pulmonary edema can occur along with bronchopneumonia.

Individuals who suffer from pulmonary diseases such as **asthma** and emphysema are more sensitive to lower levels of ozone – 0.1-0.15 ppm. There is an increase in asthmatic attacks accompanied with labored breathing. IDLH level is 10 ppm.

How ozone produces its toxic effects is uncertain. It is an oxidant and reacts quickly with organic material. It may interact with molecules within tissues in a manner similar to ionizing radiation. Ionizing radiation produces free radicals capable of disrupting chemical bonds of structural and functional molecules, thereby disrupting normal cell function. Chromosomal damage is also known to be caused by both ionizing radiation and ozone.

Particulates

Particulates arise from a variety of sources such as vehicle emissions, smokestacks, or blowing dust. The particles may be large enough to be seen with the naked eye (diameter >50 µm) or they may be microscopic in nature. The size of the particle is one of the important factors in determining the possible toxic effects on the respiratory system. Most particles 5 µm in diameter and greater are removed in the nasopharyngeal area of the respiratory tract. Particles 1-5 µm in diameter are removed from the air in the region from the trachea to the bronchioles. Only particles 1 µm or less in diameter actually reach the alveoli. These particles may cause localized damage. Asbestos fibers are capable of causing cancer, silica particles may cause a disease referred to as silicosis, and coal dust may cause pneumoconiosis. In the last two cases, alveolar tissue is damaged and is replaced with fibrous connective tissue leading to fibrosis. If the damage is severe enough, diffusion of oxygen is impaired; the resulting breathing difficulties could require supplemental oxygen on a permanent basis.

Chemical composition of the particulate also plays an important role in determining toxicity. The particulates may be composed of various types of inorganic and organic substances that may produce systemic effects in addition to localized effects within the respiratory tract. Heavy metals such as lead, cadmium, and nickel are often associated with particulates. Solubility appears to be an important characteristic in determining toxicity. Cadmium oxide and nickel oxide, for example, are less soluble in the mucosa lining the respiratory tract, while the chlorides and sulfates of these metals are more soluble.

Sulfur dioxide and chlorine are highly soluble gases and as such are usually removed in the upper portion of the respiratory tract. However, if these gases interact with particles, they form a thin film that coats the particle. If the particles are small enough, they are carried to the lower portion of the respiratory tract where they can damage the alveoli. Acidic aerosols interact with particles in a similar manner.

Indoor Air Pollution

The accumulation of indoor air pollutants in homes and offices has contributed to the development of various types of illnesses. These illnesses are collectively referred to as **building-associated illnesses (BAI)**. BAIs include **building-related illnesses (BRI)**, **tight-building syndrome (TBS)** and **sick-building syndrome (SBS)**.

Building-related illnesses have known etiological agents that are usually aerosolized and easily enter the lower portion of the respiratory tract. Examples of these types of agents include organic housedust, microorganisms, viruses, spores, pollen, and mites. Specific illnesses caused by these agents include asthma, which can be caused by pollen, mites, and spores; Legionnaire's disease caused by the bacterium *Legionella pneumophila*; and viruses, which cause diseases such as influenza and the common cold.

Heating, ventilation, and air conditioning systems (HVAC) as well as humidifiers provide the appropriate environment to produce and distribute the various agents associated with BRIs. Various types of molds, bacteria, and fungi reproduce in stagnant pools of water sometimes present in this type of equipment. The HVAC or humidifier in use can disperse these organisms in a fine mist that can be inhaled by any occupants in the building. Organisms found in these systems have been identified as the causative agents associated with asthma and Legionnaire's disease.

Tight-building syndrome and sick-building syndrome are similar in terms of the types of effects associated with each. Individuals complain of headaches; irritation of the eyes, nose, and throat; dry skin; dizziness and nausea; difficulty in concentrating and fatigue. The cause of these symptoms is thought to differ between the two syndromes. TBS is thought to be caused by reduced ventilation in well-insulated buildings. SBS, on the other hand, may be due to emission of odors, vapors, or gases from the materials used to construct the building and its furnishings. The substances responsible for causing either of these illnesses are difficult to identify. Also, the level of suspected agent may be up to a 1,000 times smaller than any of the published standards.

There are several sources and activities that contribute to what is referred to as indoor air pollution.

In many instances the level of the toxic substance is extremely low and can not be measured. As a result, only a few limits have been established for indoor air pollutants such as formaldehyde, asbestos, and **radon**. Various types of pollutants may be released from building materials used in construction. Paints and adhesives used during construction are sources of volatile organic compounds including toluene, alkylbenzenes, alcohols, methyl ethyl ketone, and styrene. Other activities or sources of potential air pollutants include the use of copy machines, humidifiers, air conditioner filters, and various types of cosmetics, as well as cleaning, polishing, and pesticide spraying.

Formaldehyde is a common indoor air pollutant. Urea- and phenol-formaldehyde glues have been used to produce laminated wood products, which in turn are used to build cabinets and furniture. Other formaldehyde-containing products are used for insulation in homes, in floor coverings, and in carpet backings. Incomplete curing of the product results in the release of formaldehyde fumes for years. Formaldehyde is also a contaminant of tobacco smoke.

Radon is another indoor air pollutant of concern. It is present in building materials, particularly concrete, but also occurs naturally in soils and drinking water. The level of radon in the home is related to the amount present in these sources and the ventilation in the home. Radioactive decay of radon results in the release of **alpha particles**, which – when inhaled in significant quantities – may cause lung cancer. Radon is also associated with smoking tobacco products.

Carbon monoxide is also a concern as an indoor air pollutant in improperly ventilated buildings, although many homes are now equipped with carbon monoxide detectors. Carbon monoxide is a byproduct of the burning of fossil fuels such as coal or natural gas and is also one of the components of tobacco smoke.

Checking Your Understanding

1. What effect does carbon monoxide have on the oxygen-carrying capacity of the red blood cells? Why?

2. What is meant by the term simple asphyxiant?

3. Explain the two stages of response in individuals exposed to nitrogen and sulfur oxides.

4. What is the source of ozone and what are some of the toxic effects produced as a result of exposure?

5. What is the diameter of particles that reach the alveoli? What illnesses may develop as a result of exposure to particulates?

6. What are building-associated illnesses? Distinguish between the various types.

7-3 Petroleum Distillates

Petroleum distillates are used in a variety of industries, such as agriculture, manufacturing, and petroleum production. The individual chemicals can be grouped into several subcategories including halogenated hydrocarbons, aromatic hydrocarbons, and other distillates such as kerosene, petroleum ether, as well as Stoddard solvent and mineral spirits. The various chemicals found in these groups have different physical, chemical, and toxicological characteristics. The following discussion provides information relative to some of the chemicals classified as petroleum distillates.

In general, petroleum distillates are fat soluble and therefore able to readily penetrate the skin causing drying and defatting. Exposure to these substances is responsible for 20 percent of the occupational **dermatitis** cases. However, the primary effects of petroleum distillates are depression of the central nervous system and pulmonary edema when aspirated into the lungs. Contact with the eye may cause irritation and burning.

Halogenated Hydrocarbons

Chloroethanes – $C_2H_nCl_n$

Monochloroethane – C_2H_5Cl
—TLV = none
—NFPA: 2H, 4F, 0R

Ethyl chloride – C_2H_5Cl
—TLV = 1,000 ppm
—NFPA: 2H, 4F, 0R

1,1-Dichloroethane – $C_2H_4Cl_2$
—TLV = 100 ppm
—NFPA: 2H, 3F, 0R

1,2-Dichloroethane (Ethylene dichloride) – $C_2H_4Cl_2$
—TLV = 10 ppm
—NFPA: 2H, 3F, 0R

Hexachloroethane – C_2Cl_6
—TLV = 1 ppm (S)

1,1,1-Trichloroethane (Methyl chloroform) – $C_2H_3Cl_3$
—TLV = 350 ppm
—NFPA: 3H, 1F, 1R

1,1,2-Trichloroethane – $C_2H_3Cl_3$
—TLV = 10 ppm (S)
—NFPA: 3H, 1F, 1R

1,1,1,2-Tetrachloroethane – $C_2H_2Cl_4$
—TLV = none

1,1,2,2-Tetrachloroethane – $C_2H_2Cl_4$
—TLV = 1 ppm (S)

Carbon Tetrachloride – CCl_4
—TLV = 5 ppm (S)
—NFPA: 3H, 0F, 0R

Dichloromethane (Methylene chloride) – CH_2Cl_2
—TLV = 50 ppm
—NFPA: 2H, 1F, 0R

Tetrachlorethylene (Perchloroethylene) – C_2Cl_4
—TLV = 25 ppm
—NFPA: 2H, 0F, 0R

Trichloroethylene – C_2HCl_3
—TLV = 50 ppm
—NFPA: 2H, 1F, 0R

The chloroethanes are chlorinated derivatives of ethane. Monochloroethane is a gas, hexachloroethane a solid, and the remaining chloroethanes listed above are liquids. Approximately five billion kg are produced annually in the United States. Several of the chloroethanes, such as 1,2-dichloroethane, tetrachloroethane, and trichloroethane, are produced and used in large quantities because of their excellent solvent properties. The rubber, plastic, and insecticide industries use 1,2-dichloroethane as a solvent. Tetrachloroethane is used as a solvent in a variety of industries and may be present as a contaminant in other chlorinated hydrocarbons. It may also be present in some household cleaners. 1,1,1-trichloroethane is an excellent degreaser, which is also used as a cleaning agent found in paint removers.

All the chloroethanes have high lipid solubility and a distinctive odor. Their primary effect is depression of the central nervous system. Symptoms include muscle weakness, intoxication, restlessness, irregular respiration, lack of muscle coordination, and unconsciousness. Chloroethanes may also cause liver and kidney damage. Two of the primary thermal decomposition products are hydrogen chloride and phosgene, which are more toxic than the chloroethanes themselves. The following information is a summary of principal characteristics associated with some chloroethanes.

Ethyl Chloride and Monochloroethane – These two names refer to the same substance, ethyl chloride being used for the liquid form and monochloroethane for the gaseous form. In both forms it is highly flammable and has an ether-like odor. In the past monochloroethane was used as a general anesthetic. Exposure to either the gas or the vapors of the liquid may result in irritation of the eyes and respiratory tract. Like other chloroethanes, high concentrations depress the central nervous system. At high doses there may be liver and kidney damage. IDLH concentration is 20,000 ppm.

1,1-Dichloroethane – 1,1-Dichloroethane is a flammable, colorless, oily liquid with a chloroform-like odor that begins to be noticeable around 100 ppm. When subjected to heat, vinyl chloride, hydrogen chloride, and phosgene are produced. Dichloroethane is mildly irritating to the eyes, skin, and respiratory tract. It depresses the central nervous system and may cause arrhythmia. Kidney and liver damage may also occur. IDLH concentration is 4,000 ppm.

1,2-Dichloroethane – 1,2-Dichloroethane is a colorless, flammable liquid that is readily absorbed through the skin. Contact with the skin and eye for prolonged periods of time may cause irritation and burns. The vapors may cause irritation to the eyes and respiratory tract. Similarly to 1,1-dichloroethane, this substance depresses the central nervous system at high concentrations. It is also capable of causing arrhythmia as well as liver and kidney damage. IDLH concentration is 1,000 ppm.

Hexachloroethane – Unlike the other chloroethanes discussed in this section, hexachloroethane is a noncombustible, white solid with a camphor-like odor. Heat causes it to decompose resulting in the production of chlorine, hydrogen chloride, and phosgene. These fumes are irritating to the eyes, skin, and mucous membranes, causing depression of the central nervous system at high doses. Kidney and liver damage may also occur. IDLH concentration is 300 ppm.

1,1,1-Trichloroethane – 1,1,1-Trichloroethane is a colorless, noncombustible liquid widely used as a solvent. Heat causes it to decompose into hydrogen chloride and phosgene. Trichloroethane was used initially for its anesthetic properties until better products were discovered. It is commonly referred to as methyl chloroform, which implies an anesthetic effect on the central nervous system. Trichloroethane is absorbed by the skin and, upon repeated exposure, may result in dermatitis and defatting reactions. It is irritating to the eyes and respiratory tract. Depression of the central nervous system occurs at high doses, but the anesthetic effect does not occur until concentrations of 10,000 to 26,000 ppm are reached in the surrounding air. While the odor is detectable at levels as low as 500 ppm, olfactory fatigue occurs, which may result in prolonged exposure due to the inability to smell the vapors. Cardiac arrhythmia may also occur along with some kidney and liver damage. IDLH concentration is 1,000 ppm.

1,1,2-Trichloroethane – This isomer is about twice as toxic than 1,1,1-trichloroethane. It is a colorless liquid that has a sweet, chloroform-like odor. 1,1,2-Trichloroethane is not combustible, but when subjected to extreme heat it will decompose producing phosgene and hydrochloric acid. The toxic effects of this isomer are similar to 1,1,1-trichloroethane. IDLH concentration is 500 ppm.

Tetrachloroethanes – There are two isomers of tetrachloroethane, 1,1,1,2-tetrachloroethane and 1,1,2,2-tetrachloroethane. 1,1,2,2-Tetrachloroethane is the more toxic of the two and the most toxic of all the chlorinated hydrocarbons. Therefore, the properties described will be primarily attributed to 1,1,2,2-tetrachloroethane. It is a colorless to light yellow liquid with a sweet, suffocating, chloroform-like smell. It is not combustible, but will decompose at high temperatures releasing hydrogen chloride and phosgene. 1,1,2,2-Tetrachloroethane is readily absorbed through the skin and may cause systemic toxicity. The vapors are irritating to the eyes and respiratory tract and may cause nausea and headaches. At high concentrations the substance depresses the central nervous system, producing a coma 1-4 hours after exposure. It may cause cardiac arrhythmia. Kidney and liver damage may be accompanied by symptoms of jaundice and **oliguria**. IDLH concentration for 1,1,2,2-tetrachloroethane is 150 ppm.

Carbon Tetrachloride – Carbon tetrachloride is a colorless liquid with an ether-like odor. The use of carbon tetrachloride has decreased dramatically because of its carcinogenic properties. Initially it was used as a solvent for dry cleaning, a degreaser and spot remover, and as an ingredient in fire extinguishers. Currently it is used primarily as a fumigant and a chemical intermediate in a variety of chemical processes. Like other chlorinated hydrocarbons, thermal decomposition results in the formation of hydrochloric acid and phosgene. Carbon tetrachloride depresses the central nervous system. Typical symptoms from inhalation or skin absorption include nausea, vomiting, dizziness, and confusion. Exposure may also cause cardiac arrhythmia. These symptoms may occur as a result of exposure to 160 ppm carbon tetrachloride for 30 minutes. Liver and kidney damage may occur as a result of exposure. Carbon tetrachloride is metabolized to chemical intermediates such as epoxides, which are more toxic than the parent compound. The chemical intermediates are responsible for the cellular damage observed as a result of exposure to carbon tetrachloride. IDLH concentration is 300 ppm.

Dichloromethane – This is a solvent used in paint and varnish removers, in degreasers, and in the polymer industry. One of its earliest uses was as an anesthetic, but it produced fatalities and was replaced. Thermal decomposition results in the formation of phosgene, chlorine, and hydrogen chloride. Dichloromethane exposure is primarily by inhalation of vapors. It may also be absorbed to some extent when in contact with the skin. The primary effect is the depression of the central nervous system. Typical symptoms of headache, dizziness, nausea, and vomiting are observed. Exposure to high concentrations causes pulmonary edema, cardiac arrhythmia, and respiratory arrest. Like other solvents its is irritating to mucous membranes, the eyes, and skin. Extensive exposure to the skin can result in chemical burns. IDLH concentration is 5,000 ppm.

Tetrachloroethylene – This is a colorless liquid used as a solvent in commercial dry cleaning and degreasing activities. Exposure can be detected as a result of the ether-like or chloroform-like odor and its eye irritating property. Exposure to vapors may result in irritation of eyes and the respiratory tract. It also depresses the central nervous system. Peripheral nerve damage may occur and is characterized by tingling, numbness, and muscle weakness. Inhalation may also produce cardiac arrhythmia. Short-term memory loss may occur as a result of chronic exposure. Skin contact will cause irritation and may result in chemical burns. IDLH concentration is 500 ppm.

Trichloroethylene – This is a colorless liquid with a sweet, chloroform-like odor. It is used as an industrial solvent and is found in household cleaning products for walls, clothing, and rugs. Also, when trichloroethylene is mixed with alkalis it releases carbon monoxide, dichloroethylene, and phosgene. Exposure results in depression of the central nervous system; the individual develops the typical symptoms of headache, dizziness, nausea, and vomiting. Exposure to trichloroethylene produces acute ventricular arrhythmia, with cardiac arrest being the eventual cause of death. Continuous exposure or repeated exposure may result in cranial and peripheral neuropathies. IDLH concentration is 1,000 ppm.

Aromatic Hydrocarbons

Benzene – C_6H_6
—TLV = 10 ppm
—NFPA: 2H, 3F, 0R

Toluene – C_7H_8
—TLV = 50 ppm (S)
—NFPA: 2H, 3F, 0R

Xylene – C_8C_{10}
—TLV = 100 ppm
—NFPA: 2H, 3F, 0R

Benzene – Benzene is a component of a variety of chemicals. As many as two million workers may be exposed to this chemical. It is used as a solvent in inks, paints, and plastics. It is a constituent of gasoline and a number of other petroleum-based products. Benzene is a clear, pleasant-smelling, volatile chemical. The odor threshold is around two ppm. Toxic effects are produced primarily as a result of inhalation. Because of the high fat solubility, one of the primary target tissues of benzene is the central nervous system. Symptoms include headache, nausea, and dizziness, which are generally noticed immediately after inhalation. Exposure to concentrations of 3,000 ppm result in a state of euphoria followed by drowsiness, fatigue, nausea, and vomiting. Exposure to 7,500 ppm of benzene for 30-60 minutes may result in respiratory paralysis and death. Aspiration of benzene into the lungs results in chemical pneumonitis or pulmonary edema. Chronic exposure affects the hematopoietic system, primarily the blood cell formation in the bone. Leukemia has been associated with exposure to benzene. Other blood disorders associated with chronic exposure include anemia and thrombocytopenia.

Toluene – Like benzene, toluene, is primarily used as a solvent in paints, cleaners, and degreasers. It is a volatile component of some glues and is responsible for the "sniffer's high" that is produced upon continual inhalation of glue vapors. Occupationally, the largest single source of exposure to toluene is during the production of gasoline. The primary route of exposure is by inhalation, diffusing from the lungs into the bloodstream and subsequently into the central nervous system. Only small amounts are absorbed through the skin. Symptoms associated with inhalation exposure are similar to those associated with exposure to benzene and include euphoria, dizziness, headache, nausea, and weakness. Higher concentrations – such as those experienced by glue sniffers – can cause delirium, coma, pulmonary edema, respiratory arrest, and consequently death due to cardiac arrhythmia. Table 7-3 lists symptoms associated with exposure to various concentrations of toluene. Exposure to 2,000 ppm of toluene is considered to be IDLH.

Xylene – This is a widely used solvent in paint removers, cleaners, inks, and degreasers. It is a clear liquid and may exist as one of three isomers: ortho-xylene (o-xylene), meta-xylene (m-xylene), or para-xylene (p-xylene). The isomer is identified by the relative position of the two sidechains associated with the benzene ring structure. Commercial grade xylene often contains toluene and benzene as contaminants. All three forms of xylene have the same toxicological properties. The primary route of exposure is by inhalation, although xylene is also absorbed through the skin. Inhalation affects the central nervous system resulting in the typical symptoms of headache, dizziness, **vertigo**, drowsiness, and unconsciousness. At high concentrations (10,000 ppm) it may cause death. Xylene has been shown to affect hematopoiesis; however, this may be attributed to benzene and toluene present in the xylene. Xylene is also a skin and eye irritant. Chronic exposure of the skin may result in defatting, cracking, dryness, and dermatitis. IDLH concentration is 10,000 ppm.

Other Petroleum Distillates

Kerosene

Kerosene is used as a weed killer, a pesticide carrier, a cleaning solvent, in thinners and varnishes, and for fuel. It is pale yellow or whitish in color, depending on the formulation. Kerosene is a mixture of hydrocarbons and in general contains 25 percent normal paraffins, 12 percent branched paraffins, 30 percent monocycloparaffins, 12 percent dicycloparaffins, one percent tricycloparaffins, 16 percent mononuclear aromatics (benzene) and five percent dinuclear aromatics. The composition percentage of kerosene is dependent on the production method and the source of the crude oil.

Kerosene is a skin irritant. The degree of irritation is determined by the type of kerosene, with the ones containing the highest percentage of paraffins being the least irritating. A burning sensation will be caused after one hour of exposure, while contin-

Exposure	Effect
37 ppm	Probably perceptible to most humans
50-100 ppm	Subjective complaints (fatigue or headache), but probably no observable impairment of reaction time or coordination
200 ppm	Mild throat and eye irritation
100-300 ppm	Detectable signs of incoordination may be expected during exposure periods up to eight hours
400 ppm	Lacrimation and irritation to the eyes and throat
300-800 ppm	Gross signs of incoordination may be expected during exposure periods up to eight hours
1,500 ppm	Probably not lethal for exposure periods of up to eight hours
4,000 ppm	Would probably cause rapid impairment of reaction time and coordination; exposures of one hour or longer might lead to narcosis and possibly death
10,000-30,000 ppm	Onset of narcosis within a few minutes; longer exposures may be lethal

Table 7-3: Acute effects in humans following short-term exposures to toluene. *Recreated with permission from Zenz C. et al.,* Occupational Medicine, *3rd edition, Mosby Co., St. Louis, 1994.*

ued exposure results in **erythema** and blistering. Continuous dermal exposure to kerosene for three years has also caused anemia and general depression of all blood cell types. However, these effects have been attributed to the benzene content of kerosene. The greatest danger associated with kerosene lies in aspiration into the lungs. Inhalation and deposition in the lungs is usually associated with the development of pneumonitis. Also, several studies have shown that as little as 0.1-0.2 ml of kerosene placed directly in the trachea may cause death.

Petroleum Ether

—TLV = 400 ppm
—NFPA: 1H, 4F, 0R

Petroleum ether is a colorless, highly flammable liquid with a kerosene-like odor, used as a component of various types of glues and adhesives. It is also used as a solvent for oils, fats, and waxes; as a fuel; in paints and varnishes; and in photography. Pentane and isohexane are the primary constituents with virtually no aromatics present. These two components have been studied to determine the toxic properties of petroleum ether. Irritation of the eyes and respiratory tract, along with depression of the central nervous system is consistent with exposure to other organic solvents. Irritation of the skin also occurs after exposure for 20 minutes, and blisters form after five hours. **Polyneuropathy** has been described in several studies as a result of workers being exposed to hexane.

Polyneuropathy is a generalized disorder of the peripheral nerves.

Hexane is metabolized to 2,5-hexanedione, which causes the peripheral neuropathy and may result in the loss of sensory perception in the fingers. A study showed that shoemakers exposed to the vapor of glue containing hexane developed sensory impairment in their extremities in a workplace with airborne concentration of hexane that ranged from 1,759-8,793 mg/m^3. In another study workers in a furniture store developed symptoms of abdominal cramps, numbness, and weakness of the extremities along with **paresthesia**.

Paresthesia is a condition characterized by a burning, itching, or tingling sensation in the skin with no apparent cause.

In this case the average airborne concentration of hexane was 2,286 mg/m^3. IDLH concentration of petroleum ether is 10,000 ppm.

Stoddard Solvent and Mineral Spirits

—TLV = 100 ppm
—NFPA: 1H, 2F, 0R

The chemical compositions of Stoddard solvent and mineral spirits are very similar. They are comprised of a mixture of hydrocarbons. Stoddard solvent contains 30-50 percent straight and branched chain paraffins, 30-40 percent naphthenes, and 10-20 percent aromatic hydrocarbons. Mineral spirits may contain 30-65 percent paraffins, 15-55 percent naphthenes, and 10-30 percent aromatic hydrocarbons. Because of their similar composition and toxicological properties these substances will be discussed as one and will be referred to as Stoddard solvent.

Stoddard solvent is a flammable, colorless liquid. It has a kerosene-like odor, and its irritating properties provide a good warning signal. There are four basic formulations of Stoddard solvent known as: regular Stoddard solvent, 140 flash solvent, odorless solvent, and low end-point solvent. The odor threshold is between 0.09 and 0.9 ppm, depending on the type of Stoddard solvent as well as on the individual exposed. It is possible to experience olfactory fatigue, which may result in prolonged exposure. Some dermal absorption may occur and dermatitis has been reported in workers after a two-week exposure. The vapors are irritating to the eyes and respiratory tract. Inhalation may result in depression of the central nervous system. Chronic exposure may cause headache, fatigue, and jaundice. IDLH concentration is 5,000 ppm.

The primary toxic effect of petroleum distillate compounds is depression of the central nervous system.

Checking Your Understanding

1. What are similarities and differences between 1,1,1,-trichloroethane and 1,1,2-trichloroethane?
2. What are the primary target tissues of solvents and the major toxic effects?
3. Why do solvents affect the tissue types identified in the previous question more readily than other tissue types?
4. What substance produces the effect known as "sniffer's high"?
5. Identify the toxic effects of breakdown products for chloroethanes. Based upon the information in this chapter choose three of the chloroethanes and determine if the breakdown products are more or less toxic than the parent compound.

7-4 Metals and Metalloids

Metals exist in rocks and soils that form the earth's crust. Natural events change their distribution in the biosphere. For example, erosion of rock and soil or volcanic activity can release the metals, which may then be carried great distances – by water or wind – from their points of origin. Also, plants absorb and bioaccumulate metals – which may be toxic – from the soil along with needed nutrients. Animals then eat these plants, metabolize them, and excrete the metal-containing waste somewhere else.

In some cases man's activities have altered these natural cycles, causing increased exposure to the toxic metals. As early as 2,000 B.C., for example, silver-smelting activities resulted in exposure to elevated levels of lead. In ancient Egypt mining of copper and tin to adorn the tombs led to the exposure of workers to arsenic. Subsequently, continued development of industry has resulted in an increased exposure to various types of metals such as cadmium, mercury, chromium, and zinc, just to mention a few.

During the early phases of the Industrial Revolution, the effects of exposure to many of these toxicants were not well known or documented. Following increased investigation and research a considerable body of knowledge has been developed concerning the effects of metals on various species of animals including man. Much of the early work addressed acute exposures and resulted in the establishment of exposure standards. Effects of chronic exposure to a single substance, but also effects of exposure to multiple toxic substances, require continued investigation to further define the possible detrimental effects.

The following discussion addresses some of the more frequently encountered metals. Information on the effects of exposure to other metals has been discussed throughout this book. Additional information can also be obtained from several of the references listed in the bibliography.

Arsenic

Arsenic – As
—TLV = 0.1 mg/m³ (arsenic compounds)
—NFPA: 3H, 2F, 0R

Arsenic trioxide – As_2O_3
—NFPA: 3H, 0F, 0R

Arsenic trifluoride – AsF_3
—NFPA: 3H, 1F, 0R

Arsenic acid – H_3AsO_4
—NFPA: 3H, 0F, 0R

Arsenic disulfide – As_2S_2
—NFPA: 3H, 1F, 1R

Arsine – AsH_3
—TLV = 0.05 ppm

Arsenic is a metalloid and the 20th most abundant element present in the earth's crust. It is present in high concentrations in areas of volcanic activity and geothermal regions. It rarely exists as a pure substance, and is normally found in a variety of different arsenic compounds. The inorganic forms include arsenic trioxide, sodium arsenite, arsenic trichloride, arsenic pentoxide, arsenic acid, lead arsenate, and calcium arsenate. Organic forms include monomethylarsonate, dimethylarsinate, and arsenobetaine.

It is estimated that worldwide 60,000 tons of arsenic are produced annually. In the United States approximately 900,000 workers may be occupationally exposed to this toxic substance. Exposure to arsenic occurs from a variety of sources. It is released into the environment as a byproduct of smelting activities; in glass manufacture; in pesticide manufacturing and application; in wood preservatives; and in the manufacture of semiconductors.

The effects of exposure to arsenic depend on the chemical state of the compound as well as the length of exposure. Exposure to arsenic trichloride can be identified from skinburns at the point of contact. Other symptoms of exposure include fatty degeneration in various organs as well as possible development of bronchopneumonia.

Arsine is a colorless gas with a slight garlic odor, used primarily in the semiconductor industry. This gas is extremely toxic causing extensive hemolysis (breakup of red blood cells). Initially, exposed individuals are lightheaded, giddy, and develop a headache. Arsine binds to the hemoglobin in the red blood cell resulting in hemolysis; kidney failure also occurs. Within a few hours after exposure blood begins to appear in the urine (hematuria). Approximately 12 hours after exposure symptoms of jaundice appear and the skin acquires a bronze-colored tone. Brief exposures to 250 ppm of arsine as well as more prolonged exposures (30 minutes) to 25-50 ppm of arsine are fatal.

Acute exposure to other arsenic compounds produces a variety of symptoms. Fever, vomiting, cramps in the extremities, **anorexia**, and **melanosis** may occur. Other symptoms include cardiac arrhythmia, convulsions, and paralysis, which may precede death.

> Melanosis is the unusual deposit of black pigment in different parts of the body.

The lethal dose for arsenic trioxide is 100 to 200 mg. If an individual survives the initial exposure other symptoms may develop such as peripheral neuropathy and changes in blood cell formation. Neurological effects may occur up to one or two weeks after exposure. Initially there may be a loss of peripheral sensory functioning, which may also be associated with some loss of motor skills. These symptoms will slowly subside after several weeks. Arsenic suppresses cellular formation in the bone marrow resulting in leukopenia, granulocytopenia, and anemia. The effects are reversible.

Chronic exposure to arsenic compounds is difficult to diagnose. The primary target tissues are the nervous system and the skin. Arsenic exposure leads to neurotoxicity of the peripheral and central nervous system. Paresthesia and muscle weakness develop over a prolonged period of time. Peripheral neuropathy of motor and sensory neurons may occur over several years.

Dermatitis, **hyperpigmentation**, and **keratosis** associated with the palms of the hands and the soles of the feet are characteristic of chronic arsenic exposure. White lines in the fingernails called **Mees' lines** are also characteristic of exposure to arsenic, both chronic and acute. Various forms of skin cancer – **basal cell carcinoma** and **squamous cell carcinoma** – are associated with chronic exposure to arsenic.

Cadmium

Cadmium – Cd
—TLV = 0.01 mg/m^3 (cadmium dust or fumes)

Cadmium chloride – $CdCl_2$
—NFPA: 3H, 0F, 0R

Cadmium oxide – CdO
—NFPA: 3H, 0F, 0R

Cadmium is a soft, white-bluish metal, a byproduct of the mining and smelting of lead- and zinc-containing ore. It is used in electroplating and galvanizing processes because it is resistant to corrosion. Cadmium and its compounds are also used in manufacturing of pigments for paints and plastics (e.g. cadmium sulfide, cadmium sulfoselenide), nickel-cadmium batteries, ceramics, different kinds of glass as well as silver solder, which may contain up to 25 percent cadmium.

Exposure primarily occurs as a result of ingestion or inhalation. Various types of meat, fish, and fruit may contain 1-50 µg/kg. Certain grains, such as rice and wheat, concentrate levels of cadmium up to 150 µg/kg. Various types of shellfish, oysters, and scallops are also efficient in concentrating cadmium (100-1,000 µg/kg). In uncontaminated areas the average daily intake is 15 µg for a person weighing 70 kg.

Airborne concentrations of cadmium in uncontaminated general areas are less than 0.1 µg/m^3, and even less in uncontaminated workplaces (0.05 µg/m^3 or less). Inhalation of cadmium is the most efficient route of exposure with up to 30 percent of inhaled cadmium being transported into the bloodstream. Cigarettes are a source of cadmium exposure: they may contain 1-2 µg of cadmium per cigarette – only 10 percent of the cadmium actually being inhaled. If a third of that inhaled cadmium is absorbed into the bloodstream, then 0.03-0.06 µg of cadmium enter the bloodstream per cigarette.

The observed effects of cadmium exposure are primarily dependent on the route of exposure and concentration. Acute ingestion of cadmium rarely occurs. Acidic foods or beverages that have been stored in cadmium-plated containers may leach cadmium from the container's surface. Consumption of these foods result in nausea, vomiting, and abdominal pain – usually occurring within minutes. Consumption of liquids containing 16 mg of cadmium per liter has been shown to initiate these symptoms. A dose of 300 mg may be fatal.

Acute exposure to airborne concentrations of cadmium may result in chemical pneumonitis and pulmonary edema. Other symptoms associated with exposure include irritation of the nose and throat, cough, dizziness, weakness, chills, fever, chest pains, and labored breathing. Inhalation of air containing 5 mg/m^3 cadmium oxide for eight hours may cause death. Chronic exposure of the lungs may cause a variety of different pathological conditions. Chronic bronchitis, fibrosis, or emphysema may develop depending on the length of exposure as well as the toxicant's concentrations.

Both chronic and acute exposure will result in concentration of cadmium in the kidneys. Cadmium is transported in the blood primarily as a cadmium-protein complex. The complex is absorbed and concentrated in the kidney tubules. Cadmium concentrations in the kidneys of 100-300 µg/g may cause damage to the cells of the kidney tubules resulting in kidney failure. An epidemiological study found that kidney failure occurred in individuals that had unknowingly ingested 140-260 µg of cadmium per day in their food for more than 50 years. Exposure for 10 years to airborne concentrations 50 µg/m^3 cadmium has produced similar results.

Cadmium may have some effect on bone metabolism disrupting normal calcium phosphate and Vitamin D metabolism. Exposed individuals may develop osteoporosis or **osteomalacia**.

> Osteomalacia is a disease that results from a deficiency of calcium or vitamin D causing softening of the bone.

The carcinogenic effects of cadmium are variable. Studies have indicated an increase of respiratory tract cancer in workers exposed to airborne cadmium oxide dust or fumes. Exposure to cadmium by ingestion does not appear to increase the number of cancer cases.

Chromium

Chromium – Cr
—TLV = 0.5 mg/m^3 (metal and Cr (III) compounds)
    0.05 mg/m^3 (water-soluble Cr (VI) compounds)
    0.01 mg/m^3 (insoluble Cr (VI) compounds)

Chromium oxychloride – CrO_2Cl_2
—NFPA: 3H, 0F, 1R

Chromic acid – CrO_3
—NFPA: 3H, 0F, 1R

Chromium is used in a number of manufacturing industries (Table 7-4). About one third of all chromium is used in the production of pigments; a fourth is used to treat metal surfaces and control corrosion; tanning activities account for another fourth of the use; and the rest is used in textile and other manufacturing.

Chromium exists in a variety of oxidation states. In nature it is found in chromite ore in the trivalent state (Cr^{3+}). In metals such as stainless steel the oxidation state is zero. In other industrial processes chromium may be found in oxidation states of Cr^{2+}, Cr^{3+}, Cr^{4+}, Cr^{5+} and Cr^{6+}. The oxidation state of chromium is important in determining whether toxic effects will be produced following exposure. The divalent, tetravalent, and pentavalent forms are unstable and pose little health risk. Most toxic effects are associated with exposure to the hexavalent and the less toxic trivalent forms.

The toxic effects observed are based on the specific types of interactions occurring between the trivalent and hexavalent forms and the tissues of the exposed individual. In terms of skin, the trivalent form of chromium does not readily penetrate it, while the hexavalent form is easily absorbed. In the case of ingestion, hexavalent chromium is converted to trivalent chromium in the stomach; the trivalent form facilitates absorption of chromium in the intestine and its clearance from the body by the kidneys. However, the hexavalent form more readily passes through the cell membrane of various tissue types such as the lungs, where intracellular absorption results in conversion of the hexavalent chromium to trivalent chromium.

Exposure occurs primarily as a result of contact with the skin or by inhalation of mists, fumes, or dust-containing chromium. Dermal contact with chromium-containing substances is the primary cause of contact dermatitis in industry. Sensitization to chromium remains for years after initial symptoms have disappeared.

Hexavalent chromium is extremely corrosive to the skin and to the mucous membranes of the nose.

Chromium-containing Materials or Objects	Job or Places of Contact	Chromium Compounds Responsible
Chromium ore	Industrial chromium production	Chromite
Chrome baths	Electro-plating industry Graphic trade Metal industry	Chromic acid, sodium dichromate Chromates Chromates, zinc chromate
Chromate colors and dyes	Painters and decorators, graphic trades, textile, rubber, glass and china industries	Chromic oxide green, chromic hydroxide green, chrome yellow (lead chromate)
Lubricating oils and greases	Metal industry	Chromic oxide
Anticorrosive agents in water system and greases	Diesel locomotive workshops and sheds, central heating and air-conditioning systems	Alkali dichromates
Wood preservation (Wolman salts)	Wood impregnation, furniture industry, carpenters, miners	Alkali dichromates
Cement, cement products, quick-hardening agents for cement (e.g., Sika 1)	Cement production, manufacture of cement products, building trades	Chromates
Cleaning materials (Eau de Javel washing and bleaching materials)	Homemakers, cleaners, laundry workers	Chromates
Textiles, furs	Textile and fur industries, everyday life	Chromates
Leather and artificial leather tanned with chromium	Leather and footwear industry everyday life	Chromium sulfate, chromium alum

Table 7-4: Examples of contact with chromium compounds. *Recreated with permission from Zenz C. et al.*, Occupational Medicine, *3rd edition, Mosby Co., St. Louis, 1994.*

Exposure results in what is referred to as **chrome ulcers** or chrome holes. The outer layers of skin undergo necrosis leaving an open wound, which allows penetration into the underlying connective tissue, but not usually all the way to the bone. Often the open wounds become infected by bacteria. Healing is a slow process and often results in the formation of scar tissue.

Acute inhalation exposure may cause perforation of the nasal septum within one week after exposure. Other respiratory tract symptoms are **sinusitis** and bronchial constriction. Chromium – primarily the hexavalent form – is considered to be a potential lung carcinogen

> Sinusitis is the inflammation of the epithelial lining of the nose.

The carcinogenic potential of hexavalent chromium appears to be related to the solubility of the substance. Insoluble hexavalent chromium compounds like thallium dichromate pose little carcinogenic risks. The substance is not readily absorbed across the cell membrane of lung tissue. The evidence supporting the role of highly soluble substances such as chromium trioxide (CrO_3) as a carcinogen is not conclusive. Initially, it was believed that highly soluble substances were rapidly absorbed into the bloodstream and excreted by the kidneys before significant cellular effects could occur. However, recent information suggests that even these types of chromium compounds may have a role in producing lung cancer. The third group of chromium compounds that are most likely to cause lung cancer are those that are only slightly soluble in tissue fluids. These compounds, like calcium chromate, dissolve slowly, thereby releasing chromium over a longer period of time, enhancing the chance that chromium will be absorbed into the pulmonary tissue cells.

Lung cancer is primarily associated with exposure to the hexavalent forms of chromium. The hexavalent form is more readily absorbed across the cell membrane than the trivalent form. However, once inside the cell the hexavalent form is converted into the trivalent form – which is believed to be involved in carcinogenesis – either in the cell cytoplasm or in the nucleus. If the hexavalent form is converted in the cytoplasm then the probability of cancer is lower than if the conversion occurs in the nucleus, because once in the nucleus it can interact with DNA and result in a mutation.

Lead

Lead – Pb
—TLV = 0.05 mg/m^3 (elemental lead and inorganic lead compounds)

Lead arsenate – $Pb(AsO_2)_2$
—NFPA: 2H, 0F, 0R

Lead arsenite – $PbAs_2O_4$
—NFPA: 2H, 0F, 0R

Lead chromate – $PbCrO_4$
—TLV = 0.05 mg/m^3 (lead), 0.012 mg/m^3 (chromate)

Lead fluoroborate – PbB_2F_8
—NFPA: 0H, 1F, 1R

Lead nitrate – $Pb(NO_3)_2$
—NFPA: 1H, 0F, 1R

Lead thiocyanate – $Pb(SCN)_2$
—NFPA: 1H, 1F, 1R

Tetramethyl lead – $(CH_3)_4Pb$
—TLV = 0.15 mg/m^3 (S)
—NFPA: 3H, 3F, 3R

Lead could arguably be characterized as one of the most widely used and broadly distributed toxic substances known. Its use may have begun as early as 4,000 B.C., with its toxic effects first being documented as early as 370 B.C. Lead is a bluish or silvery-gray soft metal. Soil concentrations from around the world average 16 ppm. Lead may be found in either the organic or inorganic form. Organic forms of lead of primary interest are tetraethyl lead and tetramethyl lead. Inorganic forms include lead acetate and lead nitrate, which easily dissolve in cold water. Lead chloride, lead chromate, and lead stearate are somewhat soluble in cold water. Other inorganic lead compounds such as lead carbonate, lead oxide, and lead sulfate are practically insoluble in water, but are soluble in some solvents and fats.

The primary use of lead is in the manufacture of lead-acid batteries, which accounts for nearly two-thirds of all the lead consumed in the United States. In the past lead was used in paint products, in pipes for home plumbing, and in the production of tanks. Paints containing inorganic lead compounds may pose a significant health problem for small children. Although the use of lead-based indoor paints has been banned in the United States, there are still more than 50 million homes that have them on interior walls. Lead-based paints are still used for exterior surfaces and may contain lead carbonate, lead sulfate, lead oxide, or lead silicates. Red lead (Pb_3O_4) is an important component of paints and is used as a metal primer or as a rust retardant. Lead compounds are also used in the manufacture of polyvinyl chloride (PVC) plastics, china, ceramics, and various types of glass.

The two forms of organic lead, tetraethyl and tetramethyl lead, were introduced as an additive in gasoline in the 1920s and 1950s, respectively. The use of lead in the production of petroleum products was second only to the use in the production of batteries, but today, with the production of unleaded gas, there has been a steady decline in the amount of lead used for this purpose.

Exposure to lead is the result of either inhalation or ingestion of lead-contaminated substances. Ingestion of lead may occur as a result of eating food from ceramic dishes, if a lead-containing glaze was applied to them and if they were improperly fired. Also, lead may be ingested when eating canned acidic foods such as tomatoes, tomato juice, pickles, or wine cider, that have leached lead from the solder used to weld the can. In Mexico and other countries in Central America lead poisoning from these sources is not uncommon.

Plants grown in lead-containing soils may concentrate the metal. Consumers usually ingest 20 μg of lead a day. Lead contained in drinking water is another possible source of ingestion. Lead leaches from pipes or solder if the water is soft, slightly acidic, or has been in contact with surfaces containing lead for a long period of time. One person typically ingests 10 mg a day in drinking water. As mentioned previously, small children may be exposed to lead by ingesting chips of lead-based paint. Lead emissions from motor vehicles are an important source of exposure by inhalation.

The toxic effects of lead are numerous and are associated with a number of different systems (Table 7-5). The most studied effects of lead exposure are those associated with the nervous system. In children, exposure to lead at an early age can have a profound effect on the development of the nervous system. The primary outcome of lead exposure is decreased mental development indicated by a decrease in the intelligence quotient (IQ) scores. Exposure also results in the development of **encephalopathy** (dysfunction of the brain) characterized by fluid accumulation in the brain (edema), damage to blood vessels in the brain, and structural changes in the neurons (loss of the myelin sheath). Most of the damage occurs in the outer layer of the brain, the **cerebral cortex**. In addition to decreased

Effect	Blood Lead Concentration (μg/dL)	
	Children	Adults
Neurological		
Encephalopathy	80-100	100-12
Hearing deficit	20	–
IQ deficits	10-15	–
In-utero effects	10-15	–
Peripheral neuropathy	40	40
Hematological		
Anemia	80-100	80-100
Renal		
Nephropathy	40	
Blood pressure (males)		30
Reproduction		40

Table 7-5: Lowest observed effect levels for lead-related health effects. *Recreated and modified with permission from Klaasen C. D. et al.*, Casarett & Doull's Toxicology, The Basic Science of Poisons, *5th edition, McGraw Hill, New York, 1996.*

mental development symptoms such as lethargy, other symptoms such as vomiting, irritability, loss of appetite, dizziness, and ataxia may occur. Coma and death may occur with acute exposure. Other long-term effects include **epilepsy**, mental retardation, and possible blindness. More subtle symptoms include poor learning ability and behavioral disorders such as aggressiveness, destructive behavior, and hostility.

In adults the primary target of lead is the peripheral nervous system. The motor axons responsible for muscle contraction are primarily affected. The neurons undergo demyelination and axonal degeneration. **Footdrop** and **wristdrop** are characteristic symptoms associated with these neurological effects. Acute exposures may cause encephalopathy characterized by convulsions and coma. Individuals who recover from this type of exposure have a decrease in intelligence and experience mood swings. Chronic exposure to lead may produce similar effects on the central and peripheral nervous systems.

Toxic effects on the kidneys may occur following both short-term acute exposures and long-term chronic exposures. Short-term acute exposures primarily affect the proximal tubules of the kidneys, where reabsorption of amino acids, glucose, and phosphate occur. Acute exposures cause these substances not to be readily reabsorbed and to be excreted in the urine. Failure to reabsorb these substances from the proximal tubules appears to be related to the effects of lead on the mitochondria in the tubule cells and the ability to produce energy (ATP) necessary to support cellular mechanisms associated with reabsorption. The acute effects are reversible. Chronic exposure has similar effects and occurs after 10 years or more of consistent exposure. Kidney damage is characterized by renal tubular degeneration, by changes in the structure of the blood vessels associated with the kidneys, and by general tissue fibrosis. These effects are generally not reversible and lead to death as a result of kidney failure. Decreased kidney function may also result in a condition known as **saturnine gout**. As a result of kidney dysfunction, **uric acid** – which is normally excreted from the blood by the kidneys – concentrates resulting in uric acid crystals being formed and deposited in the joints.

Ingestion is one of the primary routes of exposure to lead. Symptoms include loss of appetite as well as gastrointestinal tract irritation associated with diarrhea or constipation. Lead **colic** develops with continued acute exposures. Colic is characterized by severe and sharp contraction of the smooth muscle of the intestine. Similarly, the smooth muscle associated with the arterioles also contracts resulting in **pallor** of the face and a temporary increase in blood pressure.

Lead line of the gums is another well-known symptom of lead exposure. It appears as a dark line on the border of the gum and is formed by precipitation of lead sulfide. Anemia is often associated with lead exposure. The causes of anemia are twofold: 1) inhibition of heme synthesis, and 2) accelerated destruction of red blood cells. Heme is the iron-containing nonprotein portion of the hemoglobin molecule. Decreased availability of the heme molecule affects the formation of functional red blood cells. The average life of the red blood cell is usually 120 days, after which it is removed from circulation and replaced by new red blood cells. Lead accelerates the death of the red blood cell by inhibiting enzymes responsible for maintaining the integrity of the red blood cell membrane. As a result, the red blood cell undergoes premature hemolysis.

High blood pressure appears to be due to 1) the direct effects of lead on the smooth muscle found in the walls of the blood vessels and 2) the toxic effects of lead on the kidneys. In the first condition, lead appears to affect the calcium metabolism associated with smooth muscle contractions by decreasing Na^+/K^+ ATPase (an enzyme) activity and stimulating the Na^+/Ca^{2+} exchange pump. The net effect is that the smooth muscle in the walls of the arterioles seems to be more sensitive to stimuli resulting in contraction. In the second condition, exposure to lead results in the release of renin from the kidneys. Renin interacts with another chemical in the blood stream called angiotensin I. This reaction results in the formation of angiotensin II, which stimulates the smooth muscles of the arterioles causing high blood pressure.

Lead has been shown to be carcinogenic in laboratory animals. However, the doses used exceeded established exposure levels for lead. Seven thousand lead workers were studied in the United States and no statistically significant increase in cancer was identified. An analogous study in England found similar results.

Mercury

Mercury – Hg
—TLV = 0.01 mg/m^3 (alkyl mercury compounds)
    0.1 mg/m^3 (aryl mercury compounds)
    0.025 mg/m^3 (inorganic compounds including metallic mercury)

Mercury compound, solid
—NFPA: 3H, 0F, 0R

(Acetate) phenyl mercury
— TLV = 0.01 mg/m^3 (S)
—NFPA: 3H, 1F, 0R

Mercury compound, liquid
—NFPA: 3H 0F, 0R

Mercuric chloride
— TLV = 0.1 mg/m^3
—NFPA: 3H, 0F, 0R

Like lead, mercury has received much attention because of its toxic effects, which vary according to the physical state of mercury. At room temperature, elemental mercury is a silvery liquid; thus the names liquid silver and quicksilver. Liquid mercury volatilizes at room temperature, releasing mercury vapor.

Elemental mercury is found in the environment primarily as cinnabar ore. The ore is heated to 500°C producing a mercury vapor that condenses to form liquid mercury. Other natural sources of mercury in the environment are from hot springs and from volcanoes degassing. These activities may release from 25,000 to 150,000 tons per year. Mercury is also a component of fossil fuels and is released when these substances undergo combustion. Burning of fossil fuels and emissions from other industrial processes such as mining and smelting may contribute up to 20,000 tons of mercury to the environment annually.

Mercury and its compounds are used in a variety of products and manufacturing processes. Approximately 50 percent of all mercury is used in electrical products such as fluorescent tubes and mercury vapor lamps. Elemental mercury is also found in thermometers, blood pressure equipment, barometers, automatic thermostats as well as in extractive gold and silver mining operations. Inorganic mercury compounds are found in paints as an inhibitor of bacterial and mildew growth. They may also be found in floor wax, furniture polish, and fabric softeners. Organic mercury compounds have been used as seed dressings or as fungicides to prevent molding.

The toxic effects of mercury vary according to the chemical composition. In general, mercury affects primarily the nervous system and the kidneys. Exposure can be by all three routes: dermal, ingestion, and inhalation.

Metallic Mercury – Exposure results primarily from mercury vapors that are given off at room temperature. Up to 80 percent of the mercury vapors are retained in the lungs. The vapors readily diffuse across the alveolar membrane into the bloodstream. They have a high affinity for red blood cells and tissues of the central nervous system because of their high lipid solubility. Acute exposure to mercury vapors can cause corrosive bronchitis and pneumonitis, which can be fatal. Long-term effects may develop after recovery, such as tremors and hyperexcitability, which directly result from the effect of mercury on the central nervous system. Chronic exposure to mercury vapors occurs more often than acute exposure because of the lack of knowledge concerning airborne mercury vapors. Chronic mercury poisoning is often referred to as **mercurialism**. This condition may be characterized initially by any of the following symptoms: tremor, enlargement of the thyroid, tachycardia, **labile** (irregular) pulse, **gingivitis**, and increased excretion of mercury in the urine. As the disease progresses there are tremors of the muscles that control fine-motor movement. Therefore, uncontrolled tremor of the fingers, eyelids, and lips may occur. This can progress to include the larger muscle groups resulting in muscle spasms in the arms and legs. Behavioral changes indicative of the effects on the brain also occur: individuals may experience a loss of memory, increased excitability, depression, or delirium.

Mercuric Salts – Mercuric chloride has been the choice of those who have attempted suicide. It is extremely toxic and corrosive and as little as one gram may be fatal. Ingestion of this substance results in severe abdominal cramps, bloody diarrhea; and ulceration and necrosis of the gastrointestinal tract. This usually results in shock and circulatory collapse, which causes death. If the individual does recover, renal failure may occur due to destruction of the cells comprising the proximal tubular epithelium. But this condition, if not treated, may also lead to death. The cell destruction is the result of mitochondrial dysfunction and rupture of the cell membrane. Chronic exposure to inorganic mercury may also cause autoimmune disease (glomerular nephritis), which affects the glomerulus of the kidneys. This may also lead to kidney failure.

Methyl Mercury – Methyl mercury is the most toxic form of mercury. Effects include paresthesia, which is characterized by tingling and numbness in the extremities, mouth and lips; ataxia; generalized weakness; fatigue and inability to concentrate; possible loss of hearing or eyesight; tremors; and possible coma and death. Methyl mercury affects specific parts of the brain, the cerebral cortex, and the **cerebellum**.

> The cerebellum is a portion of the brain that is involved in the control and coordination of voluntary (skeletal muscle) muscular movement.

The neurons in these regions undergo necrosis. The result is destruction of the brain's gray matter and cerebral atrophy.

Exposure of the developing fetus to methyl mercury may result in impaired development of the brain. Mercury affects the migration pattern of the neurons during brain development. This produces misalignment of the neurons in various portions of the brain. Symptoms include cerebral palsy or psychomotor retardation.

Nickel

Nickel – Ni
—TLV = 1 mg/m^3 (nickel metal and insoluble compounds)
    0.1 mg/m^3 (soluble nickel compounds)

Nickel nitrate – Ni(NO$_3$)$_2$
— TLV - 0.1 mg/m^3
— NFPA: 1H, 0F, 1R

Nickel sulfate – NiSO$_4$
— TLV - 0.1 mg/m^3
— NFPA: 1H, 0F, 0R

Nickel carbonyl – Ni(CO)$_4$
— TLV = 0.12 mg/m^3
— NFPA: 4H, 3F, 3R

Nickel is an important component of a variety of products because of its hardness and resistance to corrosion. Table 7-6 lists the various types of metal alloys. Other uses of nickel are listed in Table 7-7.

Nickel is found in the environment and is also the result of man's activities. Average air concentrations of nickel are 0.02 µg/m^3 in cities and 0.006 µg/m^3 in rural areas. These levels are primarily due to burning of fossil fuels. Higher airborne concentrations of nickel are found near smelters and refineries where nickel-containing products are produced.

Ingestion of nickel-contaminated food also contributes to the exposure to this substance. Nickel can be found in tea, cocoa, and baking soda. Average daily ingestion is estimated to be between 120-60 µg/day. A cigarette contains 1-3 µg of nickel. Exposure to nickel is primarily by inhalation of nickel-containing dust or, in the case of nickel carbonyl, inhalation of vapors. Table 7-8 lists various occupations with exposure to nickel. Table 7-9 lists some of the health effects as a result of exposure.

The primary health concern associated with exposure to nickel is the development of cancer, primarily of the respiratory tract. Nasal cancer and lung cancer occur most often. But inhalation of nickel carbonyl vapors produces other effects as well. Exposure to 30 ppm nickel carbonyl for 30 minutes can be fatal. Initial symptoms include headache, fatigue, weakness, nausea, and vomiting. More severe symptoms such as shortness of breath, chest pain, and breathing difficulties may occur 12-36 hours after initial exposure. Pneumonitis may develop followed by respiratory failure. Dermatitis as a result of exposure to nickel is common. Initial sensitization may be the result of exposure by metal-containing products such as jewelry. Stainless steel cooking utensils may also be a source of initial exposure resulting in sensitization.

Types of Metals	Uses
Monels (66.5% nickel; 30% copper)	good resistance to corrosion; used for marine components, ship propellers, desalination plants
Nichromes (41.5% to 70% nickel; 15.5% to 50% chromium)	jet engine parts; furnaces; heating elements
Hastelloys (nickel/molybdenum)	resist oxidizing environments
Inconels and incoloys (nickel/iron/chromium, nickel 15.5% to 42%; 30% to 40%; 18.5% to 21.5%, respectively)	storage of liquified gases
Stainless steels (various levels of nickel)	variety of uses including storage of corrosive chemicals such as hydrofluoric acid

Table 7-6: Metal alloys containing nickel and the uses of these metals.

1. Production of stainless steel
2. Production of nickel alloys
3. Production of nickel cast iron
4. Electroplating and electroforming
5. Manufacture of alkaline batteries (nickel cadmium batteries)
6. Catalyst
7. Manufacture of coins - nickel and nickel/copper alloys are mainly used
8. Production of welding products including nickel electrodes and filler wire
9. Production of sintered components
10. Pigments
11. Electronics - nickel-containing compounds (ferrites) used in electronic and computer equipment

Table 7-7: Major uses of nickel. *Recreated with permission from Zenz C. et al.,* Occupational Medicine, *3rd edition, Mosby Co., St. Louis, 1994.*

Occupational Group	Airborne Nickel Exposure (mg/m^3)
1. Primary production of nickel	
Underground miners, sulfide mines	0.025
Open pit miners, laterite mines	0.05
Grinders and concentrating nickel	<0.05
Smelter workers	0.05-1.0
Nickel refinery workers	<1.0
Packaging nickel powders	<1.0
Discontinued calcining and sintering operations	25.30
Electrolytic tank house worker	0.02-0.3 (soluble forms)
Carbonyl refinery worker	<0.35 (gaseous form)
2. Uses of nickel	
High-nickel alloys	0.06-0.1
Manufacture and welding stainless steel	0.01-0.1
Nickel foundry workers	0.01-0.3
Electroplating of nickel	0.004-0.01 (soluble forms)
Nickel cadmium battery workers	0.4
Production of wrought nickel	1.5

Table 7-8: Occupational exposure to airborne nickel. *Recreated with permission from Zenz C. et al.*, Occupational Medicine, *3rd edition, Mosby Co., St. Louis, 1994.*

1. Increased risk of cancer of the lung
2. Increased risk of cancer of the nasal sinus
3. Increased risk of laryngeal cancer
4. Increased risk of gastric cancer
5. Increased risk of sarcoma
6. Chronic irritation of the upper respiratory tract, manifested by rhinitis, sinusitis, perforation of the nasal septum, and loss of the sense of smell (anosmia)
7. Pulmonary irritation and pulmonary fibrosis
8. Pneumoconiosis
9. Bronchial asthma
10. Increased susceptibility to respiratory infections
11. Allergic contact dermatitis
12. Acute toxic effects from exposure to nickel carbonyl

Table 7-9: Reported health effects from occupational exposure to nickel and nickel compounds. *Recreated with permission from Zenz C. et al.*, Occupational Medicine, *3rd edition, Mosby Co., St. Louis, 1994.*

Zinc – Zn

Zinc chlorate – $Zn(ClO_3)_2$
— NFPA: 2H, 0F, 2R

Zinc chloride – $ZnCl_2$
— TLV = 1 mg/m³ (Zinc chloride fumes)
— NFPA: 2H, 0F, 2R

Zinc is an essential element needed by the human body to function normally. Up to 50 percent of ingested zinc is absorbed. Absorbed zinc is required primarily in the functioning of intracellular enzymes necessary for normal cell function. The average daily intake of zinc is 12-15 mg, primarily in meat, seafood, dairy products, nuts, and legumes.

Toxic effects from ingested zinc are not observed until 200 mg or more are ingested in a single dose. Typical symptoms include vomiting, nausea, and diarrhea. Inhalation of zinc oxide fumes may produce **"metal fume fever."** The effects are primarily associated with the respiratory system. Symptoms may begin to appear 4-8 hours after exposure and last 24-48 hours. Typical symptoms include chest pain, coughing, chills, and fever (101-102°F). The effects are not fatal and usually last two days. Zinc chloride, which is corrosive, may cause ulceration of exposed skin. Solder fumes may contain zinc chloride, which may cause irritation of the mucous membranes. In addition, severe exposure may cause pulmonary edema and respiratory collapse.

Checking Your Understanding

1. What are the primary effects of exposure to arsenic?
2. How could you distinguish between exposure to arsine and other arsenical compounds?
3. Identify two tissue types affected by cadmium. What are the symptoms, and how does cadmium produce its effects?
4. What oxidation state of chromium is most important in terms of toxicity? Why?
5. What role does solubility play in carcinogenicity caused by chromium?
6. Identify three different organs affected by exposure to lead. What are the symptoms of lead exposure associated with each?
7. What is saturnine gout and how is it caused?
8. What form of mercury is the most toxic?
9. Define mercurialism and identify the symptoms associated with this illness.
10. What are the primary effects of exposure to nickel?
11. What is "metal fume fever"?

7-5 Other Important Chemicals

Ammonia – NH_3

—TLV = 25 ppm,
—NFPA: 2H, 1F, 0R

Fifteen million metric tons of ammonia are produced annually in the United States. It is the fourth most widely used chemical, ranking after sulfuric acid, lime, and oxygen. It is used in fertilizers, insecticides, as a refrigerant, and in household and commercial cleaning products. Its primary use is in the production of chemical fertilizers, which represents 20 percent of the annual use. Household cleaning agents usually contain 5-10 percent ammonia. Commercial cleaning solutions may contain 25-30 percent or more. Mixing solutions containing ammonia and chlorine or hypochlorite results in the liberation of chloramine gas, which has irritating properties similar to ammonia and chlorine.

Ammonia is a colorless gas with high water solubility, which results in the formation of a corrosive solution. This solution has direct irritating effects on the tissues of the respiratory tract, the eyes, and the skin. The odor of ammonia is usually detectable around 50 ppm. Irritation of the eyes, throat, and nose usually occurs around 100 ppm, resulting in progressively more severe effects on the respiratory tract. Inhalation of concentrations from 1,000-6,500 ppm may cause vomiting, swelling of the lips, and **conjunctivitis**, corneal irritation, dyspnea, temporary blindness, pulmonary edema, cyanosis, and tachycardia accompanied by decreased blood pressure. Bronchitis and pneumonia may also occur as a result of exposure. IDLH concentration is 500 ppm.

> Conjunctivitis is the inflammation of the mucous membrane – the conjunctiva – that lines the eyelid.

Hydrogen Fluoride (HF) Gas

—TLV = 3 ppm
—NFPA: 4H, 0F, 0R

Anhydrous hydrogen fluoride is a colorless, highly irritating gas with a pungent odor. It is used in some chemical manufacturing, uranium processing, enamel stripping, glass and quartz etching as well as electroplating. It is also a byproduct of cryolite manufacturing, which is used in the aluminum industry. Hydrogen fluoride gas is irritating to the respiratory tract and eyes because of its solubility in water. The effects may be immediate or delayed. Chronic exposure to two ppm hydrogen fluoride may cause nasal irritation as well as a slight stinging sensation in the eyes and on the skin of the face. In human subjects, conjunctival and respiratory irritation occurs after one minute of exposure to 120 ppm hydrogen fluoride. Exposure to 50-250 ppm hydrogen fluoride for five minutes may be fatal.

Symptoms usually occur in two stages. Initially, irritation of the respiratory tract and other mucous membranes occurs. This may be accompanied by coughing, choking, and chills, which can last for 1-2 hours after exposure. Then the symptoms regress and the individual may be asymptomatic for several days. The second stage may be characterized by fever, cough, tightness in the chest, cyanosis accompanied by pulmonary edema, or chemical pneumonia leading to death. IDLH concentration is 20 ppm.

Phosgene ($COCl_2$)

—TLV = 0.1 ppm
—NFPA: 4H, 0F, 0R

Phosgene is a gas used primarily in the manufacturing of isocyanates. It is also used in the manufacturing of dyes and other organic chemicals. It forms a liquid at 8°C. Approximately 400,000 kilograms are produced annually in the United States. It can also be produced as a result of the thermal decomposition of chlorinated hydrocarbons such as carbon tetrachloride, chloroform, and methylene chloride. Fatalities have been recorded from exposure to phosgene produced during fires when substances such as solvents, paint removers, and dry cleaning solutions thermally decompose. Phosgene has also been used as a chemical warfare agent.

Phosgene acts as a respiratory irritant. Once it enters the respiratory tract it is hydrolyzed and forms hydrochloric acid. The acid produces irritating symptoms and damages the cells lining the respiratory tract. Eye irritation and conjunctivitis may occur at concentrations around 1-2 ppm. Concentrations of 3-5 ppm cause coughing and irritation of the throat. Higher concentrations produce a burning sensation in the throat, vomiting, chest pains, and dyspnea. Individuals exposed to the higher concentrations of phosgene should be observed for a period of time after exposure because of the delayed onset of pulmonary edema. The symptoms may be delayed up to 72 hours depending on the length and concentration of exposure. Brief exposure to 50 ppm is lethal and is characterized by symptoms such as dyspnea, cyanosis, rapid onset of pulmonary edema, and subsequent respiratory and circulatory failure. IDLH concentration is 0.2 ppm.

Formaldehyde – HCHO

— TLV = 0.3 ppm
— NFPA: 2H, 4F, 0R

Formaldehyde is widely used throughout the United States for processing of paper, fabrics, and wood products. It is a component of adhesives used in the manufacture of construction material such as particleboard, plywood, fiberboard, and urea foam insulation. Aqueous solutions (formalin) are used as bactericides and tissue preservatives. It is estimated that about 1.5 million workers are potentially exposed to formaldehyde annually. An untold number of individuals are exposed to low levels of formaldehyde in the form of indoor air pollution. Formaldehyde is slowly released from products manufactured with formaldehyde. For instance, individuals in clothing stores are exposed to formaldehyde being released from clothing treated with formaldehyde-containing crease-resistant resins. Office buildings, houses, and mobile homes built with the construction materials mentioned above also release formaldehyde.

Formaldehyde is a flammable, colorless gas with a pungent odor. The odor threshold for some individuals is as low as 0.05 ppm. The primary effect of formaldehyde is irritation of mucous membranes and eyes at concentrations of less than one ppm. Symptoms include **lacrimation**, **rhinitis**, and **pharyngitis**.

> Lacrimation is the secretion and discharge of tears or crying. Rhinitis is the inflammation of the nasal cavity. Pharyngitis is the inflammation of the pharynx.

Formaldehyde causes precipitation of cellular protein, which can disrupt normal cellular function and result in cell death. It is soluble in water and forms formic acid, which is thought to be partially responsible for some of the toxic effects. Chronic exposure at low concentrations may be difficult to detect since it produces adaptation. On the other hand, the odor and the irritating effect of formaldehyde make it difficult to withstand concentrations of three ppm even for short periods of time. High concentrations of 50-100 ppm may result in pulmonary edema, hemorrhaging, and death.

> Formaldehyde may cause cell dysfunction and death by forming formic acid and precipitation of intracellular proteins.

The air level considered IDLH is 100 ppm. Table 7-10 summarizes the acute effects of formaldehyde at various concentrations.

Formaldehyde is a carcinogen in laboratory animals and – based on epidemiological studies – it may be a carcinogen in humans. It has been associated with the occurrence of cancer of the brain, the nasopharynx, the nasal sinus, the oral pharynx, the lungs, as well as with leukemia. There is some suggestion that formaldehyde may have some effect on the reproductive system and development of offspring. Studies have indicated an association between exposure of individuals in the garment industry and increased incidence of inflammatory diseases of the reproductive tract, sterility, and low birth weight of offspring.

Effect	Airborne Formaldehyde Concentration, ppm
Odor detection	≥ 0.05-1.0
Eye irritation	≥ 0.01-2.0
Upper respiratory tract irritation (e.g., irritation of the nose or throat)	≥ 0.10-11
Lower airway irritation (e.g., cough, chest tightness, and wheezing)	≥ 5-30
Pulmonary edema, inflammation, pneumonia	≥ 50-100
Death	>100

Table 7-10: Acute effects of airborne formaldehyde exposure. *Recreated with permission from Zenz C. et al.,* Occupational Medicine, *3rd edition, Mosby Co., St. Louis, 1994.*

Asbestos

— TLV = 2 fibers/cc

Other forms of asbestos fibers:
Amosite
— TLV = 0.5 fibers/cc

Chrysotile
— TLV = 2 fibers/cc

Crocidolite
— TLV = 0.2 fibers/cc

Asbestos has a number of unique properties. It has a high melting point; it is resistant to degradation when exposed to heat and chemicals; and it has low heat conductivity. Asbestos is composed primarily of silica, bound to various other elements such as magnesium, iron, and chromium, in the form of fibers. There are six types of asbestos: chrysotile, which represents 95 percent of all asbestos production; crocidolite; amosite; anthyophyllite, tremolite, and actinolite. It has been used in the manufacture and production of a variety of products (Table 7-11). Exposure as a result of mining, manufacturing, and handling of asbestos-containing products occurs in several occupations (Table 7-12). The use and manufacture of products containing asbestos has steadily declined in the United States, but is increasing in developing countries.

The presence of asbestos should not cause undue concern in terms of toxicity: the risk is minimal as long as the asbestos remains bound within the manufactured product. However, when airborne asbestos fibers are inhaled, they may cause development of pulmonary and **pleural** fibrosis, lung cancer, and mesothelioma.

> The term pleural refers to the thin membrane (pleura) that covers the exterior of lungs and the inner wall of the chest.

These diseases usually do not occur until several years after exposure, sometimes 15-20 years later. Toxic effects are associated with fibers that are invisible to the eye and penetrate the lung tissue. Penetration is more dependent on fiber diameter than length; however, the TLV standard is based upon particles longer than $5\mu m$ in length. Exposure to a specific type of asbestos may be associated with a specific type of disease. Crocidolite is less likely to produce fibrosis, lung cancer, and mesothelioma than amosite, but

Paper Products	Textiles
Flooring felt Roofing felt Gasket paper Pipeline wrap Rollboard and millboard Electrical insulation Thermal insulation Specialty papers	Asbestos filter pads Spray-on cement Fireproofing Acoustic tile
Asbestos Cement	**Friction Materials**
Piping and fitting Shingles Flat and corrugated sheeting	Brake linings Pads and blocks Clutch facings
Paints and Coatings	**Floor Coverings**
Adhesive Sealants Spackling	Vinyl-asbestos tile Felt-backed vinyl tile

Table 7-11: Asbestos-containing products. *Recreated with permission from Rosenstock R. and Cullen M. R.*, Textbook of Clinical Occupational and Environmental Medicine, *Harcourt Brace/W. B. Saunders, Philadelphia, 1994.*

Acoustic product installers	Electrical wire makers
Asbestos cement product makers and users	Food processing workers
	Gasket makers
Asbestos grout makers and users	Glass workers
	Iron ore (taconite) miners and millers
Asbestos millboard makers and users	Insulation workers
Asbestos millers	Loggers
Asbestos miners	Machinery makers
Asbestos paper makers and users	Maintenance and custodial workers
Asbestos plaster makers and users	Oil and gas extraction workers
Asbestos insulators	Paint makers
Asbestos tile makers and installers	Petroleum refinery workers
	Primary metal industry workers
Asphalt mixers	
Automobile repair workers	Plumbers and pipe fitters
Boiler makers	Railroad repair workers
Beer makers	Roofers
Brake lining makers	Rubber makers
Brake refabricators	Reinforced plastics makers
Chemical workers	Shipyard workers
Clay workers	Stone workers
Construction workers	Talc miners and workers
Demolition workers	Textile workers
Electric appliance workers	Transportation equipment makers and repairers
Electrical equipment workers	Transportation workers

Table 7-12: Occupations with potential asbestos exposure. *Recreated with permission from Rosenstock R. and Cullen M. R.*, Textbook of Clinical Occupational and Environmental Medicine, *Harcourt Brace/W. B. Saunders, Philadelphia, 1994.*

more likely than chrysotile asbestos. Pulmonary and pleural fibrosis is more prevalent in workers exposed to amosite asbestos. Mesothelioma is primarily associated with exposure to crocidolite.

Pulmonary and Pleural Fibrosis – The term asbestosis has been used to include both pulmonary and pleural fibrosis and will thus be used in the subsequent discussion. It is not exactly clear how asbestos causes fibrosis. Inhaled asbestos fibers are either removed by the filtration processes associated with the upper respiratory tract; or they reach the lower portions of the respiratory bronchioles and alveoli. In these regions the fibers can be engulfed by macrophages. In some circumstances the macrophages may dissolve the fiber, or they may engulf the fiber which is then removed from the respiratory tract by the mucociliary mechanism.

Fibrosis is indicative of cellular damage and stimulation of a repair process characterized by the formation of connective tissue. This process may occur when asbestos fibers are engulfed by macrophages. Phagocytosis has been shown to release chemical substances from the macrophages, which interact with the cells of the alveoli and respiratory bronchioles. The interaction results in damage to the lung tissue. As a result, a series of cellular events occurs in response to the damage, ultimately resulting in the synthesis and formation of fibrous connective tissue to replace the damaged tissue.

Some fibers are able to migrate outside the respiratory tract and affect the connective tissue surrounding the lungs and chest cavity. The asbestos fibers cause small areas of the connective tissue to thicken as a result of increased synthesis. The thickening can be an indicator of asbestos exposure, although other factors may cause a similar response. Also, there is some debate as to whether the thickening has any effect on respiratory functioning.

The primary effects of asbestosis are decreased diffusion capacity and impaired gas exchange. This results in labored breathing and a chronic persistent cough. It may also cause clubbing of the fingers, which is indicative of a decrease in the amount of oxygen reaching the tissue extremities. Smoking seems to increase the probability that an individual will develop fibrosis as a result of exposure to asbestos. This is probably related to the effects of toxins present in tobacco smoke and their effect on the mucociliary mechanism of the respiratory tract. The toxic substances decrease ciliary activity, making removal of asbestos fibers from the respiratory tract more difficult.

Lung Cancer, Mesothelioma, and Other Cancers – Studies in the 1930s, 40s, and 50s identified asbestos as a carcinogenic agent, but, it is uncertain how asbestos causes cancer. Some studies suggest it is a true carcinogen independent of any other factors, while other studies suggest it is a promoting agent that causes cancer in conjunction with other substances. One of the suggested mechanisms of action is suppression of natural killer cells of the immune system. These are specialized lymphocytes, present throughout the body, that perform the function of destroying foreign cells such as cancer. Others have suggested that various components of the asbestos fiber such as chromium and nickel – which are known carcinogens – may be involved. There is considerable evidence to suggest that fibrosis must occur first before cancer develops.

Mesothelioma is a rare form of cancer and its occurrence is indicative of exposure to asbestos. This form of cancer is associated with the tissues of the pleura and **peritoneum**.

> Peritoneum is the thin connective tissue membrane that covers the organs and the inner wall of the abdominal cavity.

Symptoms are similar to those of lung cancer: chest pain, dyspnea, and cough. Lung cancer due to asbestos exposure is treatable; however, there is no known successful treatment for mesothelioma.

Cancer of the stomach, of the liver, and of the small intestine have all been associated with asbestos exposure among other pathogens. The primary route of exposure has been attributed to clearance of asbestos fibers from the lungs and into the digestive tract through swallowing.

Chlorine – Cl_2

—TLV = 0.5 ppm
—NFPA: 3H, 0F, 0R

Chlorine is a yellowish-green gas with a distinctly irritating odor. It is highly reactive and is not found in a free form under normal conditions. Chlorine is used for disinfection in water and wastewater treatment facilities and also for bleaching in the paper and pulp industry. Production of plastics, resins, chlorinated solvents, ethylene glycol antifreeze and ethylene chloride, and antiknock additive in gasolines, all use chlorine in the manufacturing process. Household bleaches, refrigerants, pharmaceuticals, as well as the extraction of ores also use chlorine. During World War I it was used as a chemical warfare agent. Sodium hypochlorite (NaClO) is found in household bleach solutions (3-5 percent) as well as in swimming pool disinfectants and industrial strength cleaners, which contain up to 20 percent hypochlorite. When mixed with toilet bowl cleaners, solutions containing sodium hypochlorite may cause the release of and chlorine gas and other products from the hypochlorite solution.

Chlorine is 20 times more potent than hydrogen chloride and may cause extensive damage to the respiratory tract. Chlorine gas is highly soluble in water and forms HCl when it comes in contact with moisture. Therefore, exposure results in rapid onset of symptoms in the eyes, nose, throat, and lungs characterized by an immediate burning sensation, which may be accompanied by throat irritation, coughing, and development of respiratory distress. Upper respiratory irritation may occur from exposure to 0.2-16 ppm chlorine; eye irritation may occur at 7-8 ppm, and a cough may develop at 30 ppm. Death has been caused at 1,000 ppm. In addition to the symptoms described above, acute exposure results in chest pain, labored breathing, bronchitis, and possibly pulmonary edema. If respiratory damage is severe enough secondary bacterial infection may occur leading to pneumonia.

Hydrogen Sulfide – H_2S

—TLV = 10 ppm
—NFPA: 3H, 4F, 0R

Hydrogen sulfide is a colorless, flammable gas that is heavier than air. It has a distinct "rotten-egg" odor, which may be detected at concentrations as low as 0.025 ppm. It is a byproduct of petroleum refineries, tanneries, pulp-making factories, and sulfur products processing. It is also produced from a number of "natural" sources such as sulfur hot springs, pools of sewage sludge, liquid manure, or any decaying organic matter. Because hydrogen sulfide is heavier than air it can be dangerous if it is present in low-lying areas or confined workspaces.

Hydrogen sulfide is absorbed immediately from the respiratory tract and has a mechanism of action similar to cyanide. It inhibits the cytochrome oxidase system resulting in cellular asphyxia. It also directly damages the central nervous system. Exposure to 50-100 ppm hydrogen sulfide causes respiratory tract irritation. Other symptoms include conjunctivitis, headache, nausea, cough, and dizziness. Loss of smell occurs after exposure to 100-150 ppm. Pulmonary edema occurs at levels between 300-500 ppm and death at 600-800 ppm. IDLH level is 300 ppm.

Hydrogen Cyanide – HCN

—TLV = 4.7 ppm, (S)
—NFPA: 4H, 4F, 2R

Sodium cyanide (NaCN)
—TLV = 5 ppm, (S)
—NFPA: 3H, 0F, 0R

Potassium cyanide (KCN)
—TLV = 5 ppm, (S)

Calcium cyanide ($Ca(CN)_2$)
—TLV = 5 ppm, (S)

Hydrogen cyanide is a colorless, flammable, and volatile gas that smells like bitter almonds. It is used in the production of resins, fibers, plastics, and various types of chemical products such as methyl amide, hydroxynitriles, and cyanogen chloride. It is a byproduct in electroplating, metallurgy, and photographic development. It is estimated that 40,000 short tons (1 short ton equals 907.2 kg) are used annually in the United States. Ten grams of tobacco may yield up to 2 ml of hydrogen cyanide in cigarette smoke. Hydrogen cyanide is also produced as a result of burning some plastics, wool, and other synthetic and natural products.

Sodium and potassium cyanides are solids and have a weak odor similar to hydrogen cyanide. Exposure to moist air results in the release of hydrogen cyanide. Sodium and potassium cyanides are used in the extraction of gold and silver ore, in soldering of metals, in copper plating and bronzing, in photography, and in lithographic printing. They are also used as fumigants for fruit trees, in ship cargo holds containing food items, and in railway cars.

In its purest form calcium cyanide is a fine, snow-white powder, while the technical grade form is grayish-black. It readily decomposes in water and emits hydrogen cyanide when it interacts with moisture in the air. Calcium cyanide is used primarily in agriculture as a rodenticide and fumigant, as well as for leaching of gold and other ores and in the manufacture of other cyanides.

Hydrogen cyanide and the cyanide salts are absorbed through the skin. The salts act as corrosives causing irritation of the skin and some discoloration. The mechanism of action for hydrogen cyanide and for the cyanide salts is the same. All cyanides are **chemical asphyxiants** that inhibit the utilization of oxygen by tissue cells. Although sufficient oxygen is available in the bloodstream, cyanide inhibits its use by the cells' mitochondria. This produces the same consequence as low blood-oxygen levels and results in cell anoxia, which can lead to cell death. Initially, this produces an increased respiratory rate due to the effects of cyanide on chemoreceptors in the **carotid bodies** and on the respiratory center in the brain.

> The carotid body is a chemical receptor in the carotid artery that is stimulated by lack of oxygen, excess carbon dioxide, or increased concentration of hydrogen ions – which corresponds to a lower pH – in the blood.

Additional symptoms include headache, nausea, and confusion. Inhalation of 270 ppm hydrogen cyanide can be fatal to humans instantly, while 135 ppm can cause death after 30 minutes. A concentration of 50 ppm of hydrogen cyanide is considered to be IDLH. Ingestion of 150-200 mg of sodium or potassium cyanide salts may cause death.

> Cyanide is a chemical asphyxiant that prevents cells of the body from using oxygen, leading to cell death.

Nitric Acid – HNO_3

—TLV = 2 ppm
—NFPA: 3H, 0F, 0R, OX (oxidizer)

Nitric acid is used in large quantities to produce fertilizer, dyes, and plastics. It is routinely used in chemical reactions by a variety of industries. Nitric acid is a strong oxidizing agent that decomposes in the presence of light producing highly toxic nitrogen dioxide gas.

Exposure is usually the result of inhalation of mist or vapors as well as skin contact. Nitric acid vapor or mist is irritating to the eyes, the mucous membranes, and the skin. Inhalation may cause pneumonitis and pulmonary edema. These conditions may be fatal, and symptoms may not occur until 4-30 hours after exposure. Contact with the eyes or skin can cause severe burns. Concentrated solutions in contact with the skin can produce deep ulcerations and turn the skin a bright yellow or yellowish-brown color. Contact with the eye may cause permanent corneal damage and impairment of vision.

Hydrofluoric Acid – HF

—TLV = 3 ppm
—NFPA: 4H, 0F, 0R

When hydrogen fluoride gas – discussed previously – is dissolved in water it forms hydrofluoric acid. HF is used in the manufacturing of semiconductors, silicon chips, glass etching; it is a component of some rust removers and is also used in uranium recovery processes. Hydrofluoric acid has the potential to cause severe injury and even death if immediate medical assistance is not provided after exposure. Even a prompt response may not prevent serious injury. Hydrofluoric acid is considered a weak acid, which means it produces few hydrogen and fluoride ions. But even a low concentration of fluoride ions has a toxic effect. HF penetrates the skin easily, entering the underlying tissue. The highly toxic fluoride ion is released and causes dysfunction of cellular processes. Part of the cell dysfunction is related to the fluoride ion's ability to precipitate calcium. If sufficient fluoride ion reaches the bone it may cause demineralization as well as systemic **hypocalcemia** (low blood calcium levels). Hypocalcemia has a direct effect on the functioning of the heart producing arrhythmia.

Contact with the skin is the primary route of exposure. The symptoms vary according to the concentration and duration of exposure. A weak burning sensation may be noticed as a result of exposure to solutions containing up to 40 percent of hydrofluoric acid. This is primarily because HF is a weak acid so there is little immediate tissue damage. Stronger solutions (50-70 percent) will produce pain more readily. Severe effects as a result of exposure to HF often take several hours to occur. These effects are characterized by destruction and ulceration of the skin, the underlying soft tissue, and – in some cases – the bone. These injuries are extremely painful and slow to heal. Symptoms may not occur for up to 12-24 hours after exposure to 5-15 percent HF solutions. Effects as a result of exposure to 20-40 percent HF may not occur for 1-8 hours. Stronger solutions (50-70 percent) produce almost immediate effects.

> The toxic effects of hydrofluoric acid are caused primarily by the interaction between the fluoride ion and calcium.

Skin that has been exposed to hydrofluoric acid should be thoroughly washed with water to decrease the cytotoxic effects of the fluoride ion; topical applications of calcium-containing substances should be applied immediately. The calcium will bind with the fluoride decreasing its toxic properties. Gels containing calcium gluconate or calcium carbonate are available and are often used for industrial exposures.

> Calcium-containing solutions are used to decrease hydrofluoric acid toxicity.

Deaths from exposure to hydrofluoric acid have been recorded. Three chemists exposed to a 70 percent solution of HF died – even after prompt flushing of the skin – two hours after exposure due to pulmonary edema. In a separate incident, a chemist who splashed hydrofluoric acid on his face and upper extremities developed pulmonary edema three hours after exposure and died 10 hours later.

Sulfuric Acid – H_2SO_4

— TLV = 1 mg/m^3
— NFPA: 3H, 0F, 2R

Sulfuric acid is produced in greater quantity than any other chemical in the Untied States: eighty-five billion pounds are produced annually. It is used in the production of a variety of other chemicals such as alcohol, ammonium sulfate, methyl methacrylate, hydrofluoric acid, and aluminum sulfate. Sulfuric acid is also used in the manufacturing of phosphate fertilizers, batteries, cellophane, and titanium dioxide.

Sulfuric acid is a nonflammable oily liquid that is soluble in water and alcohol. When sulfuric acid is mixed with water an exothermic reaction occurs that releases heat. Interaction with various types of metal can generate hydrogen, which is flammable in the presence of oxygen. Sulfuric acid – like many other corrosive substances – irritates the eyes, the mucous membranes, and the skin as well as the gastrointestinal tract if ingested. The irritating property is due to the reaction with water and its oxidizing potential. Vapors or mist can be detected at concentrations as low as 0.5-0.7 mg/m^3. Coughing can be induced at 5-6 mg/m^3. Impairment of the respiratory system characterized by increased respiratory rate, increased airway resistance, and decreased **vital capacity** have all occurred in response to exposure to sulfuric acid mist or vapors.

> Vital capacity is the volume of air that can be expelled following full inspiration.

Additional symptoms include pulmonary edema and loss of bronchiolar and bronchiole epithelium. The mucociliary clearance mechanism of the lungs may be impaired. Bronchitis, emphysema, and fibrosis have been associated with chronic exposure.

Sodium Hydroxide – NaOH
Potassium Hydroxide – KOH

— TLV = 2 mg/m^3,
— NFPA: 3H, 0F, 1R

Although most of the following description will focus on sodium hydroxide, the effects and properties also apply to potassium hydroxide. Both chemicals may exist in either a solid or liquid form. They are extremely corrosive and may cause serious tissue damage. Caustic soda, soda lye, or white caustic are synonyms for sodium hydroxide. More than ten billion kg are produced annually in the United States. Sodium hydroxide is used in the manufacturing of soap, paper, aluminum, petroleum, rayon, and other chemicals. The liquid and solid forms cause rapid and severe tissue destruction upon contact with concentrated solutions. Mixing concentrated solutions of sodium hydroxide or the solid form with water will produce heat. Similarly to sulfuric acid, the heat generated is the result of an exothermic reaction that causes damage to the tissue. Contact with the eye may result in permanent eye damage characterized by destruction and sloughing of the corneal epithelium, clouding of the cornea, and ulceration. Recovery may not begin to occur until 7-13 days after exposure.

Contact with the skin will cause similar tissue damage unless washed from the surface immediately. Concentrated solutions, 25-50 percent, cause irritation within three minutes of exposure. Irritation caused by diluted solutions (four percent) will not be noticed until several hours after exposure.

Checking Your Understanding

1. What ion is the toxic component of hydrofluoric acid?
2. How would a topical application or an intravenous injection of a calcium solution mitigate the toxic effects of hydrofluoric acid?
3. How does formaldehyde cause cell dysfunction?
4. Identify the various lung diseases associated with asbestos exposure.
5. Would you expect more cases of gastrointestinal cancer from asbestos in nonsmokers or smokers? Why?
6. Distinguish between asphyxia and anoxia.
7. Distinguish between a simple asphyxiant and a chemical asphyxiant. What are their similarities and differences?
8. How can you distinguish between dermal contact with nitric acid and sulfuric acid?

7-6 Pesticides

More than 1,600 pesticides have been developed and used to eradicate plants and animals that are perceived to be a nuisance. Although the use of pesticides has resulted in an increase in crop production, it has also created problems. While such diseases as malaria and **encephalitis** have been eliminated in some areas, they are on the increase in other areas because pesticide-resistant mosquitoes – which carry the disease-causing organism – have developed. In addition, the environmental damage resulting from the application of pesticides like DDT has been well documented.

The tendency of DDT and other organochlorine pesticides to persist and bioaccumulate in the food chain led to the development of new classes of insecticides like organophosphates, which are not as persistent but are effective in destroying their intended target. Organochlorines, organophosphates, and carbamates were developed primarily as insecticides, while other classes of pesticides – rodenticides, herbicides, fungicides, and fumigants – were created to deal with other types of organisms.

There are 35,000 or more commercial pesticides on the market today, providing ample opportunity for exposure. Table 7-13 lists various occupational and nonoccupational exposures to pesticides. Due to the large amount of pesticides in existence it is impossible to discuss details for each one; we will therefore only discuss the best known and most commonly used compounds in each of the classes identified above.

Occupational Exposure	Nonoccupational Exposure
Research chemistry	Residues from household and lawn applications
Manufacturing active ingredient	Termite control
Formulating end product	Accidental or intentional ingestion
Transportation	Food residues
Storage/warehouse work	Water residues
Mixing and loading	Hazardous waste sites
Agricultural application	Traffic accidents
Flaggers	Industrial spills
Structural application	
Building maintenance work	
Crop duster maintenance	
Farm work	
Greenhouse, nursery, mushroom house work	
Livestock dippers and veterinarians	
Landscapers	
Park workers	
Entomologists	
Work on highway or railroad rights of way	
Wood treatment workers	
Vector control workers	
Emergency work	
Firefighters	
Sewer work	
Hazardous waste workers	
Medical personnel	
Treating contaminated workers	

Table 7-13: Occupational and nonoccupational exposures to pesticides. *Recreated with permission from Rosenstock R. and Cullen M. R.*, Textbook of Clinical Occupational and Environmental Medicine, *Harcourt Brace/W. B. Saunders, Philadelphia, 1994.*

Organophosphate and Carbamate Insecticides

Parathion – $C_{10}H_{14}NO_5PS$
— TLV = 0.1 mg/m³, (S)
— NFPA: 4H, 1F, 2R

Malathion – $C_{10}H_{19}NO_6PS_2$
— TLV = 10 mg/m³, (S)

Dichlorvos – $(CH_3O)_2POOCH=CCl_2$
— TLV = 0.1 ppm, (S)

Tetraethyl pyrophosphate – $(C_2H_5)_4P_2O_7$
— TLV = 0.004 ppm, (S)

Carbaryl – $C_{12}H_{11}NO_2$
— TLV = 5 mg/m³

Organophosphate and carbamate pesticides are widely used. Poisoning from exposure to organophosphates is common in rural areas and in developing countries. Xylene and toluene may be used as solvents for organophosphate applications and may be responsible for some of the toxic effects. Table 7-14 lists several types of organophosphates and their relative toxicity. The less toxic carbamates are used in agricultural products and commercial preparations used by veterinarians as well as in household insecticide sprays. Table 7-15 lists some examples of carbamate pesticides and their relative toxicities.

Organophosphate and carbamate pesticides produce their effects by inhibiting the enzyme acetylcholinesterase. Under normal conditions, various

Low Toxicity (LD$_{50}$ >1,000 mg/kg)	
Bromophos (3,750-8,000 mg/kg)	Primiphos-methyl (>2,000)
Etrimfos (1,800)	Propylthiopyrophosphate (2,710-5,010)
Iodofenphos (2,100) (Jodfenphos)	Temephos (4,204-10,000)
Malathion (1,000-2,800)	Tetrachlorvinphos (4,000-5,000)
Phoxim (2,000)	

Moderate Toxicity (LD$_{50}$ >50-1,000 mg/kg)	
Acephate (700-950 mg/kg)	Isoxathion (180-242)
Bensulide (271-1,470)	Leptophos (52.8)
Chlorpyrifos (96-270)	Methyl trithion
Crotoxyphos (53)	Naled (250)
Cyanophos (580-610)	Oxydemeton-methyl (30-75)
Cythioate (160)	Oxydeprofos (105)
DEF (200)	Phencapton (182)
Demeton-S-methyl (180)	Phenthoate (440)
Diazinon (300-400)	Phosalone (120)
Dichlofenthion (270)	Phosmet (147-316)
Dichlorvos (DDVP) (50)	Pirimiphos-ethyl
Dimethoate (255-310)	Profenofos (358)
Edifenphos (100-260)	Propetamphos (119)
EPBP (274)	Pyrazophos (151-778)
Ethion (21-208)	Pyradaphenthion (769-850)
Ethoprop (61.5)	Quinalphos (71)
Fenitrothion (800)	Sulprofos (150)
Fenthion (250)	Thiometon (120-130)
Formothion (365-500)	Triazophos (57-59)
Heptenophos (96-121)	Tribufos (348-712)
IBP (Kitacin) (490)	Trichlorfon (250)
Heptenophos (96-121)	

High Toxicity (LD$_{50}$ <50 mg/kg)*	
Azinphos-methyl (10 mg/kg)	Isofenphos (20 mg/kg)
Bomyl (31)	Isofluorphate
Carbophenothion (6.8-36.9)	Mephosfolan (8.9)
Chlorfenvinphos (10-39)	Methamidophos (20)
Chlormephos (7)	Methidathion (44)
Coumaphos (16-41)	Mevinphos (3-12)
Cyanofenphos (43.7)	Monocrotophos (8-23)
Demeton (2.5-6)	Parathion (2)
Dialifor (43-53)	Phorate (2-4)
Dicrotophos (17-22)	Phosfolan (8.9)
Disulfoton (4)	Phosphamidon (17-30)
EPN (26)	Prothoate (8)
Famphur (35)	Schradan (9)
Fenamiphos (3)	Sulfotep (10)
Fenophosphon (37.5)	Terbufos (4.5-9)
Fensulfothion (2)	Tetraethylpyrophosphate (1.2-2)
Fonofos (8-17.5)	Triorthocresylphosphate

* Based on oral LD$_{50}$ values in the rat.

Table 7-14: Types of organophosphate pesticides and their relative toxicities. *Obtained with permission from Olson K. R. editor,* Poisoning and Drug Overdose, *2nd edition, Appleton & Lange, Norwalk CT, 1994.*

Highly Toxic (LD$_{50}$ <50 mg/kg)	Moderately Toxic (LD$_{50}$ <200 mg/kg)	Low Toxicity (LD$_{50}$ >200 mg/kg)
Aldicarb	Thiodicarb	MTMC (Metacrate)
MIPC (Isoprocarb)	Dioxacarb	Ethiofencarb
Oxamyl	Promecarb	BPMC (Fenocarb)
Carbofuran	Bufencarb	MPMC (Meobal)
Mexacarbate	Carbosulfan	Isoprocarb
Isolan	Benfuracarb	Carbaryl
Methomyl	Propoxur	XMC (Cosban)
Methiocarb	Pirimicarb	
Aldoxycarb	Trimethacarb	
Formetanate		
Aminocarb		
Mecarbam		
Bendiocarb		
Dimetilan		

Table 7-15: Carbamate insecticides and their relative toxicities. *Obtained with permission from Olson K. R. editor,* Poisoning and Drug Overdose, *2nd edition, Appleton & Lange, Norwalk CT, 1994.*

nerves – when stimulated – release a neurotransmitter substance called acetylcholine, which binds to a receptor on the postsynaptic membrane of the affected tissue, i.e. the skeletal muscle. Acetylcholinesterase interacts with acetylcholine to break it down to acetic acid and choline, thereby preventing constant stimulation of the affected tissue. In the presence of organophosphates this reaction does not take place because the organophosphate binds with the enzyme and inactivates it, thereby producing constant stimulation of the affected tissue. This reaction is not reversible. Carbamates have a similar mechanism of action, but with a lower mammalian toxicity than organophosphates. The decreased toxicity is associated with the reversibility of the reaction between the pesticide and acetylcholinesterase.

> Organophosphate and carbamate pesticides produce continuous stimulation of affected tissue such as skeletal muscle.

Symptoms of mild exposure to these substances include headache, anxiety, and sleep disturbances. Severe exposure may result in chest tightness, seizures, loss of consciousness, bradycardia or tachycardia, and liver dysfunction. Delayed effects of acute exposure to organophosphates have been associated with polyneuropathy and specifically distal axonopathy. This condition is often associated with muscle cramps in the legs followed by paresthesia and motor weakness. Table 7-16 lists symptoms associated with exposure to pesticides that inhibit acetylcho-

Muscarinic [1] Receptors	Nicotinic [1] Receptors	Central Nervous System
Anorexia Nausea and vomiting Abdominal cramps Diarrhea Chest tightness (bronchlo constriction and increased secretions) Increased salivation and lacrimation Miosis and blurred vision (may be asymmetric) Sweating Bradycardia Involuntary defecation and urination (if severe)	Muscle twitching Fasciculations Weakness Hypertension Hyperglycemia Tachycardia (if severe)	Anxiety Irritability Altered sleep Headache Tremor Impaired cognition Seizures and coma (if severe)

[1] Muscarinic and nicotinic receptors are different types of receptors that interact with the neurotransmitter substance acetylcholine.

Table 7-16: Symptoms associated with exposure to pesticides that inhibit acetylcholinesterase. *Recreated with permission from Rosenstock R. and Cullen M. R., Textbook of Clinical Occupational and Environmental Medicine, Harcourt Brace/W. B. Saunders, Philadelphia, 1994.*

linesterase. The symptoms are identified according to the type of cholinergic receptors affected. The **muscarinic** receptors are associated with secretory glands and the postganglionic fibers in the parasympathetic nervous system, while the nicotinic receptors are associated with the skeletal muscle.

> Muscarinic receptors are types of acetylcholine receptors.

Exposure to these pesticides may occur by inhalation, ingestion, or dermal absorption. The carbamates – with the exception of aldicarb – are not as easily absorbed through the skin as the organophosphates. Symptoms are dependent on the route of exposure. Exposure by inhalation usually results in rapid onset of symptoms within 1-2 hours. Symptoms from dermal exposure may not occur for 12 hours. Repeated, moderate dermal exposure may not result in symptoms until 1-2 weeks after exposure. Fatalities have been associated with exposure to organophosphates such as Diazinon, TEPP (tetraethyl pyrophosphate), and Systox. Aldicarb poisoning has resulted from eating contaminated watermelons.

Parathion – Parathion exists in liquid or solid form. The solid form is usually mixed with a solvent for application, producing a liquid that may be a yellow to dark brown color and that has a garlic-like odor. The odor threshold for parathion is 0.04 ppm. Conditions in the environment may result in decomposition of the chemical forming sulfur, nitrogen, and phosphorous oxides. Parathion oxidizes to paraoxon, which is the chemical intermediate responsible for the anticholinesterase activity. It has been estimated that the minimum lethal oral dose for humans ranges between 10 mg and 125 mg. IDLH concentration is 20 mg/m^3.

Malathion – Malathion is a brown liquid organophosphate insecticide with a skunk-like odor. Thermal decomposition results in the production of sulfur and phosphorus oxides. Malathion is only moderately inhibiting to cholinesterase, its metabolite – malaoxon – being the primary inhibitory agent. In humans, both substances are rapidly metabolized in the liver, which contributes to its lower order of toxicity. The only reported fatalities due to this substance have been from ingestion. Although skin absorption does occur, symptoms of exposure only occur in the case of continuous dermal contact with high concentrations of the substance. Symptoms of malathion poisoning were visible in workers who absorbed an estimated 26 mg/day for one week during application. IDLH concentration is 5,000 mg/m^3.

Dichlorvos – Dichlorvos is an amber liquid with only a slight odor. Commercially it is used as a thin film on flypaper (the sticky strips hung from ceilings to catch flies). The insecticide is highly toxic to insects, but has a low order of toxicity to humans. It is well absorbed through the skin. IDLH concentration is 200 mg/m^3.

Tetraethyl pyrophosphate – Tetraethyl pyrophosphate (TEPP) is one of the most toxic organophosphates. It is produced primarily as a colorless to amber liquid with a fruity odor. Phosphoric acid mist may be produced as a result of thermal decomposition. TEPP may be absorbed through the skin, by inhalation, or injection, the first two routes being the primary routes of exposure. Poisoning as a result of inhalation of TEPP dust (one percent) has been reported. Florists exposed to TEPP residues (5-400 ppm) on flowers developed symptoms of TEPP poisoning. IDLH concentration is 10 mg/m^3.

Carbaryl – Carbaryl is a carbamate pesticide sold under the trade name Sevin. It is a white-grayish odorless solid. Decomposition of this substance results in the formation of nitrogen oxides and methylamine. Powders containing 2-5 percent carbaryl are used to kill fleas on cats and dogs. Mosquito control programs use dust or sprays containing this insecticide. Carbaryl is metabolized quickly, which contributes to its low order of toxicity to humans. Workers who were exposed for 19 months to concentrations of carbaryl ranging from 0.23-31 mg/m³ showed no signs associated with inhibition of cholinesterase activity. IDLH concentration is 600 mg/m³.

Organochlorine Insecticides

DDT (dichlorodiphenyltrichloroethane) – $C_{14}H_9Cl_5$
— TLV = 1 mg/m³

Benzene hexachloride (Lindane) – $C_6H_6Cl_6$
— TLV = 0.5 mg/m³, (S)
— NFPA: 2H, 1F, 0R (solution)
    2H, 0F, 0R (solid)

Aldrin – $C_{12}H_8Cl_6$
— TLV = 0.25 mg/m³, (S)
— NFPA: 3H, 1F, 0R (solution)
    2H, 0F, 0R (dry)

Dieldrin – $C_{12}H_8Cl_6O$
— TLV = 0.25 mg/m³, (S)

Endrin – $C_{12}H_8Cl_6O$
— TLV = 0.1 mg/m³, (S)

Chlordane – $C_{10}H_6Cl_8$
— TLV = 0.5 mg/m³, (S)

Organochlorine pesticides are often divided into groups according to their chemical structure (Table 7-17), which plays a role in determining which portion of the nervous system is affected. DDT has its primary effect on the peripheral nervous system. The cyclodienes like aldrin and endrin primarily affect the central nervous system. Chlorinated insecticides affect the rate at which repolarization of the nerve occurs, which in turn increases the sensitivity of neurons to stimuli. As a result, typical symptoms caused by exposure to organochlorine pesticides are tremors, hyperexcitability, muscular weakness, or convulsions. Other symptoms may include depression

Insecticide Class	Acute Signs	Chronic Signs

Dichlorodiphenylethanes
  DDT
  DDD (Rothane)
  DMC (Dimite)
  Dicofol (Kelthane)
  Methoxychlor
  Methiochlor
  Chlorbenzylate
	Paresthesia (oral ingestion)
	Ataxia, abnormal stepping
	Dizziness, confusion, headache
	Nausea, vomiting
	Fatigue, lethargy
	Tremor (peripheral)
		Loss of weight, anorexia
		Mild anemia
		Tremors
		Muscular weakness
		
		Hyperexcitability, anxiety
		Nervous tension

Hexachlorocyclohexanes
  Lindane (γ-isomer)
  Benzene hexachloride (mixed isomers)		

Cyclodienes
  Endrin
  Telodrin
  Isodrin
  Endosulfan
  Heptachlor
  Aldrin
  Dieldrin
  Chlordane
  Toxaphene
Chlordecone (Kepone)
  Mirex
	Dizziness, headache
	Nausea, vomiting
	Motor hyperexcitability
	Hyperreflexia
	Muscle jerking
	General malaise
	Convulsive seizures
	Generalized convulsions
		Headache, dizziness,
		hyperexcitability
		Intermittent muscle twitching and jerking
		Psychological disorders
		including insomnia, anxiety, irritability
		Loss of consciousness
		Convulsions
		Chest pains
		Skin rashes
		Ataxia, incoordination, slurred speech
		Visual difficulty, inability to focus and fixate
		Nervousness, irritability, depression
		Loss of recent memory
		Muscle weakness, tremors of hands
		Severe impairment of spermatogenesis

Table 7-17: Symptoms of acute and chronic toxicity following exposure to organochlorine insecticides. *Recreated and modified with permission from Klaasen C. D. et al., Casarett & Doull's Toxicology, The Basic Science of Poisons, 5th edition, McGraw Hill, New York, 1996.*

of the respiratory system, cardiac arrhytmia, liver and kidney damage. Table 7-17 identifies symptoms associated with chronic and acute exposure.

Commercial preparations of organochlorine insecticides can be obtained as a dry powder or as a solution. Since organochlorine insecticides are only slightly soluble in water they are usually mixed with some type of petroleum solvent such as kerosene or toluene. Aside from the fact that these solvents are toxic by themselves, they facilitate penetration of the insecticide through the skin, causing systemic effects to occur from dermal exposure. The fat-soluble organochlorine insecticides bioaccumulate in the fatty tissues of the body and are retained there for years. Inhalation and ingestion are other possible routes of exposure.

DDT (dichlorodiphenyltrichloroethane) – DDT exists in the form of solid crystals – colorless, white, or yellow – with a faint aromatic odor. Because of its long-term bioaccumulation in mammals, as well as its subsequent harm to the environment, it was banned in the United States in 1973. However, it is not banned in many other countries. Therefore, exposure may occur as a result of residues present in products imported to the United States. Typical symptoms such as tremors as well as paresthesia of extremities and dizziness occur following acute exposure. Convulsions have been induced upon ingestion of 10 mg/kg of body weight or more. The major metabolic intermediate is DDE, which is excreted in the urine.

Benzene Hexachloride (Lindane) – Lindane may exist as a white crystalline solid or as a solution. It is used as an agricultural insecticide, but also residentially to eradicate lice, fleas, ticks, flies, mosquitoes, and carpet beetles. Benzene hexachloride is a major ingredient in shampoo and sprays used to treat body lice. Lindane is used as a fumigant because of its volatility, which is 100 times greater than that of DDT. Lindane is well absorbed by all routes of exposure. Occupational exposures occur primarily by inhalation and dermal absorption. Acute exposure from ingestion or extensive dermal contamination results in symptoms appearing 1-6 hours after exposure. Vomiting and diarrhea occur initially, followed by convulsions and tremors. The organic solvent used to dissolve the solid may cause dyspnea, cyanosis, and circulatory failure. It is also irritating to the eyes and mucous membranes; continuous exposure produces nausea and headache. IDLH concentration is 1,000 mg/m³.

Aldrin, Dieldrin, Endrin – The color of aldrin solids is tan to dark brown with a mild chemical odor. Aldrin is not combustible but breaks down into hydrogen chloride gas when subjected to heat. Also, aldrin rapidly metabolizes to dieldrin. Most uses of dieldrin have been banned in the United States. Dieldrin solids are light brown in color and have a slight chemical odor. Dieldrin is the most widely used of all the chlorinated cyclodienes in the control of mosquitoes, flies, and other insects. It is fat soluble, water insoluble, and persists for years in the environment after application. Endrin is a **stereo isomer** of dieldrin.

Stereo isomers are two compounds with identical chemical constitution but different arrangements of atoms in space.

The solids may be tan, white, or colorless with a mild chemical odor. Endrin is not combustible, but decomposes into hydrogen chloride when heated.

The primary route of exposure – dermal absorption – and subsequent toxic effects are the same for all three of these insecticides. Inhalation and ingestion are secondary routes. Exposure produces increased sweating, headache, dizziness, nausea, vomiting, and convulsions. IDLH concentrations for aldrin, dieldrin, and endrin are 100 mg/m³, 450 mg/m³, and 200 mg/m³, respectively.

Chlordane – Chlordane is a viscous amber-colored liquid with a chlorine-like odor. The liquid is not combustible, but thermal degradation produces the hydrogen chloride, phosgene, and chlorine gases. In the past it was used as a general insecticide for agricultural and gardening purposes, but as of 1976 its use has been limited to the control of termites. Exposure occurs through the usual routes, with skin absorption being the primary means of exposure resulting in the appearance of systemic effects. Chlordane is a skin irritant that bioaccumulates in the fat tissue where it reaches its maximum concentration in five days. It is also removed rapidly from the fat after exposure is stopped. The primary target tissue is the nervous system, and specifically the brain. Symptoms of exposure include excitability, dizziness, headache, weakness, paresthesia, muscle twitching, convulsions, and unconsciousness. Respiratory depression may also occur. Deaths have been recorded as a result of skin absorption. A worker spilled a 25 percent mixture of chlordane and DDT and failed to remove the contaminated clothing. This resulted in convulsions within 40 minutes and death a short time later. IDLH concentration is 500 mg/m^3.

Herbicides

Herbicides are a broad class of pesticides used to eliminate unwanted plants. They perform their function by either killing the plant directly or interrupting its growth. Herbicides are widely used to kill plants along roadsides, railroad tracks, and utility right-of-ways; in deforestation programs; and in agriculture and to kill weeds in the yard. For example, 2,4-D (2,4-dichlorophenoxyacetic acid) is used to selectively kill broad-leaf weeds in cereal crops, sugar cane, grass, and pasture lands. It is also used on various fruit crops. A mixture of 2,4,5-T(2,4,5-trichlorophenoxyacetic acid) and 2,4-D known as Agent Orange was used during the Vietnam War as a defoliant. EPA has suspended the use of 2,4,5-T for crops consumed by humans, around bodies of water, in recreational areas, and in the home.

The following discussion focuses on the chlorophenoxy herbicides 2,4,5-T and 2,4-D as well as on the bipyridyl derivatives paraquat and **diquat**.

Table 7-18 lists other classes of herbicides, their trade name, and their lethal doses.

Chlorophenoxy Herbicides

2,4-Dichlorophenoxyacetic acid (2,4-D) – $C_8H_6Cl_2O_3$
—TLV = 10 mg/m^3

2,4,5-Trichlorophenoxyacetic acid (2,4,5-T) – $C_8H_5Cl_3O_3$
—TLV = 10 mg/m^3

2,4-D and 2,4,5-T – These chemicals exist in various forms. 2,4-D exists as odorless white to yellowish crystals. When heated it decomposes into hydrogen chloride and phosgene. 2,4,5-T is an odorless, light tan solid. Decomposition products include hydrogen chloride and dioxin. The chlorophenoxy herbicides have been widely used because of their apparent low toxicity to man, although suicides have occurred as a result of ingestion of 6,500 mg of 2,4-D. The LD$_{50}$ for these herbicides ranges between 300 and >1,000 mg/kg depending on species.

There are several symptoms of exposure, one of the most common being **chloracne**, an acne-like skin disorder that appears following dermal exposure. Chloracne is characterized by the formation of

Chemical Class	Generic Name	Trade Name	Oral LD$_{50}$ (mg/kg)
Acetanilides		Alachlor Metolachlor	1,200 2,780
Amides	3,4-Dichloropropionanilide	Propanil	
Arylaliphatic Acids	2-Methoxy-3,6-dichlorobenzoic acid 3-Amino-2,5-dichlorobenzoic acid	Dicamba Chloramben	3,500 5,000
Carbamates	Isopropyl carbanilate 4-chloro-2-butyny-*m*-chlorocarbanilate	Propham Barban	5,000
Dinitroanilines	*a,a,a*-Trifluoro-2,6-dinitro-*N*,N-dipropy-*p*-toluidine	Trifluralin	10,000
Nitriles	2,6-Dichlorobenzonitrile 4-Hydroxy-3,5-diidobenzonitrile	Dichlobenil Ioxynil	
Substituted ureas	3-(*p*-chlorophenyl)-1,1-dimethylurea 3-(3,4-Dichlorophenyl)1,1-dimethylurea	Diuron	
Triazines	2-Chloro-4-(ethylamino)-6-(isopropylamino)-*S*-triazine 2-Chloro-4,6-bis(ethylamino)-*S*-triazine	Atrazine Simazine	1,000 1,000

Table 7-18: Herbicidal chemicals: classes, common names, and acute toxicity. *Recreated with permission from Klaasen C.D. et al., Casarett & Doull's Toxicology, The Basic Science of Poisons, 5th edition, McGraw Hill, New York, 1996.*

comedomes (blackheads), **pustules**, and scarring of the skin.

Pustules are small inflamed swellings of the skin that are filled with pus.

Other observed symptoms of exposure to 2,4-D and 2,4,5-T include liver disorders and neurological damage. Neuropathy usually occurs as a result of either long-term chronic exposure or accidental, acute exposure to 2,4-D and 2,4,5-T. Additional symptoms are irritation of the skin, eyes, and respiratory tract; headache; nausea; dizziness; pain in the extremities; paresthesia; nervousness; irritability; dyspnea, and – in some cases – paralysis.

During the manufacturing of 2,4-D and 2,4,5-T a contaminant is produced called 2,3,7,8-tetrachlorodibenzo-para-dioxin (TCDD), which is generally referred to as dioxin. TCDD has been shown to be the primary factor responsible for chloracne in exposed workers. It is also thought to play a role in other symptoms as well, such as the occurrence of peripheral neuropathy and the formation of cancer in various types of tissues. The most controversial aspect of exposure to TCDD is the role it may play in causing cancer. In some laboratory animals TCDD is a potent carcinogen and teratogen, but whether these results can be applied to humans is still questionable. Perhaps the most widespread human exposure occurred during the Vietnam war when thousands of soldiers were exposed to Agent Orange, which was contaminated with TCDD at concentrations up to 47 mg/g of mixture. However, studies of exposed soldiers, as well as other studies performed on individuals involved in the manufacture or application of these chemicals have not helped to clarify the situation. IDLH concentrations for 2,4-D and 2,4,5-T are 500 mg/m^3 and 5,000 mg/m^3, respectively.

Bipyridyl Derivatives

Diquat – $C_{12}H_{12}Br_2N_2$
—TLV = 0.5 mg/m^3, (S)

Paraquat – $C_{12}H_{14}N_2$
—TLV = 0.5 mg/m^3 (total dust);
   0.1 mg/m^3 (respirable fraction)

Diquat and paraquat – Diquat and paraquat are used to kill a variety of nuisance plants. Diquat is used to eradicate aquatic weeds and potato hilums prior to harvesting. Paraquat is used as an herbicide on broad-leaf plants and as a desiccant (drying agent) on cotton plants to facilitate mechanical picking of the cotton balls. In the late 1970s wide scale aerial spraying of paraquat was used to kill marijuana plants in Mexico. Concerns were raised as to whether residues remaining on surviving marijuana plants would harm users, but subsequent studies indicated there were no effects.

Diquat and paraquat are usually formulated initially as white to yellowish colored solids. A solution is usually prepared to facilitate the application and to improve adherence to the targeted plants. Paraquat is rapidly altered and inactivated by sunlight but, when heated, it decomposes into nitrogen and sulfur oxides and hydrogen chloride. Paraquat is highly toxic when ingested and can be fatal; several cases of suicide have been recorded as a result of ingestion. Laboratory studies have indicated the oral LD_{50} for a variety of species to be between 22 and 262 mg/kg. Paraquat has a high affinity for lung tissue and may cause alveolar damage and fibrosis. Pulmonary edema may be seen as early as 2-4 days after exposure. Tissue damage may not occur until 10-14 days after exposure. Paraquat is actively transported by alveolar cells. Once it is absorbed, cell membrane damage occurs as a result of altered cellular metabolism. The functional integrity of the cell is compromised, affecting gas exchange and the overall respiratory process. Other observed symptoms include hypoxia, lethargy, tachycardia, diarrhea, ataxia, and convulsions.

Chronic exposure to paraquat as a result of aerial spraying can cause irritation and inflammation of mucous membranes, headaches, coughing, and skin irritation. Changes in the structure and color of fingernails have been used as an indicator of exposure to paraquat. The fingernails may exhibit discoloration or ridging; a total loss of the fingernail may also occur. Paraquat is not easily absorbed dermally; however, localized effects such as blistering and erythema have been observed. IDLH concentration of paraquat is 1.5 mg/m^3.

Diquat is less toxic than paraquat, the LD_{50} for various species of animals ranging from 100-400 mg/kg. Unlike paraquat, the primary target tissues of acute exposure are the gastrointestinal tract, the liver, and kidneys. There is no active sequestering of diquat by lung tissue. No IDLH concentration has been established for diquat.

Fungicides

Fungicides represent a diverse group of pesticides with a wide number of uses, such as the treatment of fruit trees, vegetables, seeds as well as various building supplies. Fungicides include creosote, hexachlorobenzene, organomercuric compounds, and pentachlorophenol.

Fungicides have a relatively low toxicity. Laboratory studies using rats have shown that the oral toxicity for most fungicides is relatively low, ranging from 800-10,000 mg/kg. Of more concern is the implication that many fungicides have been shown to be mutagenic and are potentially carcinogenic. Fungicides by nature are cytotoxic; they eradicate the unwanted organism either by direct lethal effects or as a result of genetic mutations. Some fungicides are now limited in use, discontinued, or have been reformulated because of these potential mutagenic effects.

Creosote

—NFPA: 2H, 2F, 0R

Creosote is an oily, dark liquid primarily used as a wood preservative, which may contain 150 or more compounds. Some of the primary classes of chemicals found in creosote are phenolic compounds. Phenols denature and precipitate cell proteins, which directly alters cell functioning. Dermal contact produces skin irritation and some corrosion. Prolonged contact may cause acne, skin pigmentation changes, and severe burns. Eye contact may result in **keratitis**, conjunctivitis, and scarring of the cornea. Vapors may cause irritation of the respiratory tract.

> Keratitis is inflammation of the cornea.

Hexachlorobenzene – C_6Cl_6

—TLV = 0.025 mg/m³, (S)

Hexachlorobenzene demonstrates many of the properties exhibited by other organochlorines. They all undergo slow degradation, they bioaccumulate in fatty tissues, and metabolize slowly. Chronic exposure has been shown to cause **hepatomegaly** (enlarged liver), **alopecia**, itching of the skin, ataxia, irritability, and tremors.

> Alopecia is loss of hair.

In hamsters, a 70-week exposure results in an increase in liver and thyroid tumors. Also, teratogenic effects as well as renal and palate malformations have been demonstrated in rats.

Organomercurial Compounds

These fungicides were often used in agriculture for the coating of seeds. While they are no longer used in the United States, they are still in use in developing countries. Because the function of these chemicals is the coating of seeds, exposure occurs primarily from eating either the seeds or the meat of animals that were fed those seeds. Signs and symptoms of intoxication from ingestion include headaches, metallic taste, paresthesia, tremors, incoordination, and constricted visual fields. Long-term effects are associated with central nervous system and peripheral nervous system neuropathy.

Pentachlorophenol – C_6Cl_5OH

—TLV = 0.5 mg/m³, (S)
—NFPA: 3H, 2F, 0R (solution)
      3H, 0F, 0R (dry)

Pentachlorophenol is a chlorinated aromatic hydrocarbon widely used as a fungicide, a general herbicide, but also as a wood preservative and in the paper industry. When subjected to heat pentachlorophenol decomposes into hydrogen chloride, chlorinated phenols, and octachlorodibenzodioxin.

Vapors of pentachlorophenol are irritating to the eyes and nose at a concentration slightly higher than the TLV. Pentachlorophenol is well absorbed by the skin, dermal exposure being the primary route of accidental exposure. Skin contact may cause irritation or burns upon prolonged contact. Exposure results in uncoupling oxidative phosphorylation in the mitochondria. As a result, energy produced from substrate metabolism is released as heat instead of forming adenosine triphosphate (ATP). This causes hyperthermia, profuse sweating, and dehydration. The elevated temperature may result in other observed symptoms of acute exposure such as tachycardia, tachypnea (rapid breathing), convulsions, and coma. Additional symptoms associated with acute exposure include headache, nausea, vomiting, and lethargy. Chronic dermal exposure may result in dermatitis. IDLH concentration is 150 mg/m³.

Fumigants

Fumigants exist either as solids, volatile liquids, or gas and are used for eradicating insects, bacteria, or rodents. They are used on fruits and vegetables, in ships, freight cars, and buildings. Because of the volatile nature of these substances they generally dissipate shortly after use except in confined spaces, where vapors, fumes, or gases may linger leading to possible exposure.

Methyl Bromide – CH_3Br

—TLV = 5 ppm, (S)
—NFPA: 3H, 1F, 0R

Methyl bromide is a colorless liquid or gas, which may be odorless or have a slight chloroform-like odor. Chloropicrin is often added as a lacrimator to provide a warning property during exposure. Methyl bromide, which is applied to soil or perishable foods, is three times heavier than air and may accumulate in low-lying areas, increasing the probability of exposure in confined spaces. This characteristic is used expressly during fumigation of buildings, where the fumigant exerts its action in a confined space because the structure is covered. The main routes of exposure are via inhalation and dermal contact. The gas or liquid causes irritation to the skin, eyes, and respiratory tract. Continuous exposure may result in dermatitis. Acute exposure to the skin may produce burns and pulmonary irritation leading to pulmonary edema.

Methyl bromide also has a neurotoxic action, which may lead to convulsions, respiratory failure, and death. Initial symptoms of exposure may be delayed for up to several hours. Typical symptoms are headache, nausea, vomiting, dizziness, and tremors. Death has been associated with exposure for a few hours to a concentration of 6,000 ppm methyl bromide, and to short and acute exposures of 60,000 ppm methyl bromide. Chronic exposure may cause peripheral neuropathy and central nervous system effects. Symptoms include **dementia** (loss of memory or concentration), **psychoneurosis**, visual disturbances, ataxia, tremors, and seizures. IDLH concentration is 2,000 ppm.

> Psychoneurosis is a mental disorder that leads to disturbances in thoughts and feelings.

Phosphine – PH_3

—TLV = 0.3 ppm
—NFPA: 3H, 4F, 1R

Phosphine is a colorless gas that is used as a fumigant for cereal crops such as wheat. It has a fishy or garlic-like smell that can be detected at an air concentration of two ppm. To produce the fumigant, aluminum phosphide is applied and allowed to react with the natural moisture in the grain, generating phosphine. Because of its density phosphine will percolate through grain stacked in a pile and provide protection for long periods of time. Phosphine is highly toxic, particularly to the lungs. It is extremely irritating to the respiratory tract and may cause pulmonary edema. Additional symptoms include dizziness, nausea, vomiting, coughing, and headache. IDLH concentration is 200 ppm.

Rodenticides

Rats, mice, rabbits, gophers, and other species of animals are often the subject of eradication programs, especially in agricultural areas. These nuisance species may damage crops either before harvest or once the crops have been placed in storage. They are also carriers of disease and – in the case of gophers – may be a danger to livestock as a result of the holes, found throughout the rangelands, created by their burrowing activity. Several of the rodenticides are non-biodegradable and non-specific in terms of their effects, which means they may harm other species of animals. Therefore, strict control has been placed on the use of some of these compounds.

Warfarin – $C_{19}H_{16}O_4$

—TLV = 0.1 mg/m³

Warfarin is a colorless, odorless solid that has been used to kill rats and mice. It is the active ingredient in the commercial product D-Con Mouse Pruf®, which contains 0.054 percent warfarin. Because rats and mice have developed resistance to it, warfarin has been replaced in some rodenticide products with more potent substances.

Warfarin contains the anticoagulant **dicumarol**. This substance inhibits liver synthesis of certain blood-coagulating factors, preventing blood clotting from taking place. This leads to bleeding gums, bleeding around the eyes, easy bruising, and life-threatening bleeding in the gastrointestinal tract and in the brain. Symptoms may not show up for 2-4 days after ingestion or dermal exposure. IDLH concentration is 200 mg/m^3.

Sodium Fluoroacetate – $CH_2FCOONa$

—TLV = 0.05 mg/m^3 (S)

Sodium fluoroacetate, also known as Compound 1080, is a highly toxic, odorless, white powder. Because of its high toxicity it has been removed from the U.S. market for general use. Exposure may occur as a result of inhalation, ingestion, or skin absorption. Toxicity is due to metabolism of fluoroacetate to fluorocitrate, which inhibits the normal functioning of the **Krebs cycle**, which in turn is important in producing ATP for all normal cell functioning. As a result, cell death may occur. Symptoms may not be observed for 30 minutes to 2 hours after exposure. Exposed individuals may experience nausea, vomiting, renal failure, confusion, seizures, coma, respiratory arrest, and arrhythmia. IDLH concentration is 5 mg/m^3.

Strychnine – $C_{21}H_{22}N_2O_2$

—TLV = 0.15 mg/m^3

Strychnine is an alkaloid that exists as a white crystal and powder. It has a bitter taste and is only slightly soluble in water. In the past strychnine was used to kill larger animals such as coyotes. Strychnine-laced carcasses or meat were left in the open rangelands to attract and kill these predators. This practice has, for the most part, been eliminated. However, strychnine is now used as an **adulterant** in illegal drugs such as cocaine and heroin.

Strychnine is a neurotoxin and is a competitive antagonist to glycine, which is a neurotransmitter in the spinal cord. Glycine causes inhibition of skeletal muscle contraction. Therefore, strychnine poisoning results in a seizure-like contraction of skeletal muscle caused by the simultaneous contraction of opposing muscle groups, the extensors and flexors. Muscular rigidity and convulsions are common symptoms. These symptoms may not occur for 10-30 minutes after exposure. Death occurs as a result of respiratory arrest. IDLH concentration is 3 mg/m^3.

Checking Your Understanding

1. Explain how organophosphate and carbamate pesticides cause their effects.

2. What characteristics distinguish carbamate pesticides from organophosphate pesticides?

3. How does the chemical structure of organochlorines affect the tissue distribution of these substances in exposed individuals?

4. What is the primary target tissue of organochlorine pesticides?

5. Identify several herbicides and the symptoms associated with exposure to each.

6. What is dioxin? What effects are associated with exposure?

7. What are the symptoms associated with exposure to warfarin? How are the symptoms produced?

Summary

Chemical, physical, and biological agents are encountered daily in our environment. The effects of exposure can range from slight irritation of the skin, eyes, or respiratory tract to the development of life-threatening diseases such as cancer. The symptoms that develop vary according to the type of substance and length of exposure.

Outdoor and Indoor Air Pollution

Sulfur, nitrogen, and carbon oxides, as well as ozone and particulates, are outdoor air pollutants with the potential to cause several toxic effects. They affect primarily the respiratory system resulting in the development of respiratory diseases such as bronchitis and asthma.

Indoor air pollution has developed, in part, as a consequence of greater energy efficiency in homes and offices. Poor ventilation and circulation of air within these structures has resulted in toxic substances accumulating and remaining indoors for longer periods of time, contributing to the development of various types of building-associated illnesses. Illnesses that occur as a result of increased exposure to known etiological agents such as molds, bacteria, and viruses are referred to as building-related illnesses. These agents may cause influenza, colds, asthma, or Legionnaire's disease. Illnesses may also be caused by poor ventilation in well-insulated buildings (tight-building syndrome) or by chemical odors, vapors, and gases from building construction material and furnishing (sick-building syndrome). Symptoms associated with these illnesses include headache, nausea, dizziness, and irritation of the eyes, nose, and throat.

Petroleum Distillates

There is a wide array of petroleum-based chemicals that are used throughout industry. In general, these chemicals are fat soluble and able to penetrate the skin. The other primary route of exposure is by inhalation of their vapors. These substances primarily affect the central nervous system and lungs. The chemicals are highly soluble in the neural tissue and cause depression of the central nervous system. Inhalation of the vapors results in damage to the lung tissue, which may cause pulmonary edema and death.

Other effects include possible liver and kidney damage. Some petroleum products are skin irritants and may cause dermatitis. Benzene has an effect on blood cell formation and may cause leukemia. The peripheral nervous system may also be affected. Polyneuropathy (impairment of the neurons in the peripheral nervous system) may produce loss of sensation in fingers and toes.

Metals

Several metals are associated with manufacturing processes and products. Exposure occurs as a result of the release of metals during manufacturing or use of the products. The metals affect different tissue types causing a variety of effects.

Arsenic compounds affect the skin causing melanosis or chemical burns, as in the case of arsenic trichloride. Cancer of the skin has also been associated with arsenic exposure. Arsenic suppresses cell formation in the bone leading to leukopenia, granulocytopenia and anemia. Exposure to arsenic compounds leads to neurotoxicity of the central and peripheral nervous system.

Cadmium affects primarily the kidneys and the liver. Acute or chronic exposure may result in kidney failure. Inhalation of cadmium-contaminated dust or vapors may result in a variety of respiratory diseases such as chronic bronchitis, fibrosis, or emphysema. Cadmium may affect normal bone metabolism, which may cause osteoporosis or osteomalacia.

Of the various forms of chromium the hexavalent form is more readily absorbed and therefore more important from the toxicological perspective. Exposure to chromium is the primary cause of dermatitis in industrial occupations. Inhalation of chromium-contaminated dusts or vapors is extremely corrosive to the upper respiratory tract leading to sinusitis or to the formation of chrome-ulcers. Lung cancer has been associated with exposure to the hexavalent form.

Lead has been used for centuries and is found in a variety of products in the organic and inorganic forms. Exposure to lead produces several effects. It affects the central and peripheral nervous system leading to nerve cell damage. In children this may lead to decreased mental development, general brain dysfunction due to structural changes in the brain, and changes in behavior. Lead may result in kidney failure and cause other symptoms including tissue fibrosis. Anemia may be caused by continuous exposure due to inhibition of heme synthesis and accelerated destruction of red blood cells.

The primary effect of mercury is on the nervous system. Chronic exposure to mercury results in mercurialism, which is characterized by tremors, increased heart rate and irregular pulse, and increased excretion of mercury in the urine. Progressive stages of the disease are characterized by tremors in the fingers, eyelids, and lips; eventually muscle spasms occur in the arms and legs. Similarly to lead, mercury affects brain development and function in children as well as some brain functions in adults.

The primary health concern with nickel is due to its carcinogenic properties. Cancer primarily develops in the respiratory tract. Nasal and lung cancers occur most often.

Inhalation of zinc oxide fumes is associated with the development of "metal fume fever." The effects are primarily associated with the respiratory system. Severe exposure may cause pulmonary edema and respiratory collapse.

Other Important Chemicals

Many chemicals like ammonia, hydrogen fluoride, phosgene, formaldehyde, and chlorine affect primarily the respiratory system. Exposure is often by inhalation of the vapors or fumes given off by these chemicals. They may cause general irritation, bronchitis, or pulmonary edema, which may lead to death.

Some chemical substances act as chemical asphyxiants. Hydrogen sulfide and cyanide compounds block the normal utilization of oxygen by cells. This is usually accompanied by symptoms of decreased oxygen in the air, such as increased respiratory rate. Both substances may lead to asphyxia and death.

Acids and bases primarily produce their effects through skin contact. Symptoms range from slight irritation to severe chemical burns resulting in loss of skin and the development of scars. Inhalation of vapors may cause respiratory tract irritation and pulmonary edema. Exposure to hydrofluoric acid can be life threatening in a different way: the fluoride ion interferes with calcium metabolism, which affects the normal functioning of the heart.

Pesticides

Many pesticides have been developed and are used to eradicate some plants and animals. Organophosphates and carbamates inhibit acetylcholinesterase resulting in constant stimulation of the affected tissue. This produces a variety of effects such as headaches, chest tightness, loss of consciousness, as well as changes in the heart rate.

Exposure to organochlorine pesticides affects the nervous system by interfering primarily with the rate of repolarization of nerve cells. This causes nerve cells to respond to weaker stimuli. Symptoms include tremors, muscular weakness, and convulsions.

Herbicides – such as the widely used paraquat and diquat – are used to kill nuisance plants. Paraquat is the more toxic of the two and has a high affinity for lung tissue. It damages the alveoli and affects gas exchange. Lethargy, tachycardia, diarrhea, ataxia, and convulsions are associated with exposure.

There are several fungicides, fumigants, and rodenticides used for a variety of purposes. Fungicides are of concern because of their potential mutagenic and carcinogenic effects. Some of them – such as hexachlorobenzene – have been shown to have teratogenic effects in laboratory animals. Effects associated with the nervous system, liver, respiratory tract, and skin are also associated with exposure to these substances.

Application and Critical Thinking Activities

1. What is the difference between TLVs and IDLH values?

2. Identify the NFPA categories and explain their usefulness.

3. Draw a graph illustrating the relationship between oxygen and carbon monoxide levels in the blood in the case of an individual being exposed to the same concentrations of each. On the same graph illustrate how this relationship affects cellular metabolism.

4. The mechanisms of action for cyanide and hydrogen sulfide are the same. If an individual were unconscious as a result of exposure, how would you determine to which chemical they had been exposed?

5. Why should you not attempt to clean a toilet bowl by using a chlorine-containing and ammonia-containing solution?

6. Why is depending on smell not a good means of determining whether exposure to a toxic substance has occurred? Give at least two examples of chemical exposure that support your answer.

7. Explain how exposure to organochlorine and organophosphate pesticides produces the same effects even though their mechanisms of actions arre different.

8. Choose an indoor air pollutant not discussed in this chapter and identify its NFPA codes (if appropriate), its TLV exposure levels, as well as its carcinogenic potential as identified by one of the organizations discussed in this text.

9. Exposure to organophosphate pesticides may result in leg cramps and paresthesia. Present a mechanism of action that would explain these effects.

10. Fungicides are sprayed on fruit trees, vegetables, and seeds where they interfere with fungal growth and development. Hypothesize how this may occur and why consuming fruits and vegetables containing these toxic substances may be of concern.

11. Several different toxic substances discussed in this chapter are neurotoxins. List the toxicants that are neurotoxic and the mechanisms of action and symptoms of each.

12. Explain why young children should not be exposed to lead and what the possible consequences of exposure are.

13. What similarities exist between the effects of exposure to mercury and lead? How could you distinguish the difference?

8

Radiation, Pathogens, and Naturally Occurring Toxins

Chapter Objectives

Upon completing this chapter, the student will be able to:

1. **Define** terms associated with radiation exposure.
2. **Describe** the effects of radiation exposure on fetal development and adults.
3. **Describe** various types of chemical and biological agents and their effects.
4. **Discuss** symptoms associated with exposure to various types of blood-borne pathogens.
5. **Identify** various types of medical and infectious wastes.
6. **Identify** types of naturally occurring toxins and toxic agents in plants and animals.

Chapter Sections

8–1 Introduction

8–2 Radiation, Radioactive Materials, and Mixed Wastes

8–3 Chemical and Biological Warfare Agents

8–4 Blood-borne Pathogens

8–5 Infectious and Medical Wastes

8–6 Naturally Occurring Toxins

8-1 Introduction

The continuing use of chemicals, high-energy sources, and the daily exposure to other toxic agents is a fact of modern life. In the last chapter, the focus was primarily on indoor and outdoor exposures to toxic chemical substances. In this chapter, the discussion on exposures to toxic substances continues with additional chemical and biological agents as well as some naturally occurring toxins.

Although exposure to background and cosmic radiation has occurred since the beginning of time, the increased use of radiation sources and radioactive materials has greatly added to the potential for biological harm. Nuclear power stations as well as the medical use of diagnostic x-rays and radioactive sources to identify and treat malignancies are just a few of the everyday examples. Depending on the type of tissue involved and the dosage received, all forms of radiation can be damaging to all living organisms.

The twentieth century has witnessed a renewed interest in the development, manufacture, and stockpiling of chemical and – more recently – biological warfare agents. Although rudimentary forms of biological warfare agents have been recorded during the Middle Ages in Europe, their use is suspected but not confirmed in modern times. The organisms that have been studied, developed, and stockpiled include bacterial, viral, and fungal agents.

Concern for and awareness of blood-borne pathogens has reached an all-time high with the worldwide spread of Acquired Immunodeficiency Syndrome (AIDS). It is transmitted through contact with infected blood in the same way as the several forms of hepatitis, syphilis, toxoplasmosis, Rocky Mountain Spotted Fever, and bacteremia. Closely associated with the concerns for blood-borne pathogens are the mounting concerns for the safe handling and disposal of medical and infectious wastes. The possibility of exposure to these materials has increased along with the use of disposable gloves, gowns, syringes, and **sharps** as well as with the decrease in the use of incineration for their disposal.

Nature has provided a dizzying array of naturally occurring toxins. These include bacteria-derived toxins causing illnesses such as botulism, Lyme disease, or tetanus; viruses such as the Ebola virus and many others; animal-derived toxins such as snake venom as well as plant-derived toxins like curare and several other toxins from plants like mistletoe, horse chestnuts, and philodendron. Effects from these toxins are generally similar to those produced by man-made toxins.

8-2 Radiation, Radioactive Materials, and Mixed Wastes

There are several types of natural radiation sources: cosmic rays; radioactive elements such as radium, thorium, and uranium, which exist in the earth's crust; and many radioactive isotopes – such as potassium-40 and carbon-14 – within the cells of the human body. There are several occupations that result in exposure to these natural as well as other artificial sources of radiation (Table 8-1). The largest source of artificial exposure is the use of x-rays in the various health-related occupations. In addition to x-rays, a variety of radioactive materials may be used for diagnostic or research purposes by hospitals. Universities and private businesses may also use radioactive materials in performing their respective activities. Most of these activities use low-level radioactive sources such as tritium, iodine-131, radium, and technetium. The estimated number of workers exposed to ionizing radiation is shown in Table 8-2.

The toxic effects of ionization radiation vary according to the type of radiation being emitted, which may be in the form of alpha particles, **beta particles**, **gamma rays**, x-rays, or **neutrons**. Each type of radiation has its own distinguishing characteristics. For example, although gamma rays and x-rays are very similar in terms of their properties, they are different in their origin. Alpha and beta particles have the potential to be more destructive in the lungs, since their external effects are primarily limited to damage to the skin tissue. Gamma rays, x-rays, and neutron-associated radiation emit high-energy electromagnetic waves that easily pass through the external barriers of the skin causing damage to internal body tissues. Table 8-3 lists various types of **radionuclides** as well as their source and type of radiation emitted.

Standards have to be set to prevent adverse health effects from exposure to radioactive substances. The **National Council of Radiation Protection (NCRP)** recommends a maximum of **0.5 rem** per year.

> A rem (roentgen equivalent man) is a unit of measure for the amount of ionizing radiation that will cause the same amount of biological injury to human tissue as one roentgen of x-ray or gamma ray dosage.

Aircraft pilots and crews	Oil well loggers
Atomic energy plant workers	Ore assayers
Cathode ray tube makers	Petroleum refinery workers
Dental assistants	Physicians
Dentists	Pipeline oil flow testers
Electron microscope makers	Pipeline weld radiographers
Electron microscopist	Plasma torch operators
Electrostatic eliminator operators	Radar tube makers
Fire alarm makers	Radiologists
Gas mantle makers	Radium laboratory workers
High-voltage television repairmen	Reactor operators and workers
High-voltage vacuum tube makers	Television tube makers
High-voltage vacuum tube users	Thickness gauge operators
Industrial fluoroscope operators	Thorium-aluminum alloy workers
Industrial radiographers	Thorium-magnesium alloy workers
Inspectors using, and workers located near, sealed gamma ray sources	Thorium ore producers
Klystron tube operators	Underground metal miners
Liquid level gauge operators	Uranium mill workers
Luminous dial painters	Uranium miners
Nuclear submarine workers	X-ray aides
	X-ray diffraction apparatus operators
	X-ray technicians
	X-ray tube makers

Table 8-1: Occupations that may result in exposure to ionizing radiation. *Recreated with permission from Rosenstock R. and Cullen M. R.,* Textbook of Clinical Occupational and Environmental Medicine, *Harcourt Brace/W. B. Saunders, Philadelphia, 1994.*

Types of Work	Number of Workers Exposed Annually	Average Dose per Year (mSv)
Nuclear energy	62,000	8.4
Naval reactor	36,000	2.2
Healing arts	500,000	1.2
Research	100,000	1.2
Manufacturing and industrial	7,000,000	0.07

Table 8-2: Estimated number of American workers who are exposed occupationally to ionizing radiation, with respective average levels of exposure. *Recreated with permission from Rosenstock R. and Cullen M. R.,* Textbook of Clinical Occupational and Environmental Medicine, *Harcourt Brace/W. B. Saunders, Philadelphia, 1994.*

Radionuclide	Typical Exposure Source	Emission	Organ of Deposition
Cesium (Cs^{137})	High-technology radiation sources	Gamma	Bone
Iodine (I^{131})	Biologic experiments, medical	Beta	Thyroid
Plutonium (Pu^{239})	Nuclear weapons	Alpha	Lung
Radium (Ra^{226})	Medical	Alpha, gamma	Bone
Radon (Rn^{222})	Mines, indoor homes	Alpha	Lungs
Strontium (Sr^{90})	Nuclear fallout	Beta	Bone
Technetium (Tc^{99})	Medical, reactors	Beta	Not stored
Thorium (Th^{232})	Industrial processes, mines	Alpha, Beta	Liver
Tritium (H^3)	Biologic research, nuclear weapons, power plants	Beta	Body water
Uranium (U^{235}, U^{238})	Nuclear weapons, nuclear power plants	Alpha, neutrons	Kidney

Table 8-3: Characteristics of commonly encountered radionuclides. *Recreated with permission from Rosenstock R. and Cullen M. R., Textbook of Clinical Occupational and Environmental Medicine, Harcourt Brace/W. B. Saunders, Philadelphia, 1994.*

The federal government has established an occupational standard of 5 rem/year to the body, the gonads, and all blood-forming organs, and 75 rem/year to the hands and feet. The NCRP also recommends that maximum exposure during pregnancy be no more than 50 millirem per month. A typical chest x-ray results in an exposure of about 15 millirem.

Radiation exerts its effects on the structural and functional molecules of the cell. To help understand the effects of radiation and radiation exposure it is useful to understand some of the associated terms. All forms of radiation are associated with energy; when absorbed by tissues, the energy is capable of causing changes in cell components and functions. The **rad** is the amount of radiation absorbed per gram of tissue.

> A rad is a unit of absorbed dose of ionizing radiation equal to an energy of 100 ergs per gram of irradiated material.

Internationally the **gray (Gy)** is used in place of the rad; however, one gray is equal to 100 rad. One rad of high penetration x-ray produces the same biological effect as one rem of ionizing radiation. The rem provides a unit of measure that allows a comparison of the biological effects of different types of radiation. The international unit called **Sievert (Sv)** is roughly equal to 100 rem.

Within the cell, the primary target of radiation that produces an effect is the DNA. If the absorbed dose is high enough, cell death occurs, which can lead to scarring of the tissue if the number of cells

Radiosensitivity	Cell Types
Very High	Lymphocytes Immature hematopoietic cells Intestinal epithelium Spermatogonia Ovarian follicular cells
High	Urinary bladder epithelium Esophageal epithelium Gastric mucosa Mucous membranes Epidermal epithelium Epithelium of optic lens
Intermediate	Endothelium Growing bone and cartilage Fibroblasts Glial cells Glandular epithelium of breast Pulmonary epithelium Renal epithelium Hepatic epithelium Pancreatic epithelium Thyroid epithelium Adrenal epithelium
Low	Mature red cells Muscle cells Mature connective tissues Mature bone and cartilage Ganglion cells

Table 8-4: Radiosensitivity of normal cells. *Recreated with permission from Mettler F. A. and Upton A. C., Medical Effects of Ionizing Radiation, 2nd edition, Harcourt Brace/W. B. Saunders, Philadelphia, 1995.*

killed is large. Cell death is associated with breaks in both strands of the DNA. Lower levels of radiation can cause single strand breaks in the DNA. The single strand breaks may be repaired or may cause mutations. The mutation may ultimately result in the formation of cancer cells.

The sensitivity of the cell to radiation varies according to the type of cell exposed (Table 8-4). Also, the phase of cellular division will determine radiosensitivity. Actively dividing cells are more radiosensitive than inactive cells that divide only sparingly or not at all. Highly differentiated cells are less sensitive, while cells that complete the mitotic cycle more slowly are more sensitive. Systemic effects may be observed if enough cells are affected as a result of acute exposure. Essentially, all cells of the body could be affected given a high enough dose. Table 8-5 lists the symptoms associated with various levels of radiation exposure.

Exposure of the embryo or the fetus to radiation is of great concern. The potential effects include lethality, cataracts, growth retardation, malformation, neuropathology, and behavioral problems. Table 8-6 illustrates the relationship between various stages of pregnancy and the occurrence of a particular effect as a result of radiation exposure. Table 8-7 lists other effects as a result of exposure to radiation.

Acute Dose Equivalent (Sv)	Incidence (%)	Latency (hours)	Syndrome or Organ Involved	Characteristic Symptoms	Critical Period after Exposure	Therapy	Prognosis	Lethality (%)	If injury is Fatal	
									Death within	Cause of Death
1-2	0-50	>3	Bone marrow	Mild leukopenia and thrombopenia	2-6 weeks	Symptomatic	Excellent	0-10	Months	Infections and/or hemorrhage
2-5	50-90	1-2	Bone marrow syndrome	Thrombopenia, leukopenia, hemorrhage, infections, **epilation**	2-6 weeks	Transfusions of leukocytes and platelets, optimal care (isolation, antibiotics, fluids), bone marrow transplantation	Uncertain depending on success of therapy	0-90	Weeks	Infections and/or hemorrhage
5-10	100	0.5-1	Bone marrow syndrome	Thrombopenia, leukopenia, hemorrhage, infections, epilation	2-6 weeks	Bone marrow transplantation, transfusions of leukocytes and platelets, optimal care (isolation, antibiotics, fluids)	Uncertain depending on success of therapy	0-90	Weeks	Infections and/or hemorrhage
10-15	100	0.5	Intestinal syndrome	Diarrhea, fever, electrolytic imbalance	3-14 days	**Palliative**	Very poor	90-100	2 weeks	**Enterocolitis**, shock
>50	100	<1	Neurologic syndrome	Cramps, fever, ataxia, lethargy, impaired vision, coma	1-48 hours	Symptomatic	Hopeless	100	1-4 hours	Cerebral edema

Table 8-5: Symptoms and effects of acute radiation syndrome. *Recreated with permission from Rosenstock R. and Cullen M. R., Textbook of Clinical Occupational and Environmental Medicine, Harcourt Brace/W. B. Saunders, Philadelphia, 1994.*

Stage	Preimplantation	Implantation	Organogenesis Early/Late	Fetal
Days since onset of menses	14-23	24-28	29-64	65-294
Days since conception (gestation)	0-9	9-14	15-50	50-280
Lethality	+++	+	+	−
Gross malformations	−	−	+++	±
Growth retardation (at term)	−	+	+++/++	+
Growth retardation (as adult)	−	+	++/+++	++
Sterility	−	±	−	++
Cataracts	−	−	+	+
Neuropathology	−	−	+++/++	+++/++

Key: +++ occurs in high incidence
++ readily apparent effect
+ demonstrated effect
± questionable effect
− no observed incidence

Table 8-6: Effects of exposure to radiation during pregnancy in relationship to different stages of egg and fetal development. *Recreated with permission from Mettler F. A. and Upton A. C., Medical Effects of Ionizing Radiation, 2nd edition, Harcourt Brace/W. B. Saunders, Philadelphia, 1995.*

1.	Microcephaly (most frequent)	16.	Nystagmus (involuntary rapid movement of the eyeball)
2.	Hydrocephalus	17.	Stillbirth
3.	Porencephaly (cysts or cavities in the brain cortex)	18.	Live birth weight (decrease)
4.	Mental deficiency	19.	Neonatal and infant death (increase)
5.	Mongolism (Down syndrome)	20.	Ear abnormalities
6.	Idiocy	21.	Spina bifida
7.	Head ossification defects	22.	Cleft palate
8.	Skull malformations	23.	Deformed arms
9.	Micromelia (small or short limbs)	24.	Clubfeet
10.	Microphthalmos (small eyes)	25.	Hypophalangism
11.	Microcornea	26.	Syndactyly
12.	Coloboma (absence or defect of some ocular tissues)	27.	Hypermetropia (farsightedness)
13.	Strabismus (deviation of the eye)	28.	Amelogenesis (defective enamel of teeth)
14.	Cataract	29.	Odontogenesis imperfecta (abnormal teeth)
15.	Chorioretinitis	30.	Genital deformities

Table 8-7: Anomalies reported after high-dose human embryonic and fetal x-irradiation. *Recreated and adapted with permission from Rugh R., "Radiology and the Human Embryo and Fetus" in Medical Radiation Biology, Harcourt Brace/W. B. Saunders, Philadelphia, 1973.*

Leukemia is most often associated with acute exposure to radiation. Other types of cancer caused by exposure to radiation include lung cancer as well as cancer of the thyroid and parathyroid glands. Also, chronic exposure to ionizing radiation from the use of radium in radiology as well as of radium-based paints in the manufacture of dials for watches and other instruments has resulted in cancer of the bone and skin. Awareness and improved protection have decreased the incidence of these types of radiation-induced cancers.

In the context of radioactivity the term mixed wastes refers to wastes containing radioactive materials and other toxic chemical substances. Waste may contain a variety of radionuclides such as plutonium, uranium, cesium, strontium, or any of the other radionuclides mentioned previously. It also contains nonradioactive materials contaminated with radionuclides. In either case, the effects of radiation are dependent on the same factors described previously: how much radiation is being emitted, how much is absorbed, and what tissue is affected.

There may be a wide variety of toxic chemicals associated with the radioactive substances. For example, metals such as cadmium, mercury, chromium, nickel, and lead are but a few of the toxic metals that can be found in mixed wastes. Their effects are discussed elsewhere in this chapter. Organic solvents such as trichloroethylene and 1,1,1-trichloroethane may also be associated with waste. Hydrofluoric, nitric, and sulfuric acids as well as caustic solutions like sodium hydroxide may be components of the mixed waste. It is also possible to have several different types of mixed wastes comprised of different combinations of these various components.

Checking Your Understanding

1. Define the terms rem and rad. Is there a relationship between the two units of measure? If so, what is it?

2. How does radiation produce its toxic effects?

3. Which cell types are the most sensitive and which are the least sensitive to radiation exposure? Why?

4. What are some of the major effects of radiation exposure on adults and on a developing fetus?

5. If a pregnant woman were exposed to radiation how might you be able to determine at what stage of her pregnancy she was exposed?

8-3 Chemical and Biological Warfare Agents

The development and use of chemical substances for the purposes of incapacitating or killing one's adversaries has occurred throughout history. In 600 B.C. the Cirrhaeans took land that belonged to the Greek temple of Apollo. War ensued and the River Pleistus, which supplied water to the Cirrhaeans, was diverted. Rather than surrender their land, the Cirrhaeans limited themselves to well and rain water. Their enemies dumped large quantities of hellebore root – which causes diarrhea – into the river and diverted it back to its original channel. Overjoyed by the abundance of water the Cirrhaeans drank their fill, but soon after had to abandon their post because of severe diarrhea, allowing the Greeks to reclaim the land. Four centuries later the Carthaginians used wine drugged with mandragora – a toxic root with a narcotic effect – to subdue their enemy. The enemy overran the Carthaginian camp, drank the wine, and fell asleep, allowing the Carthaginians to return and recapture the city.

In more recent times, chemical warfare – in the form of chlorine gas – was considered during the American Civil War, but did not occur on a large scale until World War I. Later, the production and use of chemical agents escalated, with new types being introduced to effectively counter those used by the opponent. The types of gases produced ranged from those considered harassing agents – such as bromoacetone, which causes lacrimation, and diphenyl chlorasine, which causes vomiting – to the more lethal types, such as mustard gas.

After World War I the potential devastation – in terms of human life – that could be inflicted as a result of chemical warfare was realized. Although chemical agents have not been used on the same scale in subsequent wars, the potential threat these toxic substances posed did not discourage their development. Both Germany and the United States developed and stored chemical warfare agents in Europe during World War II. Also, it is reported that the Japanese used chemical agents on the Chinese during World War II. It was during World War II that the development and possible use of biological warfare agents was first introduced.

Other wars such as Korea, Vietnam, and the Gulf War have brought accusations concerning the use of chemical agents as well as the possible use of biological agents in some circumstances. Servicemen of the Gulf War suffer from a variety of symptoms for which no apparent cause can be identified. Some evidence suggests they may have been exposed to chemical and/or biological agents.

With the end of the arms race the need to continue maintaining large stockpiles of chemical agents has been reduced, but at the same time has created a disposal problem. There are eight storage locations within the United States where stockpiles exist; the actual quantity of most of the facilities is classified. In addition, the containers holding these agents are deteriorating, which adds to the hazard potential.

The following discussion focuses on a selected few of the chemical and biological agents. A more extensive listing for each can be found in Table 8-8 and Table 8-9.

Chemical Warfare Agents

Chemical warfare agents produce different types of effects: they may be irritating to the skin and eyes, or they may cause death by acting upon the nervous system or lungs. In either case their use would be for the purpose of incapacitating the targeted individuals. The following discussion of chemical agents is divided into sections according to primary effects.

Lacrimators and Vomit-Inducing Gases – These types of gases are often used by police in riot control. These substances cause irritation to the eye, the nose, and upper respiratory tract. If enough gas reaches the upper regions of the digestive tract, vomiting occurs. The effects of these substances are immediate but not long-lasting. Two of these gases are called **CN** (chloroacetophenone) and **CS** (ortho-chlorobenzalmalononitrile) gases. CS gas produces an extreme burning sensation, coughing, tightness of the chest, nausea, and vomiting. CN gas produces effects similar to CS gas, but is more irritating to the skin.

Respiratory Irritants – In addition to the irritating effects on the respiratory system these substances may damage the respiratory epithelium as well as the alveoli, producing long-term effects. If

Disease	Organism	Infectivity	Effects	Survival	Transmission	Storage	Epidemicity	Remarks
Bacterial Diseases								
Anthrax	Bacillus anthracis	20,000 organisms of <2 μ inhaled	respiratory form normally fatal if untreated	spore forming highly stable	inhalation (also from animals) skin infection	as spores	low	one of the most stable agents
Brucellosis	Brucella melitensis	high 1,300 organisms?	long lasting recurrent severe fever; rarely fatal	stabilized with dextrin and protein products	inhalation ingestion (also from animals)	several months	low	affects both man and domestic animals
Cholera	Vibrio cholerae	low by ingestion	severe intestinal infection; sometimes fatal		ingestion (also from animals)		high	unlikely to be effective via water systems
Glanders	Malleo-myces mallei	high 3,200 organisms inhaled	in acute form, severe fever, often fatal		inhalation ingestion (also from animals)		low	affects both man and domestic animals
Melioidosis	Whitmorella pseudo-mallei	high	normally fatal fever, producing mania and delirium		inhalation ingestion (also from animals)		low	very rare and little known disease
Plague	Yersinia pestis	high 3,000 organisms	very severe; often fatal		inhalation ingestion injection by insects		high	only pneumonic (respiratory) plague likely to be of BW use
Tularaemia	Pasteurella tularensis	very high <50 organisms inhaled?	severe fever; 5-8% fatal	unstable	inhalation ingestion injection by insects		none	good BW agents apart from doubts concerning stability
Botulism	Clostridium botulinum	toxic dose 0-12 μg	severe poisoning; 60-70% mortality	decomposes in 12 hours in air	inhalation ingestion	bacteria as spores or toxin in airtight container	none	acts more quickly than any other BW agent; troops could invade after 24 hours
Viral Diseases								
Breakbone fever	dengue viruses	high single mosquito bite, 2 organisms inhaled?	most incapacitating fever known; vary rarely fatal	high	injection by mosquito, inhalation	stable eight years at 41° F	low	might be useful as an incapacitating agent
Mumps		high	incapacitating but not severe		inhalation		high	little use as BW agent because of widespread immunity
Polio-myelitis		low	severe, permanent disability; sometimes lethal	85% survive 23 hours at 21-24° C as aerosol	ingestion inhalation of moist air		low	limited by low infectivity and widespread immunity

Table 8-8: Principal agents of biological warfare. *From* Chemical and Biological Warfare *by Steven Rose, Copyright © 1968 by J. D. Bernal Peace Library and Steven Rose. Reprinted by permission of Beacon Press, Boston.*

Disease	Organism	Infectivity	Effects	Survival	Transmission	Storage	Epidemicity	Remarks
Viral Diseases (continued)								
Psittacosis		high	mild to severe fever; sometimes fatal		inhalation ingestion injection by insects		high	birds act as reservoir of disease; immunity may be fairly widespread
Smallpox	Poxvirus variolae	high a 'few' organisms	severe; often fatal		inhalation ingestion	stable when freeze dried	high	generally immunity too widespread
Yellow fever		high single mosquito bite	jaundice type fever; 30% mortality		injection by mosquito inhalation	stable at 0-4° C	none	naturally a sub-tropical disease; a strain that could survive in temperate climate might be dangerous
Rickettsial Diseases								
Q-fever	Coxiella burnetii	very high 1 organism inhaled? 10^{-6}g Infected tissue	fever for 1 week; 1% mortality	stable	inhalation ingestion injection by ticks	stable	none	very high infectivity
Epidemic typhus	Rickettsia prowazeki	high	severe; often fatal	poor	injection by louse inhalation? ingestion?	difficult	high but cannot spread man-to-man	unlikely BW agent; poor stability
Fungal Disease								
Coccidio-idomy-cosis	Coccidio-es immitis	1,350 spores	mild to severe fever; rarely fatal	spore forming highly stable	inhalation	as spores	low	highly stable; suitable agent if a vaccine were produced

Table 8-8: Principal agents of biological warfare (continued).

sufficient lung damage is caused, individuals may develop bronchitis, pneumonia, or they may die from asphyxia. Gases capable of producing these effects include chlorine and phosgene. Both were discussed in more detail earlier in this chapter.

Skin Irritants – These gases are sometimes referred to as **nettle** gases. Their effects are similar to the effects produced by lying down in a bed of nettles. At low concentrations these gases cause skin irritation and lacrimation, while at high concentrations they cause blistering of the skin and also enter the bloodstream. Once they enter the bloodstream they can produce lung damage and subsequent development of pulmonary edema, similar to the action of phosgene. **Dichloroformoxime** is an example of this type of gas.

Vesicants – While the primary target tissues of **vesicants** are eyes and skin, where they produce blistering and a burning sensation, they damage all tissues they come in contact with. Absorption through the skin may occur at high doses, producing lethal effects. Inhalation of the gases can also be lethal. These gases are not volatile and therefore persist in the field for days or weeks. There are two major classes of vesicants: mustard gases and arsenicals. The mustard gases are better known because of their use during World War I. They are effective because they have no noticeable odor and produce no initial effects. Following exposure, pain and blistering of the exposed tissue occurs which, in the case of the eyes, may cause blindness. There are several types of mustard gases, **Agent Q** being the most dangerous one. It will produce burning and blindness at low concentrations, while it may be lethal when inhaled at higher concentrations. The arsenicals are less effective as a chemical warfare agent, because they possess a sharp irritating odor and cause immediate eye pain. Therefore, protective equipment can be donned before significant exposure occurs.

Nerve Gases – Perhaps the most feared chemical warfare agents are the nerve gases. **Sarin**, tabun, and soman are examples of nerve gases. These gases are anticholinesterase agents similar to organophosphate pesticides. The symptoms vary according to the type of nerve gas an individual is exposed to as well as its concentration. Exposure to sarin at 15 mg/m^3 causes irritation to the eye along with dim vision and an inability to focus. These effects can last for several weeks. At 40 mg/m^3, difficulty in breathing can occur, coughing, nausea, heartburn, and twitching of muscles. At 55 mg/m^3 there is aching of the chest, vomiting, cramps, and tremors. Severe convulsions, paralysis, and death occur at 70 mg/m^3.

Incapacitating Gases – These substances are designed to incapacitate either the mind or the body. They may exist as part of a mixture with other chemical warfare agents. **BZ** (a derivative of lysergic acid) is an incapacitating agent. Exposure results in slowing of mental and physical functions characterized by dizziness, disorientation, hallucinations, skin flushing, urinary retention, and constipation.

Biological Warfare Agents

Biological warfare involves the use of living organisms and toxic substances produced by them. These agents are frightening and potentially devastating weapons. Like chemical warfare agents they also cause incapacitation or death. Their application or use, however, can go undetected until symptoms occur. Use of any chemical and biological warfare agents was prohibited by the 1925 Geneva Protocol and the 1972 Biological Weapons Convention.

Interest in the development of biological warfare agents was in part due to observation of the ravaging effects of the bubonic plague in Europe, which was caused by the bacterium *Yersinia pestis*. In nature, the bacterium is associated with fleas found on rodents. It is the bite of the flea that transmits the bacterium. Bubonic plague is characterized by the quick onset of chills associated with a rapid rise in body temperature to 103°F or 104°F. The increase in temperature is accompanied by an increase in heart and respiratory rate. Other symptoms include bloodshot eyes, nausea, and vomiting. The high temperature may cause delirium. **Buboes**, after which the disease is named, occur in 75 percent of the cases; they are inflammations of the lymph nodes in the armpits and the groin. The nodes reach the size of an egg and are tender to the touch. Localized internal hemorrhaging may occur resulting in large black spots. The appearance of black spots gave rise to the name "black death." The high fever may last for 2-5 days with a return to normal body temperature within two weeks. Left untreated the fatality rate is from 30-90 percent, but only 10 percent of treated victims may die.

Bacterial Agents – Other bacteria that are examples of possible biological warfare agents are the ones responsible for the diseases **anthrax** and **brucellosis**. Anthrax is caused by *Bacillus anthracis*, which is a bacterium that forms spores when exposed to air. The spores are not easily destroyed by disinfectants and can remain viable in the soil for up to 20 years. They can, however, be killed by heat and chemicals. The bacterium is transported by a variety of hosts including cattle, sheep, goats, horses, and pigs. Under normal circumstances the disease may be contracted from working with contaminated animals. The bacterium may enter the body through microscopic or large openings in the skin, by inhalation of the spores, or by ingestion of improperly cooked contaminated meat. Inhalation of the spores would be the major route of exposure as a biological warfare agent. Initially symptoms of inhalation are similar to other common upper respiratory infections. However, the disease progresses into labored breathing, cyanosis, and high fever. Death may occur within 18-24 hours if the exposed person is left untreated.

Brucellosis is caused by *Brucella melitensis*, *Brucella suis*, and *Brucella abortus*. Brucellosis is naturally occurring in cattle, pigs, goats, sheep, horses, and buffalo. Like anthrax, humans become contaminated as a result of handling contaminated products associated with these animals. The most likely route of infection is through ingestion and through damaged skin. Inhalation of contaminated dust is not as likely to result in infection. The onset of symptoms associated with brucellosis is gradual and includes headaches, loss of appetite, fever, generalized aching, chills, and stiffness in various joints. In some cases the fever undulates, usually first ap-

240 ■ Basics of Toxicology

ID Code	Name	Normal Physical State	Smell	Disseminated Form	Symptoms of Intoxication	Time of Onset of Symptoms	LD$_{50}$ Percutaneous Absorption (mg per man)	LC$_{50}$ Inhalation (mg-min/m^3)	LC$_{50}$ Percutaneous Absorption (mg-min/m^3)	Incapacitating Dosage (mg-min/m^3)	Remarks
Lethal Agents											
CG	phosgene	colorless gas	new-mown hay; imparts metallic taste to tobacco smoke	gas	**lethal** coughing, retching, frothing at mouth, cyanosis, asphyxia; pneumonia	generally delayed for several hours		3200			produced 80% of WWI gas fatalities; extensively stockpiled in WW2
AC	prussic acid	colorless liquid	bitter almonds	vapor	**lethal** giddiness, convulsions, unconsciousness, asphyxia	immediate		5000	200,000-1,000,000		stockpiled by U.S.A. from 1942 on
HD	distilled mustard	colorless to amber oily liquid	faint garlic	vapor liquid	**harassing** eyes: inflammation, photophobia, ulceration, blindness	delayed 1-48 hours	(blisters: 0.032)	1000	>10,000	7-day blindness: 200 2-week skin burns: 1,000	most widely stockpiled agent of WW2
T		oily liquid	none	liquid aerosol	skin: redness, irritation, blisters	delayed	(blisters: 0.004)	about 400			used mixed with mustard; their production methods yield such mixtures: the British stockpiled 60/40 HT during WW2
Q	sesqui-mustard	solid	none	aerosol		delayed	(blisters: 0.0003)	about 300			
HN3	nitrogen mustard	colorless to amber, oily liquid	faint geranium smell	vapor liquid aerosol	lethal resembles CG in its action on lungs; other systemic effects	delayed	(blisters: 0.06)	1000		7-day blindness: 200 2-week skin burns: 1,000	less smell than HD, so even more insidious
Nerve Gases											
GA	tabun	colorless to dark brown liquid	none to fruity	vapor liquid aerosol	**harassing** eyes: pupils constrict, vision blurs and dims, eyeballs hurt		1,500 (3 'drops')	150		20 (unmasked)	standard German tabun contained 20% chlorobenzene; to be aerosol dispersed
GB	sarin	colorless liquid	almost none	vapor liquid	respiration: chest tightness, difficulty in breathing	up to 10 min. following inhalation, or up to half an hour following percutaneous absorption	2,000 (40 'drops')	70	15,000	>20 (unmasked)	Germans stockpiled large quantities of its intermediates but could make only 1/2 ton
GD	soman	liquid	slightly fruity to camphor-like	vapor liquid aerosol	**lethal** drooling, sweating, nausea, vomiting, cramps, involuntary defecation or urination, twitching, jerking, staggering, headache, confusion, drowsiness, coma, convulsions, asphyxia		about 1,250	about 70			highly resistant to oxime therapy
GE		liquid		vapor liquid aerosol							
GF	CMPF	liquid		liquid vapor aerosol							harder to treat than GB
VE		liquid		liquid aerosol			small				
VX		liquid		liquid aerosol			small				

Table 8-9: Principal Agents of Chemical Warfare. *From* Chemical and Biological Warfare *by Steven Rose, Copyright © 1968 by J. D. Bernal Peace Library and Steven Rose. Reprinted by permission of Beacon Press, Boston.*

ID Code	Name	Normal Physical State	Smell	Disseminated Form	Symptoms of Intoxication	Time of Onset of Symptoms	LD$_{50}$ Percutaneous Absorption (mg per man)	LC$_{50}$ Inhalation (mg-min/m^3)	LC$_{50}$ Percutaneous Absorption (mg-min/m^3)	Incapacitating Dosage (mg-min/m^3)	Remarks
					Harassing Agents						
CA	BBC Camite	pinkish to brown oily liquid	soured fruit	vapor aerosol	**harassing** burning feeling in mucous membranes, severe eye irritation and lachrymation, headache	immediate		3,500		30	useful as a persistent harassing agent
CN	CAP	white crystals	apple blossom	aerosol	**harassing** burning feeling on moist skin, copious lachrymation	immediate		8,500		80	
DM	adamsite	canary yellow to brownish-green crystals	almost none	aerosol	**harassing** headache, sneezing, coughing, chest pains, nausea, vomiting	up to 3 minutes		30,000		20	
CS	OCBM	white crystals	peppery	aerosol	**harassing** stinging and burning feeling on skin, coughing, tears, chest tightness, nausea	immediate		very large		10	
					Incapacitating Agent						
BZ		solid		aerosol	slowing of physical and mental activity, giddiness, disorientation, hallucinations, occasional maniacal behavior						only standardized incapacitating agent by 1963

Table 8-9: Principal Agents of Chemical Warfare (continued).

pearing in early afternoon or evening then dissipating in the night with a drenching sweat. The temperature may reach levels as high as 104°F. Periodic episodes of fever may occur for several months. Individuals with chronic brucellosis have a low-grade fever of 100°F and general physical weakness. They may also experience headaches, vertigo, and transitory paralysis.

Viral Agents – Various types of debilitating fevers are caused by viruses. **Dengue fever** and **yellow fever** are transmitted naturally by the bite of an infected mosquito. From a biological warfare perspective the virus could be transmitted in an aerosol form. Dengue fever is an acutely disabling disease with rapid onset. It is characterized by fever, chills, intense headache, backache, joint and muscle pain, and general muscular weakness. The fever may last for 5-6 days and may reach 105°F. A rash may also occur in 3-4 days. Recovery takes several weeks after cessation of the fever. Yellow fever also has a rapid onset characterized by 102-103°F temperature that lasts 2-5 days. Symptoms are similar to Dengue fever. However, yellow fever also causes jaundice, which persists for several weeks even during time of recovery. Also, hemorrhaging from mucous membranes may occur resulting in vomiting of blood. The fatality rate may reach 40 percent of the infected population.

Rickettsial Diseases – *Rickettsias* are the microorganisms responsible for causing typhus among other things. Depending on the species of microorganism involved the symptoms may be slightly different in their degree of severity and their time of occurrence. In general, typhus is characterized by fever reaching 106°F temperature. The fever is associated with general body aches, pain, and chills. A rash appears 5-10 days after the onset of the disease. In some cases deafness may occur. Some individuals may experience delirium. Secondary infections such as bronchopneumonia may also occur and cause death. The disease may last up to 18 days with full recovery requiring several weeks to months.

Fungal Agents – A fungus called **Coccidioides immitis** occurs naturally in the dry arid soils of the southwestern United States. The organism may enter the body through abrasions of the skin, but usually it is through inhalation of airborne spores that exposure occurs. Once the fungus enters the lungs it causes symptoms similar to those associated with a cold or a mild case of the flu. More severe cases are disabling, resulting from the development of fever, restlessness, chills, cough, chest pain, headache, inflammation of the nasal passage, and night sweats. In some cases an allergic response occurs 1-2 weeks after onset of the symptoms as a result of toxic substances present in the bloodstream. This may last for 1-2 weeks or several months.

Checking Your Understanding

1. How would you distinguish between exposure to various biological agents?
2. Why would it be difficult to clearly identify exposure to a biological agent versus a non-biological agent?
3. List the various types of nerve gases and symptoms associated with exposure.
4. How are the nerve gases sarin, tabun, and soman related to organophosphate pesticides?
5. Identify the various types of biological warfare agents and some of their symptoms.

8-4 Blood-borne Pathogens

Infectious diseases may be caused by a variety of organisms. The organisms include bacteria, viruses, fungi, and parasites. These pathogenic organisms may be transmitted through air, on droplets expelled by coughing or sneezing. Tuberculosis, mumps, measles, and chicken pox are all examples of airborne-transmitted diseases. Other pathogenic organisms may be transmitted through the consumption of food. Diseases associated with this route of exposure include food poisoning, salmonella poisoning, and hepatitis A. A third possible route of transmission of disease is by blood-borne pathogens (see Table 8-10). The following discussion focuses on a few of the blood-borne pathogens.

Organism Category	Specific Organisms
Viruses	Hepatitis A virus Hepatitis B virus Hepatitis C virus Hepatitis D virus Cytomegalo virus Epstein-Barr virus Human immunodeficiency virus 1 Human immunodeficiency virus 2 Human T-cell leukemia virus 1 Human T-cell leukemia virus 2 Parvovirus B19 Colorado tick fever virus
Parasites	Malaria Babesiosis Toxoplasmosis Trypanosomiasis (Chagas disease) Leishmaniasis
Rickettsia	Rocky Mountain spotted fever Q fever
Spirochetes	Syphilis Relapsing fever
Bacteria	Yersinia enterocolitica Pseudomonas fluorescens Escherichia coli Serratis liquefaciens S. marcescens Brucella spp. Bacillus spp. Staphylococcus epidermidis

Table 8-10: Infectious agents transmitted by blood or blood products. *Recreated with permission from Mandell G. L. et al. editors,* Principles and Practices of Infectious Diseases, *Vol. 2, Churchill Livingstone, New York, 1995.*

Human Immunodeficiency Virus (HIV)

Perhaps the best known blood-borne pathogen is the virus associated with Acquired Immunodeficiency Syndrome (AIDS). AIDS exploded into the national and worldwide spotlight in the 1980s. Various segments of the United States population developed common symptoms of a weakening immune system, which ultimately resulted in death. The human immunodeficiency virus (HIV) was identified as the common causative factor of the symptoms observed in affected individuals. Although the rate of increase in AIDS-infected individuals is not as rapid as initially predicted, it is estimated that worldwide the number of infected individuals will be between 40 and 110 million by the turn of the century.

The AIDS virus can be transmitted by infected blood or by contact with other body fluids. Transmission of the virus in either of these media may occur during unprotected sexual intercourse, as a result of the sharing of infected needles by drug users, from accidental puncture wounds with contaminated needles or other sharp objects, or by transfusion with tainted blood. Perhaps the most tragic aspect of the disease is that it may be transmitted from an infected pregnant woman to the developing fetus. HIV is not transmitted through the air.

Overt symptoms may not appear for years after initial infection. Initially there may be some swelling of the lymph nodes; however, there are several other infections that may cause the same effect as well. In addition to the swelling of the lymph nodes, the number of T cell lymphocytes – which are involved in the normal immune response to infectious agents – begins to decrease. The decrease in the number of T cells continues for years; about five years after initial exposure other symptoms begin to appear that are associated with an impaired immune system. Pneumonia is usually one of the first diseases to appear in AIDS patients. Other symptoms include intermittent fever, intestinal infections coupled with weight loss and increased incidence of tuberculosis. In addition, other diseases begin to appear that are indicative of AIDS: **Kaposi's sarcoma, lymphoma, toxoplasmosis,** and **cryptococcal meningitis**. Death occurs as a result of severe infection and respiratory failure due to pneumonia.

Hepatitis A, B, C, and D

The primary target organ of the hepatitis virus is the liver. Symptoms begin to appear approximately 50 days after exposure, jaundice being the most obvious. The skin and the whites of the eyes develop a yellowish tint due to accumulation (in the blood) of bilirubin, which the liver normally removes from the bloodstream by metabolizing it. If individuals are not treated for the disease, chronic infection can lead to cirrhosis, liver failure, and liver cancer. In some cases the effects on the liver subside but the virus still exists. Although they may show no symptoms, these individuals are carriers and may infect others.

Hepatitis A (HAV) – Hepatitis A (**HAV**) is probably better known than hepatitis B (**HBV**), C (**HCV**), and D (**HDV**). Hepatitis A maybe contracted through contact with food or water contaminated with the virus. It is rarely transmitted by blood transfusion. Since the virus lives only for a short time (two weeks) in the blood of an infected individual, infectious transmission would occur only if the blood had been obtained from a recently infected individual and soon after used for transfusion. Like HIV, Hepatitis B, C, and D viruses are more readily transmitted by direct contact with blood or body fluids.

Hepatitis B (HBV) – Hepatitis B is sometimes referred to as **serum hepatitis**. One percent of the adult population in the United States carries the hepatitis B virus and generally shows no symptoms. Exposure may occur from contact with infected blood and other body fluids. The virus survives outside the body and may be present on a variety of surfaces, facilitating its transmission between individuals. High-risk individuals are intravenous drug users as well as individuals that receive transfusions, such as **hemophiliacs** and **dialysis** patients.

Hepatitis C (HCV) – Hepatitis C is very similar in nature to hepatitis B. Intravenous drug users, accidental puncture wounds with infected needles and blood transfusions are all possible means of transmitting the virus. In fact, the HCV is the primary cause for post-transfusion hepatitis. Onset of the disease is characterized by the classic symptoms of jaundice – occurring about 50 days after infection – but less severe. Similarly to hepatitis B, chronic hepatitis may either produce symptoms or exist in the carrier state.

Hepatitis D (HDV) – The hepatitis D virus occurs with one of the other hepatitis viruses. This coexistence is necessary in order for the HDV to replicate. An individual infected simultaneously with both viruses can experience more prolonged and severe symptoms than an individual infected with HBV alone. Chronic hepatitis D can either produce symptoms or simply exist in the carrier state. Individuals with chronic hepatitis D often develop cirrhosis of the liver, which is fatal.

Syphilis

Syphilis is the result of the transmission of the pathogen – the bacterium *Treponema palladium* – through sexual intercourse. The bacteria can also cross the placenta – infecting newborns – and may be transmitted by blood transfusion. Initially syphilis affects the genito-urinary system but eventually spreads throughout the body. Symptoms usually occur in three separate stages. The first stage is typically characterized by the appearance of an open sore or **ulcer** – referred to as a **chancre** – on any mucous membrane in the genital region two to four weeks after exposure. If left untreated, the chancre will disappear, and the infected individual will appear asymptomatic. The second stage – or latent stage – is characterized by the appearance of a rash on the palms of the hands and the soles of the feet. Other symptoms include fever, headache, loss of appetite, and sore throat. Symptoms of hepatitis and arthritis may appear as well. Second-phase symptoms usually begin to occur 6-8 weeks after exposure. The third stage is the most dangerous and potentially life threatening. Because it does not occur for up to two years after initial exposure, the cause of the symptoms may not be readily identified. During this stage the bacteria may attack the wall of the aorta destroying elastic tissue, predisposing to aneurysm, and producing degeneration of the aortic valves. They may also affect the central nervous system producing symptoms similar to other neurologic disorders, which can be crippling or fatal. Infection of the spinal cord leads to a condition called **tabes dorsalis**, which is characterized by a wide gait, ataxia, and slapping of the feet on the ground. The third stage may also be characterized by psychosis.

Toxoplasmosis

Pets in the home are common and normally present a low level of risk in transmitting diseases to their owners. However, in some circumstances people may become infected by organisms carried by these animals. A disease called toxoplasmosis is caused by a parasitic organism called *Toxoplasma gondii*. This organism may be transmitted by ingestion of contaminated meat, across the placenta, in blood transfusions, and in organ transplants. Of concern to pet owners is the shedding of the organism in the fecal matter of infected pets. The handling of litter from an infected pet increases the chance of contracting the disease, but good hygienic practices help avoid the problem.

One of the main concerns associated with this pathogenic organism is its potentially detrimental effect on pregnant women. Often pregnant women are advised not to handle pets or their fecal materials. Infection may result either in the birth of a healthy individual or in the premature birth of an infected individual, or also in spontaneous abortion or stillbirth. Infants born to mothers infected during the first trimester are less likely to become infected, but symptoms associated with the disease are more severe. Infants are more likely to become infected if the mother becomes infected during the last trimester of pregnancy, but they usually show little or no signs of the disease. Some of the symptoms of congenital infections are blindness, mental retardation, **microcephaly** (abnormal smallness of head) or **hydrocephaly** (excessive amount of fluid within the cranial cavity), jaundice, anemia, and **lymphadenopathy** (abnormal enlargement of lymph nodes). Acquired toxoplasmosis – as opposed to congenital – may lead to lymphadenopathy, but also malaise, rash, fever, central nervous system disorders, **myocarditis** (swelling of the heart wall), and pneumonitis.

Rocky Mountain Spotted Fever

Ticks infected with the pathogen *Rickettsia rickettsii*, which causes Rocky Mountain Spotted fever, are commonly found on dogs. The dogs serve as a means of potential contact between ticks and humans and are responsible for infecting up to 75 percent of the individuals. The disease occurs all across the United States, primarily in heavily wooded areas and coastal grasslands. It has a higher occurrence rate in children between the ages of five and nine, in men rather than women, and in Caucasians rather than African-Americans. Several symptoms may accompany the onset of the disease. Table 8-11 lists various symptoms and the percentage of their occurrence in three separate studies. Usually within one week after infection a headache, a rash, and a fever develop. Other symptoms, which may appear 7-14 days after infection, include chills, nausea, vomiting, diarrhea, muscle tenderness, respiratory symptoms such as cough, pulmonary edema, or pneumonia, as well as cardiac arrhythmia and kidney dysfunction. Death may occur during the second week due to renal failure and shock if the infected individual is left untreated.

History, Signs, or Symptoms	Sexton and Burgdorfer (N75)	Kaplowitz et al (N131)	Helmick et al (N262)
Tick bite	68	NR	60
Fever	100	100	99
Rash	100	90	88
Rash on palms and soles	NR	82	74
Headache	92	79	91
Myalgia	67	72	83
Nausea or vomiting	39	63	60
Abdominal pain	33	23	52
Conjunctivitis	15	30	30
Edema	17	20	18
Pneumonitis	NR	17	12
Any severe neurologic complications	48	23	NR

Table 8-11: Common clinical features of Rocky Mountain Spotted Fever in three case series. *Recreated with permission from Gorbach S. L. et al.,* Infectious Diseases, *Harcourt Brace/W. B. Saunders, Philadelphia, 1992.*

Bacteremia

Bacteremia is the presence of bacteria in the bloodstream (see Table 8-10) whether associated with active disease or not. The bacterium *Pseudomonas*, for example, has been shown to be responsible for up to 28 percent of transfusion-related symptoms. The bacteria may be introduced through contamination of the bag holding the blood, during processing of the blood, or through the needle. Bacteremia usually follows infection of a primary site or is associated with particularly high-risk surgical procedures that release bacteria from sequestered sites, such as the urinary or digestive tracts, into the blood, which may carry infection to remote parts of the body. Bacteremia can often be anticipated in those undergoing high-risk surgical procedures, allowing preventive antibiotic therapy to be given to forestall extensive bacterial invasion of the bloodstream. Despite precautions, however, advanced medical technology and the development of antibiotic-resistant strains of bacteria have led to an increase in the incidence of severe bacteremia since the late 1960s.

Bacteremia may also occur as a result of routine medical procedures such as the use of intravenous needles or **cannulas**.

Cannula is a tube that can be inserted into a cavity (lumen). For instance it may be inserted into a lumen of an artery.

This may result in an elevated number of bacteria or their toxic products in the blood. Of the symptoms that may appear fever is the most common. Severe cases of bacteremia may lead to death.

Checking Your Understanding

1. What virus causes AIDS? How is it transmitted?
2. What are the symptoms associated with AIDS-infected individuals?
3. What are the four types of hepatitis viruses? Describe how individuals may become infected with each type of virus and the symptoms that develop.
4. Identify the various stages and relative symptoms of syphilis infection.
5. What is toxoplasmosis?
6. What are the possible consequences of contracting toxoplasmosis during pregnancy?
7. What is bacteremia and how may it be caused?

8-5 Infectious and Medical Wastes

The American public first became aware of the problem associated with the proper management of infectious wastes during the summer of 1988 when various kinds of waste washed up on the shores and beaches of the East Coast of the United States. Among the wastes were syringes, IV tubing, and prescription bottles. Subsequent investigations discovered that the waste originated from barges traveling out to sea for several miles and then dumping their cargo containing potentially infectious medical wastes. Concern regarding the transmission of diseases by these types of wastes has been heightened by our increased awareness and knowledge of the spread of the AIDS virus by contaminated needles. However, the CDC has been able to show that the HIV concentration in liquids is reduced 10-100 times when the liquids are left to dry for several hours. Further studies on the survivability of the virus in water, sewage, and seawater have shown that the chances of becoming infected with a needle after it has passed through these media is about 1 in 15 billion. No other diseases originating from discharged medical wastes have been shown to occur. Several studies documented that routine household waste contains 100 times more potential pathogenic organisms than some types of hospital waste.

It is estimated that 3.2 million tons of infectious and medical wastes are generated annually from activities associated with hospitals, medical centers, clinics, colleges and universities, research and diagnostic laboratories, funeral homes, doctors' offices, and other facilities involved in biomedical research. There is no consensus as to what is defined as infectious waste. According to EPA the term refers to wastes that "...in all probability contain pathogenic agents that, because of their type, concentration, and quantity, may cause disease in persons exposed to the waste." For purposes of the present discussion, the categories of infectious wastes are those identified by Reinhardt and Gordon in *Infectious and Medical Waste Management*, (1991), compiled from both the Centers for Disease Control (CDC) and EPA documents.

Human Blood and Blood Products – This type of waste has a high potential of transmitting disease if not handled properly. Blood and blood products may contain a variety of blood-borne pathogens such as the HIV and the hepatitis viruses. Blood-contaminated test tubes, capillary tubes, glass slides, and blood-soaked bandages all fit into this category. To decrease the risk, the waste may be steam-sterilized or incinerated.

Cultures and Stocks of Infectious Agents – Various types of pathogenic organisms are grown in media that are conducive to their rapid reproduction. These organisms are often used for medical research and industrial purposes. Because of the high concentration of organisms present, these types of wastes pose a high risk for infection upon contact. Before disposal these wastes undergo steam sterilization.

Pathological Wastes – This type of waste includes body tissue, organs, amputated limbs, and body fluids. Potentially pathogenic organisms may become associated with this waste in two ways. First, the pathogenic organisms may be present in the waste from the beginning, since this type of waste is in many cases generated from medical activities associated with pathological conditions. Secondly, pathological wastes not properly disposed of may become a breeding ground for other types of microorganisms. Because of these concerns as well as the unappealing practice of disposing of this type of waste in landfills, this waste is incinerated.

Contaminated Sharps – One of the major concerns associated with handling medical wastes is becoming infected as a result of receiving a cut or puncture wound. Medical objects that cause punctures or cuts are referred to as sharps. Sharps include hypodermic needles, syringes, scalpel blades, intravenous needles, disposable pipettes, capillary tubes, microscope slides, cover slips, and any broken glass. Much of this type of waste is now handled in specific types of containers. Usually the containers are rigid and puncture-proof so needles will not penetrate the wall; they also have a lockable lid and a narrow opening to prevent removal of material. This waste is usually incinerated.

Contaminated Laboratory Wastes – This type of waste is generated from a variety of sources including microbiological and medical research as well as industrial laboratories. Types of waste that fit into this category are: bottles, flasks, test tubes, petri dishes, pipettes, disposable gloves, lab coats, masks, and aprons.

Contaminated Wastes from Patient Care – Wastes that belong to this category include wastes from clinics, patients rooms, emergency rooms, operating rooms, and morgues. Specific types of wastes include diapers, bed pads, intravenous tubing, catheters, and bandages contaminated with body fluids. Determination of whether the wastes are infectious or not is based on a case-by-case situation. Final treatment and disposal is dependent on the type of wastes generated.

Discarded Biologicals – Live and attenuated vaccines are included in this category. Potentially pathogenic organisms are produced for medicinal and research purposes. Other waste types such as cultures and stocks of infectious agents, sharps, animal carcasses, body tissues, and laboratory-contaminated wastes also fit into this category. This waste is generally incinerated prior to disposal.

Contaminated Animal Carcasses, Body Parts, and Bedding – This type of waste may be generated as a result of research using animals to study human pathogens. Often the animals are infected with a specific pathogen in order to study the onset and development of symptoms associated with exposure. These animals either die during the research or are terminated. Like the pathological wastes, incineration is the most appropriate means of treating the waste prior to disposal.

Contaminated Equipment – Any piece of equipment that has been used in association with the handling of pathogenic microorganisms is potentially an infectious waste. Equipment such as centrifuges or dialysis machines cannot be incinerated effectively. This type of waste is usually chemically cleaned prior to disposal.

Miscellaneous Infectious Wastes – This category includes wastes that may be potentially infectious but do not conveniently fit into the other categories. For example, materials used during cleanup of spills such as towels, spill pillows, and personal protective equipment worn during cleanup all fit into this category. These wastes can generally be incinerated.

Checking Your Understanding

1. What is the definition of infectious waste?
2. Identify the various types of infectious and medical wastes. Give examples of each.
3. For each type of infectious and medical waste identified in question 2, explain how the waste is disposed of.

8-6 Naturally Occurring Toxins

Before man introduced artificially produced toxic substances into the environment, many toxins were – and still are – existing in nature in a variety of animals, plants, bacteria, and viruses. These toxins may have evolved as a protective mechanism against predators or as tools to immobilize prey. Their effects on humans take place in different tissues and vary from producing mild to debilitating illness or death. Their effects are generally the same as those produced by the man-made chemical toxins: they all are capable of disrupting normal cell function. Following are some examples of naturally occurring toxins in the four main groups: bacteria, viruses, animal, and plants.

Bacteria-Derived Toxins

Botulism

Botulism is associated with the presence of the organism *Clostridium botulinum*. There are four distinct types of botulism: 1) food-borne botulism, 2) infant botulism, 3) adult enteric (intestinal) infectious botulism, and 4) wound botulism. Regardless of the type, it is the production of a toxin by the organism that is responsible for the illness.

Food-borne botulism is probably the most recognized form of this type of illness. In the past it was most often associated with consumption of foods that were poorly preserved such as smoked hams and canned foods. More recently botulism has been associated with freshly prepared foods – such as potato salad – that are not sufficiently refrigerated.

Infant botulism is the most common form of botulism reported in the United States. When it occurs, it is most likely to happen during the second month after the child is born. *Clostridium* spores colonize the intestinal tract and begin to produce the toxin once they become established. The cause of adult enteric botulism is not clearly known, but is believed to be similar to infant botulism

Wound botulism is rare. This form of botulism occurs when spores enter a wound via contaminated soils or via infected needles during drug use. Symptoms of this type of botulism appear about seven days after contamination occurs.

Toxic Effects – Effects of *Clostridium botulinum* infection may be seen as early as six hours after infection or not until 14 days after infection. The toxin is absorbed into the bloodstream and the lymphatic system and distributed through the circulatory system to the neuromuscular junctions. The toxin blocks the release of the neurotransmitter acetylcholine. This blocking action appears to be irreversible, which explains the prolonged effects experienced by individuals who contract botulism. The central nervous system is not affected by the toxin, possibly because it may not able to cross the blood-brain barrier.

Individuals who contract a mild form of botulism experience muscular weakness, abdominal pain, nausea, vomiting, and diarrhea. More severe cases of botulism can result in peripheral muscular weakness or respiratory paralysis, which can lead to death.

Lyme Disease

Lyme disease was first identified in 1975. It is transmitted to humans through bites from ticks infected with *Borrelia burgdorferi*. The first symptoms occur between three and 30 days after the tick bite. The disease usually follows three stages of development, each characterized by different symptoms.

Stage I is characterized by the dissemination of the infectious organism throughout the body via the bloodstream and lymphatic system. A skin rash appears, sometimes accompanied by fever, chills, fatigue, and headaches. In addition, some patients may develop pneumonitis, hepatitis, and early symptoms of meningitis.

During stage II and III neurological symptoms begin to appear, possibly due to the infection spreading to the central nervous system. Patients may develop meningitis and encephalitis. Patients with meningitis often have symptoms of poor memory, confusion, irritability, and disorientation. During stage II, effects of Lyme disease on the heart may also be observed. The most common effect is on the conduction of the electrical signal between the atrium and right ventricle of the heart, producing an irregular heartbeat. The left ventricle may not function properly as well. These symptoms may last up to six weeks.

Development of arthritis is common during stage III. The arthritis usually occurs in large joints such as the knee. This may lead to bone and cartilage erosion in some patients.

Tetanus

Tetanus is the result of infection with the bacterium *Clostridium tetani*. This bacterium produces two exotoxins: tetanolysin and tetanospasmin. Tetanolysin is a **hemolysin** toxin, which is a substance that breaks up the red blood cells. Tetanospasmin is a powerful neurotoxin and is primarily responsible for the symptoms associated with tetanus.

Symptoms develop following contamination of wounds with *Clostridium tetani* spores, which release a toxin that is absorbed into the central nervous system. Symptoms first become visible between three days and three weeks after infection. Symptoms are mediated by the inhibitory effect of the toxin at various neuron-neuron and neuromuscular junctions. The classic tetanic muscle spasms are due to disruption of the inhibitory neurons associated with skeletal muscle contraction. As a result, the skeletal muscles only receive excitatory stimuli, which forces them to remain in a state of contraction. The disease can be prevented by obtaining routine immunization shots.

Toxic Shock Syndrome

In the 1970s and 1980s deaths from **Toxic Shock Syndrome (TSS)** became widely publicized. Most of the early cases were associated with menstruating females. However, additional investigation has shown that TSS may occur in children and men as well.

It is believed that TSS is caused by the common bacterium *Staphylococcus aureus*, which produces a **pyrogenic** toxin (a substance that generates heat or fever). Symptoms usually occur one to three days after infection. Initially patients may experience a sore throat, fever, fatigue, and headache, followed by vomiting or diarrhea, signs of **hypotension** (low blood pressure) and development of a rash. Temperatures of 104-106°F are common. The effects of hypotension vary from slight dizziness to shock, which may be lethal. Hypotension may also cause kidney failure. Antibacterial therapy during early stages of TSS has been shown to be effective in treating and alleviating the symptoms.

Virus-Caused Illnesses

Ebola (African Hemorrhagic Fever)

The Ebola and Marburg viruses are distinguished from other viruses that cause diseases such as yellow fever or Lassa fever. Ebola and Marburg are responsible for producing acute hemorrhagic fever in humans and other relatively unique symptoms. Early symptoms include headache, progressive fever, sore throat, and diarrhea. Hepatitis develops four to five days after infection. By the fifth or sixth day bleeding from various sites is observed, usually from areas covered by thin mucous membranes such as the interior of the nose or of the gastrointestinal tract and other internal organs. Blood does not coagulate and shock may develop. If the patient progresses to this level, death usually occurs by the ninth day.

Hantavirus

Deaths caused by infection with the Hantavirus in the Southwestern part of the United States during the 1990s have renewed attention to this organism. The virus was first described in the late 1970s. It is found in rodents and shrews and is spread by contact with excreta from these animals.

Symptoms first begin to appear 10-35 days after infection. During the initial phase of the disease there may be abrupt chills and fever associated with headache, backache, weakness, and general malaise. Blood platelets decrease significantly. During the following phase the fever declines, but a general state of hypotension occurs, caused by the decreased blood volume due to gastrointestinal bleeding and loss of blood protein through the capillaries. This phase is followed by what is referred to as an **oligouric** phase (reduced excretion of urine). As a result, metabolic wastes build up in the bloodstream producing lethal effects if they are not removed. This phase is then followed by a phase characterized by increased amounts of urine being produced, resulting in the loss of excess amounts water and of electrolytes – such as Na^+, K^+, and Ca^{++} – which can be life threatening.

There are no vaccinations to prevent infection or to fight the virus once infection occurs. To avoid death, patients must be watched continuously in order to provide medical care, which helps maintain fluid and electrolyte balance, blood volume, and in order to promptly detect and treat possible secondary bacterial infection.

Animal-Derived Toxins

Animal toxins are produced by a variety of species throughout the animal kingdom. The toxins are produced for the purpose of defense or of obtaining prey, or – as in the case of poisons – as a byproduct of metabolism. Animals capable of producing toxins can be divided into two categories: venomous and poisonous animals. Venomous animals produce the toxin in highly specialized secretory glands. Contact with the toxin is the result of being bitten or stung. Poisonous animals are characterized by having various tissues that contain the toxin. Exposure to the toxin requires ingestion of the poisonous tissue.

Snakes are probably the most recognized types of venomous organisms. There are approximately 400 different types of snakes that can deliver dangerous amounts of toxin to humans. The venom of snakes is composed primarily of several types of enzymes and polypeptides (proteins), which are responsible for the observed effects.

The types of effects observed vary according to the type of snake venom. Symptoms may include changes in blood vessel permeability; changes in the cellular components of the blood and the mechanism associated with clotting; effects on the heart and respiratory system; and dysfunction of the nervous system. Death – usually as a result of hypotension or shock – may occur anywhere from less than one hour to several days after being bitten. Shock and hypotension occur as a result of increased permeability of the capillaries to proteins and other large molecules resulting in fluid loss from the circulatory system. In addition, there may be some hemolysis and development of acidosis (low blood pH), which affects the respiratory system and the heart.

Poisonous and venomous organisms are found throughout the marine environment. There are approximately 1,200 marine organisms ranging from single cell protista – which include various algae, diatoms, bacteria, and fungi – to more complex organisms such as the various types of poisonous and venomous fish. The types of toxins vary and produce a range of effects from localized erythema to dysfunction of the nervous system and cardiovascular effects.

"Red tide" is the massive accumulation of single-cell dinoflagellatae in various regions of the marine environment. These dinoflagellatae can be lethal to large marine animals including fish. Human consumption of shellfish that have eaten the dinoflagellatae may produce gastrointestinal or respiratory distress. The shellfish become poisonous as a result of the formation of a toxic substance referred to as saxitoxin.

There are up to 700 marine fish that are either poisonous or toxic to humans. Many of these fish are commonly used as food sources and include snapper, sea bass, eel, tuna, skipjack, and bonito. Snapper, sea bass, and eel are the most common types of marine fish associated with poisoning. Symptoms include increased salivation, lethargy, respiratory difficulties, ataxia, and loss of reflexes. In tuna, skipjack, and bonito a toxin may form in the muscle if the fish is not well preserved. The toxic substance is referred to as saurine. Ingestion of the contaminated fish can cause nausea, vomiting, and diarrhea. The symptoms may first appear two hours after ingestion, but they usually disappear within a day.

Plant-Derived Toxins

Like the animal kingdom, the plant kingdom has evolved a variety of toxins used primarily as a defensive mechanism against predators. Over the centuries humans have used these naturally occurring toxins for many purposes, including medical preparations and arrow poison for hunting. Unfortunately, they were also used for less than noble purposes in deliberate poisoning of political adversaries.

The effects of plant toxins are as varied as those of animal toxins. It seems that all major organ systems can be affected by some type of plant toxin. Ingestion of various types of plant toxins result in the common symptoms of nausea, vomiting, and diarrhea. Some plants, such as the castor bean – which is an ornamental plant grown in temperate regions – can cause additional and sometimes lethal symptoms such as bloody diarrhea and dehydration.

Horse chestnuts and the Ohio buckeye are common trees that produce nuts containing a toxin. Ingestion by humans causes inflammation of the gastrointestinal tract. Poisoning in cattle is more prevalent as a result of eating the nuts as well as new leaves and buds. The poison has a neurotoxic effect in cattle, which is characterized by a stiff-legged gait or by seizures.

There are several species of plants that cause contact dermatitis. In addition to poison ivy, common household plants such as narcissus, hyacinths, daffodils, tulips, and *Philodendron* also produce this effect. Dermatitis is caused by the release of chemicals in sap or other fluids from the damaged leaves or stems.

A number of plant toxins produce cardiovascular effects. Perhaps the best known of these is mistletoe. Mistletoe contains a chemical substance called phoratoxin. Exposure results in hypotension, bradycardia (slow heart rate), decreased strength of contraction and vasoconstriction of blood vessels in the skin and muscle.

Curare is well known for its neurological effects. It is obtained from two tropical species of plants, *Strychnos* and *Chondrodendron*. Curare causes respiratory paralysis by inhibiting the transmission of nerve impulses across the neuromuscular junction of respiratory skeletal muscle.

Tall larkspur is commonly found in the Western United States and may be consumed by grazing cattle. This plant contains the toxin methyl caconitine. Ingestion of the substance causes muscle tremor, ataxia, and respiratory failure. Like curare, methyl caconitine blocks acetylcholine at the neuromuscular junction.

Cyanogens are present in apple seeds as well as in cherry and peach pits. Although apple seeds pose little risk, there is enough cyanogen in peach pits to poison a small child if several are eaten. Laetrile – a drug derived from apricot pits – was initially touted to be effective in fighting certain types of cancer; however, its effectiveness was never substantiated.

Checking Your Understanding

1. How is botulism caused, and what are the toxic effects associated with it?
2. What is Lyme disease and what symptoms are associated with its various phases?
3. What are the symptoms associated with tetanus? Briefly explain how these symptoms may develop.
4. What is Toxic Shock Syndrome? How is this disease caused and what are the symptoms?
5. What is the difference between venomous animals and poisonous animals? Give examples of each.
6. What is "Red tide"? Why is it considered dangerous?
7. Identify various plants that produce toxins, the route of exposure, and their effects.

Summary

Chemical, physical, and biological agents are encountered daily in our environment. The effects of exposure can range from slight irritation of the skin, eyes, or respiratory tract to the development of life-threatening diseases such as cancer. The symptoms that develop vary according to the type of substance and length of exposure.

Radiation, Radioactive Material, and Mixed Waste

Ionizing radiation may be in the form of alpha particles, beta particles, gamma rays, x-rays, and neutrons. Alpha and beta particles have their greatest effect in the lungs. Gamma rays, x-rays, and neutrons penetrate the skin and cause damage to internal tissues.

Radiation exerts its effects on the structural and functional molecules of the cell. If the absorbed dose is high enough then cell death occurs. Cell death is associated with breaks in both strands of the DNA. Low levels of radiation cause single-strand breaks in DNA, which may either be repaired or cause mutations.

The effects produced by radiation exposure are dependent on the sensitivity of the cell. Actively dividing cells are more sensitive, while highly differentiated cells are less sensitive. In the fetus exposure can result in lethality, growth retardation, malformation, and neuropathology. In adults the symptoms may be severe radiation burns or cancer.

Chemical and Biological Warfare Agents

A large number of chemical agents has been developed and stored around the world. These agents may either produce an irritating effect to the skin and eyes; or cause death by interacting with the nervous system and lungs. The various types of chemical agents include lacrimators and vomit-inducing gases, respiratory irritants, skin irritants, vesicants, nerve gases, and incapacitating gases. Biological agents may consist of bacterial agents, viral agents, species of *Rickettsia*, and fungal agents.

Blood-borne Pathogens and Infection and Medical Waste

Transmission of pathogens via contact with contaminated blood or blood-contaminated articles is one way of spreading disease. Pathogens commonly transmitted are the AIDS virus, the hepatitis virus, bacteria associated with syphilis, and various other organisms.

Pathogenic organisms may also be transmitted by contact with infectious medical wastes. This waste may include human blood and blood products; cultures and stocks of infectious agents; pathological wastes; contaminated sharps; contaminated laboratory wastes; contaminated wastes from patient care; discarded biologicals; contaminated animal carcasses, body parts; contaminated bedding and equipment; and miscellaneous infectious wastes. Proper handling of these types of wastes is important in preventing infection and the spreading of pathogenic organisms.

Naturally Occurring Toxins

There are many microorganisms in the plant and animal kingdom that produce toxins, poisons, or that cause illness in other organisms with their presence. Bacteria and viruses are well known for their ability to cause illnesses or diseases such as botulism, Lyme disease, and hepatitis. Poisons and toxins are produced in a wide range of animals from the single-cell dinoflagellatae that cause "red tide" to venomous snakes and poisonous fish. Many species of plants such as *Philodendron*, Mistletoe, and Ohio buckeye are toxic to a variety of species.

Application and Critical Thinking Activities

1. Leukemia is frequently associated with acute exposure to radiation. Propose an explanation for this association.

2. In this chapter the use of the hellebore root was described as an example of a chemical warfare agent. What is the active ingredient in hellebore root and what are its toxic effects? Are there any other known uses for this substance?

3. A variety of chemical and biological warfare agents were discussed in this chapter. Select a substance from the list of naturally occurring toxins discussed and propose how it might be used as a warfare agent. What would be its advantages and disadvantages?

4. There are claims that cellular telephone users may develop health problems from being exposed to damaging electromagnetic radiation. How much radiation is produced by a cellular telephone and, based on your research, are the claims warranted or unwarranted?

5. Several naturally occurring toxins are discussed in this chapter. Obtain references that discuss these types of organisms, identify at least one other species from each of the categories identified in this chapter, and discuss the effects of exposure to the toxins.

9

Mutagens, Teratogens, and Carcinogens

Chapter Objectives

Upon completing this chapter, the student will be able to:

1. **Define** the term mutation.
2. **Identify** and distinguish between various types of mutations.
3. **Define** the terms teratogen, carcinogen, and neoplasm.
4. **Distinguish** between a benign and a malignant tumor.
5. **Understand** the terminology associated with identification of cancer based on tissue origin.
6. **Identify** and discuss the types and sources of carcinogens.
7. **Identify** and explain the stages of tumor development.
8. **Recognize** the different agencies involved in classifying carcinogens.
9. **Describe** the different types of ionizing radiation.
10. **List** and explain the effects of ionizing radiation.
11. **Identify** the types of nonionizing radiation.
12. **List** and explain the effects of nonionizing radiation.

Chapter Sections

9-1 Introduction

9-2 General Chromosome Structure and Function

9-3 Mutagens

9-4 Teratogens

9-5 Carcinogens

9-6 Radiation

9-1 Introduction

One of the most feared effects of exposure to a particular chemical or physical agent is cancer. Although the fear is well founded, cancer is not always lethal. In addition, cancer is not the only concern, since other illnesses may occur as a result of exposure that may be as lethal or as debilitating as cancer.

The symptoms associated with cancer are due to changes in the normal functioning of cells. These changes are a result of exposure to toxic substances affecting either the DNA in the cell nucleus or other cell functions. Changes in the DNA, referred to as mutations, are caused by substances referred to as **mutagens**. During fetal development changes in the DNA may result in severe and disabling birth defects or death. The particular agents causing birth defects are referred to as **teratogens**.

In order to understand the mechanism of action of mutagens, teratogens, and carcinogens, it is important to have a basic understanding of their primary interaction with the DNA of chromosomes. The following discussion provides a general overview describing the basic structure of DNA and its role in protein synthesis.

9-2 General Chromosome Structure and Function

Chromosomes are composed primarily of DNA and **histones**. The DNA consists of long strands of **nucleotides**, each of which is composed of a sugar (deoxyribose), a base, and a phosphate group linked together (Figure 9-1).

> A nucleotide is the portion of the DNA molecule that consists of deoxyribose, a base, and a phosphate group.

The histones are proteins that play a role in regulating the expression of genes located on the chromosomes. DNA is composed of two strands of nucleotides bound to each other and forming a double-helix structure. The binding occurs as a result of hydrogen bonds between the bases in each strand. There are four bases found in DNA: adenine and guanine (referred to as purines), and thymine and cytosine (referred to as pyrimidines) (Figure 9-2). These bases are arranged in a sequence in the DNA molecule; bases in one strand of the DNA bind to bases in the opposite strand in a very specific manner: adenine binds to thymine; cytosine binds to guanine. In somatic cells there are two sets of chromosomes, whereas sex cells contain only one set of chromosomes.

The genes, which are portions of the DNA molecule and are located along the chromosome strands, may be composed of tens to thousands of nucleotides. They are responsible for initiating the synthesis and production of structural and functional proteins important in normal cell function. There is a very specific sequence of events that occurs during the process of protein production. The double helix of the chromosome unravels and the two strands separate exposing a sequence of bases. The exposed bases serve as a code – or source of information – for production of a specific protein. The DNA code directs the manufacture of a molecule referred to as **messenger RNA (mRNA)**.

> mRNA "interprets" the information in the DNA necessary for protein synthesis and carries it to the site of protein synthesis.

The mRNA is made up of bases that are the mirror image of – or complementary to –the bases in the exposed portion of the DNA molecule. The messenger RNA "carries" the genetic information from the nucleus to the ribosome, located in the cell cytoplasm, and attaches to it.

Amino acids are the building blocks of proteins. The sequences of amino acids in the protein are determined by base triplets on the mRNA and are referred to as **codons**. Each codon specifies the amino acid that will be incorporated into the protein. Therefore, the sequence of codons in the mRNA will determine the sequence of amino acids in the protein. Another kind of RNA, **transfer RNA (tRNA)** carries a single amino acid and positions it according to the code possessed by the mRNA.

> tRNA transports the building blocks – the amino acids – to the site of protein production.

The interactions between the mRNA and the tRNA result in the binding of the individual amino acids in a specific sequence to form a protein molecule. Figure 9-3 provides an overview of the process and components involved.

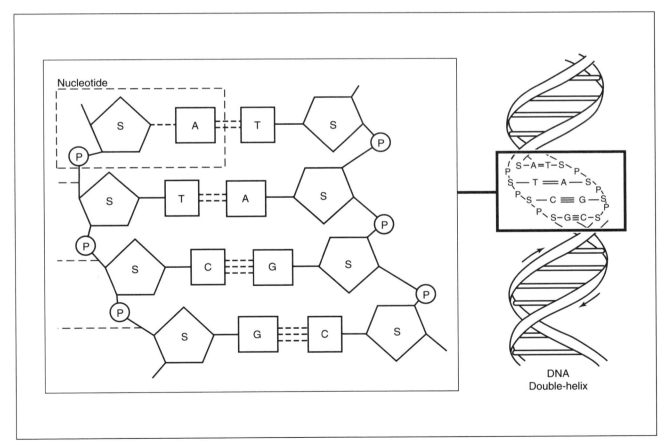

Figure 9-1: Generalized illustration of the DNA molecule showing the individual strands arranged in a double-helix configuration. The strands are held together as a result of bonds between the base pairs.

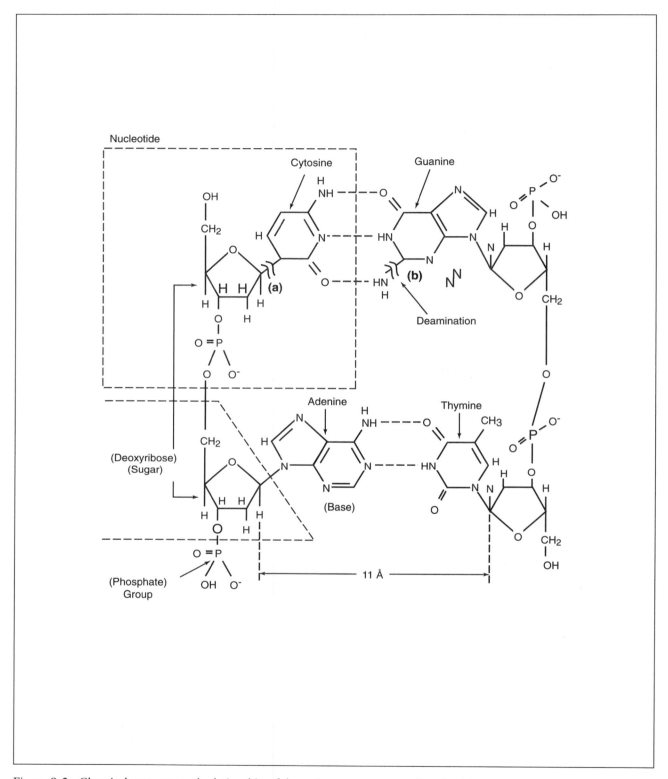

Figure 9-2: Chemical structure and relationship of the various components of nucleotides.
a. Breaking the bond between the base (cytosine) and the sugar (deoxyribose) may result in a spontaneous mutation if the bond is not repaired.
b. Removal of the amine group ($-NH_2$) by deamination may also result in spontaneous mutation.

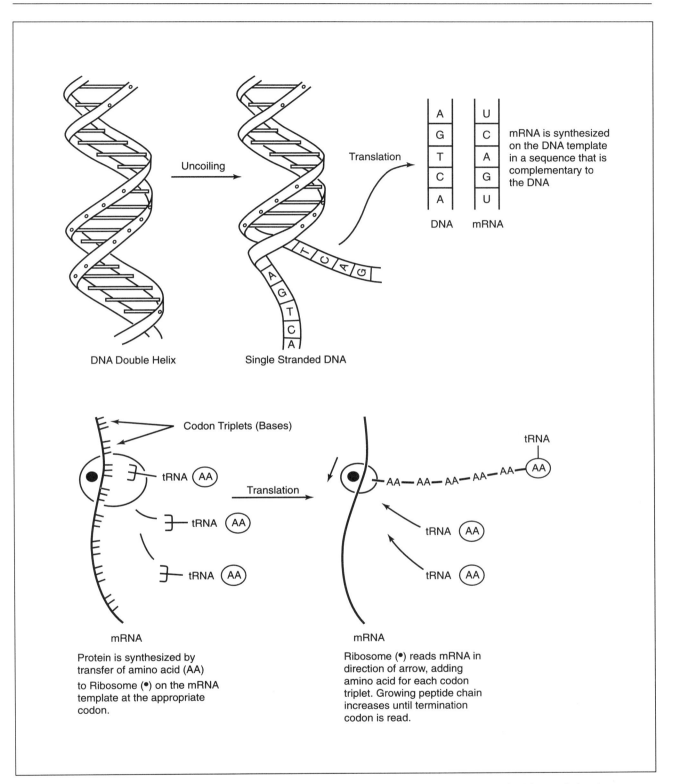

Figure 9-3: Simplified illustration of the role of DNA and RNA in protein synthesis. *Recreated with permission from Williams P. L. and Burson J. L.*, Industrial Toxicology: Safety and Health Applications in the Workplace, *Van Nostrand Reinhold, New York, 1985.*

Checking Your Understanding

1. Define the terms mutagen and mutation.
2. Provide a general description of the components of the DNA molecule and their arrangement within the molecule.
3. Which DNA base pairs bind to each other?
4. What are the basic components of a gene? What do they produce?

9-3 Mutagens

Mutations have occurred since the beginning of life, leading to the development of new species of bacteria, algae, fungi, protozoa, plants, and animals. Changes in the DNA allowed some organisms to adapt to their environment. Other mutations not only did not enhance an organism's chance of survival but, on the contrary, may have hastened its demise.

In many textbooks, discussions on DNA focus primarily on its structure and function as it relates to protein synthesis and cell division. Except during cell division, where changes in segments of chromosomes can occur, the DNA molecule is portrayed as a relatively "inert" and intact molecule that maintains its integrity on a continuous basis. However, DNA interacts with a variety of endogenous (inside) and exogenous (outside) toxic substances on a daily basis. Some of these interactions result in mutations.

> Mutations are changes in the base sequence of DNA that may be either beneficial, harmful, or have no effect.

Categories of Mutations

Spontaneous Mutations

Mutations may occur without the introduction of an exogenous mutagenic agent. These types of changes are referred to as **spontaneous mutations**. Various types of chemical reactions that normally occur in the cell may cause these type of mutations. For example, the bond between the sugar – deoxyribose – and any of its four bases – cytosine (C), thymine (T), guanine (G), or adenine(A) – can be broken (Figure 9-2) by chemical intermediates such as superoxides (O_2^-), hydroxyl-free radicals (OH*), and hydrogen peroxide (H_2O_2) produced during cell metabolism. It is estimated that breaks in the DNA due to interaction with these chemical intermediates occur at a rate of 100 per cell per day. If the DNA is not repaired, a blank in the DNA code occurs at this location. A second type of chemical reaction, called **deamination**, may cause a spontaneous mutation. This process removes an amino ($-NH_2$) group from the bases, thus affecting the bonding between base pairs (Figure 9-2). Removal of an amine group from cytosine produces uracil, which is a base normally found in mRNA, not DNA. If left uncorrected, uracil will bind with adenine instead of guanine when the DNA is replicated. A third type of intracellular chemical reaction that can cause mutations is methylation (addition of a methyl group, $-CH_3$). Methylation results in the change of the chemical structure of guanine and adenine, thus affecting the bonding with their complementary base pairs, cytosine and thymine, respectively.

There are several endogenous enzymes that are involved in DNA repair. These enzymes can make a mistake during replication – such as substituting the wrong base – producing a mutation. In addition, DNA bases, such as thymine, may exist in multiple isomeric forms that are in equilibrium. The predominant isomer – referred to as the keto form – binds with adenine, which is considered to be the normal base pairing (T–A). The less prevalent isomer – the enol form–binds with guanine (Figure 9-4). Therefore, a small number of abnormal thymine-guanine (T–G) base pairs will exist initially in a DNA molecule during replication.

Spontaneous mutations may also occur as a result of addition or deletion of base sequences during DNA replication. These types of mutations may occur when there are short complementary base sequences located in the same DNA strand (Figure 9-5). Recall that during DNA replication the hydrogen bonds are broken and the two strands separate, and also that the complementary bases (A–T and G–C) normally bind to each other. This binding will occur if the bases are in opposite strands of a single DNA molecule (normal) or if they are in the same strand of DNA (abnormal). For example, in Figure 9-5(a) the adenine (A) and thymine (T) in the template strand bind to each other during DNA replication. As a result only the cytosine bases are "exposed and read" during formation of the new complementary DNA strand. Therefore, nine bases are deleted in the new DNA strand. In 9-5(b) the opposite occurs: nine new bases are inserted in the new DNA strand.

> Spontaneous mutations occur without the introduction of exogenous mutagenic agents.

Induced Mutations

Induced mutations are associated with the introduction of exogenous chemical agents or physical

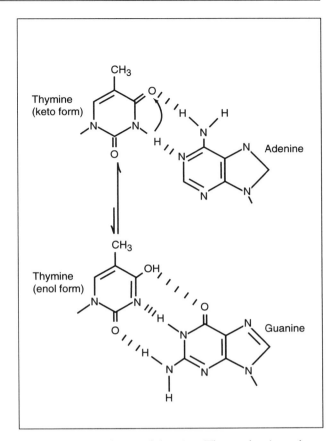

Figure 9-4: Two forms of thymine. The predominant keto form binds with adenine. The enol form binds with guanine, which results in a mutation that can be passed on during replication. *From* Principles of Cell and Molecular Biology, *2nd edition, by Lewis J. Kleinsmith and Valerie M. Kish. Copyright © 1995 by HarperCollins College Publishers. Reprinted by permission of Addison Wesley Educational Publishers.*

agents – such as radiation – into the cell. Chemical mutations occur by any of the following three general mechanisms: 1) base-analog substitution, 2) base modification, and 3) base intercalation. Base-analog mutations occur when a chemical substance similar in chemical structure replaces one of the bases. These chemicals (base analogs) are capable of forming bonds with the normal bases found in DNA, producing a mutation. For example, 5-bromouracil is similar in structure to thymine. During DNA replication thymine is replaced by 5-bromouracil, which does not bind with thymine's normal base pair adenine. Rather, 5-bromouracil binds with other bases such as guanine.

Base-modifying agents are chemicals that interact with the DNA bases changing their structures.

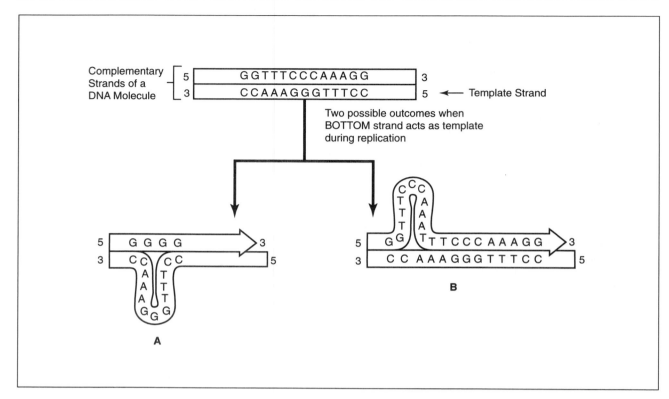

Figure 9-5: Spontaneous mutations during DNA replication. Deletions and insertions occur when complementary bases in the same strand bind to each other.
a. Loop forms in template strand during replication, causing nine bases to be deleted in the new strand.
b. Loop forms in the replicating strand during replication, causing nine bases to be inserted in the new strand.
From Principles of Cell and Molecular Biology, *2nd edition, by Lewis J. Kleinsmith and Valerie M. Kish. Copyright © 1995 by HarperCollins College Publishers. Reprinted by permission of Addison Wesley Educational Publishers.*

These agents may either remove the amino group of the base, or add a hydroxyl group (OH$^-$), or add a hydrocarbon group (C_nH_n) to the base. These types of changes are similar to those caused by interaction with endogenous chemical agents discussed previously. Addition or removal of these groups affects the binding properties of the bases by changing their chemical structure. Examples of chemicals that may cause these types of changes include chemotherapeutic agents such as chloroethylnitros-ureas, nitrogen mustards, epoxides, and aldehydes.

Intercalating agents are substances that are inserted between the bases in the DNA and that form a hydrogen bond with one of the bases. For example, Actinomycin D intercalates (inserts) between guanine and cytosine (G-C) base pairs and forms hydrogen bonds with guanine. This overall process can cause unraveling of the DNA helix by breaking the hydrogen bonds between base pairs. This type of chemical interaction may result in a base either being deleted from or inserted into the DNA.

Physical agents such as radiation can also cause mutations. **Ultraviolet radiation (UV)** from the sun is known to cause skin cancer as a result of prolonged exposure. UV radiation affects the bonds between adjacent pyrimidines located on the same or opposite DNA strands. When this occurs, thymine, for example, will bond to thymine forming a T–T dimer called cyclobutyl thymine (Figure 9-6). Radiation also causes single strand breaks, double strand breaks, and various types of DNA base damage. Of these, single strand breaks are more easily repaired by the intracellular DNA repair mechanism. On the other hand, when a double strand break in the DNA molecule occurs, there are three possible outcomes: 1) the molecule will be connected with no errors; 2) the molecule is repaired incorrectly and produces a mutated DNA molecule; 3) the DNA molecule is not repaired. A DNA molecule with a double strand break results in fragmentation of the chromosomes followed by an abnormal distribution of chromosomes during cell division.

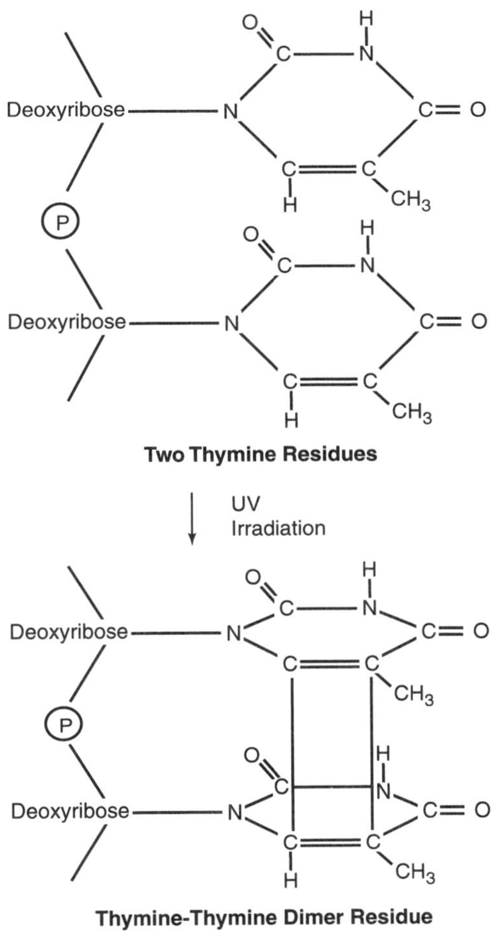

Figure 9-6: Formation of a thymine-thymine dimer in DNA. UV radiation can cause adjacent thymine bases to bind to each other (T–T) resulting in a mutation. If not repaired it may initiate events associated with skin cancer.

> Induced mutations are caused by exogenous agents that alter the base of the DNA.

Large and Point Mutations

Mutations are also categorized according to the size – large and point – of the DNA segment undergoing mutation. Deletion, inversion, and translocation are processes that involve hundreds to thousands of base pairs and several different genes, producing changes in large segments of the DNA. These are called **large mutations** and have been associated with the development of tumors in somatic cells. Deletions – which occur at any point along the chromosome – generally result in fewer bases in a chromosome. This is caused by a break in the DNA strand that is not repaired. Inversion is a chromosomal aberration in which a segment of the DNA is inverted 180 degrees. Translocation occurs when segments of DNA are exchanged between different chromosomes, or when segments are transferred to a different part of the same chromosome. Exposure to radiation, for example, can cause double-strand breaks in the DNA molecule resulting in the translocation of large sections of the DNA between chromosomes.

On the other hand, a change in a single base is a **point mutation**. The substitution of cytosine for thymine, adenine for guanine, or vice versa for either pair, is referred to as a **transitional point mutation**. Another type of point mutation is the **transversion point mutation**, which occurs when purine bases are substituted for pyrimidine bases. For example, adenine might be replaced by cytosine, or guanine for thymine. Point mutations are caused by endogenous metabolic processes or by the exogenous agents discussed previously.

Somatic and Germinal Mutations

Mutations may also be categorized according to their location. If they occur in cells other than the germinal cells of the reproductive system, they are referred to as **somatic mutations**. Mutations occurring in the reproductive cells are referred to as **germinal mutations**. The mechanisms causing the changes in these cell types are the same. The differentiating factor is that somatic mutations are limited to the individual in which the mutation occurs, whereas mutations of the germinal (sex) cells may be passed on to the offspring, where its effects may be observed.

> Mutations may also be categorized according to the size of change in the DNA molecule and the type of cell where the mutation occurs.

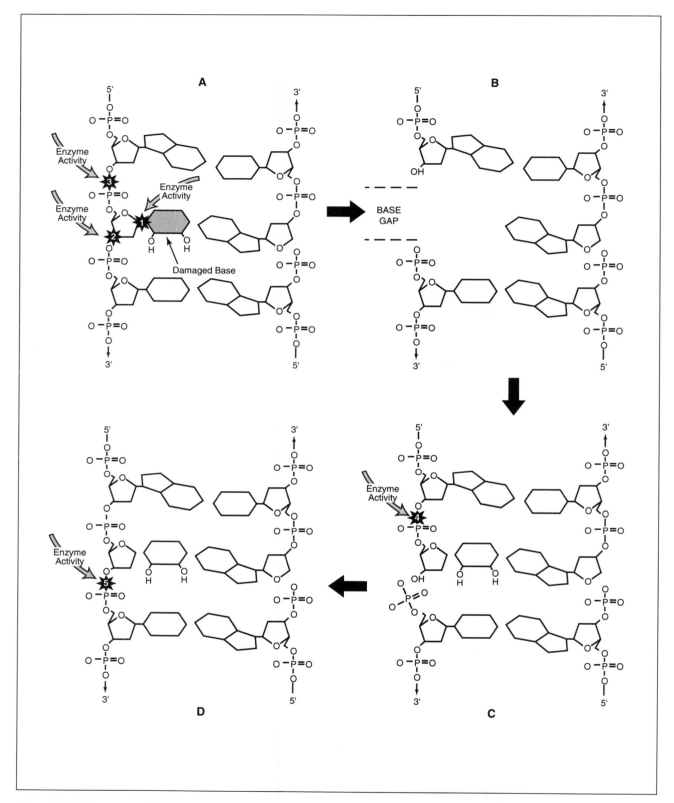

Figure 9-7: Illustration of the events associated with the base excision repair (BER) mechanism. The damaged base and associated sugar and phosphate group are removed through enzyme activity (A and B). The gap is filled by a new nucleotide through enzyme activity (C), which promotes nucleotide synthesis (D).

DNA Repair and Mutations

If all mutations caused by the various events described above were left intact, the cell, the tissue, or perhaps the whole organism, would probably not survive very long. Fortunately, there are intracellular DNA mechanisms that are able to recognize changes in the DNA molecule and repair them. Two of them are the **base excision repair (BER)** and the **nucleotide excision repair (NER)**.

As mentioned previously, reactions such as methylation or deamination can damage DNA bases. Figure 9-7 illustrates the sequence of events associated with the BER process involved in replacing a damaged base. A specific enzyme breaks the bond between the damaged base and the deoxyribose sugar (A). A different set of enzymes breaks the bond between the sugar and the phosphate group. This will leave a single base gap in the DNA molecule (B). Nucleotide synthesis occurs and fills the gap (C-D). This type of repair occurs frequently in response to spontaneous mutations.

The NER repair mechanism is more common and also requires several different types of enzymes to perform the excision (cutting out) and subsequent repair. There are five steps involved in the NER process (Figure 9-8). The damaged portion of the DNA molecule is identified (A) and interacts with enzymes, resulting in incision of the altered portion of the DNA strand at both ends (B). A separate group of enzymes is then involved in the removal of the damaged portion of the DNA strand, leaving a gap (C-D). The sequence of bases of the undamaged strand serves as a template for synthesis of new DNA, which will fill the gap in the damaged strand (D). The new DNA segment is then joined at both ends to the existing DNA molecule (E).

These mechanisms are two of the many DNA repair mechanisms that are involved in limiting the potential mutation damage. It is estimated that there are hundreds of genes involved in the synthesis of the enzymes required for recognition and repair of altered DNA. In addition, mutations of the genes involved in the DNA repair mechanism itself may hinder the repair of altered DNA, allowing mutated genes to survive and affect normal cell functioning. For example, individuals who have the disease **Xeroderma pigmentosum** are extremely sensitive to UV radiation and readily develop skin cancer. This disease has been associated with mutation of the genes involved in DNA repair. Here the cross-linking of the pyrimidine and thymine discussed earlier is not repaired and may ultimately lead to the development of skin cancer. Other types of diseases associated with DNA-repair defects are, for example, **Cockayne's syndrome** – caused by UV irradiation and characterized by dwarfism, retinal atrophy, photosensitivity, and deafness – and **Fanconi's anemia** – due to cross-linking agents and potentially leading to leukemia.

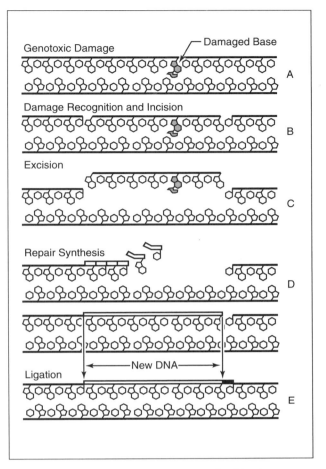

Figure 9-8: Illustration of the steps involved in the nucleotide excision repair pathway (NER). There are several enzymes involved in recognition, removal, and synthesis of DNA. *Recreated with permission from Craighead J. E.*, Pathology of Environmental and Occupational Disease, *Mosby Co., St. Louis, 1995.*

If the DNA repair mechanisms do not have sufficient time to repair the DNA damage, or if the number of mutations overwhelms the repair mechanisms, then the mutation can be passed on to the new cells during cell division. Base-pair mutations or DNA strand breaks may lead to the dysfunction of cell processes such as active transport and ATP production. As a result, production of enzymes involved in these cell functions is decreased or nonexistent. This can cause a loss of cell membrane integrity and subsequent cell death. Death of a single cell does not normally result in an adverse effect. In fact, it may be advantageous for mutated cells to be eliminated from the cell population. However, if the number of mutated dysfunctional cells is large for a given tissue type, then it is possible to have more adverse effects in the organism as a whole. Cancer, nerve degeneration, and premature aging are some of the illnesses associated with mutations affecting the whole organism. Table 9-1 lists various diseases associated with genetic mutations and chromosome abnormalities. In germinal cells, mutations that are not repaired may be severe enough to prevent embryonic development and lead to miscarriage. On the other hand, the mutation may be passed on to the offspring where it may subsequently express itself.

Checking Your Understanding

1. List and describe the chemical reactions that may be associated with spontaneous mutations.

2. Briefly explain how addition and deletion of nucleotides result in spontaneous mutations.

3. Distinguish between spontaneous and induced mutations. Explain how each may occur.

4. What is the difference between a large mutation and a point mutation?

5. Differentiate between somatic and germinal cell mutations.

6. Is it possible for somatic cell mutations to be passed on to an individual's offspring? Why or why not?

7. What is the role of the base-excision repair and the nucleotide repair pathway in preventing mutations?

Chromosome Abnormalities

a **Cri-du-chat syndrome** (partial deletion of chromosome 5)
b **Down's syndrome** (triplication of chromosome 21)
 Klinefelter's syndrome
c (XXY sex chromosome constitution; 47 chromosomes)
d **Turner's syndrome** (X0 sex chromosome constitution; 45 chromosomes)

Gene Mutations

e **Chondrodystrophy** (bone growth disorder)
f **Huntington's chorea**
g **Retinoblastoma**
h **Albinism**
i **Cystic fibrosis**
j **Diabetes mellitus**
k **Fanconi's syndrome**
l **Hemophilia**
 Xeroderma pigmentosum
m **Anencephaly**
n **Club foot**
o **Spina bifida**

a Congenital abnormality so named because the infant's cry resembles the cry of a cat. Characterized by mental retardation, small brain, dwarfism, and defects associated with voice box.
b Congenital abnormality characterized by moderate-to-severe mental retardation.
c Congenital abnormality characterized by abnormal development of the testes resulting in development of individuals with long legs an subnormal intelligence.
d Congenital hormone disorder caused by failure of the ovaries to respond to pituitary hormone stimulation. Individuals are short in stature and do not attain sexual maturation.
e A disturbance in bone growth resulting in dwarfism. Individuals have shorter than normal arms and legs, but the rest of the bones grow normally.
f An inherited disease of the central nervous system that usually has its onset between 30 to 50 years of age. Disease characterized by decline in mental capabilities, and irregular, spasmodic contraction of limb and facial muscles.
g A cancer associated with the retina of the eye.
h A congenital condition characterized by partial or total absence of pigment in the skin, hair and eyes. Individual will appear white.
i An inherited disease that affects various glands of the body, but primarily the respiratory system. Usually begins during infancy and is characterized by chronic respiratory infections.
j An inherited disease in which insufficient amounts of insulin are produced by the pancreas resulting in elevated blood sugar levels, excessive thirst, and urine production and slow healing of wounds.
k An inherited disease characterized by decreased production of blood cells.
l A hereditary blood disorder characterized by a greatly prolonged clotting time. The blood fails to clot and bleeding occurs. Caused by deficiency in specific plasma proteins involved in the clotting mechanism.
m Abnormality caused by lack of development of the brain and skull bones resulting in a condition that is incompatible with life.
n A congenital abnormality associated with the structure of the foot. Several conditions may exist: exaggerated normal arch of the foot; foot is flexed and only the heel touches the ground; or foot is extended and individual only walks on the toes.
o Abnormality caused by incomplete development and joining of vertebrae (spinal column) resulting in the spinal cord protruding through this area.

Table 9-1: Examples of human chromosome and genetic disorders. *Recreated with permission from Williams P. L. and Burson J. L.*, Industrial Toxicology: Safety and Health Applications in the Workplace, *Van Nostrand Reinhold, New York, 1985.*

9-4 Teratogens

The most tragic demonstration of the effects of exposure to a toxic substance occurs when it impacts fetal development and the birth of a child. At birth, the number of children identified as having either a major or minor malformation is approximately five percent. However, all congenital defects are not visible at birth and may develop later. During the first year after birth an additional six percent of the children may display evidence of some congenital defect. Several types of birth defects are listed in Table 9-2. Some of these defects can be caused by exposure to toxic substances during embryonic and fetal development. Toxic agents that cause abnormal development resulting in birth defects are referred to as teratogens (Table 9-3).

Mechanisms of Action

Sixty-five percent of all birth defects have an unknown cause, in twenty percent the cause is genetic, in five percent it is specifically chromosomal, in four percent it is due to maternal illness (including alcoholism), in three percent it is due to infection, in two percent it is mechanical, and in only one percent it is due to toxicants (from R. L. Brent and D. A. Beckman "Principles of Teratology" in *Reproductive Risks and Prenatal Diagnosis*, M. I. Evans editor, Appleton & Lange, 1992). Birth defects may be associated with various mechanisms such as: mutation, chromosomal damage, mitotic interference, altered nucleic acid integrity and/or function, altered energy sources, changes in cell membrane characteristics, and enzyme inhibition (from J.C. Wilson "Current Status of Teratology" in *Handbook of Teratology*, J. G. Wilson and F. C. Fraser editors, Plenum Press, 1977). Birth defects are most likely to occur if exposure takes place during organ development. Organogenesis (development of organs) begins to occur about the third week after conception and continues until the definitive characteristics of the organ are achieved. During this time there is extensive cellular replication and differentiation of cells leading to the formation of the major organ systems. It is during this time that the cells are very sensitive to teratogenic agents. The type and severity of birth defect that occurs is often dependent on the time of exposure and the toxic agent involved. Manifestations of exposure may appear as physical malformations, growth impairment, death of the developing embryo/fetus (miscarriage), or functional impairment of organs. The causes of several birth defects and the estimates of their frequency are identified in Tables 9-4 and 9-6.

For many teratogenic substances high doses are fatal, and the developing embryo/fetus may be aborted. While lower doses do not kill they may al-

Malformation	Rate (per 10,000 infants)
Anencephaly [a]	0.8-18.4
Spina bifida [b]	1.8-17.5
Encephalocele [c]	0.3-3.2
Hydrocephaly [d]	2.1-8.5
Microtia [e]	0.1-6.4
Hypoplastic left heart syndrome [f]	0-3.4
Cleft palate	2.6-10.1
Cleft lip	5.2-16.3
Esophageal atresia [g]	0.4-3.6
Renal agenesis/dysgenesis [h]	0.2-4.1
Limb reduction defects	3.1-8.0
Abdominal wall defects	1.9-6.8
Conjoined twins	0-0.3
Down syndrome [i]	5.2-15.3

[a] Abnormality caused by lack of development of the brain and skull bones, resulting in a condition that is incompatible with life.
[b] Abnormality caused by incomplete development and joining of vertebrae (spinal column) resulting in the spinal cord protruding through this area.
[c] Abnormality characterized by protrusion of the brain between bones of the skull that have not closed at birth.
[d] Abnormality characterized by excess accumulation of fluid in the brain, which may cause enlarged or misshapened head.
[e] Small external ear.
[f] Left ventricle is smaller than normal and is composed primarily of connective tissue instead of muscle.
[g] Abnormality characterized by lack of complete development of the esophagus so there is no passageway to the stomach
[h] Lack of formation of the kidneys or imperfect development of some portion of the kidneys.
[i] Abnormality characterized by varying degrees of mental retardation; individuals have an extra chromosome.

Table 9-2: Rates of occurrence for specific abnormalities, 1974-1988. *Recreated with permission from Craighead J. E.*, Pathology of Environmental and Occupational Disease, *Mosby Co., St. Louis, 1995.*

Radiation
Therapeutic
Radioiodine
Atomic fallout
Infections
Rubellus virus
Herpes virus
Toxoplasmosis
Syphilis
Maternal Metabolic Imbalances
Alcoholism
Diabetes
Folic acid deficiency
Hyperthermia
Rheumatic disease
Drugs and Chemicals
Androgenic chemicals
Antibiotics
Tetracyline
Anticancer drugs
Aminopterin, methylaminopterine, cyclophosphamide, busulfan
Chlorobiphenyls
Cigarette smoke
Cocaine
Warfarin
Diethylstilbesterol
Ethanol
Ethylene oxide
Iodides
Lithium
Metals
Mercury (organic)
Lead
Thalidomide

Table 9-3: Human teratogenic factors. *Recreated with permission from Klaasen C. D. et al.*, Casarett & Doull's Toxicology, The Basic Science of Poisons, *5th edition, McGraw Hill, New York, 1996.*

Drugs of Abuse	
Cigarette smoking	
Ethanol	
Heroin	
Pharmaceutical Agents	
Glucocorticoids	A general class of hormones produced by the adrenal glands that protect against stress and affect protein and carbohydrate metabolism; also used as an anti-inflammatory agent.
Propranolol	A drug used to treat various heart disorders and high blood pressure.
Phenytoin	A drug used to treat involuntary muscle contractions (convulsions). Used to treat patients with epilepsy.
Warfarin	A drug used to prevent blood clotting.
Infectious Agents	
Herpes simplex	An infectious disease caused by a herpes simplex virus.
Cytomegalovirus	A group of viruses that cause enlargement of cells of various organs. Infection of the embryo may result in malformation and fetal death.
Rubella	Acute infectious disease characterized by drowsiness, fever, sore throat, rash and swollen glands in the neck. Infection during the first 2 or 3 months of pregnancy can cause fetal abnormalities such as congenital (ocourring at birth) cataracts.
Toxoplasmosis	A disease due to infection with the protozoa *Toxoplasma gondii*. Infection of pregnant women can result in babies being born with abnormalities such as microcephalus (small heads) or hydrocephalus (fluid on the brain).

Table 9-4: Examples of toxicants associated with impairment of fetal growth. *Recreated with permission from Craighead J. E.*, Pathology of Environmental and Occupational Disease, *Mosby Co., St. Louis, 1995.*

ter the development of the embryo. Exposure to lower doses of the same teratogenic substance on different days may produce different effects. This may occur even if the dose administered is the same for each day. For instance, exposure to a particular substance may produce cleft palate one day and development of a heart defect on the next day.

Growth impairment is more difficult to identify and to associate with exposure to teratogens since there is no obvious malformation. Table 9-4 lists toxic substances that impair fetal growth. Cell death (discussed below) during development is thought to be the cause of the impaired growth. Miscarriages and stillbirths occur in many pregnancies. Data suggest that up to 30 percent of pregnancies spontaneously abort. Stillbirths occur at the rate of approximately eight per 1,000. Twenty percent of the stillborn fetuses have birth defects. Functional impairment is difficult to assess, since it may not occur until several years after birth. During this time exposure to various other toxicants may also occur, making it difficult to discern whether the dysfunction is the result of exposure before or after birth.

There are several proposed mechanisms by which teratogenic substances may cause their effects. These mechanisms include cell death, genotoxicity, impaired cell function, maternal toxicity, and placental toxicity.

Cell Death

The cells of a developing embryo interact with each other by cell-to-cell contact and by production of biochemical messengers. These interactions provide a means of coordinating the differentiation and development of the various cell types within the embryo or fetus. The death of cells as a result of exposure to toxicants may affect the normal development by disrupting the cell-to-cell communication. Brain development is a good example of the importance of cell-to-cell communication. Migration of nerve cells to certain parts of the developing brain requires them to follow a path established by previously migrated neurons (nerve cells) and glial cells (support cells). Death of these initial migrating cells can cause the remaining cells to either not migrate or follow improper paths. The end result may be impaired development of the brain. This type of effect is observed in animals exposed to x-rays during pregnancy. The fetus has a decreased brain size and a thin cerebral cortex (outer portion of brain), both associated with neurological dysfunction.

As mentioned previously, some stillbirths are attributable to exposure to toxic substances. If large numbers of cells are killed by exposure to toxicants then it is possible that a spontaneous abortion will occur. However, if exposure during early pregnancy results in the death of only a small number of cells in the developing embryo, it is possible for the embryo to replace the dead cells and continue its development. For example, mouse embryos exposed to methylmethane-sulfonate – which breaks the DNA strand – spontaneously abort or produce small but normally formed offspring. Table 9-5 lists several toxic substances that may impair fetal growth.

Exposure to toxicants during the later stages of pregnancy usually results in the development of malformations rather than fetal death. Hydroxyurea – which inhibits DNA synthesis – causes cell death; in rabbit embryos the cell death occurs in the tissue associated with limb development, thereby producing limb malformation in newborn rabbits.

During normal embryonic development cells are not only preprogrammed to replicate and differentiate; some cells are preprogrammed to die at a specific time. It is important that these events occur as scheduled to assure normal development. Interference with this preprogrammed cell death – which is part of the process of developing fetuses in the same way as cell replication – can cause birth defects. For example, rats exposed to aspirin are born with more toes than normal, because the aspirin interferes with the preprogrammed death of cells associated with toe development.

Genotoxicity

Genotoxicity refers to the ability of a substance to damage the DNA of a cell. As discussed previously, there are several chemical and physical agents capable of altering the structure of the bases found in DNA or of causing breaks in the DNA strands, possibly leading to cellular dysfunction. These mutations are often associated with the development of various types of diseases or illnesses such as cancer.

Exposure of embryonic and fetal DNA to toxic substances may result in cell death, discussed previously, causing spontaneous abortions or stillbirths. If cell death is not extensive, offsprings are smaller at birth but have no defects. Late exposure can cause malformations associated with the arms and legs. Exposure of female mice to ethylene oxide a few hours after mating had multiple effects ranging from early spontaneous abortions to fetal death and mal-

Physical Agents	
Irradiation	
Social and Personal Habits	
Alcohol ingestion	Mineral deficiency
Cigarette smoking	Hallucinogens
Vitamin deficiency	
Metals	
Tellurium	Lead
Therapeutic Agents	
Chemotherapeutic cancer agents	Large doses of steroid hormones
Streptomycin	Thiazide diuretics
Tetracyclines	Quinine
Sulfonamides	Prednisolone
Novobiocin	Antihistamine antiemetic
Chloramphenicol	Narcotics
Erythromycin	Thalidomide
Anticoagulants	Excess vitamin K
Antidiabetics	Chloroquine
Aminopterin	
Chemicals	
Alkylating agents	Methyl mercury
Anesthetic gases	
Biological Agents	
Rubella virus	Syphilis
Toxoplasmosis	
Pesticides	
Organophosphates	Carbamates
Other	
Acute hypoxia	

Table 9-5: Physical and chemical agents affecting fetal development. *Recreated with permission from Williams P. L. and Burson J. L.*, Industrial Toxicology: Safety and Health Applications in the Workplace, *Van Nostrand Reinhold, New York, 1985.*

formation. Ethylene oxide affects the structure of the bases found in the DNA molecule.

Impaired Cell Function

Some agents do not kill cells but inhibit their normal functions. The dysfunction of these metabolic processes may lead to symptoms similar to those associated with cell death. Lead and cadmium are known animal teratogens capable of inhibiting protein synthesis by acting on important enzymes – such as DNA polymerase – involved in DNA synthesis. Dysfunction of enzymes involved in energy production and maintenance of intracellular homeostasis can result in malformation of the developing fetus. For example, 6-aminonicotinamide is an antagonist of niacin, which is one of the B vitamins that serves as an enzyme cofactor involved in the body's oxidation-reduction reactions. Exposure of experimental animals during development results in an increase in cleft palate and limb defects.

Maternal and Placental Toxicity

Factors associated with the general health of the mother or with the physiology and functioning of the placenta may also affect development. In some instances the observed effects may be the result of both direct exposure to the toxicant and indirect toxic conditions related to the mother's overall health.

Maternal conditions affecting development include: decreased uterine blood flow, maternal anemia, nutritional status, diabetes, electrolyte balance, stress, and viral or bacterial infection during pregnancy. The importance of vitamins in the normal development of the fetus has been demonstrated in experiments using vitamin A and folic acid. A deficiency in vitamin A produces a higher rate of cleft palate in pigs, while an increased intake of folic acid during pregnancy reduces the incidence of neural tube defects in newborn children. Offspring of rats and mice exposed to stress developed various types of malformations including cleft palate, fused ribs, or an increased number of ribs. There was also an increase in the number of fetal deaths. These effects may be associated with an increase in circulating steroid hormones produced during a period of stress.

Infections of the mother with a virus or bacteria may result in malformations as well. Viral infections have been associated with fetal death, microencephaly, mental retardation, blindness, and deafness. Infections due to *Toxoplasma gondii* cause malformation of the head and eye. Fused, missing or extra ribs; fused vertebrae; **exencephaly**; eye abnormality (one eye missing or smaller than the other); and bent or shortened bones are all conditions that have been observed in test animals exposed to toxicants during pregnancy.

> Exencephaly is a congenital abnormality in which part of the brain forms outside of the skull.

Normally the placenta is responsible for providing nutrition, gas exchange, and waste removal for the developing fetus. Toxic substances that impair these functions may produce the effects described previously. Several substances have been shown to be toxic to the placenta, including arsenic, mercury, and cadmium. Cadmium has been shown to produce placental toxicity by decreasing blood flow, tissue necrosis, and inhibition of nutrient transport across the placenta. Injection of pregnant rats with cadmium results in fetal death even though only small amounts of cadmium actually enter the fetus.

Teratogenic Agents

Radiation

Soon after their discovery and use for diagnostic purposes it was observed that x-rays had the potential to harm the developing fetus. High doses resulted in children being born with small heads and varying degrees of mental retardation. As discussed previously, ionizing radiation causes breaks in both strands of the DNA molecule and alters the chemical structure of the nitrogen bases. Because the cells of embryos are undergoing rapid division, it is possible that DNA repair is not completed; or that translocation of DNA segments between different chromosomes occurs, which could be associated with the symptoms mentioned above.

Metals and Solvents

The Minamata disease – described in Chapter 1 – is a tragic example of how metals can become teratogenic agents. The inorganic mercury – discharged to the ocean and converted to methyl mercury – concentrated in fish and shellfish, which was used as a food source by the local population. Pregnant women who ate the contaminated fish had children with a higher incidence of neurological disorders such as cerebral palsy.

Other metals such as lead and cadmium inhibit enzyme functions associated with energy production; this may lead to cell death or delayed cell reproduction, both of which cause developmental abnormalities. For example, women working in copper smelters in Sweden were exposed to airborne levels of copper, lead, arsenic, and cadmium, and had children with increased numbers of birth defects. Women working in chemical factories have – in some cases – an increased incidence of children born with malformations associated with the mouth. Central nervous system malformations were observed in children from women exposed to organic solvents.

Thalidomide

Thalidomide was used both as a sedative and to reduce nausea and vomiting during pregnancy. It was first associated with teratogenicity following an increase in the number of limb malformations in German newborns. In a hospital in Hamburg the number of newborns with limb malformations went from one in 1959 to 30 in 1960 and 154 in 1961. The newborns were characterized by absence of limbs (**amelia**) and/or a shortening of the long bones of the arms and legs known as **phocomelia**. The arms are typically affected more than the legs. Other symptoms that occurred were congenital heart disease, eye, kidney, and intestinal abnormalities, as well as malformation of the external ear.

Testing of thalidomide illustrates the inherent difficulties of trying to extrapolate information from animal studies to establish safe exposure levels in humans. Prior to widespread distribution and use of thalidomide animal toxicity tests were performed using mice. Results from these studies indicated no adverse effects; however, they were not designed to investigate possible teratogenic effects. After adverse symptoms began to appear in the offspring of individuals using thalidomide, additional animal toxicity tests using different species of animals were performed, producing varying results. No teratogenic effects were observed in hamsters and certain strains of mice. Increased numbers of malformed offspring and resorption of young occurred in rats exposed to thalidomide. Various strains of rabbits as well as eight of nine primate strains developed anomalies similar to humans. The concentration at which effects were seen in the various species varied widely from as low as one to as high as 100 mg/kg of body weight.

Diethylstilbestrol

Diethylstilbestrol (DES) stimulates the production of estrogen and progesterone in the placenta, which decreases the probability of a miscarriage. However, male and female offspring of women who used DES develop various abnormalities. Females develop cancer of the vagina at an earlier age than is typically seen. Male offspring have a higher incidence of epididymal cysts, small testes, and decreased quantity and viability of semen.

Ethanol

Fetal alcohol syndrome (FAS) was first identified in the 1970s when studies clearly indicated that there was a correlation between alcohol consumption and developmental abnormalities. Symptoms include behavioral, neurological, and skeletal problems. Offspring may be characterized by malformation of the head and face, slowed growth prior to and after birth, mental retardation, and poorly developed psychomotor skills. These symptoms may be caused by the consumption of as little as 3-4 ounces of alcohol per day although a definitive dose-response relationship has been difficult to establish.

Cocaine

Cocaine-babies have become a major health problem since the 1980s following an increase in the use of cocaine during pregnancy. Both the mother and the fetus are affected. In the mother the usual symptoms are premature labor and delivery. The offspring may be born microcephalic and with decreased birth weight; they may be addicted at birth exhibiting tremors, irritability, seizures, as well as abnormal sleep patterns due to disruption of normal neurological function; or be more prone to **Sudden Infant Death Syndrome (SIDS)**.

Checking Your Understanding

1. What is a teratogen?
2. Is a teratogen a mutagen? Why or why not?
3. Identify different mechanisms of action that may explain how teratogenic agents cause their effects.
4. Identify factors that cause maternal or placental toxicity and describe the possible teratogenic effects.
5. How does radiation cause teratogenic effects? List other types of teratogens and their possible effects.

9-5 Carcinogens

It has been estimated that three out of four Americans will develop some form of cancer during their lifetime. The role chemical substances and physical agents play in the development of cancer is of considerable debate. It has been estimated that up to 90 percent of all cancers are the result of exposure to environmental agents. However, many feel that cancer is a normal result of the aging process.

Cancer is the uncontrolled division and growth of cells. Cancer cells divide rapidly, invading tissues and replacing normal living cells. These rapidly dividing cells consume large amounts of energy and displace cells that perform the normal functioning of the tissues. What triggers this process is still not totally understood.

Definitions

A simple definition of a carcinogen is any cancer-causing substance – chemical or physical – such as benzene, cadmium, asbestos, or certain types of radiation. The term cancer refers to a type of **neoplasm**. A neoplasm is defined as a new (neo) growth (plasia). Therefore, cancer is a type of neoplasm characterized by tissue undergoing rapid and abnormal growth. Tissues that grow faster than the surrounding tissue form an enlarged mass of cells referred to as a tumor.

There are two basic types of tumors, **benign** and **malignant**. Benign tumors are usually encapsulated by fibrous tissue; they are characterized by uncontrolled but slow growth of cells that do not **metastasize** and do not invade surrounding tissues.

> Metastasis refers to the ability of tumor cells to break off, enter the bloodstream, and be transported to other locations in the body where they can begin to grow. This is a characteristic of malignant tumors.

Malignant tumors are not encapsulated and invade healthy surrounding tissue. The cell growth is rapid, and metastasis to other parts of the body occurs. Because metastasis leads to multiple sites of tumor growth – in addition to its original site – it is more difficult to successfully treat malignant tumors than benign tumors, which are localized. Table 9-6

Benign	Malignant
Encapsulated	Nonencapsulated
Noninvasive	Invasive
Highly differentiated	Poorly differentiated
Rare cell division	Cell division common
Slow growth	Rapid growth
Little or no **anaplasia**[a]	Anaplastic to varying degrees
No metastases	Metastases

[a] Anaplasia is the loss of cellular differentiation and function. Characteristic of most malignancies.

Table 9-6: Characteristics of benign and malignant tumors. *Recreated with permission from Williams P. L. and Burson J. L.*, Industrial Toxicology: Safety and Health Applications in the Workplace, *Van Nostrand Reinhold, New York, 1985.*

compares the characteristics of benign and malignant tumors.

Benign and malignant forms of cancer can be further classified according to the type of tissue in which they develop. Benign tumors have the suffix "oma" attached to the term identifying where the tumor originates. For instance, a benign tumor that develops in the dermal papillae is referred to as an epidermal papilloma. In this case, the epidermal papillae indicate the tissue type, while the suffix "oma" indicates it is a benign tumor.

Table 9-7 lists various types of benign and malignant tumors associated with various tissue types. A malignant tumor that has its origin in epithelial cells is referred to as a **carcinoma**. A malignant tumor of the epidermis is referred to as an epidermal carcinoma. The term epidermal indicates the tissue type is the epidermis, while the term carcinoma is used because it had its origin in epithelial cells of the epidermis and it is a malignant tumor. A malignant tumor that originates in bone or muscle is referred to as **sarcoma**.

> Cancer represents uncontrolled growth of cells that have their origin in various types of tissue, subsequently forming tumors that are either benign or malignant.

Tissue	Benign Tumor	Malignant Tumor
Epidermis	Epidermal papilloma	Epidermal carcinoma
Biliary tract	Cholangioma	Cholangiocarcinoma
Adrenal cortex	Adrenocortical adenoma	Adrenocortical carcinoma
Stomach	Gastric polyp	Gastric carcinoma
Liver	Hepatoma	Hepatocellular carcinoma
Fibrous tissue	Fibroma	Fibrosarcoma
Cartilage	Chondroma	Chondrosarcoma
Fat	Lipoma	Liposarcoma
Muscle	Rhabdomyoma	Rhabdomyosarcoma
Bone	Osteoma	Osteosarcoma
Kidney	Renal Adenoma	Renal adenocarcinoma
Lung	Pulmonary adenoma	Pulmonary adenocarcinoma

Table 9-7: Examples of tumors based on tissues of origination. *Recreated with permission from Williams P. L. and Burson J. L.,* Industrial Toxicology: Safety and Health Applications in the Workplace, *Van Nostrand Reinhold, New York, 1985.*

Mechanisms of Action

How toxic chemicals or physical agents cause cancer is not clearly understood. The toxic substance or agent may interact directly with the DNA or affect DNA replication. These types of toxic substances are referred to as **genotoxic carcinogens**.

The **somatic cell mutation theory** of cancer describes how genotoxic carcinogens produce their effects. The rate at which cells divide and multiply is determined by the genetic information of the cell. Genotoxic carcinogens interact with the DNA and produce mutation of the genes controlling cell division. As a result, uncontrolled growth of nondifferentiated cells occurs, forming a tumor in the affected tissue.

There is some evidence to support the idea that mutations are involved in cancer development. For instance, some mutagens have been shown to cause cancer in animals; and chromosome abnormalities are found in some types of cancer. Individuals with Xeroderma pigmentosum disease have a higher incidence of skin cancer and, as discussed previously, this disease has been associated with mutation of genes involved in the repair of damaged DNA. It is important to remember that not all mutagens cause cancer. Some mutations are beneficial or have no effect at all.

Another theory that has been associated with exposure to genotoxic carcinogens is the **single-hit theory** or **one-hit theory** of carcinogenesis. This theory considers a worst-case scenario in which cancer is caused by a single genotoxic event, resulting in irreversible damage to a critical segment of the DNA responsible for controlling cell division. The mutation results in the development of cancer shortly after exposure or several years later. According to this theory, the damage may occur at any exposure level. This theory has played an important role in establishing acceptable levels of exposure to carcinogenic substances.

Development of cancer several years after exposure to the carcinogenic agent can also occur as a result of continuous exposure to **epigenetic carcinogens**. These types of carcinogens promote the development of cancer without directly interacting with the DNA of cells. Epigenetic carcinogens affect the physiology of the cell in such a way that there may be loss of cell communication, hormone imbalances, or cellular damage and excessive regeneration. These abnormal events stimulate abnormal cell growth.

Stages of Development

In general, the development and growth of tumors involves three distinct stages: 1) initiation; 2) promotion; and 3) progression.

Initiation Stage – The initiation stage is characterized by an irreversible change in the DNA structure caused by genotoxic agents. These agents are referred to as initiators. They begin the change in the cell that ultimately leads to tumor growth.

The DNA change may disrupt normal cell function and cause cell death. If this happens the mutated cell is eliminated and no further effects are observed. However, DNA mutation may result in a dysfunctional cell that will divide rapidly. Most cells have a limit as to the number of divisions they can undergo. This is partially due to the cell-to-cell contact that occurs during cell division and the subsequent tissue formation. As a result of this contact a message is sent by individual cells comprising the tissue that limits the number of times a cell can divide. Cancer cells, on the other hand, divide repeatedly and without limit because they have lost the ability to control division by cell-to-cell contact. Examples of initiating carcinogens include nitrosamines, polycyclic aromatic hydrocarbons (PAHs), dimethyl sulfate, nitrosureas, cadmium, chromium, and nickel. In addition to these toxic substances, several of the endogenous agents discussed previously can cause spontaneous mutations or produce errors in DNA replication. These events are also considered initiating factors.

> The initiation stage is characterized by irreversible mutation of cellular DNA.

Promotion Stage – Promotion occurs when tumor growth is stimulated without the carcinogen interacting with DNA. Promotion may either enhance the expression of the altered genes formed during the initiation phase or occur without DNA involvement. The carcinogen that causes the genetic change during the initiation state may or may not be the carcinogen that promotes the development and growth during the promotion stage.

Promotion differs from initiation because removal of the promoter carcinogen can result in stopping tumor growth, whereas the genetic change associated with the initiation phase is irreversible. Tumor development related to the concentrations of the promoting agent typically shows a dose-response relationship. For instance, the frequency of epidermal papillomas (skin tumors) occurring in mice decreases significantly once the application of the promoting agent decreases. The decreased number of tumors that follows is caused by the death of tumor cells or by the return of tumor cells to a normal cell type after removal of the promoting agent.

Many substances that are consumed as a normal part of our diet are thought to be promoters. Four possible mechanisms have been proposed by which dietary factors may be associated with the development of cancer (Table 9-8). One of the chief culprits thought to promote cancer is fat. How this

1.	Ingestion of complete carcinogens, initiating agents, or progressor agents	Aflatoxin
2.	Exogenous or endogenous production of carcinogens from dietary constituents	Heterocyclic amines N-nitrosation
3.	Alteration of transport, metabolic activation, or inactivation of carcinogens	Ethanol, vitamins A and E
4.	Promoting agents that act on spontaneously initiated cells	Calories, dietary fat

Table 9-8: Possible mechanisms of dietary carcinogenesis. *Recreated with permission from Klaasen C. D. et al., Casarett & Doull's Toxicology, The Basic Science of Poisons, 5th edition, McGraw Hill, New York, 1996.*

happens is not clearly understood, but several possible mechanisms have been proposed. Fat may alter hormone levels; affect cell membrane composition; increase the level of fatty acids in the circulation, thereby inhibiting the immune system; or stimulate the synthesis of bile acids in the liver. Cancer rates are 4-6 times higher in animals fed a high-calorie diet than in those fed a restricted low-calorie diet. People who consume a fatty diet have a higher incidence of prostate, colon, and breast cancer. To avoid the fat associated with frying foods, many people broil their food. However, broiling can cause the formation of carcinogenic polyaromatic hydrocarbons or of mutagenic proteins.

Ingestion of substances referred to as complete carcinogens may be another mechanism. A complete carcinogen is a substance capable of triggering all three stages of cancer development: initiation, promotion, and progression, without requiring any other substance. Aflatoxins – produced by a strain of mold *Aspergillus flavus* – produce cancer of the liver in rats. Foods contaminated by this mold and consumed by individuals have been associated with increased frequency of occurrence of liver cancer. Heterocyclic amines are produced in foods when they are cooked, and nitrosamines are produced by endogenous metabolic processes within the digestive tract. Both of these substances have carcinogenic properties.

In addition to substances that promote cancer, there are other dietary factors that have the capability of inhibiting cancer formation. Retinoids such as vitamin E, beta-carotene (vitamin A precursor), and selenium have been shown to inhibit carcinogenic development. The retinoids have their greatest effect at the promotion stage. Vitamin E, beta-carotene, and selenium seem to inhibit the metabolic conversion of noncarcinogenic substances to carcinogens.

Hormones such as estrogen and progesterone promote development of cancer. These hormones normally result in proliferation of various cell types of the reproductive tract. In women they promote the rapid development of the epithelial lining of the uterus during the menstrual cycle. Elevated levels of these substances may either result in abnormal growth of the cells and development of cancer or function as promoters for cells that were initiated by other carcinogens.

Carcinogens that are promoting agents demonstrate a classic dose-response curve. There is a threshold concentration below which no cancer is detected, and there is a maximum concentration above which the frequency no longer increases. Like noncarcinogenic chemicals, the promoting agent's dose-response relationship is affected by factors such as the rate of chemical metabolism, the size of the molecule, the electrical charge, and the lipid solubility. All of these factors will affect the shape of the dose-response curve.

Additional examples of promoting carcinogenic agents are saccharin – which affects the lining of the urinary bladder – and phenobarbital – which may cause cancer of the liver. Tetrachlorodibenzo-p-dioxin (TCDD), androgens, and estrogens may also promote cancer development in the liver. Tumors in the renal tubular cells of rats can be caused by exposure to 2,2,4-trimethylpentane and unleaded gasoline.

> The promotion stage is associated with tumor growth and demonstrates a dose-response relationship.

Progression Stage – Like the initiation stage, progression is characterized by being irreversible. Once the affected cells enter this stage, removal of the carcinogen has no effect on cellular growth and development. Progression is characterized by rapid cell growth, invasiveness of normal tissue, increased frequency of metastases, lack of response to other environmental factors, and lack of cellular differentiation.

The transition between promotion and progression may be caused by exposure to chemical or physical substances referred to as **progressor agents**. These agents may affect the DNA by inhibiting DNA repair. Examples of progressor agents include arsenic salts, asbestos fibers, benzene, benzoyl peroxide, and 2,5,2',5'-tetrachlorobiphenyl.

> The progression stage is irreversible and possibly initiated by a progressor agent, producing a malignant tumor.

Sources of Carcinogens

Almost every aspect of life provides an opportunity to become exposed to chemical and physical agents that are carcinogens (Figure 9-9). We may encounter them in the workplace, in the household products we use, in the food we eat, or in our general environment. Only some of these substances are known to be carcinogenic; some others are potentially carcinogenic, while the carcinogenic potential for the rest is unknown.

Lifestyle and Exposure to Carcinogens

A great deal of research and discussion has occurred concerning the relationship between diet and development of cancer. Some researchers suggest diet plays a significant role in the development of various types of cancer. Many others make counterclaims suggesting a normal diet poses no additional risk for the development of cancer.

Consumption of alcoholic beverages seems to be associated with both positive and negative health effects. Recent studies have suggested that the consumption of a glass of wine a day reduces heart disease. However, excessive consumption of alcoholic beverages may be associated with development of cancer. Specifically, ethanol – which is an important constituent in all alcoholic beverages – may be the primary agent associated with development of cancer. The ethanol is metabolized to acetaldehyde, which is mutagenic. But, not all mutagens are car-

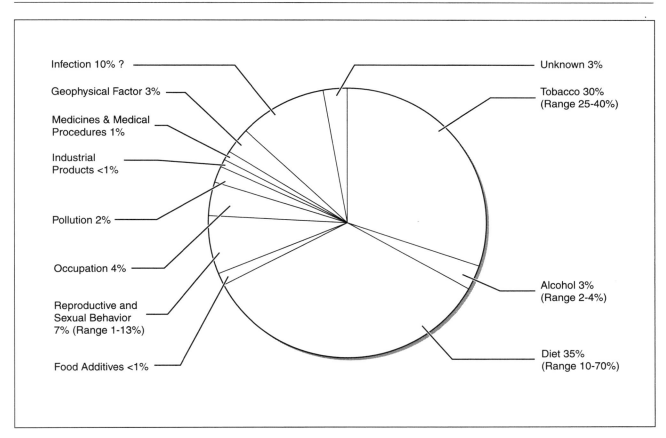

Figure 9-9: Proportions of cancer deaths attributed to various environmental factors. *Recreated with permission from Klaasen C. D. et al.*, Casarett & Doull's Toxicology, The Basic Science of Poisons, 5th edition, McGraw Hill, New York, 1996.

cinogens, and ethanol may be a promoting agent. Cancer of the mouth and pharynx increases in individuals who are chronically exposed to both alcohol and tobacco compared to those who only participate in one of these activities. Similarly, individuals who consume excessive alcohol and have been infected with the hepatitis B virus develop cancer of the liver more readily.

Tobacco products are probably some of the best studied and documented substances in relation to their carcinogenesis. The number of individuals who develop some form of cancer as a result of the use of tobacco varies according to the study. However, it has been generally estimated that around 90 percent of all lung cancers are the result of tobacco use and around 30 percent of all cancer deaths can be attributed to use of tobacco products. It appears that tobacco products, like ethanol, are promoting agents. Cessation of use results in decreased frequency of cancer occurrence.

Occupational Exposure to Carcinogens

A number of chemical substances found in the workplace have been associated with the development of cancer (Table 9-9). Exposure to asbestos fibers causes two kinds of cancer: bronchogenic carcinoma – a type of lung cancer – and malignant mesothelioma – a type of cancer associated with connective tissue, the latter being directly associated with inhalation of asbestos fibers. The frequency of occurrence is affected by other lifestyle factors such as smoking: workers smoking cigarettes and exposed to asbestos have a higher incidence of bronchogenic carcinoma. However, there appears to be a decrease in the frequency of cancer occurrence in individuals who stop smoking compared to those who continue to smoke. This suggests that under these circumstances asbestos fibers may be a promotion agent. Other potential workplace carcinogens include fiberglass, wood dust, various metals, benzene, and

Established

Agent	Industries and Trades with Proved Excess Cancers and Exposure	Primary Affected Site
Para-aminodiphenyl	Chemical manufacturing	Urinary bladder
Asbestos	Construction, asbestos mining and milling, production of friction products and cement	Pleura, peritoneum, bronchus
Arsenic	Copper mining and smelting	Skin, bronchus, liver
Alkylating agents (mechloro-ethamine hydrochloride and bis[chloromethyl]ether)	Chemical manufacturing	Bronchus
Benzene	Chemical and rubber manufacturing, petroleum refining	Bone marrow
Benzidine, beta-naphthylamine, and derived dyes	Dye and Textile production	Urinary bladder
Chromium and Chromates	Tanning, pigment making	Nasal sinus, bronchus
Isopropyl alcohol manufacture	Chemical manufacturing	Cancer of paranasal sinuses
Nickel	Nickel refining	Nasal sinus, bronchus
Polynuclear aromatic hydro-carbons from coke, coal tar, shale, mineral oils, and creosote	Steel making, roofing, chimney cleaning	Skin, scrotum, bronchus
Vinyl chloride monomer	Chemical manufacturing	Liver
Wood dust	Cabinetmaking, carpentry	Nasal sinus

Suspected

Agent	Industries and Trades	Suspected Human Sites
Acrylonitrile	Chemical and plastics	Lung, colon prostate
Beryllium	Beryllium processing, aircraft manufacturing, electronics, secondary smelting	Bronchus
Cadmium	Smelting, battery making, welding	Bronchus
Ethylene oxide	Hospitals, production of hospital supplies	Bone marrow
Formaldehyde	Plastic, textile, and chemical production; health care	Nasal sinus, bronchus
Synthetic mineral fibers (e.g., fibrous glass)	Manufacturing, insulation	Bronchus
Phenoxyacetic acid	Farming, herbicide application	Soft tissue sarcoma
Organochlorine pesticides (e.g., chlordane, dieldrin)	Pesticide manufacture and application, agriculture	Bone marrow
Silica	Casting, mining, refracting	Bronchus

Table 9-9: Exposures to chemical carcinogens in the workplace. *Recreated with permission from Cullen M. R., Cherniak M. G., Rosenstock L., "Occupational Medicine," New England Journal of Medicine 322: 675-682, 1990.*

formaldehyde. Organophosphate pesticides are thought to be carcinogenic primarily because of the presence of dioxins (i.e., TCDD) as a contaminant. Vinyl chloride – used in the manufacture of polyvinylchloride (PVC) plastic – produces cancer of the liver.

Identification and Classification of Carcinogens

The National Toxicology Program of the U.S. Public Health Service produces a regularly updated list of carcinogens. The list includes known human carcinogens for which there is sufficient evidence to establish a cause-effect relationship as well as chemicals that are "suspected or reasonably anticipated to be human carcinogens." These suspected carcinogens are listed separately either because there is limited evidence on their carcinogenic effects in humans or because the data from animal experimentation are insufficient to suggest a potential effect in humans.

The International Agency for Research on Cancer (IARC) – which is part of the World Health Organization – was one of the first organizations to identify and classify chemicals according to their carcinogenicity. Table 9-10 summarizes the classification scheme used by IARC.

Other organizations such as the Environmental Protection Agency (EPA) and the Chemical Manufacturers Association also have developed lists and classification schemes for carcinogenic substances. The American Conference of Governmental Industrial Hygienists (ACGIH) and the Occupational Safety and Health Administration (OSHA) are concerned with exposure of individuals to carcinogens in the workplace. Both the ACGIH and OSHA have established lists of chemicals that are either known or suspected human carcinogens. In this case, known human carcinogens are substances for which a carcinogenic effect either has been demonstrated or is considered potential.

Because there are several organizations involved in identifying and classifying carcinogens, there are multiple lists. Often these lists vary in length for the same category (i.e., known carcinogen) but they may not contain the same substances. This complicates the process of effectively communicating the level of risk associated with chemical exposure.

Group	Evidence	Examples
1. Agent is carcinogenic	Sufficient (human)[a]	Arsenic, aflatoxin, benzene, estrogens, vinyl chloride
2A. Agent is probably carcinogenic	Limited (human)[b] Sufficient (animal)	Benz(a)anthracene, diethylnitrosamine (DEN), polychlorinated biphenyls (PCB), styrene oxide
2B. Agent is possibly carcinogenic	Limited (human) Inadequate (human)[c] Sufficient (animal)	TCDD, styrene, urethane
3. Agent is not classifiable as to carcinogenicity		5-azacytidine, diazepam
4. Agent is probably not carcinogenic	Inadequate (human) Inadequate (animal)	Caprolactam

[a] "*Sufficient evidence* of carcinogenicity, which indicates that there is a casual relationship between the agent or agents and human cancer."
[b] "*Limited evidence* of carcinogenicity, which indicates that a causal interpretation is credible but that alternative explanations such as chance, bias, and confounding variables could not be completely excluded."
[c] "*Inadequate evidence*, which indicates that one of three conditions prevailed: (a) there were few pertinent data, (b) the available studies, while showing evidence of association, did not exclude chance, bias, or confounding variables, (c) studies were available that did not show evidence of carccinogenicity."

Table 9-10: IARC classification of the evaluation of carcinogenicity for the human species. *Recreated with permission from Klaasen C. D. et al., Casarett & Doull's Toxicology, The Basic Science of Poisons, 5th edition, McGraw Hill, New York, 1996.*

Checking Your Understanding

1. What is a neoplasm?

2. Differentiate between the terms malignant tumor, benign tumor, carcinoma, and sarcoma. Give an example of each.

3. Briefly explain the somatic cell mutation theory and the single-hit or one-hit theory of carcinogenesis.

4. How does the mechanism of action for a genotoxic carcinogen differ from that of an epigenetic carcinogen?

5. What are the three stages of tumor growth and the characteristics of each?

6. What characteristics can be used to differentiate carcinogens that are considered to be a promoting agent, an initiating agent, and a progressor agent?

7. Identify several lifestyle or dietary factors that may contribute to the development of cancer.

8. Which agencies are involved in identifying and categorizing carcinogenic substances?

9-6 Radiation

There are two major types of radiation: ionizing and nonionizing. The following is a brief discussion of their sources and effects.

Ionizing Radiation

The sources of ionizing radiation are **radioactive** elements –called radionuclides – that undergo spontaneous disintegration. The transfer of energy from one source and its absorption by a body is referred to as radiation. Ionizing radiation can produce mutagenic, teratogenic, and carcinogenic effects. In order to understand the effects of ionizing radiation on tissues, it is important to define the terms associated with this phenomenon. The energy is either in the form of particles – alpha particles, beta particles, and neutrons – or in the form of electromagnetic waves, such as x-rays and gamma rays. The type of energy emission is dependent on the radionuclide.

> Radioactivity is often confused with the term radiation. Radioactivity is the spontaneous emission of energy – radiation – which is characteristic of certain elements referred to as radionuclides.

Energy emitted from a radionuclide and colliding with matter, such as a cell, instigates changes in cell molecules. When radiation has sufficient energy to cause electrons in the cellular atoms to be displaced, ions are formed. This process is referred to as ionization. The ions, in turn, react with cellular molecules such as DNA.

> Ionizing radiation produces effects by ionizing atoms that, in turn, interact with structural and functional cellular molecules.

Types of Ionizing Radiation

There are five main types of ionizing radiation: 1) alpha particles, 2) beta particles, 3) gamma rays, 4) neutrons, and 5) x-rays. The specific characteristics of each type contribute to the differentiation between the various observed biological effects.

Alpha particles (α) – Alpha particles have the largest mass of all the types of ionizing radiation. The particle consists of two protons and two neutrons that are emitted from the nuclei of certain radionuclides, such as radium-226 and plutonium-

239. Alpha particles are large, slow-moving particles that are unable to penetrate the keratinized layer of the skin. Nonkeratinized tissues may be penetrated to a depth of about 50 μm, which roughly corresponds to the thickness of five cells. Alpha particles, therefore, produce the greatest injury when they are introduced internally, as for example in the lungs.

Beta particles (β) – Beta particles are high-energy electrons emitted from the nuclei of radionuclides such as strontium-90 or iodine-131. Because these particles have a smaller mass than alpha particles they are able to penetrate several centimeters of tissue. They therefore have a greater potential to cause injury by external exposure.

Gamma rays (γ) – Gamma rays are electromagnetic waves that originate from within the nuclei of certain radionuclides. Unlike alpha and beta particles they are composed of **photons**, which have no mass. Gamma rays can easily penetrate tissues causing cellular damage and injury. They are high energy and can pass through the human body.

Neutrons (n°) – The nuclei of some radionuclides emit neutrons. They are readily absorbed by the nuclei of surrounding atoms and therefore may not travel very far from their source. However, when they do, their penetrating power is similar to gamma rays and may have significant biological effects.

X-Rays – X-rays are similar to gamma rays: they are also composed of photons, but they have a longer wavelength and, therefore, lower energy. X-rays are produced by the transition of an electron from a high to a low energy level. The difference in energy between the two levels is emitted as photons with x-ray wavelength. In commercial and medical occupations x-rays are generated by bombarding a metallic target with a high-energy electron source. The x-rays are used for a variety of diagnostic activities. Excessive exposure to x-rays produces significant tissue damage and fetal abnormalities.

Sources of Ionizing Radiation

Several sources of exposure to ionizing radiation are associated with different occupations and industries. Perhaps the best known source of radioactive materials is within the nuclear power industry. Nuclear reactors use uranium or plutonium as fuel to generate electricity, to perform research, to provide propulsion, or to produce other radionuclides. People working within this industry can be exposed to radiation while mining the uranium, manufacturing nuclear fuel rods, or working on the reactors.

Uranium miners are exposed to radon gas – an alpha particle emitter – since it is one of the byproducts of the natural decay of uranium. Other sources of radon exposure include soils, rock, and building materials. Of some concern to health professionals are increased levels in homes with poor ventilation if the underlying rock and soil contain radon.

X-ray machines are used for a variety of purposes. They are routinely used in hospitals and dental offices to identify anatomical abnormalities. They are also used by various industries to inspect metal castings or welds for cracks or other deformities.

Astronauts, pilots, and airplane passengers are all exposed to cosmic radiation. The amount of exposure is dependent on the altitude and duration of flight. Astronauts could potentially be exposed to higher radiation levels because of the lack of an atmosphere, which provides some level of shielding to pilots and passengers of airplanes. Exposure under these circumstances has not been shown to have any adverse effects.

Effects of Ionizing Radiation

Cellular Effects – The initial effects of radiation on the tissues of the body occur at the cellular level. If enough cells are affected, symptoms such as radiation burns or cataracts may be observed. Other symptoms of radiation exposure – such as the development of leukemia – may not be observed until after a long period of latency.

Effects at the cellular level may take on several different forms. DNA in the nucleus is responsible for providing information to synthesize important structural and functional proteins necessary to maintain normal cell function. Exposure to radiation can affect the DNA structure by changing the base sequence of the DNA molecule or breaking the bonds between the individual DNA strands, which can cause cross-linkage between the strands and subsequent DNA dysfunction. All of these changes can impact cell functions.

One of the initial effects of radiation exposure is the inhibition of cell mitosis (cell division), which occurs for a period of minutes or hours depending on the level of exposure. Once mitosis resumes, it initially occurs at a rate higher than normal until the cell population returns to normal. This effect appears to be associated with irradiation of the nucleus and possible effects on the DNA, since exposure of the cell cytoplasm alone does not have similar effects.

In addition to inhibition of cell mitosis, radiation may lead to chromosomal aberrations. During normal mitosis duplication and division of the chromosomes occurs in such a way that each cell gets exactly the same number of chromosomes with the same genetic information. If damage to the DNA caused by radiation exposure is not repaired, the normal distribution of chromosomes to the daughter cells can be disrupted. As a result, there is an increased chance that the dividing cells will not receive the same number of chromosomes with the same information. Increased occurrence of chromosome aberrations have been observed in lymphocytes of atomic bomb survivors, radiation workers, victims of radiation accidents, and in individuals residing in areas with elevated, natural background levels of radiation.

Cells that undergo cell division are more likely to be killed than nondividing cells, the latter being more resistant and requiring a larger dose to be destroyed. Blood-forming cells as well as cells involved in spermatogenesis (sperm formation) and oogenesis (egg formation) are particularly susceptible.

> Effects of radiation exposure are primarily the result of effects on the DNA molecule.

Tissue Effects – The various tissues of an organism respond differently to radiation. The effects may be observed only within a few cells, or in the organism as a whole when major organ functions are disrupted. The following list describes some of the effects on major tissues.

a. Skin
   Erythema – a reddening of the skin – is one of the first observable symptoms associated with radiation exposure. This occurs as a result of the release of histamine in response to cell damage. Depending on the level of exposure, erythema may disappear within days or weeks. With higher levels of exposure the erythema may continue for weeks and produce cell death and sloughing of the skin. After many months or years of exposure there may be changes in skin pigmentation (color); atrophy of the epidermis, the sweat glands, the sebaceous glands, and the hair follicles; fibrosis of the dermis; and increased ulceration of the skin.

b. Bone Marrow and Lymphoid Tissue
   The stem cells in the bone are sensitive to radiation exposure and are easily destroyed. Radiation exposure produces a decrease in all types of circulating blood cells and particularly the leukocytes, leading to leukopenia (decreased number of white blood cells). If the dose is large enough, the organism may die within 4-6 weeks after exposure as a result of bacterial or viral infection due to the decrease in white blood cells, or as a result of hemorrhaging due to decreased numbers of platelets. At lower doses the blood-forming components may recover. However, after a latent period leukemia may develop.

c. Gastrointestinal Tract
   Along with erythema, one of the first symptoms associated with radiation exposure is the development of digestive tract disorders such as diarrhea. The precursor cells responsible for producing the epithelial cells that line the small intestine are extremely sensitive to radiation. Under normal conditions the epithelial cells are continuously replaced. If the precursor cells are damaged by radiation, then replacement of the cells is impaired and the lining is not replaced. This interferes with the normal digestive processes of the small intestine and eventually leads to ulceration of the tissue and intestinal bleeding. Extensive damage may be lethal.

d. Gonads
   Mature sperms are relatively resistant to radiation exposure except at large doses. However, developing sperm cells are more susceptible and exposure may cause decreased sperm cell numbers. Mature oocytes are very sensitive to radiation exposure. Depending on the level of exposure, temporary or permanent sterility can occur.

e. Lens of the Eye
   The epithelium of the eye undergoes cellular division like other epithelial layers. Therefore, it is susceptible to injury from radiation exposure. If the exposure is severe, the cells of the cornea and lens can be damaged and become opaque. This leads to impaired vision or cataracts.

f. Respiratory Tract
The lungs are extremely sensitive to radiation. The alveoli can be easily damaged: the cells die and the alveoli are replaced by connective tissue. If lung damage is significant enough, it can interfere with normal breathing and conditions similar to pneumonitis (inflammation of the lungs), pulmonary edema, (illness similar to pneumonia), or pulmonary fibrosis may develop. These conditions affect the exchange of gases in the lung, and are characterized by labored breathing, possibly leading to death.

g. Effects on Developing Fetus
The developing fetus is a bundle of rapidly dividing and multiplying cells that are extremely sensitive to radiation exposure during certain critical stages. In humans, exposure to radiation has been shown to increase the frequency of severe mental retardation. In laboratory animals a wide variety of malformations have been observed such as congenital blindness, limb malformation, and polydactyly (additional toes).

h. Carcinogenic Effects
The cancer-causing effects of radiation exposure are well documented. X-ray workers develop skin cancer, radiologists have increased frequency of leukemia, individuals who painted the dials of clocks with radium were more susceptible to developing bone cancer, and hard rock and pitchblende (uranium ore) miners had increased frequency of lung cancer.

Radium has been used for a variety of purposes since its discovery. As mentioned above, it was used on watch dials to produce glow-in-the-dark timepieces. Radium dial painters licked the tips of their paintbrushes to shape them, unknowingly ingesting hundreds and thousands of micrograms of the substance. Radium therapy for rheumatism and for the treatment for mental disorders was accepted by the American Medical Association during the early part of the twentieth century. Patients could either receive injections or drink solutions for treatment of symptoms. In Europe radium was used to treat diseases such as tuberculosis.

The primary effect of internal exposure to radium was the development of bone cancer. This is not surprising since radium acts very much like calcium and is incorporated into the crystalline matrix of bone. However, there appears to be no effect on the hematopoietic tissue (blood-cell forming) of the bone since leukemia is rarely found as a result of internal exposure.

The increased frequency of lung cancer in uranium miners is well documented. The primary substance responsible for the increased rate is radon, which is a gas produced by the decay of radium, which, in turn, is a decay product of uranium. The decay product can be inhaled and deposited in the respiratory tract, primarily in the areas where it begins to branch. This is where most pulmonary tract cancers related to radon exposure are discovered.

Whether radon exposure in the home causes an increased occurrence of lung cancer is still being debated. Currently there is not sufficient evidence to positively establish a cause-effect relationship. Only a study in Sweden suggests that there may be an increased chance of developing lung cancer from radon exposure, particularly if the individual smokes cigarettes.

Nonionizing Radiation

Solar radiation, microwave radiation, lasers, and extremely low-frequency (long wavelength) radiation are all possible sources of nonionizing radiation, with varying effects on humans. There is extended debate about the possible effects of extremely low-frequency radiation on humans and it is analogous to the debate on the effects of exposure to low-level ionizing radiation. In each case, much work still needs to be done to determine if either of these two types of radiation produces a deleterious effect.

> The various forms of nonionizing radiation target the same organs, particularly skin and eyes.

Nonionizing radiation differs from ionizing radiation. First, exposure does not result in the generation of ions in the exposed material. Secondly, the source of the radiation is not the radioactive decay of specific radionuclides. As mentioned above, there are several different sources of nonionizing radiation, each having different effects.

> Nonionizing radiation does not cause its effects by producing ions but by thermal damage.

Types of Nonionizing Radiation

Solar Radiation – Solar radiation is subdivided by wavelength into infrared, visible, and ultraviolet radiation (UV). Each type produces different effects on tissues.

1. Infrared Radiation

    Infrared radiation can be emitted from hot objects such as furnaces, metals, lasers, drying equipment for paints, and welder's torches, causing a variety of effects. Infrared radiation only penetrates the upper layers of the skin and is involved in elevating skin temperature. Continuous exposure may result in skin burns and increased pigmentation. Prolonged exposure of the eye can cause damage to the retina and lead to the development of cataracts. To avoid tissue damage individuals usually wear special goggles and some form of protective or reflective clothing over the surface area exposed.

2. Visible Radiation

    Exposure to visible radiation does not normally produce injury unless there are continuous bright pulses like those produced by electronic flash lamps or pulse lasers. Continual glare from video display terminals may be a potential source of high-intensity visible radiation. These types of exposures may damage the retina.

3. Ultraviolet Radiation (UV)

    Perhaps the best known source of UV radiation is the sun. However, UV radiation is also produced from hot surfaces, welder's torches, and fluorescent lights, and is used in hospitals for its germicidal properties. There are three types of UV radiation: UVA, UVB, and UVC. Only the first two types of UV radiation reach the earth from the sun. The ozone that surrounds the earth absorbs all of the UVC.

    UV radiation has received a great deal of attention because of its association with the development of skin cancer. Individuals who have light skin, burn easily, and have difficulty tanning are more susceptible to the two forms of skin cancer – basal cell carcinoma and squamous cell carcinoma – caused by exposure to UV radiation. UVB seems to be the primary cause of skin cancer and is considered a complete carcinogen.

    UV radiation does not pass entirely through the skin: it is either reflected, or scattered, or absorbed by the various layers. The absorbed radiation excites molecules in the tissue causing them to undergo various types of chemical reactions such as oxidation, reduction, and the breaking of chemical bonds. These changes alter the functioning of the cell, producing mutations when they occur within the DNA molecule. If these mutations are not repaired, then changes in normal cell functioning can result in the development of skin cancer.

    Other effects of UV radiation include sunburn and some immune system responses. Sunburn is associated with reddening of the skin, heat, pain, and sometimes blistering. The reddening is caused by dilation of the blood vessels found in the skin. Pain is the result of cell damage, which releases chemicals that stimulate pain nerve fibers. The damaged cells also release substances that facilitate the dilation of the blood vessels. When the dilation is accompanied by increased capillary permeability to circulating blood fluids then blisters occur in the damaged area.

    One mechanism used to protect the skin against further damage is increased skin pigmentation i.e. tanning, during which melanocytes (pigment-containing cells) in the skin increase in number and become larger. The pigment in the cell absorbs the UV radiation, decreasing the damage to the underlying skin. The protective effect of increased skin pigmentation is not unlimited. Continuous exposure to UV radiation may result in damage to the underlying connective tissue of the skin causing the formation of wrinkles and the development of a leathery texture to the skin.

    UV radiation seems to have some effect on the immune system. The major effect seems to be on the suppression of cell-mediated immunity, which is important in destroying tumors. UVB-exposed individuals have an increased susceptibility to tumor development.

    On the other hand, a positive effect of UV radiation is the production of vitamin D. The skin contains a chemical called 7-dehydrocholesterol that is converted to previtamin D_3 upon exposure to UVB. The previtamin D_3 undergoes a series of chemical reactions and is converted to vitamin D, which is important for the absorption of calcium from the small intestine.

Microwave Radiation – Microwave radiation is generated by radar and communications equipment as well as from industrial and consumer ovens. Studies on the effects of radiation from these sources are inconclusive. Localized body heating seems to occur in some cases. Depending on the area of exposure there may be some harmful effects observed. The eye may be particularly susceptible to this type of radiation, which may cause cataracts as well as corneal and retinal damage. Some effects have been observed in animal experimentation, but their relevance to humans is unclear. The symptoms include effects on the nervous, reproductive, and immune systems. Additional effects include behavioral changes, decreased birth weight of offspring, and changes in the proportions of the various cellular components of the blood.

Extremely-Low-Frequency Radiation – This type of radiation has been the source of great controversy during the last twenty years. It is associated with power transmission lines and various types of appliances and, specifically, with the electromagnetic field (EMF) generated by this type of equipment.

Research has focused on trying to link exposure to this type of radiation to the development of leukemia. There is no definitive proof that there is a cause-effect relationship between the two; however, other effects have been observed. In tissue cultures, effects have been observed on DNA, RNA, and protein synthesis; on cell proliferation; on immune responses; and on cell membrane. Studies on live organisms have indicated effects on the nervous and reproductive systems, on blood chemistry, and on bone growth and repair.

Lasers – Lasers are high-energy coherent light sources used in cutting metal, micro-welding, surveying, guidance equipment, communications technology, medical and surgical equipment. There are four different classes of lasers. Class I lasers pose no danger from radiation. Class II lasers have a limited potential to cause damage. If the beam is stared at for prolonged periods of time, eye damage may occur. Class III lasers may cause eye damage. Class IV lasers may damage the eye and skin requiring various controls to be implemented to avoid injury.

Checking Your Understanding

1. Briefly describe the characteristics of the five types of ionizing radiation.
2. What is the difference between ionizing and nonionizing radiation?
3. Explain the difference between radioactivity and radiation.
4. What are some of the effects of radiation on cell structure and function?
5. How does ionizing radiation affect various tissues of the body?
6. Identify various types of nonionizing radiation and any effects of exposure.
7. Briefly explain how ionizing and nonionizing radiation may result in mutations, teratogenic effects, or cancer.

Summary

Toxic substances may either be mutagenic, teratogenic, carcinogenic, or they may produce no effect at all. Mutagens cause a change in the DNA of cells. Spontaneous mutations are caused by endogenous substances, while induced mutations occur as a result of exposure to exogenous agents and are difficult to distinguish from spontaneous mutations. Point source mutations involve a single base, while large mutations involve many genes. The mutations may be repaired by the DNA repair mechanisms such as the base-excision repair (BER) mechanism or by the nucleotide excision repair process (NER). These mechanisms remove the damaged portion of the DNA and replace it with a new DNA segment.

Substances that cause mutations may also be teratogens. Exposure to teratogenic substances can cause birth defects. In order to produce its effects the teratogen either interacts with DNA directly or interferes with normal cell metabolism. Teratogenic agents include metals, solvents, radiation, and various types of drugs.

Carcinogens cause uncontrolled growth of normal cells in a variety of different tissues. They are classified as either genotoxic or epigenetic.

Genotoxic carcinogens affect the DNA of cells. Epigenetic carcinogens promote cancer by changing the cell physiology and without directly interacting with the DNA. The tissues grow abnormally and faster than the surrounding tissue, forming tumors, either benign or malignant. They appear to develop and grow in three stages: 1) the initiation stage, 2) the promotion stage, and 3) the progression stage. The initiation stage is characterized by irreversible mutation of the DNA. Exposure to a promoting agent may enhance the expression of altered genes formed during the initiation stage and result in tumor growth. The promotion stage can be reversed if the promoting agent is removed. The progression stage is again irreversible and continues even with the removal of the progressor agent. Sources of carcinogens include occupational exposure, diet, and lifestyle choices.

Ionizing radiation is capable of producing mutagenic, teratogenic, and carcinogenic effects. Exposure results in changes in the DNA structure affecting normal cell functioning. It can inhibit cell mitosis, resulting in the abnormal distribution of chromosomes in dividing cells or causing cell death. There are several symptoms associated with radiation exposure. They include reddening of the skin, a decrease in all types of circulating blood cells, gastrointestinal and reproductive effects, cataracts, birth defects, respiratory illnesses, and cancer.

Nonionizing radiation is capable of producing mutagenic and carcinogenic effects primarily by ultraviolet radiation. UV radiation is capable of altering the structure of the DNA molecule. If left unrepaired, the change may lead to skin cancer.

Application and Critical Thinking Questions

1. There are three different agencies referenced in the chapter. Each is involved in the classification of physical and chemical agents as carcinogens. Obtain references from two of the agencies, identify the criteria used to classify carcinogenic agents, and compare the agents listed in each category. How many substances are the same? Are the same substances listed under different categories?

2. What is Xeroderma pigmentosum? Give two possible mechanisms of action to explain its origin.

3. Various populations around the world have been exposed to radiation as a result of nuclear accidents or as a result of warfare. The Japanese were exposed as a result of the explosion of atomic bombs; the Russians and some Europeans were exposed to fallout from Chernobyl; and some citizens of the United States were exposed to radiation from the accident at Three Mile Island. Obtain references that describe these events and identify the type of radiation exposure that may have been encountered as well as the observed effects on human and animal populations. Based upon what you know provide a brief explanation as to why the observed effects occurred. Can they be correlated with the type of radiation exposure that occurred?

10

Risk Assessment and Acute Exposure Treatment

Chapter Objectives

Upon completing this chapter, the student will be able to:

1. **Define** risk.
2. **Identify** and discuss the steps involved in performing a risk assessment.
3. **Describe** the relationship between risk assessment and risk management.
4. **Discuss** the relation among safety factors, acceptable daily intake, reference dose, and reference concentration.
5. **Identify** and explain how a safe human dose is calculated.
6. **List** general signs and symptoms associated with chemical exposure.
7. **Identify** sources of chemical information to assist in first aid response.
8. **List** and explain first aid steps implemented in responding to chemical exposure.
9. **Describe** the steps involved in responding to acute chemical exposure.

Chapter Sections

10–1 Introduction

10–2 Risk Assessment

10–3 Administering First Aid

10–4 Toxic Substance Exposure: Symptoms and Treatment

10-1 Introduction

Paracelsus' statement "All substances are poisons, there is none that is not a poison. The right dose differentiates a poison from a remedy" describes the central theme that underlies the processes of **risk assessment** and **risk management**. Any chemical given in excess can cause harm, while in moderation it can be of benefit. For instance, aspirin in the right dose can relieve pain and symptoms associated with inflammation as well as providing some level of protection against heart disease. Conversely, excess consumption of aspirin can cause serious health problems; there are even cases of attempted suicide by aspirin overdose. Other chemicals, such as the pesticide DDT, have no direct beneficial health effect on people, but may contribute indirectly to the general well-being by eliminating unwanted insects. DDT is not highly toxic to humans, but it does bioaccumulate in the fatty tissues and is retained in the body for years. DDT has a weak estrogen-like effect; as a result, tissues of the female reproductive tract, such as the ovaries, may be constantly stimulated. The estrogenic effects may explain the appearance of ovarian cancer in some exposed individuals. Another example is penicillin, an important antibiotic for preventing and curing illnesses. However, at a given dose one individual may be cured, another will see no effect, and a third one may die because of hypersensitivity to the substance. A variety of other effects may be seen between the extremes of death and no response.

Determining what constitutes a "safe" level of exposure becomes a difficult task when we consider the wide array of chemicals in existence as well as the variability of individual response to them. To determine acceptable exposure levels, two major pieces of information need to be obtained: 1) the toxic effects associated with exposure to the substance, and 2) the level of risk the public or the user is willing to accept.

All human activities are associated with some degree of risk. Risk is defined as the probability (chance) that an adverse effect will occur under a specific set of circumstances. Table 10-1 identifies the risk associated with a variety of activities. Notice that the list contains a number of activities that individuals are involved in on a daily basis. For instance, many people drink soda, eat charbroiled steaks, and smoke cigarettes. There are risks associated with each of these activities, yet individuals

Action	Nature of Risk
Smoking 1.4 cigarettes	Cancer, heart disease
Drinking 0.5 liter of wine	Cirrhosis of the liver
Spending one hour in a coal mine	Black lung disease
Spending three hours in a coal mine	Accident
Living two days in New York or Boston	Air pollution/heart disease
Traveling six minutes by canoe	Accident
Traveling ten miles by bicycle	Accident
Traveling 30 miles by car	Accident
Flying 1,000 miles by jet	Accident
Flying 6,000 miles by jet	Cancer caused by cosmic radiation
Living two months in Denver on vacation	Cancer caused by cosmic radiation
Living two months in average stone or brick building	Cancer caused by natural radioactivity
One chest x-ray taken in a good hospital	Cancer caused by radiation
Living two months with a cigarette smoker	Cancer, heart disease
Eating 40 tablespoons of improperly stored peanut butter	Liver cancer caused by aflatoxin B
Drinking heavily chlorinated water (e.g., Miami) for one year	Cancer caused by chloroform
Living five years at site boundary of a typical nuclear power plant in the open	Cancer caused by radiation
Drinking 1,000 24-oz. soft drinks from recently banned plastic bottles	Cancer from acrylonitrile monomer
Living 20 years near PVC plant	Cancer caused by vinyl chloride (1976 standard)
Living 150 years within 20 miles of a nuclear power plant	Cancer caused by radiation
Eating 100 charcoal broiled steaks	Cancer from benzopyrene
Living within five miles of a nuclear reactor for 50 years	Cancer caused by radiation

Table 10-1: Actions increasing risk of death by one in a million. *Recreated with permission from MIT's Technology Review Magazine, Copyright © 1997.*

continue to participate in them. However, the same group of individuals may find the idea of living near a nuclear reactor an unacceptable risk. Indeed for every individual who enjoys diet soda, a charbroiled steak, or smoking cigarettes, there may be many individuals who think there is too much risk involved in participating in these same activities. However, these people may be quite willing to live near a nuclear power facility.

All of the activities in Table 10-1 present the same level of risk. There is a one in a million chance that by participating in any one of these activities an individual will develop some type of disease or have an accident. Whether an individual becomes involved is based primarily on his/her perception of the risk associated with the activity. This is generally an individual decision in which the benefits derived from the activity are determined to outweigh the perceived risk. We can draw an analogy to individuals who enjoy skydiving. The thrill of floating effortlessly through the sky outweighs the apparent risk of the parachute not opening, resulting in the ultimate "toxic effect" – death. The relief of stress and the taste enjoyment are perceived as immediate benefits by individuals who smoke cigarettes. The potential lung and heart disease derived from smoking, on the other hand, develop over a long period of time; the effects are therefore not immediately realized. If the negative effects were felt immediately, would the individual's perception of the risk change? Many patients who undergo chemotherapy treatment for various forms of cancer, using experimental or proven drugs with known toxic side effects, are willing to accept the risk associated with use because the outcome is deemed to be beneficial and worth the discomfort (adverse side effects). In the final analysis an individual will undertake a particular activity based upon his/her perception of the risk involved weighed against the apparent benefits that will be derived.

Regulatory agencies (Table 10-2) such as the Environmental Protection Agency (EPA), the Food and Drug Administration (FDA), and the Occupational Safety and Health Administration (OSHA) are involved in determining risk associated with chemical exposure. EPA, through a variety of environmental laws, regulates exposure to toxic substances that may be released into the environment (soil, water, and air). The FDA is concerned with toxic substances added to foods and cosmetics. OSHA evaluates and regulates toxic substances in the workplace. One of the primary functions of these agencies is to establish guidelines and safe levels of exposure for toxic substances.

Often the use of the word "safe" is interpreted by the public to mean that there is no risk (zero) associated with a particular activity or – in the context of toxic chemicals – that there are no adverse effects as a result of exposure. In reality, there is always some risk because of the differences in susceptibility and sensitivity of individuals to a particular toxic substance. Therefore, it is not possible for regulatory agencies to establish absolute, safe levels guaranteeing that no individual will ever develop an adverse response. However, the agencies can determine the circumstances and concentrations where the risk of an adverse response is extremely low, thus defining the level of exposure that may be considered safe. There are two phases in the establishment of acceptable exposure levels: risk assessment and risk management. Information derived from these activities is used to determine the risk involved in exposure to toxic substances and how to best manage the risk.

When individuals become exposed to toxic substances, the symptoms may be a result of both long-term chronic or acute exposure. Often it takes months and sometimes years for symptoms to develop. For example, various types of lung disease such as cancer, emphysema, and bronchitis occur as a result of smoking. Liver disease may occur from prolonged consumption of alcohol or exposure to various types of organic solvents. Chronic exposure to mercury may cause renal failure.

Symptoms associated with acute exposure are easily recognized and are often immediate, or occur within hours or a few days after exposure. For example, splashing hydrochloric acid in the eye or on the skin will result in an immediate burning sensation to both. If left untreated tissue damage and scar formation will occur. Inhalation of noxious, irritating gases like chlorine or ammonia can cause coughing and breathing difficulties.

Acute chemical exposures may occur as a result of inhalation, ingestion, injection, or contact with the skin or eyes; the type of response implemented to assist and treat exposed individuals varies accordingly. However, there are some basic steps that should be followed when attempting to help any exposed individual regardless of the route of exposure. The first things to do are: 1) assess the situation; 2) determine the nature and severity of injuries or problems; 3) administer first aid; and 4) if possible, obtain help. Some circumstances may not be conducive to following the steps in the order listed. In some situations it may not be possible to admin-

290 ■ Basics of Toxicology

Legislation	Administering Agency	Regulated Products
Food, Drug and Cosmetics Act	FDA[a]	Food, drugs, cosmetics, food additives, color additives, new drugs, animal and feed additives, and medical devices
Federal Insecticide, Fungicide and Rodenticide Act	EPA[b]	Pesticides
Dangerous Cargo Act	DOT[c], USCG[d]	Water shipment of toxic materials
Atomic Energy Act	NRC[e]	Radioactive substances
Federal Hazardous Substances Act	CPSC[f]	Toxic household products
Poultry Products Inspection Act	USDA[g]	Food, feed, color additives, and pesticide residues
Occupational Safety and Health Act	OSHA[h], NIOSH[i]	Workplace toxic chemicals
Poison Prevention Packaging Act	CPSC	Packaging of hazardous household products
Clean Air Act	EPA	Air pollutants
Hazardous Materials Transportation Act	DOT	Transport of hazardous materials
Clean Water Act	EPA	Water pollutants
Marine Protection, Research and Sanctuaries	EPA	Ocean dumping
Consumer Product Safety Act	CPSC	Hazardous consumer products
Lead-Based Paint Poison Prevention Act	CPSC, HEW[j], HUD[k]	Use of lead paint in federally assisted housing
Safe Drinking Water Act	EPA	Drinking water contaminants
Resource Conservation and Recovery Act	EPA	Solid waste, including hazardous wastes
Toxic Substances Control Act	EPA	Hazardous chemicals not covered by other laws, includes pre-market review
Federal Mine Safety and Health Act	DOT, NIOSH	Toxic substances in coal and other mines
Comprehensive Environmental Response, Compensation, and Liability Act	EPA	Hazardous substances, pollutants, and contaminants at waste sites

[a] Food and Drug Administration
[b] Environmental Protection Agency
[c] Department of Transportation
[d] United States Coast Guard
[e] Nuclear Regulatory Commission
[f] Consumer Product Safety Commission
[g] United States Department of Agriculture
[h] Occupational Safety and Health Administration
[i] National Institutes of Occupational, Safety, and Health
[j] Department of Health, Education and Welfare
[k] Department of Housing and Urban Development

Table 10-2: Federal laws related to exposures to toxic substances.

ister first aid until additional help arrives. For example, an individual exposed to toxic gases may not receive assistance until the arrival of the proper breathing apparatus for first aid personnel. The proper sequence of events should be determined in accordance with the evaluation of the overall circumstances The choice of a course of action should always allow for the quickest treatment to the victim without endangering individuals attempting to provide assistance.

10-2 Risk Assessment

Risk assessment is the process of gathering all available information on the toxic effects of a chemical and evaluating it to determine the possible risk associated with exposure. The process of gathering and evaluating the information can be divided into four steps: 1) hazard identification, 2) hazard evaluation or dose-response assessment, 3) exposure assessment, and 4) risk characterization. Although each of these steps will be discussed individually, the information associated with each step is interrelated and may be evaluated to different degrees in each of the steps.

Hazard Identification

Initiation of the risk assessment process presumes that exposure to a toxic substance may occur, therefore creating a need to determine if there is a chance of causing adverse effects in exposed individuals. Hazard identification, which is the first step in the risk assessment process, consists of compiling and examining data primarily from toxicological and epidemiological studies. Information from these studies should provide an answer to the questions: "Does exposure to the substance produce any adverse effects? And, if adverse effects are produced, what are the circumstances associated with the exposure?"

Table 10-3 lists the various types of information that should be identified and examined when performing the hazard identification. This type of information is important in initially determining whether adverse effects produced in one set of circumstances will produce adverse effects under a different set of circumstances. This is true for data derived from both toxicological and epidemiological studies.

The advantages and disadvantages associated with toxicological and epidemiological studies were discussed in Chapter 2. The major advantage of the epidemiological study is that it represents "real-world" exposure situations. Any adverse effects observed are the result of performing daily routines that result in exposure to the toxic substance at concentrations occurring in the environment. The major disadvantage is that the data collected and evaluated are often incomplete or of questionable accuracy. This is particularly true when retrospective epidemiological studies are performed. In many cases a direct cause-effect relationship is hard to establish.

1. Identify substance (name)
2. Physical/chemical properties of the toxic substance
    A. Solubility, chemical reactivity, molecular size, ion/non-ion state
3. Source of toxicity information
    A. Epidemiological studies
        (1) prospective study
        (2) retrospective study
    B. Toxicological study
        (1) Acute toxicity studies
            – Thresholds, no-observed-effect-level (NOEL), no-observed-adverse-effect-level (NOAEL), LD_{50}, LC_{50}
        (2) Chronic toxicity study
            – Mutagenic, teratogenic, carcinogenic, other biological effects (heart failure, liver disease, skin rash, etc.)
        (3) Species of test animal
        (4) Other variables affecting toxicity
            – Age, sex, health
4. Exposure to toxic substance
    A. Route of exposure
        (1) Skin contact
        (2) Inhalation
        (3) Ingestion
        (4) Injection
    B. Duration of exposure
    C. Frequency of exposure
    D. Exposure to other toxic substances
5. Other confounding factors
    A. Diet, lifestyle choices, occupation

Table 10-3: Types of information collected and considered when performing the hazard identification phase of a risk assessment. *Obtained from the U. S. Environmental Protection Agency,* Risk Assessment Guidelines and Information Directory, *Government Institute, Rockville, MD, 1988.*

Chemical	Carcinogenic Site(s) in Humans	Carcinogenic Site(s) in Animals
Aflatoxins	Liver	Liver, colon, and kidneys in mice, rats, fish, duck, marmosets, tree shrews, and monkeys
4-Aminobiphenyl	Bladder	Urinary bladder in mice, rats, rabbits, and dogs
Asbestos	Lung, mesothelium	Lung and mesothelium in mice, rats, hamsters, and rabbits
Benzidine	Bladder	Liver in rats and hamsters; bladder in dogs
DES	Vagina	Estrogen-responsive tissues in mice, rats, hamsters, frogs, and squirrel monkeys
2-Naphthylamine	Bladder	Urinary bladder in hamsters, dogs, and nonhuman primates; liver in mice

Table 10-4: Sites of carcinogenic activity in humans and animals for a variety of chemicals. *Information derived from International Agency for Research on Cancer (IARC),* Chemicals and Industrial Processes Associated with Cancer in Humans, *IARC, Lyon, September, 1979.*

Toxicological studies have the advantage of being able to establish a cause-effect relationship. Therefore, these types of studies may be more useful in determining whether a risk exists. However, there are two major disadvantages associated with obtaining data from toxicological studies. First, various types of test animals are used, since humans are seldom intentionally exposed to toxic substances. Second, low-dose values can only be obtained by extrapolation from information obtained at high-dose levels.

In general, it is assumed that toxic effects observed in test animals will also be observed in humans. Table 10-4 lists several examples where the toxic effect (cancer) that occurs in the test species also occurs in humans. However, in some circumstances the toxic effects observed in the test species are not observed in humans and vice versa. Different species of animals may metabolize toxic substances by different metabolic pathways, which may result in either bioactivation or deactivation of the substance. For example, acetylaminofluorene (AAF) is a carcinogen in many species of animals. However, it is not a carcinogen in guinea pigs because the carcinogenic metabolite, 2-hydroxy-AAF, is not produced in the guinea pig. This example points out the importance of understanding the difference between species when evaluating toxicological data. Compilation of toxicity data should encompass as many different test protocols as possible in order to understand all possible effects of exposure to a toxic substance. Differences in the toxicity of various substances in relation to different species of test animals are discussed in more detail in Chapters 4 and 5.

Toxicological studies are characterized by test animals being exposed to various doses of a toxic substance followed by measurement of the response. The information is used to develop a dose-response curve. The test animals are exposed to high concentrations in order to observe a response in a short period of time. Therefore, response information at the lower concentrations must be extrapolated from test results and represented by the dose-response curve. This approach presents a level of uncertainty due to the assumption that the exposed test animal will respond in the same manner at the low doses as at the high doses. For instance, it is assumed that the rate of metabolism, bioactivation, or deactivation of the toxic substance is the same at all levels of exposure.

In addition to considering the type and source of the toxicological information, data needs to be obtained on the various physical/chemical properties of the toxicant, such as solubility, molecular size, and electrical charge, which affect toxicity. Lipid-soluble toxicants (xylene, benzene) are easily absorbed through the skin, the digestive tract, and the lungs. Water-soluble toxicants are more readily absorbed when inhaled or ingested than with skin contact. If the primary route of exposure is determined to be contact with the skin and the substance is lipid insoluble (and therefore water-soluble), then it may be concluded that the risk associated with exposure is minimal. The toxicant will not be able to easily penetrate the keratinized outer layer of the skin. On the other hand, if the primary route of exposure is by inhalation, then the associated risk of exposure is high because the solubility of the toxicant in the moist lining of the respiratory tract is high.

In some cases the risk assessment ends with the collection of toxicological information and data concerning the physical/chemical properties of the toxicant, because the data indicate there are no potential hazards. However, when the data are inadequate, the decision may be to initiate further testing to obtain additional data. When further testing is not possible, the hazard determination may be merely based on scientific judgment.

> Hazard identification consists of the collection of data – from various sources – that are necessary to determine whether a substance is toxic.

Hazard Evaluation and Dose-Response Assessment

If the hazard identification process produces evidence of a hazard, then a hazard evaluation is performed. The purpose of this step is to quantify, if possible, the dose at which an effect will occur. Results from dose-response curves, if available, can be used to provide this type of information. There may be several dose-response relationships defined for a specific toxic chemical. The differences may be due to different exposure conditions or to the measurement of different responses. Both of these factors may yield different dose-response curves for the same toxic substance. Dose-response curves can be used to identify a threshold level, a no-observed-effect-level (NOEL), or a no-observed-adverse-effect-level (NOAEL) for noncarcinogenic substances. This information can, in turn, be used to determine safe exposure levels.

As mentioned previously, test animals are often exposed to toxic doses much higher than those encountered by humans. To address this disparity, high

Chemical	LD_{50} (mg/kg body weight)
Sucrose (table sugar)	29,700
Sodium chloride (common Salt)	3,000
Vanillin	1,580
Aspirin	1,000
Copper sulfate	960
Chloroform	908
Caffeine	192
DDT	113
Nicotine	53
Strychnine	16
Sodium cyanide	6
Aflatoxin B1	5

1. Most values were taken from NIOSH Registry of Toxic Effects of Chemical Substances, 1979. Results reported elsewhere may differ, as will results for species other than rats.
2. Chemicals are listed in order of increasing toxicity in rats, i.e., sucrose is the least and aflatoxin B1 is the most toxic.

Table 10-5: Approximate oral LD_{50} in Rats[1,2]. *Information obtained from the NIOSH Registry of Toxic Effects of Chemical Substances, 1979.*

doses are extrapolated to doses that may be considered safe or acceptable. Perhaps the easiest method used is that associated with applying a safety factor. Although their range is anywhere from 10 to 2000, safety factors of 10, 100, or 1,000 are routinely used. The choice of the safety factor is determined by the data available. A safety factor of 10 is used when there is human exposure data available or when the substance has a relatively low toxicity. Low toxicity can be indicated by a large LD_{50} (see Table 10-5), or LC_{50}. A safety factor of 100 is used when there are reliable data from animal studies but no human exposure data are available. Because there may be variability between species, the observed toxic effects may not be similar to those that may occur in humans at a given dose. A safety factor of 1,000 is used when there are no human data and data from animal studies are scarce, when there is great interspecies variability, and the substance is extremely toxic.

> Safety factors are used to account for the variability in response of test animals and differences in toxicity. They are most often used for non-carcinogenic chemicals where a threshold can be established.

Suppose that information derived from animal studies indicated that the no-observable effect-level (NOEL) was 50 ppm. If the information was derived primarily from animal studies, the exposure level considered to be safe would be 50 ppm ÷ 100 = 0.5 ppm. A safety factor of 100 is used because the supporting data are derived primarily from animal studies.

The concentration of a substance derived from applying a safety factor to a no-observable effect-level can be used to establish the **Acceptable Daily Intake (ADI)** for that substance. The ADI is the daily intake of a chemical, which during an entire lifetime appears to be without appreciable risk on the basis of all known facts at the time. It is expressed in milligrams of the chemical per kilogram of body weight (mg/kg). ADIs are often established for substances such as food additives and pesticides. It is important to understand that ADIs do not imply that there will be no adverse effects under any circumstances. ADIs are based on the best available data at the time they are established. Additional testing may uncover other toxicological effects not previously considered resulting in a change in the ADI value.

Other terms utilized to indicate acceptable levels of exposure are the **reference dose (RfD)** for exposures from ingestion or skin contact; and **reference concentration (RfC)** for inhalation exposures. RfDs and RfCs are estimates of a daily exposure to an agent that is assumed to be without an adverse health impact on the human population.

RfDs and RfCs are calculated by dividing the no-observed-adverse-effect-level (NOAEL) or the lowest-observed-adverse-effect-level (LOAEL) by uncertainty factors (value of 10). Uncertainty factors are essentially safety factors that take into account the differences in response between species (extrapolation of results from animals to humans) and within species (variation in human sensitivity). An additional uncertainty factor of 10 is used for LOAEL (see formula below).

The general formula for calculating an RfD or RfC is:

$$\text{RfD or RfC} = \frac{\text{NOAEL}}{10 \times 10}$$

Where:

10 = Variation in human sensitivity
10 = Extrapolation of results from animals to humans

$$\text{RfD or RfC} = \frac{\text{LOAEL}}{10 \times 10 \times 10}$$

Where:

10 = Variation in human sensitivity
10 = Extrapolation of results from animals to humans
10 = Use of LOAEL and not NOAEL

Suppose you wanted to determine the reference dose for mercury that would not cause kidney damage. Remember that the LOAEL represents the lowest concentration at which the toxic effect is first observed; in this case kidney damage. If the LOAEL was 1.5 mg/kg/day for rats fed mercury, then the RfD would be:

$$\text{RfD} = \frac{1.5 \text{ mg/kg/day}}{10 \times 10 \times 10} = 0.0015 \text{ mg/kg/day}$$

Where:

10 = Variation in human sensitivity
10 = Extrapolation of results from animals to humans
10 = Use of LOAEL and not NOAEL

A more complex, but perhaps more exact, method for determining a safe human dose incorporates several other variables such as absorption factors, body weight, exposure protocol, the half-life of the toxicant, in addition to the safety factor. The numbers used for the absorption factor, exposure protocol, and half-life all represent ratios that attempt to compensate for the differences between the test animals and humans. As a result, the overall equation for determining the safe human dose for a given toxic substance is:

$$\text{SHD} = \frac{\text{ThD} \times \text{BW} \times \text{AF} \times \text{ER} \times t_{1/2}}{\text{SF}} = \text{mg/day}$$

Where:

SHD = safe human dose;

ThD = threshold in the test species;

BW = body weight of exposed individual usually considered to be 70 kg;

AF = absorption factor which is the ratio of absorption in test animals/absorption in humans;

ER = ratio of exposure concentration in test animal versus exposure in humans;

$t_{1/2}$ = ratio of half-life of substance in test animal over half-life in humans; and

SF = safety factor, which is dependent on source of the data and relative toxicity of the substance.

Suppose you wanted to determine the safe human dose for cadmium-containing food that would not cause liver damage, and you know that the threshold dose causing liver damage in rats is 15/mg/kg/day (ThD). To determine the safe human dose, additional information would need to be gathered concerning: 1) the absorption rate of cadmium from the digestive tract of humans and rats (AF), 2) the exposure regime for both (ER), and 3) any differences in metabolism ($t_{1/2}$). If you assume that the exposure regime and the metabolic rate are the same for both rats and humans and that standard body weight for humans is 70 kg (BW), then the only other variables will be the difference in the absorption rate and whatever safety factors need to be included. If it is further assumed that only 10 percent of the cadmium is absorbed from the digestive tract of humans and 30 percent is absorbed from the digestive tract of rats, the AF value would be 3 (30 percent divided by 10 percent). The safe human dose then would be calculated as follows:

$$\text{SHD} = \frac{15/\text{mg/kg/day} \times 70 \text{ kg} \times 3 \times 1 \times 1}{10 \times 10}$$

$$= \frac{15 \text{ mg/kg/day} \times 70 \text{ kg} \times 3}{100}$$

$$= \frac{3{,}151 \text{ mg/day}}{100}$$

$$\text{SHD} = 31.51 \text{ mg/day}$$

Two safety factors were used in this example assuming that no human exposure data were available (safety factor of 10) and that data were available from chronic toxicological studies (safety factor of 10).

Use of Statistical Models in Hazard Evaluation

Chemicals suspected or known to produce cancer often use statistical models. Animal studies routinely use only two doses to determine if a toxic substance produces cancer; therefore, a dose-response curve is drawn using only these two points. The statistical models are utilized to extrapolate the remainder of the curve for the lower doses. This method is used because it is not possible to accurately measure chemical exposures at the lower doses.

Information derived from these dose-response curves is utilized to determine the acceptable risk of exposure to potential cancer-causing substances. An acceptable level of exposure is often expressed in terms of the probability – above background – of causing an increase in the occurrence of cancer in a given population. This is normally expressed numerically by regulatory agencies that consider the acceptable level of risk to be in a range between 10^{-5} to 10^{-7}. Therefore, exposure to a specific cancer-causing substance would be acceptable only if daily exposure – at a given dose for a lifetime – increased the rate of cancer occurrence by 1 in 100,000 (10^{-5}) or 1 in 10,000,000 (10^{-7}) above background.

296 ■ Basics of Toxicology

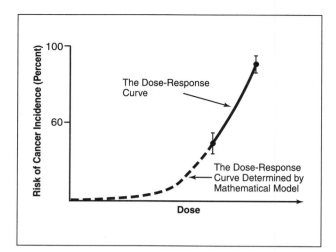

Figure 10-1: Dose-response curve illustrating extrapolation from high to low doses. The solid line represents the dose-response curve derived from toxicity (animal) studies. *Recreated with permission from Williams P. L. and Burson J. L.*, Industrial Toxicology: Safety and Health Applications in the Workplace, *Van Nostrand Reinhold, New York, 1985.*

Probit Model
Logit Model
Weibull Model
Linearized Multistage Model
One-hit Model
Gamma Multi-hit Model

Table 10-6: Models routinely used to extrapolate information from high doses to low doses.

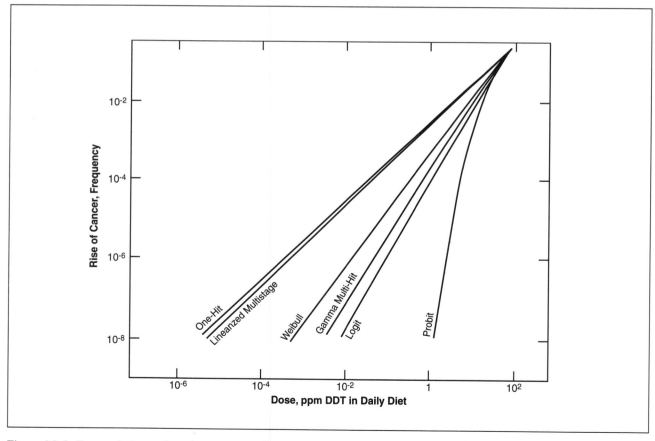

Figure 10-2: Extrapolation to low doses using different mathematical models. *Obtained from the U. S. Environmental Protection Agency*, EPA Toxicology Handbook, *Government Institute, Rockville, MD, 1994.*

Very complex statistical evaluations as well as different approaches are used to extrapolate results from high to low doses. The different approaches – called models – take into account variability of test procedures and results. Each model produces a different dose-response curve. Figure 10-1 is a generalized illustration of this concept. In this case, the graph represents the results from a toxicological animal study using a carcinogen. It is assumed that there is no threshold, therefore the dose-response curve intersects the X-axis and the Y-axis at zero. Table 10-6 lists six of the models routinely used. Each of the models is based on a different set of assumptions; therefore different results are obtained at the lower concentrations when extrapolated from the results obtained from exposure at higher concentrations. Figure 10-2 illustrates this point. The most conservative model – the one providing the largest margin of safety – is the one-hit model, which takes into account the highest margin of risk. The least conservative is the probit model.

> A probit is a statistical unit of measurement of probability based on deviations from the mean of a normal frequency distribution.

There is approximately one chance in 10 million that exposure to 10^{-6} ppm DDT will result in development of liver tumors using the one-hit model. Using the probit model, a higher concentration of approximately one ppm is needed to produce the same effect.

Table 10-7 illustrates the same concept for exposure to methylene chloride. In this example, the multistage model is more conservative. According to this model, the projected risk of developing cancer is higher than the ones projected by the other models.

Exposure Assessment

The exposure assessment is performed to identify the affected population and, if possible, quantify the magnitude, frequency, duration, and route of exposure. Table 10-8 lists various factors that should be considered when performing an exposure assessment. In general, consideration should be given to the physical/chemical properties of the toxic substance, source, exposure pathways and environmental fate, concentration, the exposed population, and the calculation of the total exposure from all pathways. Each of these areas is briefly discussed below.

General Information for Each Chemical

The physical/chemical properties of the toxic substance affect how it is transported, how it is accumulated in the environment and in tissues, and how it is transformed when it is released into the environment. Some of the characteristics that should be identified and considered are: vapor pressure (volatility), solubility in water, and sorption to soil or sediments. Collectively, the interactions between these factors will affect the dose and route of exposure.

Air Concentration µg/m³	Multistage Model	One-Hit Model	Weibull Model	Log-Probit Model
1	1.8×10^{-7}	2.0×10^{-7}	4.8×10^{-10}	3.5×10^{-31}
10	1.8×10^{-6}	2.0×10^{-6}	1.7×10^{-8}	1.6×10^{-22}
100	1.8×10^{-5}	2.0×10^{-5}	6.1×10^{-6}	2.5×10^{-15}
1,000	1.8×10^{-4}	2.0×10^{-4}	2.0×10^{-5}	1.3×10^{-9}
10,000	1.9×10^{-3}	2.0×10^{-3}	6.1×10^{-4}	2.4×10^{-1}

Table 10-7: Estimates of lifetime risk to humans from exposure to methylene chloride based on salivary gland region sarcomas in male rats derived from four different models. *Information obtained from* USEPA Health Assessment Document for Dichloromethane (Methylene Chloride), Final Report, *February, EPA/600/8-82/004F.*

1. General Information for Each Chemical
    a. Identity
        (1) Molecular formula and structure
        (2) Other identifying characteristics
    b. Chemical and Physical Properties

2. Sources
    a. Characterization of Production and Distribution
    b. Uses
    c. Disposal
    d. Summary of Environmental Releases

3. Exposure Pathways and Environmental Fate
    a. Transport and Transformation
    b. Identification of Principal Pathways of Exposure
    c. Predicting Environmental Distribution

4. Measured or Estimated Concentrations
    a. Uses of Measurements
    b. Estimation of Environmental Concentrations

5. Exposed Populations
    a. Human Populations
        (1) Population size and characteristics
        (2) Population location
        (3) Population habits

6. Integrated Exposure Analysis
    a. Calculation of Exposure
        (1) Identification of the exposed population
        (2) Identification of pathways of exposure

Table 10-8: Suggested outline for an exposure assessment. *Obtained and modified from the U. S. Environmental Protection Agency,* Risk Assessment Guidelines and Information Directory, *Government Institute, Rockville, MD, 1988.*

Vapor pressure is defined as the pressure exerted by a vapor that is in equilibrium with a solid or liquid. It is a measure of the ease with which a substance can change from a solid or liquid to a gas. The higher the vapor pressure, the greater the tendency of the substance to enter the gaseous state. In general, the lower the molecular weight, the higher the vapor pressure. Exposure by inhalation is more likely to occur when a toxic substance has a high vapor pressure. Substances with high vapor pressures may or may not persist long in the environment. Also, the sorption of these substances to surfaces such as soils and sediments will slow their volatilization rate.

Some substances are more soluble in water than others. Therefore, it is possible that exposure may occur via ingestion (drinking water). Even though many of the organic toxic substances are considered to be not soluble in water, they still dissolve in water to a small degree. This fact must be considered in the exposure assessment.

Adsorption to soil and sediments can affect the distribution of toxic substances in the environment, thus affecting exposure, which may then occur when the soil is handled, or if food crops are grown in the contaminated area. The toxic substance may be absorbed by the plants or aquatic organisms and subsequently consumed. Metals such as cadmium and lead have been shown to be absorbed by plants from contaminated soil. The accumulation of organochlorine pesticides such as DDT in the sediments of rivers and lakes is well documented. The sediments provide a potential, continuous source of the toxicant to various aquatic species. Ultimately, these substances are bioaccumulated in the food chain, resulting in human exposure. The degree to which a toxic substance will be adsorbed is dependent on the type of soil/sediment and the properties of the toxic substance. In general, soils/sediments with a high organic content will have a higher affinity for water-insoluble toxic substances.

Sources of Exposure

Exposure to toxic chemicals may occur almost anywhere. In the home, for example, there are many sources: cleaning products, medicines, various types of paints, pesticides, solvents, and adhesives. In addition, lifestyle choices such as smoking and consuming alcoholic beverages may result in exposure to other toxic substances. Table 10-9 lists some of the possible household sources and the chemicals they contain. Outside the home, exposure to airborne chemical pollutants occurs by inhalation. Oxides of nitrogen and sulfur are common air pollutants that result from various industrial activities. Occupational exposure may involve a variety of toxic agents (see Table 10-10).

Table 10-11 lists possible sources of exposure from various manufacturing organizations. Although they may not apply to every situation, all possible means of exposure should be considered.

Products	Constituents
Insecticides and Pesticides	All ant, roach, and moth poisons, animal flea collars, Chlordane, lindane, toxaphene, arsenic, Malathion, diazinon, fumigants, nicotine, moth balls (naphthalene, paradichlorophenol), Sodium fluoroacetate, phosphorus, thallium, barium, strychnine, methyl bromide, cyanides
Plants	Foxglove - digitalis; cherry seed - cyanide; thorn apple - atropine; mushrooms, oleander, poison ivy
Flammables and Extinguishers	Kerosene and gasoline Fire lighter - Methanol, petroleum hydrocarbons, denatured alcohol Fire extinguishers - Carbon dioxide
Cleaning Supplies	Chlorinated hydrocarbons, solvent distillate, lye (sodium hydroxide), ammonia, Bleach - Sodium hypochlorite, oxalic acid Drain cleaner - Lye, sodium acid sulfate Rug cleaner - Chlorinated hydrocarbons Wallpaper cleaner - Kerosene
Medicines	Salicylates - Aspirin, methyl salicylate Acetaminophen Sedatives - Barbiturates, bromides, benzodiazepines, ethanol Antidepressants - Imipramine, amitriptyline, doxepin Antiepileptic agents - Barbiturates, hydantoins Antihistamines - "Cold" tablets and seasickness pills Cathartic pills - Strychnine, atropine Cough mixtures - Codeine, methadone, other opiates Nose drops - Ephedrine, Privine, Neo-Synephrine, etc. Reducing or slimming tablets - Amphetamines, thyroid, digitalis Cardiac drugs - Digitalis, quinidine Antiseptics - Boric acid, mercuric chloride, iodine, phenol Liniment - Methyl salicylate, alcohols Drugs of abuse - Amphetamines, PCP, methadone, LSD.
Cosmetics	Hair - Silver salts, anilines, potassium bromate, selenium sulfide
Paints and Painting Supplies	Paint - Thinner, lead, arsenic, chlorinated hydrocarbons Paint remover - Chlorinated hydrocarbons, acids, alkalies Lacquer - Ethyl acetate, amyl acetate, methanol Shellac - Methanol Wood bleach - Oxalic acid
Hobbies: Painting	Toxic pigments, e.g., arsenic (emerald green), cadmium, chromium, lead, mercury; acrylic emulsions; solvents
Ceramics	Raw materials - colors and glazes containing barium carbonate; lead, chromium, uranium, cadmium Firing - Fumes of fluoride, chlorine, sulfur dioxide Gas-fired kilns - Carbon monoxide
Sculpture and Casting	Grinding Silica-containing stone - Silica (silicon dioxide) Serpentine rock with asbestos - Asbestos Woodworking - Wood dust Metal casting - Metal fume, sand (silica) from molding, binders of phenol formaldehyde or urea formaldehyde
Welding	Metal fume, ultraviolet light exposure, welding fumes, carbon dioxide, carbon monoxide, nitrogen dioxide, ozone or phosgene (if solvents nearby)
Plastics	Monomers released during heating (polyvinyl chloride), methyl methacrylate, acrylic glues, polyurethane (toluene 2,4-diisocyanate), polystyrene (methyl chloride release), fiber glass, polyester, or epoxy resins
Woodworking	Solvents, especially methylene chloride
Photography	Developer - Hydroquinone, metal Stop bath - Weak acetic acid Stop hardener - Potassium chrome alum (chromium) Fixer - Sodium sulfite, acetic acid, sulfuric acid Hardeners and stabilizers - Formaldehyde

Table 10-9: Household products and some of their constituents. *Obtained and modified with permission from Dreisbach R. H. and Robertson W. D.*, Handbook of Poisoning, *12th edition, Appleton & Lange, Norwalk CT, 1987 – McCunney R. J.,* Occupational and Environmental Medicine, *2nd edition, Lippincott-Raven, Philadelphia, 1994.*

Occupation	Toxic Agent
Bakers	Flour, grain dust (asthma)
Battery manufacturing or repair	Lead, cadmium
Butchers	Vinyl plastic fumes (asthma)
Cement workers	Potassium chromate, dichromate (asthma)
Coal miners	Coal dust, (lung disease), methane, hydrogen sulfide
Dentists	Mercury, waste anesthetic gases
Dry cleaners	Solvents (liver, skin, and neurologic disease (neuropathy)
Electronics workers	Toluene diisocyanate (asthma); solvent exposure, hydrofluoric acid (skin burns)
Electroplating	Cyanide, chromium, nickel
Explosives manufacturing	Nitrates (headache and rebound vasoconstriction)
Farm workers	Pesticides; infectious agents; e.g., nitrogen dioxide, pentachlorophenol, hydrogen sulfide
Fire fighters	Carcinogens, smoke products; e.g., carbon monoxide, hydrogen cyanide, nitrogen dioxide
Foundry workers	Silica (lung disease), resins (asthma), metals
Fumigators	Methyl bromide (encephalopathy, neuropathy), sulfuryl fluoride (pulmonary edema), organophosphates and carbamates
Hospital workers	Infectious agents, radiation, cleansers, ethylene oxide
Insulation industry	Formaldehyde, asbestos, fiberglass
Insulators	Asbestos, fiberglass
Meat packing or weighing	Polyvinylchloride or papain (asthma), infectious agents (Q fever)
Miners	Dust (lung disease), hydrogen sulfide, carbon monoxide
Painters	Solvents, lead, isocyanates
Pathology technicians	Fluorocarbons (palpitations), formaldehyde, phenol
Petroleum industry	Benzene (leukemia), ammonia and hydrogen sulfide, polycyclic aromatic hydrocarbons
Pottery industry	Silica, lead
Poultry workers	Infectious agents, dusts (asthma)
Radiator repairmen	Lead
Rubber manufacturing	Ethylenediamine (asthma)
Seamen	Asbestos
Sewer workers	Hydrogen sulfide, methane
Shoe repairmen	Benzene (leukemia), benzidine (bladder cancer)
Sterilizer operators	Ethylene oxide (nerve diseases)
Vintners	Arsenic (cancer and dermatitis)
Welders	Metal and polymer fume fever, ultraviolet radiation (skin and corneal burns), lead
Woodworkers	n-hexane (nerve disease), other solvents, dust (asthma), wood preservatives

Table 10-10: Toxic agents associated with certain occupations. *Recreated with permission from Olson K. R. editor,* Poisoning and Drug Overdose, *2nd edition, Appleton & Lange, Norwalk CT, 1994.*

To Surface Water
Treated process wastewater
Leaks and spills to unrecirculated (once-through) cooling waters
Treated washwaters, condenser waters, scrubber waters
Deep ocean disposal
To Atmosphere
Process stack and vent emissions
Fugitive emissions, e.g., imperfect connections
Storage tank vents
Volatilization from wastewater treatment operations
Volatilization from land farming, land filling, and impoundments
Stack emissions from incineration of process wastes
To Soil
Solid wastes, liquid residues, and wastewater treatment sludges to landmarks (soil cultivation) and landfills
Incinerator and boiler ash to landfill
Deep well injection
Spills and leaks directly to soil
Important Intercompartmental Transport
Storm water washing chemicals and contaminated silt from soil to surface water
Groundwater flow to surface water, and vice versa
Washout and fallout of chemicals from atmosphere to surface water and soil
Volatilization from surface water and soil

Table 10-11: Various sources of toxic substances from manufacturing operations. *Obtained and modified with permission from Conway R. A. editor,* Environmental Risk Analysis for Chemicals, *Van Nostrand Reinhold, New York, 1982.*

Exposure Pathways and Environmental Fate

Once the sources have been identified it is necessary to determine the route and the nature of the exposure. Table 10-12 lists factors that may need to be considered when identifying the exposure pathway and environmental fate of a toxic substance. All of the factors identified may not need to be considered in each case. For instance, evaluating the occupational hazard of an individual in a coal mine would be different than evaluating it for a farmer who is going to be applying a pesticide. The primary concern for the coal miner would be from exposure to airborne toxicants. For the farmer, exposure could occur from inhalation during application or from skin exposure if the substance is lipid soluble.

Measured or Estimated Concentrations

When performing an exposure assessment it is preferable to obtain actual samples from the source of exposure and quantify the toxic substance of concern. If actual measurements are not available then estimates can be used based on distribution models. These models attempt to estimate the concentration of a substance at the point of exposure. Modeling is most often used when evaluating the distribution of substances released into the air. Other models may be used to evaluate the distribution of substances in aquifers or in lakes. Concentrations of the toxic substance must be obtained for all possible sources of exposure using either measured quantities, values derived from distribution models, or a combination of both.

Exposed Population

It is important to identify and characterize the exposed population in terms of variables such as sex and age distribution, number of small children, pregnant women, and the number of chronically ill individuals. In addition, personal traits such as eating habits, workplace habits, recreational habits, and exercising habits should be identified. All of these factors affect toxicity of a substance. It may be advantageous to consider some of these groups of people individually because of the possible differences in susceptibility to the toxic substance being evaluated. For example, small children and the chronically ill are more susceptible than healthy adults to airborne pollutants such as ozone and particulates.

Calculation of Exposure

Once all the information has been obtained and evaluated, possible exposure can be calculated. Exposure may occur by single or multiple pathways. When exposure occurs by more than one pathway, the total exposure is determined by totaling the contributions of all pathways. When specific exposure

I. Original Input to Environment During Manufacturing, Distribution, Use, and Disposal

A. To Water
  1. Treatment plant effluents at manufacturing and/or formulating plants
  2. Spills during manufacturing (original and formulating) and distribution
  3. Disposal after use
B. To Soil
  1. Direct applications as an agricultural chemical or for vegetation control
  2. land disposal, e.g., landfill or land cultivation operations
  3. Spills
C. To Atmosphere
  1. Stack emission during manufacture
  2. Fugitive volatilization losses, e.g., from leaks, storage tank vents, and wastewater treatment
  3. Losses during use and subsequent disposal

II. Mechanisms for Transformation and Transport Within and Between Environmental Compartments

A. Water
  1. Transport from water to atmosphere, sediments, or organisms
     a. Volatilization
     b. Adsorption onto sediments; desorption
     c. Absorption into cells (Protista, plants, animals); desorption
  2. Transformations
     a. Biodegradation (effected by living organisms)
     b. Photochemical degradation (nonmetabolic degradation requiring light energy), direct or via a sensitizer
     c. Chemical degradation (effected by chemical agents), e.g., hydrolysis, free-radical oxidation
B. Soil
  1. Transport to water, sediments, atmosphere, or cells
     a. Dissolution in rain water
     b. Adsorbed on particles carried by runoff
     c. Volatilized from leaf or soil surfaces
     d. Taken up by Protista, plants, and animals
  2. Transformation
     a. Biodegration
     b. Photo degradation on plant and soil surfaces
C. Atmosphere
  1. Transport from atmosphere to land or water
     a. Adsorption to particulate matter followed by gravitational settling or rain washout
     b. Washout by being dissolved in rain
     c. Dry deposition (direct absorption in water bodies)
  2. Transport within atmosphere
     a. Turbulent mixing and diffusion within troposphere
     b. Diffusion to stratosphere
  3. Atmospheric transformation
     a. Photochemical degradation by direct absorption of light, or by accepting energy from an excited donor molecule (sensitizer), or by reacting with another chemical that has reached an excited state
     b. Oxidation by ozone
     c. Reaction with free radicals
     d. Reactions with other chemical contaminants

Table 10-12: Mechanisms affecting environmental fate of chemicals. *Reprinted in large part with permission from Environmental Science and Technology Review 1978, 12 (4), 398-407 Copyright 1978 American Chemical Society.*

Exposed Population	Exposure Pathway	Daily Intake Rate	Exposure Frequency	Exposure Duration	Body Weight
Residential	Ingestion of potable water	2 liters	350 days/yr	30 yr	70 kg
	Ingestion of soil and dust	200 mg (child) 100 mg (adult)	350 days/yr	6 yr 24 yr	15 kg (child) 70 kg (adult)
	Inhalation of airborne particulates and gases	20 m³ (total) 15 m³ (indoors)	350 days/yr	30 yr	70 kg
Commercial/Industrial	Ingestion of potable water	1 liter	250 days/yr	25 yr	70 kg
	Ingestion of soil and dust	50 mg	250 days/yr	25 yr	70 kg
	Inhalation of airborne particulates and gases	20 m³ (8-hr workday)	250 days/yr	25 yr	70 kg
Agricultural	Ingestion of potable water	2 liters	350 days/yr	30 yr	70 kg
	Ingestion of soil and dust	200 mg (child) 100 mg (adult)	350 days/yr	6 yr 24 yr	15 kg (child) 70 kg (adult)
	Inhalation of airborne particulates and gases	20 m³ (total) 15 m³ (indoors)	350 days/yr	30 yr	70 kg
	Consumption of homegrown produce	42 g (fruit) 80 g (vegetables)	350 days/yr	30 yr	70 kg
Recreational	Consumption of locally caught fish	54 g	350 days/yr	0 yr	70 kg

Table 10-13: Standard default exposure factors. *Obtained and modified from U. S. Environmental Protection Agency (1991), Human Health Evaluation Manual – Supplemental Guidance: "Standard Default Exposure Factors," OSWER Directive 9285.6-03, Office of Solid Waste and Emergency Response, Washington D.C. March 25.*

data are not available, certain assumptions are made; Table 10-13 lists some of the default values (assumptions) used when assessing exposure.

Calculation of exposure by various routes may be performed by using the following equations:

Equation 1 – Inhalation:

Daily Exposure = air level (concentration) × 1-2 m³/hour × hours/day

Equation 2 – Ingestion:

Daily Exposure = soil level (concentration) 25 mg soil/day × percent absorption

Equation 3 – Dermal Exposure:

Daily Exposure = Surface level (concentration) × 1,500 cm² (estimated surface area exposed) × fraction of day exposed × percent absorption

Several assumptions are associated with each of the calculations. It is assumed that during rest the average inhalation rate for an adult is one cubic meter per hour. If the individual is involved in moderate activity it is assumed that the inhalation rate is two cubic meters per hour. Regardless of the respiratory rate it is assumed that there is 100 percent absorption by this route.

The second equation listed above addresses specifically those situations related to ingestion of soil or from airborne soil particles. Ingestion may occur as a result of either acute or chronic exposure such as eating or smoking. In these situations the amount of toxic substance should be measured or – as generally is the case – at least estimated.

Dermal exposure is difficult to calculate relative to the other routes of exposure. It is assumed that the actual concentration of the substance in contact with the skin is the same as the concentration in the environment immediately adjacent to the skin.

Exposures by inhalation, ingestion, or dermal contact may also be calculated by using the following general equation:

Equation 4:

$$\text{Intake} = \frac{\text{Contaminant Concentration} \times \text{Intake Rate} \times \text{Exposure Time} \times \text{Exposure Frequency} \times \text{Exposure Duration}}{\text{Body Weight} \times \text{Averaging Time}}$$

$$I = \frac{C \times IR \times ET \times EF \times ED}{BW \times AT}$$

Where:

C = Concentration at point of exposure;
　　Units of Measure:
　　　Soil – mg/kg
　　　Water – mg/liter
　　　Air – mg/m^3

IR = Intake rate;
　　Units of Measure:
　　　Ingestion – soils (mg/kg), water (liter/day), food (g/day)
　　　Inhalation – air (m^3/day)
　　　Dermal Contact – variable depending on substance

ET × EF × ED = Represents the total duration of exposure;
　　ET = Exposure time (hours/day or hours/ exposure event)
　　EF = Exposure frequency (days/year or events/year)
　　ED = Exposure duration in years

BW = Body weight of exposed individual

AT = Averaging time in days
　　For carcinogens – 25,550 days (70 years)
　　For noncarcinogens – 365 days/year × ED

To determine the total exposure of an individual drinking water contaminated with 0.05 mg/liter methyl mercury, the following calculations would be performed using the information from Table 10-13:

$$I = \frac{0.05 \text{ mg/liter} \times 2 \text{ liter/day} \times 350 \text{ days/yr} \times 30 \text{ years}}{70 \text{ kg} \times 25,550 \text{ days}}$$

I = 0.0006 mg/kg/day

If exposure to methyl mercury occurred as a result of eating contaminated fish, the same equation would be used to calculate exposure by ingestion. Both values would then be added together to determine the total exposure.

Statistical models are available that address various pathways of exposure. These models may calculate exposure to humans as a result of the deposition of airborne substances on the surface of plants that are subsequently eaten by people or by animals that are, in their turn, used as food. Exposure from the consumption of fish or plants that have bioaccumulated toxic substances by absorption from water or soil, respectively, can also be calculated. These calculations are often performed by individuals experienced in modeling and risk assessment.

> Exposure assessment attempts to identify actual exposure levels and characterize the affected population.

Risk Characterization

The final step in the risk assessment is the compilation of all the information gathered from the previous steps and the determination of the actual risk of exposure to a specific toxic substance. This step becomes more subjective in nature because it relies on the assessor's analysis concerning the validity of the data provided. Consideration must be given to the assumptions associated with data derived from models. Are the routes of exposure analogous to those experienced by the affected population? Are the test animals and parameters measured the most sensitive, and are they representative of individuals in the exposed population?

Based upon evaluation of the available information the assessor must determine what is an acceptable risk, which is usually in the range of 1 in 10,000 to 1 in 1,000,000. This implies that whatever the effect being measured, it may be acceptable that its chance of occurring in an exposed population is 1 in 10,000, 1 in 1,000,000, or somewhere in between. As indicated previously, the acceptable risk for cancer-causing substances established by various regulating agencies is 1 in 100,000 to 1 in 10,000,000.

It is important to clearly list and explain all the assumptions utilized in establishing the limits of exposure. The impact of these assumptions on the final decision must be clearly identified so as not to mislead individuals who may utilize the information to establish risk management programs and policies.

Generally, assumptions, particularly those associated with models, are conservative resulting in exposure levels that may be considered overly protective. However, other factors may offset these conservative assumptions resulting in a more realistic exposure level and risk factor: data may be obtained using test animals that are less sensitive to the toxicant of concern than humans, or the route of exposure may different than in the actual exposed population.

Risk Management

The information derived from the risk assessment is based primarily on quantitative data. This information is then evaluated along with certain economic, social, legal, and policy issues to determine the final approach to be utilized. This approach is subjective in nature and may result in a policy that establishes an exposure level that is different from the one derived from the data used in the risk assessment. Often the policies established are influenced by risks perceived by specific social, economical, or political interest groups. A more detailed discussion of risk management is beyond the scope of this book.

Checking Your Understanding

1. What are the four steps involved in performing a risk assessment?

2. List the types of information compiled during the hazard identification phase of a risk assessment.

3. Briefly explain why and how safety factors are used to establish safe exposure levels.

4. What is the difference between a reference dose and a reference concentration?

5. What factors should be considered when performing an exposure assessment? Why?

6. What is the relationship between risk characterization and risk management?

10-3 Administering First Aid

The decision concerning the appropriate type of treatment for an exposed individual is complicated by the hundreds-of-thousands of chemicals in use and the variability of individual response. Information concerning the type of chemical, the route of exposure, and the status of the exposed individual will determine the type of first aid to be provided. Table 10-14 provides a list of general symptoms that may appear as a result of acute exposure to a toxic substance.

Assessing Nature and Severity of Circumstances

A broad range of circumstances can be associated with acute exposure to toxic substances. For instance, a small child may inadvertently swallow aspirin or other medication. An individual spraying pesticides on trees or shrubs may become exposed by inhalation or skin contact. Chemical spills at work or during transport may involve large quantities of chemicals. Before attempting any first aid to the individual exposed, it is important to correctly assess all of the circumstances associated with the exposure. Observations should be made concerning the presence or absence of containers, the type of containers, information that may be present on their labels, posted signs, placards, and the presence of any spilled material. Most chemical containers have labels giving information about the constituents and the associated hazards. The format used to present this information lists such things as the type of chemical, the container size, as well as the quantity and concentration of the chemical. Additional information concerning hazardous household products can be obtained from regional poison control centers or from local medical facilities.

Industrial chemicals have labels that identify hazards in addition to a warning statement. For example, the National Fire Protection Association (NFPA) has developed a labeling system that identifies the hazards associated with a particular chemical (Figure 10-3). The label consists of a color-coded diamond, on which information is provided on flammability, reactivity, health hazards, and other hazards that may be associated with the chemical. Table 10-15 provides further information related to the NFPA codes.

The Department of Transportation (DOT) uses a variety of warning placards on vehicles that transport chemicals. The placards usually have a 4-digit substance identification code and a single-digit hazard classification code (Figure 10-4). The 4-digit substance identification number may be either on the placard or on an orange, rectangular panel located on the truck. The number provides the identification of the chemical: in the example shown in Figure 10-4 the number 1090 corresponds to acetone. The single-digit hazard classification code identifies a specific class of chemicals. The number 3 in the example indicates the material belongs to the class of flammable liquids. Information concerning the substance identification code and the single-digit hazard classification code may be obtained from the DOT Emergency Response Guidebook, the Regional Poison Control Centers, the Chemical

1. Severe nausea, vomiting, or diarrhea.
2. Muscle twitching, convulsions, collapse, paralysis.
3. Delirium, drowsiness, unconsciousness.
4. Restlessness, agitation, signs of fear or panic.
5. Unusual flushing, paleness, or blueness of skin.
6. Signs of burns about the mouth or on the skin.
7. Corrosion, swelling, bleaching of the skin.
8. Pain, tenderness, cramps - especially sudden pain with no apparent cause.
9. Pain or burning in the throat area.
10. Characteristic odor on the breath.
11. Unusual urine color (red, dark green, bright yellow, black).
12. Slow or labored breathing.
13. Unusual contraction or dilation of pupils.
14. Shock, cardiac arrest.
15. Presence of an empty bottle or other signs that would indicate possible poisoning.

Table 10-14: General signs and symptoms of acute exposure. *Obtained and modified with permission from Lefevre M. J., First Aid Manual for Chemical Accidents, Van Nostrand Reinhold, New York, 1989.*

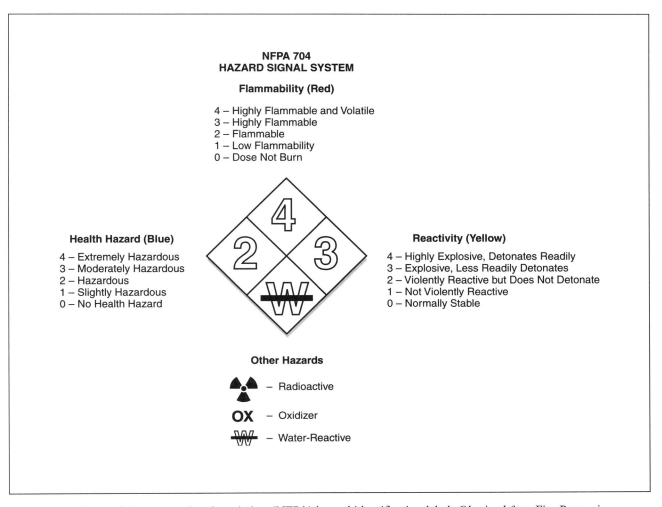

Figure 10-3: National Fire Protection Association (NFPA) hazard identification label. *Obtained from* Fire Protection Guide on Hazardous Materials, *9th edition, National Fire Protection Association, 1986.*

Transportation Emergency Center (CHEMTREC), or the National Response Center. CHEMTREC and the National Response Center work together to provide information – through toll-free phone numbers – to individuals involved in handling hazardous chemicals.

Odors or vapor clouds may indicate the presence of toxic gases. Circumstances involving toxic gases or gases that act as asphyxiants may be difficult to evaluate properly. In some circumstances there may be no warning signs that a gas is present, as is the case for carbon monoxide, which has no odor. Gases in a confined space may also present a problem, since there is little or no circulation that allows the gas to be dispersed. In this situation, gases that may not generally be treated as hazardous can be toxic as a result of reaching a higher concentration level.

After carefully assessing the physical conditions existing at the scene of exposure, the exposed individual should be examined. If the individual is conscious and is able to talk, then all appropriate information should be obtained. However, if the individual is unconscious, then other signs need to be used to provide information concerning the type of exposure.

There are some general signs and symptoms that can be used to indicate the possible route of exposure. For instance, ingestion of chemicals may cause nausea, vomiting, diarrhea, chest or abdominal pain, breathing difficulties, sweating, loss of consciousness, and seizures. Commonly ingested toxic substances include aspirin, ammonia, detergents, bleaches, cleaning solutions, insecticides, and gasoline. Other symptoms include burns on the lips, tongue, and mucous membranes of the mouth,

National Fire Protection Association (NFPA) Identification of the Hazards of Materials					
Identification of Health Hazard **Color Code: BLUE**		**Identification of Flammability** **Color Code: RED**		**Identification of Reactivity (Stability)** **Color Code: YELLOW**	
Signal	Type of Possible Injury	Signal	Susceptibility of Materials to Burning	Signal	Susceptibility of Release of Energy
4	Materials that on very short exposure could cause death or major residual injury even though prompt medical treatment was given	4	Materials that will rapidly or completely vaporize at atmospheric pressure and normal ambient temperature, or that are readily dispersed in air and will burn readily.	4	Materials that in themselves are readily capable of detonation or of explosive decomposition or reaction at normal temperatures and pressures.
3	Materials that on short exposure could cause serious temporary or residual injury even though prompt medical treatment was given.	3	Liquids and solids that can be ignited under almost all ambient temperature conditions.	3	Materials that in themselves are capable of detonation or explosive reaction but require a strong initiating source, or that must be heated under confinement before initiation, or that react explosively with water.
2	Materials that on intense or continued exposure could cause temporary incapacitation or possible residual injury unless prompt medical treatment is given.	2	Materials that must be moderately heated or exposed to relatively high ambient temperatures before ignition can occur.	2	Materials that in themselves are normally unstable and readily undergo violent chemical change but do not detonate. Also materials that may react violently with water or that may form potentially explosive mixtures with water.
1	Materials that on exposure would cause irritation but only minor residual injury even if no treatment is given.	1	Materials that must be preheated before ignition can occur.	1	Materials that in themselves are normally stable, but can become unstable at elevated temperatures and pressures, or that may react with water with some release of energy but not violently.
0	Materials that on exposure under fire conditions would offer no hazard beyond that of ordinary combustible material.	0	Materials that will not burn.	0	Materials that in themselves are normally stable, even under fire exposure conditions, and are not reactive with water.

Table 10-15: National Fire Protection Association (NFPA) identification of the hazards of materials. *Obtained from the National Fire Protection Association.*

which would be indicative of ingestion of corrosive chemicals. Symptoms associated with inhaled chemicals include dizziness, headache, breathing difficulties, unconsciousness, and pale or bluish skin color. Toxic gases such as chlorine and ammonia as well as fumes from insecticides, solvents, and automobile exhaust may cause these symptoms. Exposure to the skin may be characterized by a rash, itching, burning, by changes in breathing and pulse, or by headaches. Various types of chemicals such as acids, bases, insecticides, pesticides, herbicides, and solvents may produce these effects. Exposure to the eye may result in pain, irritation, and tearing, which can be caused by many of the chemicals already mentioned.

Spiders, snakes, and various types of insects can also inject toxic substances. Symptoms generally associated with this type of exposure are pain, tenderness, and swelling at the site of injection. Additional symptoms may be nausea, vomiting, depression or failure of the respiratory and circulatory systems.

Basic First Aid Steps

When assisting an exposed individual there are two general rules that should be followed: 1) anyone providing assistance should ensure that they don't also become a victim, and 2) the most serious symptom should always be treated first. For example, if a person has stopped breathing as a result of a skin exposure, then cardiopulmonary resuscitation should be started immediately. Once the individual is breathing then any skin contamination should be addressed. It is important to clearly assess the circumstances in order to determine the appropriate steps to implement. A hasty response can endanger the responder and complicate attempts to assist the exposed individual.

The following information summarizes the general steps that should be performed when providing first aid. The first aid steps listed below are adapted from *First Aid for Health Emergencies* (Hafen and

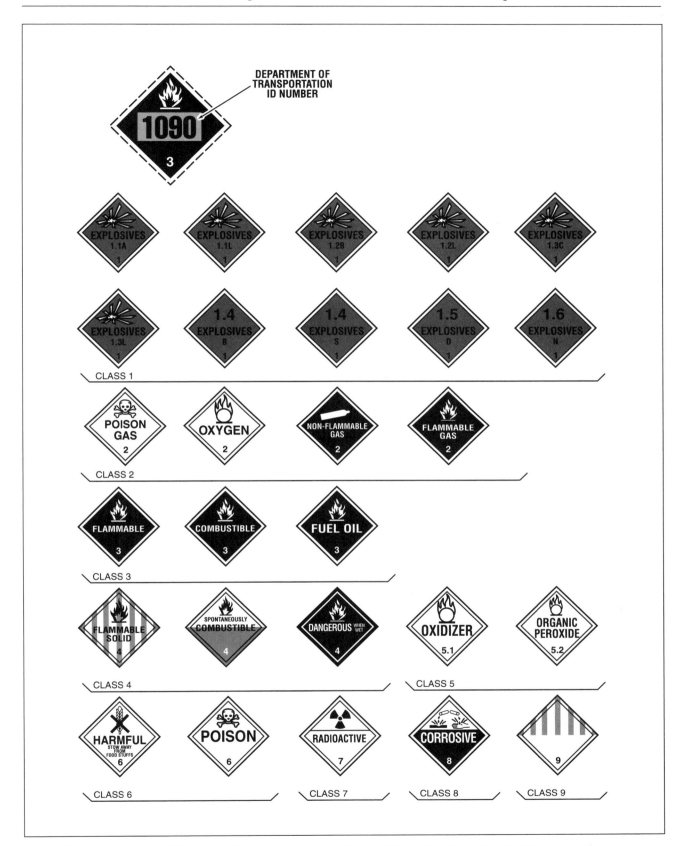

Figure 10-4: Department of Transportation hazard placards. *Obtained from U. S. Department of Transportation.*

Class 1	Explosives
Division 1.1	Explosives with a mass explosion hazard
Division 1.2	Explosives with a projection hazard
Division 1.3	Explosives with a predominantly a fire hazard
Division 1.4	Explosives with no significant blast hazard
Division 1.5	Very insensitive explosives
Class 2	**Gases**
Division 2.1	Flammable gases
Division 2.2	Nonflammable gases
Division 2.3	Poison gases
Division 2.4	Corrosive gases (Canadian)
Class 3	**Flammable Liquids**
Division 3.1	Flashpoint below −18° C (0° F)
Division 3.2	Flashpoint −18° C and above but less than 23° C (73° F)
Division 3.3	Flashpoint of 23° C and up to 61° C (141° F)
Class 4	**Flammable Solids; Spontaneously Combustible Materials; and Materials That are Dangerous When Wet**
Division 4.1	Flammable solids
Division 4.2	Spontaneously combustible materials
Division 4.3	Materials that are dangerous when wet
Class 5	**Oxidizers and Organic Peroxides**
Division 5.1	Oxidizers
Division 5.2	Organic peroxides
Class 6	**Poisonous and Etiologic (infectious) Materials**
Division 6.1	Poisonous materials
Division 6.1	Etiologic (infectious) materials
Class 7	**Radioactive Materials**
Class 8	**Corrosives**
Class 9	**Miscellaneous Hazardous Materials**

Table 10-16: United Nations classification system used by the Department of Transportation. *Obtained from the U. S. Department of Transportation,* Emergency Response Book: Guideline for Initial Response to Hazardous Materials Incidents, *Publication No. P5800.4.*

Peterson, 1977) and *First Aid Manual for Chemical Accidents* (Lefevre, 1989); they can generally be performed by a lay-person. More advanced treatment should be performed by trained medical personnel. Certain activities are identified as **contraindications**, which means that a particular treatment or activity should not be undertaken in response to a particular set of circumstances.

Inhalation – Acute Exposure

Examples: carbon dioxide, carbon monoxide, chlorine, ammonia, nitrogen, hydrogen sulfide.

First Aid Steps:
1. If possible, move the exposed individual from contaminated area to a cool, well ventilated area. If the exposed individual cannot be reached initially because of continued danger, then appropriate rescue personnel should be contacted immediately for assistance.
2. Loosen tight-fitting clothing.
3. If the victim has stopped breathing, begin artificial respiration immediately.
4. Contact appropriate medical personnel.

Contraindications:
1. Do not give any substance orally to a person who is unconscious or is experiencing convulsions, or if the nature of the toxic substance is unknown.
2. Do not use direct mouth-to-mouth resuscitation if the nature of the toxic substance is unknown.

Ingestion – Acute Exposure

Examples: pesticides, aspirin, lighter fluid, bleach, detergent, metals.

First Aid Steps:
1. Examine the mouth, lips, and tongue to determine if there are any burns. Burns or inflamed (swollen) tissue may indicate ingestion of alkali or acids.
2. If the victim has stopped breathing, begin artificial respiration immediately.
3. If victim is unconscious but breathing, lay the individual on his side, or with his head lowered. This will help avoid choking and the entry of toxin into the respiratory tract if the victim should vomit.

4. If victim is conscious and the poison is known, then induce vomiting, if appropriate. **Syrup of ipecac** should be used, if available; a soapy water solution may be used if syrup of ipecac is not available.

5. Contact appropriate medical personnel.

Contraindications:
1. Do not give water or any other substances if the patient is unconscious, experiencing convulsions, or if the chemical ingested is unknown.

2. Do not induce vomiting if the patient: a) is unconscious, b) ingested an unknown substance, c) ingested acid or alkali, d) ingested petroleum products (kerosene, gasoline, paint thinner, lighter fluid), or e) ingested substances that cause coma or seizure (antidepressant drugs, cocaine, strychnine). Acids and alkali will cause further irritation or perforation of the esophagus if vomiting is induced. Vapors may be aspirated into the respiratory tract, possibly causing pneumonitis, if vomiting is induced after ingestion of petroleum products.

3. Do not give mouth-to-mouth resuscitation if the nature of chemical ingested is unknown.

Other types of treatment may be performed as a result of ingestion of toxic substances. Activated charcoal is a highly absorbent material and is administered for a variety of poisons. The charcoal is mixed with a liquid, usually water, and swallowed. The amount of charcoal given is about ten times the weight of the ingested substance. Additional doses may be given at 1-2 hour intervals to assure removal of most of the substance.

Cathartics may also be administered to hasten the elimination of the toxin-containing activated charcoal.

> A cathartic is an orally administered substance that enhances the elimination of the contents in the digestive tract.

This will decrease the probability of desorption of the toxic substance from the charcoal while in the intestinal tract. Also, it may enhance the elimination of substances that are poorly or not at all absorbed by charcoal, such as alkalis, ethanol, methanol, and mineral acids. Magnesium citrate (10 percent) and 70 percent sorbitol are often used as cathartics. Sorbitol is often added to commercially prepared charcoal used for this purpose.

Most emergency rooms use gastric **lavage** to remove ingested poisons. It is most effective in removing ingested liquids, but less effective in removing solids, such as a pill overdose. The process involves the insertion of a flexible tube through the mouth and into the stomach. The contents of the stomach are withdrawn, followed by administration of activated charcoal – when appropriate – to absorb any remaining substance. Water or saline solution (200-300 ml) is added to the stomach and then removed. This portion of the process is repeated until the ingested substance has been removed.

Another type of treatment that has met with varying degrees of success is the use of **chelating agents**. These are used to absorb toxic substances to facilitate their elimination from the body. They are most effective in binding with toxic metals such as lead, mercury, copper, nickel, zinc, cadmium, cobalt, beryllium, and arsenic. Interaction between the metal and the chelating agent results in the formation of an organo-metallic complex that is eliminated from the body.

Common chelating agents include: **BAL (2,3-dimercaptopropanol), DMPS (2,3-dimercapto-1-propanesulfonic acid), EDTA (ethylene diaminetetraacetic acid)**, and **penicillamine (β,β-dimethylcysteine)**.

BAL interacts and forms stable organo-metallic complexes with inorganic mercury, cadmium, chromium, and nickel. Although it can be used to remove toxic metals from the body, it has a number of side effects that limit its usefulness. For example, it removes cadmium from the body, but it also causes cadmium to concentrate in the kidney. Other symptoms associated with the administration of BAL are stimulation of the central nervous system, anorexia, restlessness, generalized aches and pains, itching, salivation, and increased blood pressure. DMPS was developed because of the toxic side effects caused by BAL. It is effective in removing mercury and lead without side effects.

In the past EDTA was used to remove lead from the body. However, EDTA is poorly absorbed from the gastrointestinal tract and must be given **parenterally**.

> The term parenteral refers to the introduction of a substance into the body other than by the digestive system.

The major side effect associated with administration of EDTA is renal toxicity. DMPS, on the other hand, is easily absorbed from the gastrointestinal tract and is the preferred choice for removal of lead.

Penicillamine is used primarily for exposures to lead, mercury, and copper. It is readily absorbed from the gastrointestinal tract and rapidly excreted. Side effects include nausea, skin rash, fever, leukopenia, thrombocytopenia, aplastic anemia, and kidney disease.

Skin Contact

Examples: hydrochloric acid, sulfuric acid, nitric acid, sodium hydroxide.

First Aid Steps:

1. Chemicals that are splashed on the skin should be flushed with large volumes of water and later soap, if available. It is recommended that the area be flushed for 15 minutes.
2. Remove all contaminated clothing while flushing the contaminated area.
3. Contact appropriate medical personnel.

Contraindications:

1. Do not touch open wounds.
2. Do not apply chemical antidotes to the skin unless indicated by a physician.

Flushing of the skin is an effective treatment for exposure to most acids and bases. Large volumes of water dilute the substance and decrease the extent of potential injury. Exposure to hydrofluoric acid requires a two-step process because it presents both a corrosive and a toxic hazard. Once the fluoride ion enters the circulation it can produce systemic effects including heart arrhythmias and even death. For skin exposures, the exposed area should be thoroughly washed and a calcium gluconate gel applied. The calcium in the gel binds with the fluoride, therefore negating its toxic effects.

Eye Contact

Examples: acids, sodium hydroxide, solvents, particulates.

First Aid Steps:

1. Gently wash eyes with plenty of liquid for 15 minutes holding the eyelids open.
2. If contact lenses are present, remove them and wash eyes again.
3. Contact appropriate medical personnel.

Contraindications:

1. Do not rub eyes.
2. Do not apply chemical antidotes into the eyes unless indicated by a physician.

Checking Your Understanding

1. If you saw a toxic chemical container with a DOT label that indicated a Class 2 substance, what is the most probable route of exposure?
2. What type of information is collected when assessing the nature and severity of an exposure to a toxic substance?
3. What symptoms would be indicative of an individual who has ingested a corrosive?
4. What are the two general rules that should be followed when administering first aid?
5. Define the term contraindication.
6. What is a cathartic substance? In general, under what circumstances is it used?
7. Activated charcoal is effective in treating what type of chemical exposure? In which type is it not effective?

10-4 Toxic Substance Exposure, Symptoms, and Treatment

Tables 10-17 and 10-18 list several chemicals and drugs that may be encountered. They also list symptoms resulting from acute or chronic exposure as well as a summary of the first aid steps that may be implemented in case of acute exposure. Specific routes of exposure and the particular organ system directly affected are provided whenever possible. Symptoms that may be the result of toxic effects on several different organ systems are indicated under the category of "general symptoms." Also, similar symptoms, which are the result of different routes of exposure, are identified as "various routes." The substances identified and the effects listed in the tables are not all inclusive. They are indicative of the types of symptoms that may occur as a result of exposure to a specific substance or to chemicals which comprise a class of chemicals such as organochlorine pesticides. Not all symptoms may be observed in every exposed individual, and all substances in a given class may not cause the symptoms identified. Exposure to toxic substances produced by various types of snakes, insects, and plants are presented in Tables 10-19, 10-20, and 10-21.

Chemical Compound	Route of Exposure	Symptoms		Treatment
		Acute Exposure	Chronic Exposure	
Acetaminophen *Examples:* Anacin-3, Liquiprin, Panadol, Tylenol, Comtrex, Nyquil, Vicks Formula 44-D	Ingestion	*General Symptoms:* nausea, vomiting, drowsiness, confusion, liver tenderness, low blood pressure, cardiac arrhythmias, jaundice, acute liver and renal failure	*General Symptoms:* liver damage (rare)	Acute Exposure *Ingestion:* induce vomiting
Acetylsalicylic acid (aspirin)	Ingestion or Skin Absorption	Mild Poisoning *General Symptoms:* burning pain in mouth, throat or abdomen; increased rate of breathing, lethargy; vomiting, hearing loss and dizziness Moderate Poisoning *General Symptoms:* increased rate of breathing, excitability, delirium, fever, sweating, incoordination, restlessness Severe Poisoning *General Symptoms:* increased rate of breathing, coma, convulsions, cyanosis, pulmonary edema, respiratory failure	*General Symptoms:* ringing in the ear, gastric ulcer, weight loss, mental deterioration	Acute Exposure *Ingestion:* induce vomiting

Table 10-17: Common chemicals encountered: symptoms and treatment. *The information has been obtained, modified, and adapted for use in this table from the following sources: Dreisbach R. H. and Robertson W. D.*, Handbook of Poisoning, *12th edition, Appleton & Lange, Norwalk CT, 1987; Lefevre M. J.*, First Aid Manual for Chemical Accidents, *Van Nostrand Reinhold, New York, 1989; Olson K. R. editor,* Poisoning and Drug Overdose, *2nd edition, Appleton & Lange, Norwalk CT, 1994; Zenz C. et al.,* Occupational Medicine, *3rd edition, Mosby Co., St. Louis, 1994; Rosenstock R. and Cullen M. R.,* Textbook of Clinical Occupational and Environmental Medicine, *Harcourt Brace/W. B. Saunders, Philadelphia, 1994.*
NOTE: The information contained in this table is for general knowledge purposes only. No effort should be made to diagnose or administer first aid to an exposed individual based upon this information.

314 ■ Basics of Toxicology

Chemical Compound	Route of Exposure	Symptoms		Treatment
		Acute Exposure	Chronic Exposure	
Aluminum Compounds *Examples:* aluminum acetate, aluminum chloride, aluminum oxide, aluminum hydroxide	Various Routes		*Bone:* osteomalacia *Hematopoietic System:* anemia *Nervous System:* dysfunction of the brain characterized by tremors, weakness, ataxia, learning and memory disorders; some association with Alzheimers Disease	Acute Exposure *Ingestion:* dilute with water and remove by gastric lavage *Inhalation:* remove from source and treat symptoms *Skin:* flush with large quantities of water; remove contaminated clothing *Eye:* flush with large quantities of water Chronic Exposure Reduce exposure or remove from source
	Inhalation (particulate)	*Respiratory System:* chronic inflammation of bronchi and trachea; pulmonary edema; fibrosis; asthma-like response	*Respiratory System:* pulmonary disease such as fibrosis, **alveolitis**, pneumoconiosis (respiratory disease), asthma, emphysema	
	Ingestion	*General Symptoms:* rigidity of limbs and convulsions, jaundice, hypotension; collapse and convulsion, vomiting, burning pain in mouth *Respiratory System:* respiratory distress, profuse salivation		
	Dermal		*Skin:* sensitization	

Table 10-17: Common chemicals encountered: symptoms and treatment (continued).

		Symptoms		
Chemical Compound	Route of Exposure	Acute Exposure	Chronic Exposure	Treatment
Arsenic Compounds *Examples:* arsenic trioxide, arsenic acid, arsenates, arsenites	Inhalation (dust)	*Respiratory System:* pulmonary edema, restlessness, dyspnea, cyanosis	*Respiratory System:* lung cancer *Blood:* anemia *Cardiovascular System:* cardiac failure *Nervous System:* parasthesias (tingling, burning sensation, pins and needles)	Acute Exposure *Inhalation:* remove from source and treat symptoms *Ingestion:* gastric lavage followed by administration of chelating agents dimercaptol (BAL); D-penicillamine or dimercaptosuccinic acid (DMSA)
	Ingestion	*General Symptoms:* restlessness, spasms; convulsions, paralysis precede death; headache, dizziness, chills, irritability *Cardiovascular System:* hypotension (decresed blood pressure), ventricular arrythmias *Gastrointestinal Tract:* acute abdominal cramps, vomiting, diarrhea	*General Symptoms:* weight loss, carcinogen of lungs and liver *Cardiovascular System:* cardiac failure *Gastrointestinal Tract:* nausea, vomiting, abdominal cramps *Nervous System:* muscular weakness, motor dysfunction, and paresthesia *Skin:* hyperpigmentation and hyperkeratosis, carcinogenic (skin cancer), dermatitis *Blood:* anemia	*Skin:* flush with large quantities of water; remove contaminated clothing *Eye:* flush with large quantities of water Chronic Exposure Reduce exposure or remove from source
—Arsenic trichloride	Dermal	*Skin:* burns	*Skin:* dermatitis	See information for arsenic compounds
	Inhalation	*Respiratory System:* bronchopneumonia		
—Arsine (gas)	Inhalation	*General Symptoms:* low doses produce giddiness, headache, light headiness, jaundice *Blood:* hemolysis, anemia *Gastrointestinal Tract:* nausea and vomiting; abdominal pain *Kidneys:* hematuria, renal failure *Nervous System:* vertigo		No specific antidote; maintain kidney function

Table 10-17: Common chemicals encountered: symptoms and treatment (continued).

Chemical Compound	Route of Exposure	Symptoms		Treatment
		Acute Exposure	Chronic Exposure	
Barium Compounds *Examples:* barium (soluble salts), barium acetate, barium carbonate, barium chloride, barium fluoride, barium hydroxide, barium nitrate, barium oxide, barium sulfide	Inhalation	*General Symptoms:* irritation of nose and eyes (barium fluoride, barium hydroxide, and barium oxide); tendency to fatigue; cold sweat, headaches, dizziness *Respiratory System:* laryngitis, bronchitis, emphysema		**Acute Exposure** *Ingestion:* gastric lavage followed by administering orally sodium sulfate solution *Inhalation:* remove from source and treat symptoms
	Ingestion	*General Symptoms:* anxiety; convulsions; bluish face and lips; state of shock (weak and rapid pulse, cold sweat, pale complexion, tendency to fainting, cold hands and feet); paralysis of lower limbs, spreading to upper limbs *Digestive Tract:* disagreeable taste; muscular contractions; nausea and vomiting; stomach pains and diarrhea *Respiratory System:* difficulty in breathing; cyanosis and death by fainting and partial or complete temporary respiratory failure or respiratory paralysis		*Skin:* flush with large quantities of water; remove contaminated clothing *Eye:* flush with large quantities of water **Chronic Exposure** Reduce exposure or remove from source
	Dermal	*Skin:* irritation of skin and mucous membranes; prolonged contact of dusts with moist skin causes ulcerations, necrosis (barium fluoride only)		
	Eye	*Eye:* mechanical and chemical irritation; watering of eyes; inflammation of conjunctivas (barium fluoride, barium oxide, and barium hydroxide); burning sensation		

Table 10-17: Common chemicals encountered: symptoms and treatment (continued).

		Symptoms		
Chemical Compound	**Route of Exposure**	**Acute Exposure**	**Chronic Exposure**	**Treatment**
Cadmium Compounds *Examples:* cadmium sulfate, cadmium nitrate, cadmium oxide	Various Routes	*General Symptoms:* necrosis of internal organs such as liver and brain; edema in brain	*KidneyS:* proteinuria, increase excretion of calcium and phosphorus, glucouria *Bone:* osteomalacia, osteoporosis and spontaneous bone fractures *Cardiovascular System:* hypertension	Acute Exposure *Inhalation:* remove patient from exposure, treat symptoms *Ingestion:* administer milk or beaten egg to reduce irritation; activated charcoal may be administered. **DO NOT INDUCE VOMITING** because of irritant nature of cadmium salts
	Inhalation (dust or fumes)	*General Symptoms:* weakness, chills, fever, dizziness, headache *Respiratory System:* bronchial and pulmonary irritation characterized by irritation and dryness of nose and throat, cough, chest pain and breathlessness; pulmonary edema	*General Symptoms:* weight loss, irritability, yellow-stained teeth; liver and kidney damage *Respiratory System:* emphysema, pulmonary fibrosis, chronic bronchitis, cough, dyspnea *Blood:* anemia	*Skin:* flush with large quantities of water; remove contaminated clothing *Eye:* flush with large quantities of water
	Ingestion	*Cardiovascular System:* cardiopulmonary depression leading to shock *Gastrointestinal Tract:* nausea and vomiting; abdominal cramps, diarrhea *Kidneys:* may experience renal failure	*Cardiovascular System:* cardiopulmonary depression leading to shock *Gastrointestinal Tract:* nausea and vomiting; abdominal cramps, diarrhea *Kidneys:* may experience renal failure	Chronic Exposure Reduce exposure or remove from source

Table 10-17: Common chemicals encountered: symptoms and treatment (continued).

		Symptoms		
Chemical Compound	Route of Exposure	Acute Exposure	Chronic Exposure	Treatment
Caustic (Corrosive) Compounds, Acids and Bases *Examples of Acids:* acetic acid, formic acid, hydrochloric acid, nitric acid, oxalic acid, perchloric acid, phosphoric acid, sulfuric acid, sulfurous acid, trichloroacetic acid, hydrogen peroxide	Inhalation	*General Symptoms:* irritation of mucous membranes, (nose, mouth, eyes, throat); watering of eyes, nasal discharge, sneezing, coughing; headache, bluish face and lips, salivation, giddiness, nausea, muscular weakness, ulceration of mucous membranes (nose) *Respiratory System:* difficulty in breathing, rapid breathing, chemical bronchitis, acute pulmonary edema, secondary chemical pneumonia, death	*General Symptoms:* dental erosion with loss or etching of enamel *Respiratory System:* bronchitis, emphysema, fibrosis	Acute Exposure *Inhalation:* (fumes, vapors): remove from source and treat symptoms *Ingestion:* dilute with water; **DO NOT INDUCE VOMITING** *Skin:* flush with large quantities of water for a least 15 minutes; remove contaminated clothing *Eye:* flush with large quantities of water for at least 15 minutes; hold eyelids apart during washing Chronic Exposure Reduce exposure or remove from source
	Ingestion	*General Symptoms:* state of shock (weak and rapid pulse, cold sweat, pale complexion, tendency to fainting, cold hands and feet); convulsions, coma, death *Digestive Tract:* irritation and burning sensation of lips, mouth, and throat; pain in swallowing; abundant salivation; ulceration of the mucous membranes and color change of the tongue (gray with hydrochloric acid, yellow with nitric acid, white to black with sulfuric acid, white with hydrogen peroxide, oxalic acid and phenol); intense thirst; burning sensation at the esophagus, back of throat, and stomach; painful abdominal cramps; nausea and vomiting, occasionally with blood; risk of perforation of the stomach *Respiratory System:* difficulty in breathing		
	Dermal	*General Symptoms:* shock can occur as a result of pain, weak and rapid pulse, cold sweat, pale complexion, tendency to fainting, cold hands and feet *Skin:* burning sensation, inflammation; burns that may be very painful and become white (in the case of hydrochloric acid) or yellow (in the case of nitric acid); painful blisters; extensive damage to tissues;		
	Eye (splashing of liquid in, or contact of vapor with, eyes)	*Eye:* stinging or burning sensation; watering of eyes; conjunctivitis (inflammation of eyes); burning sensation in eyelids and eyes, with ulceration of the tissues; yellow eyes (nitric acid only); opaqueness of the cornea; loss of sight		

Table 10-17: Common chemicals encountered: symptoms and treatment (continued).

		Symptoms		
Chemical Compound	Route of Exposure	Acute Exposure	Chronic Exposure	Treatment
Caustic (Corrosive) Compounds, Acids and Bases *(continued)* –Hydrofluoric Acid	Inhalation	*General Symptoms:* irritation of mucous membranes (nose, eyes, mouth, throat); watering of eyes, salivation, ulceration of mucous membranes (nose and throat); pain in throat, headache, fatigue, giddiness, pale or bluish face; state of shock (weak and rapid pulse, cold sweat, pale complexion, tendency to fainting, cold hands and feet); *Digestive Tract:* nausea, stomach pains *Respiratory System:* chest congestion, difficulty in breathing, bronchitis, chemical pneumonia; death		Acute Exposure *Inhalation:* remove from source and treat symptoms *Ingestion:* dilute with water; **DO NOT INDUCE VOMITING** *Skin:* flush thoroughly with water for 15-60 minutes; administer calcium gluconate gel to bind fluoride; remove contaminated clothing *Eye:* flush with large quantities of water for at least 15 minutes Chronic Exposure Reduce exposure or remove from source
	Ingestion	*General Symptoms:* state of shock (weak and rapid pulse, cold sweat, pale complexion, tendency to fainting, cold hands and feet); convulsions, coma, death *Digestive Tract:* irritation and painful burning sensation of lips, mouth, and throat; ulceration of mucous membranes; pain in swallowing; intense thirst; violent burning sensation in the esophagus and stomach; painful stomach cramps with distension of stomach; difficulty in breathing; muscular fatigue and general weakness; coughing; acute pulmonary edema; perforation of stomach; nausea and vomiting, occasionally with blood; pale or bluish face		
	Dermal	*General Symptoms:* shock may occur as a result of pain (weak and rapid pulse, cold sweat, pale complexion, tendency to fainting, cold hands and feet) *Skin:* deep penetrating acid causing burning sensation and inflammation; irreparable damage to skin, which gradually turns white and becomes very painful; blisters; extensive damage to the tissues; can cause death		
	Eye (splashing of liquid in, or contact of vapor with, eyes)	*Eye:* intense stinging and burning sensation; watering of eyes; conjunctivitis (inflammation of eyes); burning sensation in eyelids and eyes, with ulceration of the tissues; irreparable damage to cornea; loss of vision		

Table 10-17: Common chemicals encountered: symptoms and treatment (continued).

Chemical Compound	Route of Exposure	Symptoms		Treatment
		Acute Exposure	Chronic Exposure	
Caustic (Corrosive) Compounds, Acids and Bases *(continued)* *Examples of Bases:* Potassium Hydroxide and Sodium Hydroxide	Inhalation	*General Symptoms:* irritation of nose, eyes, and throat; *Respiratory System:* burning sensation in nose; difficulty in breathing; sneezing; fits of coughing; risk of acute pulmonary edema; risk of chemical bronchitis		Acute Exposure *Inhalation:* remove from source and treat symptoms *Ingestion:* dilute with large quantities of water *Skin:* flush with large quantities of water until disappearance of soapiness; remove contaminated clothing *Eye:* flush with large quantities of water for at least 15 minutes Chronic Exposure Reduce exposure or remove from source
	Ingestion	*General Symptoms:* state of shock (weak and rapid pulse, cold sweat, pale complexion, tendency to fainting, cold hands and feet); loss of consciousness; death *Digestive Tract:* immediate, intense burning sensation in mouth, throat, and stomach; immediate effect on buccal (cheeks and mouth cavity) membranes, which become white; intense pain in swallowing; strong salivation, which increases the pain; vomiting - vomit is brown and contains pieces of mucous membrane of the stomach (digestive hemorrhage); stomach cramps; diarrhea, possibly blood-stained; perforation of stomach *Respiratory System:* rapid breathing		
	Dermal	*General Symptoms:* state of shock (weak and rapid pulse, cold sweat, pale complexion, tendency to fainting, cold hands and feet) *Skin:* itching and tingling sensation; painful sensation; painful ulceration; extensive, irreparable damage of tissues	*Skin:* dermatitis	
	Eye (splashing of liquid in, or contact of vapor with, eyes)	*Eye:* highly painful, instantaneous irritation of eyes and eyelids; intense watering of eyes; victim keeps eyelids tightly closed; burns and irreparable damage of mucous membranes; ulceration of eyes; perforation of eyes; loss of eyes		

Table 10-17: Common chemicals encountered: symptoms and treatment (continued).

	Symptoms			
Chemical Compound	**Route of Exposure**	**Acute Exposure**	**Chronic Exposure**	**Treatment**
Chromium Compounds *Examples:* chromic acid, chromium chloride, potassium chromate, potassium, dichromate, sodium chormate, sodium dichromate	Inhalation	*General Symptoms:* irritation of nose and eyes *Respiratory System:* tingling and burning sensation in respiratory tract; sneezing; coughing, sometimes violent; difficulty in breathing; bluish face and lips; risk of pulmonary edema	*General Symptoms:* conjunctivitis, lacrimation; hepatitis with jaundice *Respiratory System:* painless ulceration, bleeding, and perforations of the nasal septum; carcinogen	Acute Exposure *Inhalation:* remove from source and treat symptoms *Ingestion:* dilute with water. **DO NOT INUDCE VOMITING**; some chromium compounds are corrosive in nature
	Ingestion	*General Symptoms:* state of shock (weak and rapid pulse, cold sweat, pale complexion, tendency to fainting, cold hands and feet); possible convulsions; risk of paralysis (chromium chloride only); coma; death *Digestive Tract:* burning sensation in mouth and throat; salivation, burning sensation in stomach; stomach cramps; nausea and vomiting, occasionally with blood; general weakness, dizziness; diarrhea, possibly blood-stained		*Skin:* flush with large quantities of water; remove contaminated clothing *Eye:* flush with large quantities of water
	Dermal (contact with moist skin)	*Skin:* itching; irritation and burning sensation; inflammation; ulceration and possibly profound necrosis (destruction of tissue)	*Skin:* dermatitis associated with edema, ulceration	Chronic Exposure Reduce exposure or remove from source
	Eye	*Eye:* stinging and burning sensation; watering of eyes; inflammation of conjunctivas; risk of serious lesions		

Table 10-17: Common chemicals encountered: symptoms and treatment (continued).

		Symptoms		
Chemical Compound	Route of Exposure	Acute Exposure	Chronic Exposure	Treatment
Cyanide Compounds *Examples:* potassium cyanide, sodium cyanide	Inhalation	Very Acute Exposure Wheezing; foaming at mouth; violent convulsions; loss of consciousness; almost immediate death Acute Exposure *Excitement Phase:* headache; breath smells of bitter almond; dizziness; nausea, occasionally vomiting; rapid breathing; anguish *Depression Phase:* difficulty in breathing; pain in area around heart; anguish *Convulsion Phase:* convulsions; jaws clenched together; foaming at mouth; loss of consciousness *Paralysis Phase:* prolonged coma; dilated pupils; weak and irregular pulse; breathing stops; death Slight Exposure Headache; dizziness; anguish		Acute Exposure (for cyanide poisoning) *Inhalation:* remove from source and give amyl nitrite by inhalation *Ingestion:* administer activated charcoal; induce vomiting if more than 20 minutes from medical facility and activated charcoal not available; give amyl nitrite inhalation *Skin:* flush with large quantities of water; remove contaminated clothing *Eye:* flush with large quantities of water Chronic Exposure Reduce exposure or remove from source
	Ingestion	See symptoms described under inhalation route of exposure - acute and slight poisoning		
	Dermal	*Skin:* the gaseous and liquid compounds are quickly absorbed by the skin and cause symptoms described under inhalation route of exposure - acute and slight poisoning; depending on their nature, they can be very or only slightly irritating		
	Eye	*Eye:* when absorbed by mucous membranes of the eyes, these compounds can cause the same symptoms described in inhalation route of exposure - acute and slight poisoning; irritation, watering of eyes		
–Acrylonitrile	Inhalation	*General Symptoms:* nausea, vomiting, diarrhea, weakness, headache, jaundice	*General Symptoms:* carcinogenic	
–Cyanogen Chloride	Inhalation		*General Symptoms:* dizziness, weakness, conjunctivitis, loss of appetite, weight loss, mental deterioration	

Table 10-17: Common chemicals encountered: symptoms and treatment (continued).

		Symptoms		
Chemical Compound	Route of Exposure	Acute Exposure	Chronic Exposure	Treatment
Gases (non-irritating, possible asphyxiant) *Examples:* carbon dioxide, carbon monoxide, ethylene, freon depressant of central nervous system (types of freon: 11,12, 13, 14, 21, 22, 116, 142b, 143, 151a, 152a),methane, nitrogen, propane, propylene	Inhalation Ingestion (liquid) Dermal (liquid and gases) Eye	*General Symptoms:* need for fresh air; headache; fatigue; mental confusion; nausea and vomiting; giddiness; exhaustion; loss of consciousness; convulsions; death *Cardiovascular System:* arrythmias *Respiratory System:* rapid, occasionally irregular breathing *General Symptoms:* breath has ether smell; rapid intoxication; drowsiness *Skin:* feeling of intense cold; insensitivity to pain; congestion of blood vessels, rapidly becoming painful (freezing); defatting, erythema *Eye:* stinging pain; watering of eyes; inflammation of conjunctivas; cloudiness to opaqueness of eyes		Acute Exposure *Inhalation:* remove from source and treat symptoms *Ingestion* (liquid, Freon): administer activated charcoal. **DO NOT INDUCE VOMITING** *Skin:* flush with large quantities of water; remove contaminated clothing *Eye:* flush with large quantities of water Chronic Exposure Reduce exposure or remove from source
Gases (irritating, noxious) *Example:* Ammonia, Chlorine, Nitrogen oxides, Sulfur dioxide	Inhalation Ingestion (aqueous solutions, ammonium hydroxide, hypochloride) Dermal Eye	*General Symptoms:* irritation of nose, throat, and eyes; burning sensation; fits of coughing; sudden death *Respiratory System:* irritation of respiratory tract, hoarseness; difficulty in breathing; chemical pneumonia or bronchitis; acute pulmonary edema; tingling sensation in respiratory tract *General Symptoms:* state of shock (weak and rapid pulse, cold sweat, pale complexion, tendency to fainting, cold hands and feet) *Digestive Tract:* immediate burning sensation in mouth, throat, and stomach; pain in swallowing; strong salivation; nausea and vomiting of blood; stomach cramps; diarrhea, occasionally blood-stained; risk of stomach perforation *Respiratory System:* rapid breathing *Skin:* itching and tingling sensation; painful burning; painful ulcerations *Eye:* immediate, highly painful irritation of eyes and eyelids; intense watering of eyes; burns and irreparable damage to mucous membranes of eyes; serious lesions in eyes	*Respiratory System:* bronchial irritation with chronic cough, bronchial pneumonia	Acute Exposure *Inhalation:* remove from source and treat symptoms *Ingestion:* (for liquids only) dilute with water. **DO NOT INDUCE VOMITING** *Skin:* (for liquids only) flush with large quantities of water; remove all contaminated clothing *Eye:* (for liquids only) flush with large quantities of water Chronic Exposure Reduce exposure or remove from source

Table 10-17: Common chemicals encountered: symptoms and treatment (continued).

		Symptoms		
Chemical Compound	Route of Exposure	Acute Exposure	Chronic Exposure	Treatment
Gases (irritating, noxious) *(continued)* —Hydrogen Sulfide	Inhalation	*General Symptoms:* headache, nausea, vomiting, confusion, dizziness, seizures, coma, death *Respiratory System:* upper airway irritation, chemical pneumonitis, pulmonary edema, respiratory arrest *Cardiovascular System:* cardiovascular collapse	*General Symptoms:* low blood pressure, nausea, loss of appetite, weight loss, impaired balance, conjunctivitis, chronic cough	Acute Exposure *Inhalation:* remove from source and treat symptoms *Eye:* flush with large quantities of water Chronic Exposure Reduce exposure or remove from source
	Dermal	*Skin:* dermatitis		
	Eye	*Eye:* irritation and burning sensation		
Lead Compounds	Inhalation, Ingestion and Dermal Absorption	*General Symptoms:* moody, restless, easily excited, headaches, jaundice, weight loss, lethargy, joint pain *Blood:* **hemolytic anemia** *Gastrointestinal Tract:* abdominal colic, constipation, abdominal pain, persistent metallic taste *Nervous System:* vertigo, epileptic fits, transient disorders with speech and vision, clouding of consciousness, coma	*General Symptoms:* moody, restless, easily excited, headaches, jaundice, weight loss, lethargy, joint pain *Blood:* anemia *Cardiovascular System:* arrythmia, tachycardia or bradycardia, hypertension *Gastrointestinal Tract:* intermittent abdominal pain, nausea, pigmentation of gums and teeth, diarrhea *Nervous System:* headache, vertigo, easily fatigued, irritable, sleep disturbances, memory deterioration, decreased sensory perception, narrowing of visual field; dementia, reduced visual acuity; tremors of hands and legs	Acute Exposure *Inhalation:* remove from source and treat symptoms *Ingestion:* induce vomiting *Skin:* flush with large quantities of water; remove contaminated clothing *Eye:* flush with large quantities of water Chronic Exposure Reduce exposure or remove from source
	Eye	*Eye:* irritant		

Table 10-17: Common chemicals encountered: symptoms and treatment (continued).

		Symptoms		
Chemical Compound	Route of Exposure	Acute Exposure	Chronic Exposure	Treatment
Mercury Compounds				
–Mercury Vapor	Inhalation	*Gastrointestinal Tract:* gingivitis, dark line forms on inflammed gums, loose teeth, nausea, vomiting, diarrhea *Nervous System:* muscle tremors, poor memory, headaches *Respiratory System:* chemical pneumonitis, bronchitis, **brochiolitis**, pain and tightness of chest, difficulty in breathing, coughing, pulmonary edema	*Gastrointestinal Tract:* gingivitis, excessive salivation *Nervous System:* initial tremor of muscle of fingers, eyelids and lips progressing to tremor of extremities, loss of memory, increased excitability, severe depression, possible delirium and hallucinations	Acute Exposure *Inhalation:* remove from source and treat symptoms *Ingestion* (for inorganic mercuric salts): administer activated charcoal, if possible **DO NOT INDUCE VOMITING** because of possible corrosive injury; (for organic mercury) administer activated charcoal and perform gastric lavage *Skin:* flush with large quantities of water; remove contaminated clothing *Eye:* flush with large quantities of water Chronic Exposure Reduce exposure or remove from source
	Dermal	*Skin:* dermatitis and irritation	*Skin:* dermatitis and irritation	
–Inorganic Mercury Compounds *Example:* mercuric chloride	Ingestion	*General Symptoms:* fluid loss, renal failure and shock; intestinal necrosis *Gastrointestinal Tract:* irritation, nausea, abdominal pain, vomiting, diarrhea, metallic taste, gingivitis, dark line forms on inflammed gums, loose teeth	*Nervous System:* poor concentration and recall, emotional instability, intellectual deterioration, numbness and tingling fingers, toes, mouth, lips and tongue; ataxia, tremor	
–Organic Mercury Compounds *Examples:* Methylmercury, Ethylmercury, Phenylmercury	Inhalation, Ingestion, Dermal		*Nervous System:* poor concentration and recall, emotional instability, intellectual deterioration, numbness and tingling in fingers, toes, mouth, lips and tongue; ataxia, tremors	
Nitrobenzenes *Example:* nitrobenzene	Inhalation	*General Symptoms:* face, lips, and hands are deep blue; headache; dizziness; mental confusion; general weakness; convulsions; coma; death *Respiratory System:* rapid and difficult breathing;	*General Symptoms:* weight loss, anemia, weakness, irritability	Acute Exposure *Inhalation:* remove from source and treat symptoms *Ingestion:* induce vomiting; administer activated charcoal if available *Skin:* flush with large quantities of water; remove contaminated clothing *Eye:* flush with large quantities of water Chronic Exposure Reduce exposure or remove from source
	Ingestion	*Gastrointestinal Tract:* irritation of mouth and stomach; stomach cramps and diarrhea		
	Dermal	*Skin:* irritation; dermatitis; ulceration and necrosis	*General Symptoms:* weight loss, anemia, weakness, irritability	
	Eye	*Eye:* irritation; inflammation of conjunctivas; photophobia (abnormal intolerance of light); severe lesions		

Table 10-17: Common chemicals encountered: symptoms and treatment (continued).

		Symptoms		
Chemical Compound	Route of Exposure	Acute Exposure	Chronic Exposure	Treatment
Phenol Compounds *Examples:* dinitrophenols, pentachlorophenol	Inhalation	*General Symptoms:* irritation of nose; nausea and vomiting; coughing; high fever; profuse sweating; rapid, bluish face and lips (yellow in the case of dinitrocresols only); intense fatigue; dizziness; convulsions; loss of consciousness; death due to heart failure *Respiratory System:* difficult breathing; risk of severe pulmonary edema if victim survives; methemoglobinemia	Similar symptoms as acute exposure	Acute Exposure *Inhalation:* remove from source and treat symptoms *Ingestion:* administer activated charcoal, if available. **DO NOT INDUCE VOMITING** because of risk of seizures
	Ingestion	*General Symptoms:* dizziness; high fever; profuse sweating; bluish face and lips (yellow in the case of dinitrocresol only); intense fatigue; convulsions; lack of consciousness; death *Digestive Tract:* burning sensation in mouth and throat; salivation; nausea and vomiting *Respiratory System:* rapid and difficult breathing; risk of pulmonary edema if victim survives		*Skin:* flush with large quantities of water; remove contaminated clothing *Eye:* flush with large quantities of water Chronic Exposure Reduce exposure or remove from source
	Dermal	*Skin:* irritation; inflammation; blisters	*Skin:* contact dermatis	
	Eye	*Eye:* burning pain in eyes; irritation; watering of eyes; inflammation of conjunctivas	*Eye:* cataracts, glaucoma	
Organic Solvents (which in high concentration have narcotic effects) *Examples:* benzene, cyclohexane, gasoline, hexane, pentane, petroleum ethers, toluene, white spirit, xylenes	Inhalation (by vapors) (By mist or fine droplets)	*General Symptoms:* excitability, with intoxication and mental confusion; staggering; headache; fatigue; nausea with vomiting; dizziness; drowsiness; narcosis (stupor and unconsciousness); loss of consciousness, convulsions; coma and death *Respiratory System:* rapid breathing; *General Symptoms:* coughing; bluish face and lips; nausea and vomiting; fatigue *Respiratory System:* difficulty in breathing; bronchitis and fever; chemical pneumonia; acute pulmonary edema	*Nervous System:* memory problems, concentration difficulties, depression or aggression, vertigo, sleeping problems, headaches *Blood:* anemia, aplasia, leukemia	Acute Exposure *Inhalation:* remove from source and treat symptoms *Ingestion:* administer activated charcoal, if available. **DO NOT INDUCE VOMITING** because of risk of onset of coma or convulsions *Skin:* flush with large quantities of water; remove contaminated clothing *Eye:* flush with large quantities of water Chronic Exposure Reduce exposure or remove from source

Table 10-17: Common chemicals encountered: symptoms and treatment (continued).

		Symptoms		
Chemical Compound	Route of Exposure	Acute Exposure	Chronic Exposure	Treatment
Organic Solvents (which in high concentration have narcotic effects) *(continued)*	Ingestion	*General Symptoms:* dizziness; fatigue; loss of consciousness; coma and death *Gastrointestinal Tract:* slight gastrointestinal irritation *Respiratory System:* if subject recovers, there is a risk of coughing, bronchitis, fever and pneumonia		
	Dermal	*Skin:* dryness; cracks causing irritation; dermatitis		
	Eye (splashing of liquid in, or contact of vapor with, eyes)	*Eye:* stinging sensation; watering of eyes; inflammation of the conjunctivas		
Organic Solvents (which have a depressant effect) Examples: acetone, aliphatic alcohols, butanol, ethyl acetate, ethyl alcohol, ethylene glycol, methyl alcohol, pentanol, propyl alcohols, propylene glycol	Inhalation	*General Symptoms:* slight irritation of nose and eyes; head feels hot and face is flushed; excitability and talkativeness; intoxication; staggering and lack of co-ordination; headache; mental confusion and visual disturbance; tiredness; nausea and vomiting; paleness of complexion; dizziness; eyes are sensitive to and painful in direct light (methyl alcohol only); drowsiness; stupor; loss of consciousness; coma and death	*Eye:* blurring of vision progressing to contraction of visual fields and sometimes blindness (methyl alcohol only)	Acute Exposure *Inhalation:* remove from source and treat symptoms *Ingestion:* administer activated charcoal, if available; induce vomiting except for isopropyl alcohol *Skin:* flush with large quantities of water; remove contaminated clothing *Eye:* flush with large quantities of water Chronic Exposure Reduce exposure or remove from source
	Ingestion	*General Symptoms:* symptoms described in inhalation route of exposure *Gastrointestinal Tract:* gastrointestinal irritation;	*General Symptoms:* cirrhosis (ethanol) *Gastrointestinal Tract:* **gastroenteritis**, diarrhea (ethanol) *Eye:* optic atrophy (ethanol) *Nervous System:* **polyneuritis** (inflammation of nerves) with pain, sensory and motor loss in extremities; memory loss, tremor, impaired judgement (ethanol)	
	Dermal	*Skin:* highly volatile products (methyl alcohol and acetone) produce a feeling of cold; alcohols, acetates, and acetone remove oils from the skin, which becomes dry and eventually develops cracks or dermatitis; glycol derivatives have little effect on healthy skin		

Table 10-17: Common chemicals encountered: symptoms and treatment (continued).

328 ■ Basics of Toxicology

Chemical Compound	Route of Exposure	Symptoms		Treatment
		Acute Exposure	Chronic Exposure	
Organic Solvents (which have a depressant effect) *(continued)*	Eye	*Eye*: vapors are slightly uncomfortable; splashes are irritating; irritation with painful burning or stinging sensation; watering of eyes; inflammation of the conjunctivas; eyes are sensitive to and painful in the light		
Organic Solvents: Chlorinated and Chloro-fluorinated *Examples:* carbon tetrachloride, chloroethanes (mono-, di-, tri-, tetra-), chloroform, chloromethanes (di-, tri-, tetra-), dichloro-benzenes, methyl-chloroform, tetrachloro-ethylene, tetrachloro-ethane, trichloroethylene, vinyl chloride	Inhalation	*General Symptoms:* irritation of eyes, nose, and throat (not in the case of methyl chloride); overexcitement; headache; intoxication (staggering, loss of equilibrium); loss of consciousness; narcosis (deep unconsciousness); cardiac arrythmias *Respiratory System:* pulmonary edema	*General Symptoms:* fatigue, anorexia, anemia, nausea, memory loss, paresthesias, tremors, blurring vision; liver and kidney damage; anesthetic effect	Acute Exposure *Inhalation:* remove from source and treat symptoms *Ingestion:* administer activated charcoal, if available. **DO NOT INDUCE VOMITING** *Skin:* flush with large quantities of water; remove contaminated clothing *Eye:* flush with large quantities of water Chronic Exposure Reduce exposure or remove from source
	Ingestion	*General Symptoms:* drowsiness; loss of consciousness; narcosis (deep unconsciousness); state of shock (weak and rapid pulse, cold sweat, pale complexion, tendency to fainting, cold hands and feet) *Digestive Tract:* irritation of lips and mouth; gastrointestinal irritation; nausea, vomiting; diarrhea, possibly blood-stained *Respiratory System:* risk of acute pulmonary edema		
	Dermal	*Skin:* dry skin, freezing, inflammation, blisters that later become painful	*Skin:* defatting, dermatitis	
	Eye: (splashing of liquid in, or contact of vapor with, eyes)	*Eye:* irritation of eyes; watering of eyes; occasionally inflammation of conjunctivas		

Table 10-17: Common chemicals encountered: symptoms and treatment (continued).

	Symptoms			
Chemical Compound	Route of Exposure	Acute Exposure	Chronic Exposure	Treatment
Organochlorine Pesticides *Examples:* DDT, methoxychlor, dieldrin, aldrin, chlordane, endrin, heptachlor, toxaphene	Various Routes Inhalation Ingestion Dermal Eye	*General Symptoms:* malaise, increased sweating, headache, dizziness, causes convulsions, arrhythmias *General Symptoms:* convulsions *Respiratory System:* nose and throat irritation *Gastrointestinal Tract:* convulsions, vomiting, severe gastritis (inflammation of stomach lining) *Respiratory System:* respiratory distress and cyanosis *Nervous System:* seizures, followed by death *Skin:* irritation *Eye:* irritation	*General Symptoms:* malaise, headache, dizziness *Nervous System:* paresthesia, tremors, confusion	Acute Exposure *Inhalation:* remove from source and treat symptoms *Ingestion:* administer activated charcoal; **DO NOT INDUCE VOMITING** because of risk of abrupt onset of seizures *Skin:* flush with large quantities of water; remove contaminated clothing *Eye:* flush with large quantities of water Chronic Exposure Reduce exposure or remove from source
Organo-phosphorus Compounds *Examples:* diazinon, malathion, parathion, phosdrin	Inhalation Ingestion Dermal Eye	*First phase:* profuse salivation; nausea and vomiting; stomach cramps; diarrhea; headache and dizziness; troubled breathing (asthmatic); very small pupils (myosis); anxiety and agitation *Second Phase:* muscular disorders (muscle contractions that first occur in the face and subsequently spread throughout the body; convulsions; decreasing muscle power); hyper-secretion of the bronchia, weakness of respiratory muscles); disorders of the heart (weak and slow pulse, irregular pulse) *Third phase:* high fever; loss of consciousness; death See symptoms described under Inhalation route of exposure; dermatitis See symptoms described under Inhalation route of exposure *Eye:* irritation; watering of eyes; these are quickly followed by the symptoms described under Inhalation route of exposure	Symptoms similar to acute exposure can occur as a result of chronic exposure	Acute Exposure *Inhalation:* remove from source and treat symptoms *Ingestion:* induce vomiting *Skin:* flush with large quantities of water; remove contaminated clothing *Eye:* flush with large quantities of water Chronic Exposure Reduce exposure or remove from source

Table 10-17: Common chemicals encountered: symptoms and treatment (continued).

		Symptoms		
Chemical Compound	**Route of Exposure**	**Acute Exposure**	**Chronic Exposure**	**Treatment**
Quaternary Ammonium Pesticides *Examples:* Paraquat and Diquat	Inhalation	*General Symptoms:* irritation of nose and eyes; sneezing; nose bleed; painful obstruction of throat *Respiratory System:* fibrosis *Cardiovascular System:* tachycardia		Acute Exposure *Inhalation:* remove from source and treat symptoms *Ingestion:* administer activated charcoal, if available; induce vomiting *Skin:* flush with large quantities of water; remove contaminated clothing *Eye:* flush with large quantities of water Chronic Exposure Reduce exposure or remove from source
	Ingestion	*Gastrointestinal Tract:* burning sensation in mouth and throat; irritation of mucous membranes; difficulty in swallowing; burning sensation in stomach; nausea and vomiting, possibly of blood; painful abdominal cramps; diarrhea, may be blood-stained *Respiratory System:* chemical pneumonia, cyanosis		
	Dermal	*Dermal:* irritation; inflammation; painful blisters		
	Eye	*Eye:* painful burning sensation; watering of eyes; inflammation; sensitiveness to bright light		
Particulates *Example:* asbestos	Inhalation	*Respiratory System:* sneezing; slight irritation of nose	*Respiratory System:* pulmonary fibrosis (asbestosis), mesothelioma, lung cancer; labored breathing *Gastrointestinal Tract:* increased incidence of gastrointestinal cancer	Acute Exposure *Inhalation:* remove from source and treat symptoms *Ingestion:* remove from source *Skin:* flush with large quantities of water; remove contaminated clothing *Eye:* flush with large quantities of water Chronic Exposure Reduce exposure or remove from source
	Ingestion	no symptoms		
	Dermal	no symptoms		
	Eye	*Eye:* mechanical irritation; watering of eyes; inflammation of conjunctivas		

Table 10-17: Common chemicals encountered: symptoms and treatment (continued).

Chapter 10 – Risk Assessment and Acute Exposure Treatment ■ 331

Chemical Compound	Route of Exposure	Symptoms		Treatment
		Acute Exposure	Chronic Exposure	
Silica (crystalline)	Inhalation	*Respiratory System:* sneezing; slight irritation of nose; pneumoconiosis	*Respiratory System:* silicosis, labored breathing, dry cough *Gastrointestinal Tract:* increased incidence of gastrointestinal cancer	Acute Exposure *Inhalation:* remove from source and treat symptoms *Ingestion:* remove from source *Skin:* flush with large quantities of water; remove contaminated clothing *Eye:* flush with large quantities of water Chronic Exposure Reduce exposure or remove from source
	Ingestion	no symptoms		
	Dermal	no symptoms		
	Eye (splashing in eyes)	*Eye:* mechanical irritation; watering of eyes; inflammation of conjunctivas		
Strychnine	Ingestion	*General Symptoms:* muscular stiffness and painful cramps; seizurelike contraction of skeletal muscles; renal failure, respiratory arrest causing death		Acute Exposure *Ingestion:* administer activated charcoal, if available. **DO NOT INDUCE VOMITING**

NOTE: The information contained in this table is for general knowledge purposes only. No effort should be made to diagnose or administer first aid to an exposed individual based upon this information.

Table 10-17: Common chemicals encountered: symptoms and treatment (continued).

		Symptoms		
Chemical Compound	**Route of Exposure**	**Acute Effects**	**Chronic Effects**	**Treatment**
Stimulants				
–Amphetamine ("bennies", pep pills, "wakeups", "copilots") *Examples:* Benzedrine, dexedrine, methedrine –Methamphetamine ("meth", "speed", "crystal") *Example:* Methedrine –Methylphenidate *Example:* Ritalin	Ingestion Injection	Agitation, talkativeness, insomnia, anorexia lasting 4-8 hours, abnormal heart rhythms, nausea, vomiting, sweating, abnormal cramps, psychosis, decrease in hunger	Possibility of infection, weight loss, high blood pressure, schizophrenic-like reaction	Maintain body heat, give artificial respiration if necessary; give water or milk, induce vomiting, get to emergency facility
–Cocaine ("snow", "coke") *Examples:* Methylester of benzoylecgonine, made from leaves of cocoa bush	Inhalation Ingestion Injection	Feelings of great mental and physical prowess, courage, and invincibility; excitation, loss of sense of time, exaggerated reflexes lasting a few minutes to 4 hours	Vertigo, confusion, exhaustion, convulsions, paralysis of respiratory center	Maintain body heat, give artificial respiration if necessary; give water or milk, induce vomiting, get to emergency facility
–Inhalants - fast-drying glue or cement (such as model glue), many paints and lacquers and their thinners and removers, gasoline, kerosene, lighter fluid and dry cleaning fluid, nail polish remover; chief active ingredient toluene	Inhalation (usually from inside plastic bag)	Blurred vision, slurred speech, lack of coordination, hallucinations lasting 1-4 hours; appears to be drunk	Nausea, vomiting, blood damage, paralysis of vocal cords, convulsions, respiratory arrest	If a person is found with a paper bag or other apparatus over his head, remove it immediately; if breathing stops, administer artificial respiration, obtain medical assistance immediately
Depressants				
–Barbiturates - phenobarbital ("goofballs"), pentobarbital ("yellow jackets"), amobarbital ("blue devils"), secobarbital ("red devils"); made from barbituric acid, often used to induce sleep	Ingestion Injection	Slurred speech, staggering, confusion, mood shifts lasting 4-8 hours, relief of anxiety and excitement, tendency to reduce mental and physical activities, slight decrease in breathing, relaxation	Allergic reactions, convulsions, coma, death	Dilute with water, induce vomiting, give oxygen, get victim to physician immediately, maintain body heat
–Tranquilizers - meprobamate, diazepam, reserpine, phenothiazine; used to relieve mental tension, anxiety, and relax muscles without producing sleep or seriously affecting mental physical capacity	Ingestion	Slurred speech, staggering and confusion, mood shifts lasting 4-8 hours, relief of anxiety and excitement, tendency to reduce mental and physical activities, slight decrease in breathing, relaxation	Psychic and physical dependence	Arouse the victim, if possible, by lightly slapping him with a cold, wet towel and try to get him on his feet; maintain an open airway and apply artificial respiration if indicated; maintain body temperature; get the victim to a physician, hospital

Table 10-18: Common drugs: symptoms and treatment. *Obtained and modified with permission from Hafen B. Q. and Peterson B.,* First Aid for Health Emergencies, *Brooks/Cole, Pacific Grove, CA, 1977.*

		Symptoms		
Chemical Compound	Route of Exposure	Acute Effects	Chronic Effects	Treatment
Depressants (continued)				
–Narcotics - codeine (methylmorphine); weakest derivative of opium	Ingestion	Drowsiness, difficulty concentrating, dulled perceptions lasting about 4 hours	Reduction in awareness of pain, decrease in activity, tendency to sleep more, decrease in breathing and pulse rate, reduction of hunger and thirst. Some unpleasant reactions to narcotics include sweating, dizziness, nausea, vomiting, and constipation, exposure to infection, malnutrition, respiratory arrest	Arouse the victim, if possible, by lightly slapping him with a cold, wet towel and try to get him on his feet; maintain an open airway and administer artificial respiration if indicated; maintain body temperature; induce vomiting if taken orally and seek medical assistance as soon as possible
–Heroin (diacetylmorphine), "smack", "horse", "scag"; synthetic alkaloid formed from morphine	Injection Inhalation	Euphoria, drowsiness, and confusion lasting 4-8 hours, circulation and respiration slowed, blood pressure lowered	Reduction in awareness of pain, decrease in activity, tendency to sleep more, decrease in breathing and pulse rate, reduction of hunger and thirst. Some unpleasant reactions to narcotics include sweating, dizziness, nausea, vomiting, and constipation, exposure to infection, malnutrition, respiratory arrest	Arouse the victim, if possible, by lightly slapping him with a cold, wet towel and try to get him on his feet; maintain an open airway and administer artificial respiration if indicated; maintain body temperature; induce vomiting if taken orally and seek medical assistance as soon as possible
–Morphine (morphine sulfate); principal derivative of opium	Ingestion Injection	Euphoria, difficulty in concentrating lasting 4-8 hours	Reduction in awareness of pain, decrease in activity, tendency to sleep more, decrease in breathing and pulse rate, reduction of hunger and thirst. Some unpleasant reactions to narcotics include sweating, dizziness, nausea, vomiting, and constipation, exposure to infection, malnutrition, respiratory arrest	Arouse the victim, if possible, by lightly slapping him with a cold, wet towel and try to get him on his feet; maintain an open airway and administer artificial respiration if indicated; maintain body temperature; induce vomiting if taken orally and seek medical assistance as soon as possible
–Alcohol (ethyl); usually from beers, wines, or distilled beverages; similar effects may be produced by use of methyl alcohol, canned heat, etc.	Ingestion	Affects higher reasoning areas of the brain, feeling of relaxation, exhilaration, release of inhibitions, coordination affected, disturbed speech, dilation of blood vessels causing heat loss. Eventually, if enough is taken in, the area of the brain that controls respiration is depressed and the individual will "pass out"	Nutritional deficiencies, brain damage, fatty or cirrhotic liver, and possibly cancers of the mouth, pharynx, and larynx	Usually sleep and time will relieve symptoms. In severe cases, treat for shock; if no response, get medical aid immediately, give artificial respiration if necessary, maintain body heat. If the person is unconscious, try to arouse him, if possible, and induce vomiting. If the person cannot be aroused, do not induce vomiting, seek medical help. If the person's respiration becomes very shallow give artificial respiration

Table 10-18: Common drugs: symptoms and treatment (continued).

334 ■ Basics of Toxicology

Chemical Compound	Route of Exposure	Symptoms		Treatment
		Acute Effects	Chronic Effects	
Hallucinogens				
–LSD ("acid", d-lysergic acid diethylamide); synthetic derivative of lysergic acid	Ingestion Injection	Euphoria, rapid mood shifts, depersonalization lasting 6-10 hours, increased activity through its action on the central nervous system, increased heart rate, increased blood pressure, increased body temperature, dilated pupils, flushed face	Recurrence of initial effects, panic or paranoid psychosis, hallucinations, delusions of grandeur, possible chromosome damage	Get to a physician immediately. Limit external stimulation such as intense light or loud noises, have person lie down and relax
–Psilocybin (phosphoryl-tryptamine); extract of Mexican mushroom	Ingestion	Euphoria, confusion, hallucinations lasting 3-14 days	Sudden violence, nerve impairment, respiratory arrest, possible chromosome damage	Get the victim to a physician or hospital as soon as possible
–Mescaline; extract of peyote cactus	Ingestion Injection Inhalation	Exhilaration, anxiety, hallucinations lasting 6-12 hours	Gastric distress, paranoid psychosis	Get the victim to a physician or hospital as soon as possible
Marijuana (cannabis sativa); mild hallucinogen, made from the flower tops of leaves of the hemp plant; chief active ingredient, tetrahydrocannabinol	Smoked Inhalation Ingestion	Euphoria, mood swings, loss of inhibitions, distortion of time and space lasting 2-4 hours, throat irritation, increased heart rate, red eyes, dizziness, poor coordination, sleepiness, increased appetite	"Drop-out" syndrome, possible panic or psychotic reactions	First aid will ordinarily not be required for marijuana users, but some users may get frightened and exhibit paranoid tendencies in emergency situations

Table 10-18: Common drugs: symptoms and treatment (continued).

Snake	Description	Mechanism of Exposure	Symptoms	Treatment

Copperhead	Found in hilly, rocky country and in lowlands from Massachusetts to northern Florida, west to the Mississippi River and across to Texas	Lightning-like strike, small amount of venom injected from two forward fangs	Circulatory system poison. Extensive local tissue damage as venom actually digests the tissues. Contains hemorrhagic factor that destroys blood vessels. Severe pain, swelling, dark purplish marks (one or more), growing weakness, shortness of breath, increasing lassitude leading to unconsciousness, dimness of vision, rapid pulse, nausea, vomiting	**First aid treatment for snakebite**
				1. Immobilize the bite area in a position slightly lower than the heart. Do not allow the victim to walk or move.
				2. Do not take unnecessary risks that may result in another bite. If the snake is not poisonous, treat the bite as you would a puncture wound.
				3. Apply a flat constricting band two to four inches above the bite, between the bite and the heart if the bite is on an arm or leg. It should be tight enough to restrict venous blood and lymph flow but not arterial blood blow. The band should be loose enough so that a finger can be inserted beneath it without force.
				4. If possible, cleanse the bitten area with alcohol sponges. With a sterile scalpel or knife make one straight incision that connects both fang marks. The incision should go no deeper than through the skin so that no underlying vessels, nerves, or tendons are cut. Do not make cross incisions because they heal poorly, cause scars, and are contradictory to surgery principles.
				5. Squeeze venom gently from the incision with the fingers for 30 minutes or until a doctor takes over. Mechanical suction should be applied in a vigorous but not forceful manner.
				6. Antivenin can be administered in the field in an emergency, but instructions contained in each package must be followed rigidly. The required skin test for allergy must have proven negative. Antivenin counteracts or neutralizes the snake venom.
				7. Get the victim to a doctor or hospital as quickly as possible, but without exertion on his part.
				8. Treat victim for shock.
				9. Give artificial respiration if needed.
				10. Even if the bite has been inflicted by a nonpoisonous snake, a physician should be consulted about antibiotic therapy and prevention of tetanus.
				11. If the victim is without help and must walk, he should move slowly.
				12. Some "don'ts" in treating snakebite:
				   a. Don't give alcohol to the victim. This causes dilation of the blood vessels.
				   b. Don't pack the wound in ice. This can cause serious damage.
				   c. Don't exercise the bite area. This should be done by a physician, if at all.
				   d. Don't inject antivenin around the wound site. This does not counteract the venom and increases local tissue damage. Antivenin can also cause allergic reactions.

Water Moccasin	Found in hilly, rocky country and in lowlands from Massachusetts to northern Florida, west to the Mississippi River and across to Texas	Lightning-like strike, small amount of venom injected from two forward fangs	Circulatory system poison. Extensive local tissue damage as venom actually digests the tissues. Contains hemorrhagic factor that destroys blood vessels. Severe pain, swelling, dark purplish marks (one or more), growing weakness, shortness of breath, increasing lassitude leading to unconsciousness, dimness of vision, rapid pulse, nausea, vomiting	
Coral Snake	Found in Northern Carolina south to Florida, west to Texas, and along the Mississippi Valley to Indiana. Particularly located in coastal regions	Injects poison through small teeth with chewing motion, usually while victim is asleep. Coral snake bite is rare, but its venom is the most toxic of American snakes	Nerve poison, little local tissue reaction. Great drowsiness, sinking into unconsciousness, little local pain or swelling but extensive bruising	
Nonpoisonous snakes	Variety	Bite resembles scratches. No fang marks	Small puncture wounds, local minor pain, and inflammation	

Table 10-19: Symptoms associated with snakebites. *Obtained and modified with permission from Hafen B. Q. and Peterson B., First Aid for Health Emergencies, Brooks/Cole, Pacific Grove, CA, 1977.*

336 ■ Basics of Toxicology

Insect or Spider	Description	Mechanism of Exposure	Symptoms	Treatment
Brown Recluse Spider	Oval body with eight legs. Light yellow to medium dark brown. Has distinctive mark shaped like a fiddle on its back. Body from 3/8 to 1/2-inch long, 1/4-inch wide, 3/4-inch from toe-to-toe	Bites producing an almost painless sting that may not be noticed at first.	In two to eight hours pain may be noticed followed by blisters, swelling, hemorrhage, or ulceration. Some people experience rash, nausea, jaundice, shills, fever, cramps or joint pain.	Summon doctor. Bite may require hospitalization for a few days. Full healing may take 6-8 weeks. Weak adults and children have been known to die.
Black Widow Spider	Color varies from dark brown to glossy black. Densely covered with short microscopic hairs. Red or yellow hourglass marking on the underside of the female's abdomen. Male does not have this mark and is not poisonous. Overall length with legs extended is 1-1/2-inch. Body is 1/4-inch wide.	Bites	Local redness, two tiny red spots may appear. Pain follows almost immediately. Larger muscles become rigid. Body temperature rises slightly. Profuse perspiration and tendency toward nausea follow. It's usually difficult to breathe or talk. May cause constipation, urine retention. Venom is more dangerous than a rattlesnake's but is given in much smaller amounts. About 5 percent of bite cases result in death. Death is from asphyxiation due to respiratory paralysis. More dangerous for children, to adults its worst feature is pain. Convulsions result in some cases.	Use an antiseptic such as alcohol or hydrogen peroxide on the bitten area to prevent secondary infection. Keep victim quiet and call a doctor. Do not treat as you would a snakebite since this will only increase the pain and chance of infection; bleeding will not remove the venom.
Scorpion	Crablike appearance with clawlike pincers. Flashy post-abdomen or "tail" has 5 segments, ending in a bulbous sac and stinger. Two poisonous types: solid straw yellow or yellow with irregular black stripes low on back. From 2-1/2 to 4 inches.	Stings by thrusting its tail forward over its head.	Swelling or discoloration of the area indicates a nondangerous, though painful, sting. Excessive salivation and facial contortions may follow. Temperature rises to more than 104 degrees. Convulsions, in waves of increasing intensity, may lead to death from nervous exhaustion. First 3 hours most critical.	Apply tourniquet. Keep victim quiet and call a doctor immediately. Do not cut the skin or give pain killers. They increase the killing power of the venom. Antitoxin, readily available to doctors.
Bee	Winged body with yellow and black stripes. Covered with branched or feathery hairs. Different species vary from 1/2 to 1 inch in length.	Stings with tail when annoyed.	Burning and itching with localized swelling occur. Usually leaves venom sac in victim. It takes between 2 and 3 minutes to inject all the venom. If a person is allergic, more serious reactions occur - nausea, shock, unconsciousness. Swelling may occur in another part of the body. Death may result.	Gently scrape (don't pluck) the stinger so venom sac won't be squeezed. Wash with soap and antiseptic. If swelling occurs, contact doctor. Keep victim warm while resting.
Tick	Oval with small head; the body is not divided into definite segments. Gray or brown. Measures from 1/8-inch to 1/4-inch when mature.	Attaches itself to the skin and sucks blood. After removal there is danger of infection, especially if the mouthparts are left in the wound.	Sometimes carries and spreads Rocky Mountain spotted fever, tularemia, Colorado tick fever. In a few rare cases, causes paralysis until removed.	Seek medical help for treatment of symptoms associated with specific disease.

Table 10-20: Symptoms associated with insect bites and stings. *Obtained and modified with permission from Hafen B. Q. and Peterson B.*, First Aid for Health Emergencies, *Brooks/Cole, Pacific Grove, CA, 1977.*

Poisonous Plants	Route of Exposure	Symptoms	Treatment
Hemlock	Ingestion	*General Symptoms:* abdominal pain, nausea, vomiting, sweating, diarrhea, convulsions, cyanosis, respiratory failure	Remove poison by gastric lavage or emesis
Mushrooms (Various Types)	Ingestion	*General Symptoms:* severe nausea, vomiting, diarrhea, blood in stools, jaundice pulmonary edema, headache, mental confusion and depression, coma or convulsions; cardiac arrhythmias; ataxia	Induce vomiting
Poison Ivy (Poison Oak) and Poison Sumac	Dermal, Ingestion, or Inhalation of smoke of burning plants	*General Symptoms:* generalized edema, pharyngeal or laryngeal edema, weakness, malaise, fever *Local Effects (Dermal):* itching, swelling, papule formation, blisters, oozing, and crusting	*Ingestion:* Induce vomiting *Skin:* Wash thoroughly with soap and water *Inhalation:* Remove from source

Table 10-21: Symptoms as a result of contact with poisonous plants. *Obtained, modified, and adapted for use from Dreisbach R. H. and Robertson W. D.,* Handbook of Poisoning, *12th edition, Appleton & Lange, Norwalk CT, 1987.*

Checking Your Understanding

1. How do you treat an individual who has been acutely exposed to sulfur dioxide? Hydrogen sulfide? Which is more toxic and why?

2. Describe the symptoms associated with exposure to carbon monoxide and hydrogen cyanide. Are the first aid steps implemented in the same way or not? Why?

3. Based upon the symptoms produced could you distinguish between an acute exposure to organochlorine and organophosphate pesticides? Why or why not?

Summary

Risk is defined as the probability, or chance, that an adverse effect will occur. To determine the level of risk involved in exposure to toxic substances two activities should be performed. The first one is risk assessment, which includes four steps: 1) hazard identification, 2) hazard evaluation and dose-response assessment, 3) exposure assessment, and 4) risk characterization. Hazard identification consists of compiling and examining data from toxicological and epidemiological studies concerning potential hazards associated with exposure. The hazard evaluation is performed to quantify the dose at which an effect will occur. The exposure assessment will identify the affected population and quantify the magnitude, frequency, duration, and route of exposure. The final step in the risk assessment is risk characterization. This step involves determining what the actual risk may be when exposure occurs. The purpose of this process is to accumulate and evaluate all objective information relative to a given toxic substance and determine its relevancy and usefulness in establishing an exposure level that has an acceptable risk. The second activity consists in the actual development and implementation of a risk management policy based on the information derived from the risk assessment. The policy is developed as a result of evaluating the information from the risk assessment along with certain economic, social, and legal issues to determine the final approach.

Exposure to toxic chemicals occurs on a daily basis. They are encountered in the home, the workplace, and the environment. Chronic exposure may result in symptoms appearing months or years after initial exposure. Acute exposure usually results in symptoms that are noticed within a few seconds, minutes, hours, or days after contact with the toxic substance.

In response to acute exposures, there are some basic steps that should be followed to assist and treat exposed individuals. The first step is to assess the overall situation and obtain as much information as possible. Identification of the type of chemical, route of exposure, and toxic properties should be obtained, which is necessary information for appropriate first aid. The second step is to identify the general signs and symptoms demonstrated by the exposed individuals. Third, the appropriate first aid must be administered according to the previously collected information. The fourth step is to obtain assistance. This step may occur earlier if approaching the exposed individual would place rescue personnel in danger.

Application and Critical Thinking Activities

1. What does "acceptable risk" mean to you?

2. Suppose you are assessing information to determine the policy concerning exposure to a toxic substance in the workplace. Explain why species, dose, route of exposure, and age are important in assessing the toxicological information. How would each of these factors affect both the safety factor chosen and the determination of a safe exposure level?

3. Speculate why a safety factor of 100 is used when calculating a reference dose using a no-observed-adverse-effect-level versus a safety factor of 1,000 when using the lowest-observed-adverse-effect-level.

4. Calculate the reference dose if the no-observable-adverse-effect-level (NOAEL) for rats fed lead-containing water is 0.027 mg/kg/day.

5. List and explain the variables involved in determining a safe human dose for a given toxic substance. How would exposure by different routes affect the final value?

6. Joe Smith lives in a large city and works outdoors as a landscape contractor. His normal workday is eight hours long. After work he takes an easy jog for an hour, while the rest of the evening he remains inside his air-conditioned home. If the concentration of lead in the air is 0.005 mg/m^3, what is Joe's daily exposure using Equation 1 provided in this chapter? Explain any assumptions you may use to calculate your answer.

7. The Department of Transportation (DOT) labels and placards address some of the same classes of chemicals as the National Fire Protection Association (NFPA) diamond. For instance, both address flammable chemicals. What chemical characteristics are used by each of these organizations to categorize them? If they differ from one organization to the other would they alter your response in providing assistance? Identify one other chemical class that is addressed by both of these organizations and answer the same questions.

8. Identify ten toxic household chemical products in your home. List the chemical constituents identified, any warning labels or statements, and the treatment suggested as a result of chemical exposure. How would you treat an acute exposure to each of these?

Glossary

Absorption – The taking up of a substance by capillary, osmotic, chemical, or solvent action, such as the movement of a toxic substance into a cell or into the circulation.

Acceptable Daily Intake – The daily intake of a chemical that, during a lifetime, appears to be without appreciable risk based on all known facts at the time; expressed in milligrams of chemical per kilogram of body weight (mg/kg).

Acetylation – A Phase II metabolic reaction involved in either activating or deactivating aromatic amines.

Acetylcholine – Neurotransmitter released from motor neurons that innervate skeletal muscle fibers; a neurotransmitter associated with most neurons in the sympathetic and parasympathetic nervous systems.

Acetylcholinesterase – An enzyme that breaks down acetylcholine into acetic acid and choline.

ACGIH (American Conference of Governmental Industrial Hygienists) – Professional organization involved in evaluating chemical exposure information; updated information is published annually in "*Threshold Limit Values (TLVs) for Chemical Substances and Physical Agents and Biological Exposure Indices.*"

Acidosis – A condition of the blood characterized by a pH that is lower than normal (pH < 7.35).

Action Potential – An electrical charge that is the result of the movement of sodium and potassium ions across the cell membrane of a neuron in response to a stimulus, such as pain or the sensation of heat.

Active Transport – The movement of materials against a concentration gradient by the expenditure of metabolic energy.

Acute – Occurring over a short period of time; used to describe brief exposures and effects that appear promptly after exposure.

Acute Exposure – Exposure to a substance over a very short period of time, usually 24 hours; usually refers to a single exposure; however, repeated exposure may also occur.

Acute Toxicity – Adverse effects caused by a toxic agent and occurring within a short period of time following exposure.

Additive Effect – The combined effect of two chemicals, which is equal to the sum of each of the individual effects.

Adenosine Triphosphate – The primary source of energy required for cellular chemical reactions.

ADI – See Acceptable Daily Intake

Adrenal Gland – Gland located on top of the kidneys that secretes steroidal hormones, epinephrine, and norepinephrine, which affect various tissues of the body.

Adulterant – A substance that debases or makes another substance impure, resulting in an undesirable effect.

Afferent Nerves – Nerves that carry stimuli from the periphery to the central nervous system.

Agent Q – One of the mustard gases that causes severe pain and blistering of exposed tissue; blindness may occur as a result of exposure; exposure to high concentrations may result in death.

Agranulocyte – Leukocyte with no visible granules in the cytoplasm; lymphocytes and monocytes.

AIDS – Acquired Immunodeficiency Syndrome; disease caused by a virus that attacks the human immune system.

Albinism – A recessive genetic condition characterized by the inability to produce pigment. Individuals typically have a milky or translucent skin, white or nearly colorless hair, and eyes with pink or blue iris, and deep-red pupils.

Albumin – A blood protein that binds to and transports various types of materials, including toxic substances, through the blood.

Aldrin – An organochlorine insecticide containing mostly (95 percent or more) 1,2,3,4,10,10-hexachloro-1,4,4a,5,8,8a-hexahydro-1-4,5,8-endo,exo-dimethanonaphthalene.

Alopecia – A loss of hair; baldness.

Alpha Particle – A charged particle, α^{2+}, composed of two neutrons and two protons and emitted from the nuclei of certain radionuclides.

Alveolitis – An inflammation of the alveoli.

Alveolus (Alveoli, pl.) – One of the terminal air sacs of the lung; site of gas exchange between the lungs and the circulatory system.

Alzheimer's Disease – An illness of the brain associated with degeneration of mental capacities, loss of orientation, and forgetfulness.

Amelia – A developmental abnormality characterized by the absence of one or more arms and legs in newborns; newborns exposed to thalidomide during pregnancy have increased probability of developing this condition.

ALS – See Amyotropic Lateral Sclerosis

Amyotrophic Lateral Sclerosis – A syndrome characterized by muscular weakness and atrophy due to degeneration of motor neurons in the spinal cord and various regions of the brain.

Analgesia – A loss of pain sensation.

Anaplasia – A loss of cell differentiation and function; characteristic of most malignancies.

Anemia – Any condition that results in less than normal hemoglobin in the blood or a lower than normal number of erythrocytes.

Anencephaly – An abnormality caused by lack of development of the brain and skull bones resulting in a condition that is incompatible with life.

Angiotensin – A protein synthesized in the liver that interacts with renin resulting in vasoconstriction of arteriole smooth muscle and increasing blood pressure.

Anoxia – An extreme condition of oxygen deficiency that may result in permanent tissue damage.

ANS – See Autonomic Nervous System

Antagonistic – Two or more chemicals interacting in such a way that the actions of one of them is modified by the action of the other; interference with the uptake or by an opposing physiological reaction.

Anterior Chamber – The chamber of the eye bounded in front by the cornea and in the back by the iris and lens.

Anterior Pituitary – A portion of the pituitary gland that releases the hormones LH and FSH.

Anthrax – An acute, infectious disease caused by an increase in the concentration in the bacterium *Bacillus anthracis*, usually found in cattle, sheep, horses, and goats. It may attack the lungs producing respiratory distress, cyanosis, shock, and coma; death may occur if it is left untreated. The disease is contracted by man as a result of handling contaminated products from these animals.

Antibody – A protein found in the plasma that interacts with antigens (foreign substance) resulting in their destruction; produced by lymphocytes.

Aplasia – A cessation of the usual regenerative process in an organ or tissue.

Apnea – The absence of breathing.

Aqueous Humor – A transparent fluid-like substance found in the anterior chamber of the eye between the cornea and the lens.

Arrhythmia – Any irregular beating of the heart; atrial and ventricular fibrillation are examples of arrhythmia.

Arteriole – One of the small terminal twigs of an artery that joins the artery to its capillary bed.

Aspartame – An artificial sweetener formed by combining two amino acids, aspartic acid and phenylalanine.

Asphyxia – A condition in which decrease in the amount of oxygen in the body is accompanied by an increase of carbon dioxide and leads to loss of consciousness or death.

ASTDR (Agency for Toxic Substances and Disease Registry) – Established in 1980 to produce health-based information in support of chemical waste disposal sites.

Asthma – A condition of the lungs characterized by widespread narrowing of the airways due to spasm of the smooth muscle, edema of the mucosa, the presence of mucus in the lumen of the bronchi and bronchioles and characterized by wheezing and difficulty in expiration.

Astrocyte – A star-shaped cell involved in forming the blood-brain barrier.

Ataxia – A loss of muscular coordination.

ATP – See Adenosine Triphosphate

Atria – The chambers of the heart that receive blood from veins and pump blood into the ventricles prior to ventricular contraction.

Atrial Fibrillation – Characterized by rapid, irregular unsynchronized contraction of the atria, which produces irregular filling of the ventricles and a decreased transport of blood to the body.

Atrioventricular (AV) Node – A small collection of specialized cardiac muscle fibers located in the right atrium; its function is to delay propagation of the action potential to the ventricles of the heart.

Atrioventricular Bundle – Bundle of modified cardiac muscle fibers that projects from the AV node through the interventricular septum; it conducts action potentials from the AV node rapidly through the interventricular septum. It is also called bundle of His.

Attributable Risk – The number of individuals who actually develop a particular illness or disease that is associated with exposure to an etiological factor.

Autoimmune Response – Disorder resulting from the production of antibodies that attack normal body tissue; the immune system erroneously identifies the cells of the body as foreign substance that must be destroyed.

Autonomic Nervous System – That part of the peripheral nervous system composed of efferent fibers that reach from the central nervous system to smooth muscle, cardiac muscle, and glands.

Axon – The main process of a neuron; usually conducts action potentials away from the neuron cell body.

Axonopathy – Disorder caused by the degeneration and loss of the axon of a neuron; impairs nerve cell transmissions.

Bacillus anthracis – A bacterium, found in soils, that is responsible for causing the disease anthrax in cattle, goats, sheep, and horses.

Bacteremia – The presence of bacteria in the blood.

BAI – See Building-associated Illness

BAL – 2,3-Dimercaptopropanol, a chelating agent capable of binding a variety of toxic metals such as inorganic mercury, antimony, bismuth, cadmium, chromium, cobalt, and nickel. It is used as an antidote for ingested toxic substances such as heavy metals.

Baroreceptor – The sensory nerve endings in the walls of the atria of the heart, aorta and carotid sinus; sensitive to stretching of the wall caused by increased blood pressure.

Baroreceptor Reflex – Process in which baroreceptors detect changes in blood pressure and produce changes in heart rate, force of heart contraction, and blood vessel diameter that return blood pressure to normal levels.

Basal Cell Carcinoma – A slow-growing, locally invasive, neoplasm of the skin derived from basal cells of the epidermis.

Base Excision Repair – Like nucleotide excision repair, BER is a DNA molecule repair mechanism in which the damaged base and associated sugar and phosphate group are first removed through enzyme activity, then the gap is filled by a new nucleotide.

Basophil – A leukocyte with granules; promotes inflammation and prevents clot formation.

BEI – Biological Exposure Indices; recommended maximum concentrations of various types of toxic substances used as guidelines to evaluate the potential hazards associated with exposure.

Benign Tumor – A neoplasm (new growth) composed of cells which are proliferating but not invasive i.e., do not spread to surrounding tissues; usually encapsulated within fibrous connective tissue.

BER – See Base Excision Repair

Beta Particle – High energy electron emitted from the nuclei of certain radionuclides.

Bias – An error introduced in experimental design that can result in misleading conclusions which are not truly representative of the population being studied.

Bile – A fluid secreted from the liver, stored in the gall bladder, and released into the duodenum where it facilitates the digestion of fats and oils.

Bilirubin – A by-product formed mainly by the breakdown of hemoglobin during destruction of old, dead erythrocytes; normally it is converted to a water-soluble substance by the liver and excreted in the bile.

Biliverdin – A green bile pigment formed from hemoglobin.

Bioaccumulation – The retention and concentration of a substance by an organism, such as the bioaccumulation of organochlorine pesticides in fatty tissue.

Biotransformation – The conversion of a substance into a closely related substance, or to a new, chemically distinct substance.

Blood-brain Barrier – A barrier, composed of capillary endothelial cells and cellular processes of astrocytes, which prevents certain toxic substances from reaching brain tissue.

Botulinus Toxin – Toxic substance produced by bacterium *Clostridium botulinum* that is associated with food poisoning.

Botulism – Food poisoning caused by the bacterium *Clostridium botulinum*.

Bowman's Capsule – The enlarged end of the nephron; Bowman's capsule and the glomerulus make up the renal corpuscle.

Bradycardia – The slowing of the heart rate; stimulation of the parasympathetic nerves will decrease heart rate.

BRI – See Building-related Illness

Bronchiogenic Carcinoma – A form of cancer centering on the bronchial passages.

Bronchiole – A minute thin-walled branch of the primary bronchi, in which the diameter becomes progressively smaller and terminates in one or more pulmonary alveoli.

Bronchiolitis – Inflammation of the bronchioles, often associated with bronchopneumonia.

Bronchitis – An acute or chronic inflammation of the bronchial tubes or any part of them.

Bronchopneumonia – Pneumonia involving acute inflammation of the walls of the many relatively small areas of the lung adjacent to the smaller bronchi.

Brucella abortus – A bacterium that causes brucellosis leading to abortion in cows, horses, and sheep as well as undulant fever in man.

Brucella melitensis – A bacterium that causes brucellosis in man, abortion in goats, and a wasting disease in chickens; it may infect cows and hogs and be excreted in their milk.

Brucella suis – A bacterium that may cause brucellosis in man; it may also infect horses, dogs, cows, monkeys, and goats.

Brucellosis – A widespread infectious disease affecting primarily cattle, pigs, and goats, but sometimes affecting other animals, including humans.

Bubo – An enlarged, inflamed lymph node, usually in the armpit and groin region.

Bubonic Plague – An acute infectious disease caused by the bacterium *Yersinia pestis*, which is transmitted to humans from the bite of the rat flea. The high mortality rate in this disease is due to pneumonia. Bubonic plague was called black death in the Middle Ages.

Building-associated Illness – An illness related to indoor air pollution, the cause of which is not always known.

Building-related Illness – Illness associated with indoor air pollution in which the causative agent is known, i.e., viruses, spores, pollen.

Bulbourethral Gland – One of two glands located near the prostate; it secretes an alkaline fluid.

Bundle Branches – The left and right branches of the atrioventricular bundle, also known as the bundle of His, after it divides; they give rise to many small conducting Purkinje fibers.

Bundle of His – See Atrioventricular Bundle

BZ – An incapacitating chemical warfare agent; symptoms include dizziness, hallucinations, and disorientation.

CAA – See Clean Air Act

CaBP – See Calcium-binding Protein

Calcium-binding Protein – A protein involved in the absorption of calcium from the small intestine; it can bind with toxic heavy metals such as cadmium and lead.

Cannula – A small tube made for insertion into a body cavity, duct, or vessel.

Carbaryl – A pesticide sold under the trade name Sevin; it is used to control fleas on cats and dogs and in mosquito-control programs.

Carbohydrate – A class of organic molecules containing only carbon, hydrogen, and oxygen, with a 2:1 ratio of hydrogen and oxygen; it includes all sugars such as glucose.

Carbon Tetrachloride – A nonflammable, sweet-smelling, colorless liquid; it is used as a degreasing agent and a solvent.

Carbonic Acid – A weak colorless acid formed by dissolving carbon dioxide in water.

Carboxyhemoglobin – A compound formed in red blood cells when carbon monoxide binds to hemoglobin.

Carcinogenic – Any substance or agent having the ability to produce cancer.

Carcinoma – A malignant tumor originating in epithelial cells; for example, a malignant tumor of the epidermis is referred to as an epidermal carcinoma.

Cardioregulatory Center – A specialized area within the medulla oblongata of the brain that receives sensory input and functions to control the sympathetic and parasympathetic stimulation of the heart.

Carotid Body – A small epithelioid structure – located just above the bifurcation of the common carotid artery – that serves as a chemoreceptor organ responsive to lack of oxygen, excess carbon dioxide, and increased concentration of hydrogen ions.

Carrier Molecule – A specialized molecule associated with the cell membrane that can attract and transport substances across the cell membrane.

Catabolic – Refers to one of the two metabolisms (the other one being anabolic) characterized by the enzymatic breakdown of large nutrient molecules such as carbohydrates, lipids, and proteins.

Catalyst – A substance that changes the rate of a chemical reaction without undergoing a chemical change.

Catecholamines – Any of various amines (as epinephrine, norepinephrine, and dopamine) that function as hormones or neurotransmitters or both.

Cathartic – A substance – administered orally – that enhances the elimination of gastrointestinal contents.

CDC – See Centers for Disease Control and Prevention

Cells of Leydig – Testosterone-secreting cells located between the seminiferous tubules of the testes, also referred to as interstitial cells.

Centers for Disease Control and Prevention – A branch of the U.S. Public Health Service under the Department of Health and Human Services involved in the study and prevention of infectious diseases and in establishing standards, guidelines, and recommendations relative to toxic chemicals.

Central Nervous System – The portion of the nervous system comprised by the brain and the spinal cord.

Central Vein – A vein that passes through the lobules of the liver and collects blood from the hepatic sinusoids.

CERCLA – See Comprehensive Environmental Response, Compensation, and Liability Act

Cerebellum – A portion of the brain involved in the control of skeletal muscles and the coordination of voluntary muscular movements.

Cerebral Cortex – The outer portion of the cerebrum (the largest part of the brain), which is involved in the interpretation of sensory information, voluntary muscular activity, memory, learning, intelligence, and emotions.

CFR – See Code of Federal Regulations

Chancre – A lesion (ulcer) associated with syphilis, usually appearing 2-3 weeks after infection.

Chelating Agent – A chemical substance capable of binding to a toxicant, inactivating it or facilitating its removal from the body. Chelating agents are used to remove heavy metals, such as lead or mercury, from the blood.

Chemical Antagonism – An interaction between two chemicals resulting in a less toxic product; often referred to as chemical inactivation.

Chemical Asphyxiant – A substance, such as a toxic gas, that induces asphyxia.

Chemical Transportation Emergency Center – A chemical database that provides various types of information on manufactured chemicals.

Chemoreceptor – A sensory nerve ending, like a taste bud, that is stimulated by and reacts to certain chemicals. It is found in large blood vessels like the aorta.

CHEMTREC – See Chemical Transportation Emergency Center

Chloracne – An acne-like skin disorder caused by prolonged exposure to chlorinated hydrocarbons.

Chloroacetophenone – See CN Gas

Chloroform – A sweetish, colorless, volatile liquid used at one time as an anesthetic; also a solvent and fumigant.

Cholera – An acute infectious disease caused by the bacterium *Vibria cholera*, characterized by profuse watery diarrhea and vomiting resulting in loss of fluid, cramps, and dehydration.

Chondrodystrophy – A disturbance in bone growth resulting in dwarfism; individuals have arms and legs shorter than normal, but the rest of the bones grow normally.

Choroid – The layer of tissue located between the sclera and the retina of the eye; it contains blood vessels and connective tissue.

Chrome-ulcers – An open sore or lesion of the skin or mucous membrane caused by exposure to chromium VI-containing compounds.

Chromosome – A gene-carrying threadlike strand of DNA and associated proteins located in the nucleus of cells; it functions in the transmission of hereditary information.

Chronic – Occurring over a long period of time (weeks, months, or years) either continuously or intermittently; the term is used to describe ongoing exposures and effects that develop only after a long exposure.

Chronic Exposure – Exposure to a substance continuously or intermittently over a long period of time (weeks, months, or years).

Chronic Toxicity – Adverse effects occurring after a long period of exposure to a toxic substance; effects are usually considered to be permanent and irreversible.

Cilia – Small, bristle-like projections on the free surface of cells; in the respiratory system they beat in a coordinated fashion to move substances from the lower regions of the respiratory tract to the back of the throat where they may be either expectorated or swallowed.

Ciliary Body – A part of the middle layer of the eyeball that – along with the choroid and the iris – make up the vascular layer. The ciliary body includes the ciliary muscles and the ciliary processes; it secretes the aqueous humor. Ciliary muscles are responsible for changing the shape of the lens.

Ciliostasis – The cessation of the rhythmic beating of cilia.

Cirrhosis – A chronic disease of the liver characterized by replacement of normal liver tissue with connective tissue.

Clean Air Act – A federal act that implements regulations intended to protect the public health and welfare from the harmful effects of air pollution by setting maximum acceptable levels of pollution in ambient air.

Clean Water Act – A federal act that regulates discharge of pollutants to surface waters.

Clinical Toxicologist – An individual involved in the study of diseases and illnesses that are a result of exposure to toxic substances.

Clonic – Relating to clonus, which is a series of alternating contractions and partial relaxations of a muscle that in certain nervous diseases occurs in the form of convulsive spasms involving complex groups of muscles.

Clostridium botulinum – A bacterium that grows in improperly processed food and that produces a toxin causing botulism.

Clubfoot – A congenital abnormality associated with the structure of the foot. Several conditions may exist: exaggerated normal arch of the foot; foot is flexed and only the heel touches the ground; or foot is extended and individual only walks on the toes.

CN Gas – A chemical warfare agent that causes lacrimation and severe irritation to the skin; additional symptoms of exposure may include tightness of the chest, nausea, and vomiting.

CNS – See Central Nervous System

Coccidioides immitis – A fungus found in the soil of semi-arid areas of the southwestern United States and Central and South America.

Cockayne's Syndrome – A condition characterized by dwarfism, retinal atrophy, photosensitivity, deafness.

Code of Federal Regulations – The federal government's official publication of federal regulations. It contains the implementing language and regulatory requirements for federal laws.

Codon – A sequence of bases in a strand of DNA that provides the genetic code for a specific amino acid.

Colic – A painful spasm of the muscle comprising the wall of the intestine.

Collecting Duct – A straight tubule that extends from the cortex of the kidneys to the tip of the renal pyramid.

Comedo (Comedones, pl.) – A thickened secretion plugging a duct of the skin, especially of a sebaceous gland; it is usually associated with dermatitis or acne.

Common Bile Duct – The duct formed by union of the left and right hepatic ducts that empties into the duodenum.

Comprehensive Environmental Response, Compensation, and Liability Act – A federal act established for the cleanup of hazardous waste sites and for the definition of requirements for response to hazardous waste spills. Also called Superfund.

Concentration Gradient – The result of an unequal distribution of molecules with the tendency for the molecules to move from an area of higher concentration to an area of lower concentration.

Confounding Factor – Factors, other than the ones being tested in an epidemiological study, that have the potential to influence the study results. Examples include sex, age, health, social habits, or history of past exposure of the individual participating in the study.

Conjugation – The joining of two substances to form a single molecule; conjugation of toxic substances can increase water solubility and excretion.

Conjunctiva – The mucous membrane that lines the eyelid.

Conjunctivitis – An inflammation of the conjunctiva, characterized by redness and often accompanied by discharge.

Consumer Product Safety Commission – An independent federal regulatory agency that helps protect the public from unreasonable risk of injury by developing uniform safety standards, and promoting research into the causes and prevention of product-related illnesses, injuries, and death.

Contraindication – An indication, symptom, or condition that makes inadvisable a particular treatment or procedure.

Cornea – The clear, transparent anterior portion of the eye.

Corneal Reflex – The closing of eyelids as a result of irritation of the outer layer of the eyeball (cornea).

Corneal Stroma – The thickest part of the cornea composed primarily of connective tissue and a small number of corneal cells.

Corpus Luteum – A yellowish body formed in the ovary in the site of a ruptured follicle immediately after ovulation; it secretes progesterone and estrogen.

Cortex – The outer or superficial part of an organ or structure, such as the kidney, the adrenal gland, or the gray matter of the cerebrum and cerebellum.

CPSC – See Consumer Product Safety Commission

Cri-du-chat Syndrome – A congenital abnormality characterized by mental retardation, small brain, dwarfism, and defects associated with the voice box. The infant's cry resembles the cry of a cat (French: cri du chat) and gives the syndrome its name.

Cryptococcal Meningitis – An inflammation of the membranes of the spinal cord or brain caused by the fungus *Cryptococcus*.

CS Gas – A chemical warfare agent that causes lacrimation and irritation to the skin; additional symptoms of exposure may include tightness of the chest, nausea, and vomiting.

CWA – See Clean Water Act

Cyanosis – A bluish discoloration of the skin and mucous membranes resulting from inadequate oxygenation of the blood.

Cystic Fibrosis – An inherited disease that affects primarily the respiratory system but also various glands of the body; it usually begins during infancy and is characterized by chronic respiratory infections.

Cytoplasm – The fluid, translucent substance located outside of the cell nucleus and bound by the cell membrane. It is composed of fats, proteins, and other types of molecules; intracellular structures such as ribosomes and endoplasmic reticulum are found in this substance.

DBCP (1,2-Dibromo-3-chloropropane) – A nematocide soil fungicide.

DDOH (2,2-Bis (p-chlorophenyl) ethanol) – A metabolite of DDT.

DDT (Dichlorodiphenyltrichloroethane) – An organochlorine pesticide used to kill insects.

Deactivation – The process of making a substance inactive; rendering a toxic substance less toxic.

Deamination – The removal of an amine group from a molecule, usually by hydrolysis.

DEHP (Di(2-ethylhexyl)phthalate) – A plasticizer for many resins and elastomers.

Dementia – A deterioration of intellectual faculties, such as memory, concentration, and judgement, often accompanied by emotional disturbances and personality changes.

Dendrite – Any of the usual branching protoplasmic processes that conduct impulses toward the body of a nerve cell.

Dengue Fever – An acute disease characterized by sudden onset, headache, fever, joint and muscle pain, and the appearance of a rash.

Deoxyribonucleic acid – A nucleic acid, localized in the cell nuclei, that is the molecular basis of heredity and is constructed of a double helix held together by hydrogen bonds between purine and pyrimidine bases, which project inward from two chains containing alternate links of deoxyribose and phosphate.

Dermis – Layer of tissue located below the epidermis and above the subcutaneous fatty layer of the skin; it is composed primarily of connective tissue that is responsible for the elasticity of the skin.

DES – See Diethylstilbestrol

Descriptive Toxicologist – Toxicologist involved in obtaining information concerning the effects of chemical exposure on animals.

Diabetes Mellitus – An inherited disease in which insufficient amounts of insulin are produced by the pancreas resulting in elevated blood sugar levels, excessive thirst, urine production, and slow healing of wounds.

Dialysis – The process of diffusing blood across a semipermeable membrane to remove toxic materials and maintain normal fluid, pH, and solute balance.

Diaphragm – A muscular separation between the chest and abdominal cavities; its contraction assists inhalation.

Dichloroformoxime – A chemical warfare agent that is primarily a skin irritant; at high concentrations it may cause blistering and enter the bloodstream resulting in death.

Dicumarol – An anticoagulant that inhibits the formation of prothrombin in the liver.

Dieldrin – An organochlorine insecticide.

Diethylstilbestrol – A synthetic estrogen that stimulates production of estrogen and progesterone in the placenta.

Differential Leucocyte Count – A test to determine the number of white blood cells present in the circulatory system; it is calculated by counting the number of white blood cells in a milliliter of blood.

Diffusion – The net movement of molecules from a region of high concentration to one of lower concentration.

Dioxin – Any of several carcinogenic or teratogenic heterocyclic hydrocarbons that occur as impurities in petroleum-derived products.

Diquat – A type of herbicide used to kill nuisance plants.

Dispositional Antagonism – Absorption, biotransformation, distribution, or excretion of a toxic chemical such that a less toxic product is formed or the duration at the target organ is shortened.

Distal Convoluted Tubule – The twisted tubule of the nephron that extends from the ascending limb of the Loop of Henle and ends in a collecting duct.

DMPS (2,3-Dimercapto-1-propanesulfonic Acid) – A chelating agent used in the treatment of mercury and lead poisoning.

DNA – See Deoxyribonucleic Acid

Dopamine – A monoamine that occurs especially as a neurotransmitter in the brain and is a precursor of epinephrine and norepinephrine.

Dose – The quantity of a chemical administered at one time.

Dose-response Curve – A graphical presentation of the relationship between the degree of exposure to a chemical (dose) and the observed biological effect or response.

Dose-response Relationship – A quantitative relationship between the dose of a chemical and an effect caused by the chemical.

Down's Syndrome – A congenital abnormality characterized by moderate-to-severe retardation.

d-tubocurarine – A drug used to treat poisoning from black widow spider bite, also used on tips of poison arrows; it can produce skeletal muscle relaxation; at certain doses it may be lethal.

Ductus Deferens – A duct of the testes; running from the epididymis to the ejaculatory duct; also called vas deferens.

DUI – An acronym for driving under the influence of alcohol.

Duodenum – The first portion of the small intestine, originating at the stomach and extending 10-12 inches until it merges with the jejunum.

Dyspnea – Difficulty in breathing, often associated with lung or heart disease and resulting in shortness of breath.

EDB – Ethylene dibromide, used as a grain and tree crop fumigant and as a scavenger for lead in gasoline.

Edema – An excessive accumulation of fluid.

EDS – Ethylene dimethane sulfonate; it interacts with the Leydig cells and decreases testosterone levels, which reduces spermatogenesis.

EDTA – See Ethylene Diaminetetraacetic Acid

Efferent Ductule – A small duct that leads from the testes to the epididymis.

Efferent Nerves – The nerve fibers going from the central nervous system outward toward the peripheral nervous system and muscles – motor fibers.

Electromagnetic Fields – The coupled energy fields (electric and magnetic) that are generated by electric current passing through a conductor.

Emergency Planning and Community Right-to-Know Act – A federal act that requires companies to report inventories of hazardous chemicals and toxic releases. It also requires state and local governments to develop plans for responding to emergency releases.

EMF – See Electromagnetic Fields

Emphysema – A pathological condition of the lungs marked by an abnormal increase in the size of the air spaces, resulting in labored breathing and increased susceptibility to infection; it can be caused by destruction of alveolar walls.

Encephalitis – An inflammation of the brain.

Encephalocele – An abnormality characterized by protrusion of the brain between bones of the skull that did not close at birth.

Encephalopathy – Any of a variety of diseases of the brain.

Endometrium – The mucous membrane comprising the inner layer of the uterine wall.

Endoplasmic Reticulum – A membrane system located within the cell associated with intracellular transport, storage, and secretion.

Enterocolitis – An inflammation of the intestine and colon.

Enterohepatic Circulation – The recurrent cycle in which the bile salts and other substances excreted by the liver are reabsorbed through the intestinal mucosa, returned to the hepatic cells, and then excreted again.

Environmental Protection Agency – A government agency responsible for establishing and implementing laws and regulations to protect people and the environment.

Environmental Toxicologist – An individual studying the interaction of toxic substances with various components of the ecosystem (environment).

Enzyme – Any of numerous complex proteins that are produced by living cells and catalyze specific biochemical reactions within the body.

Eosinophil – A leukocyte or other granulocyte with cytoplasmic inclusions that stain red with eosin. The sodium or potassium salts are used as a biological stain for cytoplasmic structures.

EP – See Erythrocyte Protoporphyrine

EPA – Environmental Protection Agency

EPCRA – See Emergency Planning and Community Right-to-Know Act

Epidemiological Studies – Studies that investigate the causative factors associated with the incidence and distribution of disease and illness in a specific population.

Epidermis – The outer nonsensitive and nonvascular layer of the skin of a vertebrate that overlies the dermis.

Epididymis – The elongated structure connected to the posterior surface of the testes; it is the site of storage and maturation of sperm cells.

Epigenetic Carcinogen – A chemical or physical agent that causes the development of cancer, but does not interact with the DNA of cells.

Epilation – The loss of hair due to exposure to ionizing radiation.

Epilepsy – A disorder of the brain characterized by brief and sudden attacks of altered consciousness and motor activity.

Epinephrine – A neurotransmitter, also known as adrenaline, that is the principal blood-pressure raising hormone secreted by the adrenal medulla. It is used medicinally as a heart stimulant, vasoconstrictor in controlling hemorrhages of the skin, and a muscle relaxant in bronchial asthma.

Epithelial Cells – The cells that compose the layer of tissue covering the internal and external surfaces of the body, including the lining of blood vessels and other small cavities. It consists of cells that are joined by small amounts of a cementing substance.

EPN – See o-ethyl o,p-nitrophenyl phenylphosphorothionate

Equilibrium – The state in which the movement of molecules, ions, or atoms in any one direction is equal to the movement of molecules in the opposite direction.

ER – See Endoplasmic Reticulum

Erythema – A reddening of the skin caused by an increased blood flow through the capillaries located in the dermis.

Erythrocyte – A red blood cell that carries oxygen from the lungs to the tissues of the body.

Erythrocyte Count – A test to determine the number of red blood cells present in the circulatory system; calculated by counting the number of red blood cells in a milliliter of blood.

Erythrocyte Protoporphyrine – A derivative of hemoglobin found in red blood cells.

Erythropoietin – A hormone that stimulates erythrocyte formation in red bone marrow.

Escherichia coli **(E. coli)** – A bacterium normally found in the colon; presence of this microorganism in milk or water is an indicator of fecal contamination; food contaminated with E. coli may be lethal.

Esophageal Atresia – An abnormality characterized by lack of complete development of the esophagus in a way that there is no passageway to the stomach.

Esophagus – A long muscular tube that is posterior to the larynx and extends from the back of the mouth and terminates in the upper portion of the stomach.

Estrogen – A steroid hormone secreted primarily by the ovaries; it is involved in the maintenance and development of female reproductive organs.

Ether – A volatile, colorless, highly inflammable liquid with an aromatic odor; used as an anesthetic and solvent.

Ethylene Diaminetetraacetic Acid – A chelating agent used primarily in the treatment of lead poisoning.

Etiology – The study of causes; in the toxicological context, the study of causes of disease or illness.

Excretion – The process of expelling or discharging a substance from the body, an organ, or a cell.

Exencephaly – A congenital abnormality in which the brain is located outside the skull.

Extravascular Space – The space outside of blood vessels or lymphatics.

Exudate – A fluid that has exuded out of a tissue or its capillaries due to injury or inflammation.

Facilitated Diffusion – The movement of a substance by a carrier molecule across the cell membrane without expenditure of energy.

Fanconi's Anemia – A condition characterized by decreased production of blood cells, various congenital anomalies, leukemia.

FAS – See Fetal Alcohol Syndrome

Fat – See Lipid

FDA – See Food and Drug Administration

Federal Insecticide, Fungicide, and Rodenticide Act – A federal act that requires pesticide manufacturers to register and test pesticides produced; it also regulates their use, sale, and distribution.

FEL – See Frank-effect Level

Fenestrae – The large pores in the endothelial lining of the hepatic sinusoids; they facilitate the absorption of large molecular weight substances.

Fetal Alcohol Syndrome – A syndrome characterized by developmental abnormalities in the fetus as a result of consumption of alcohol during pregnancy.

Fibrinogen – A globular protein in plasma that gives rise to fibrin to form a blood clot.

Fibrosis – A disease in which alveolar cell death occurs and is replaced with fibrous connective tissue; it decreases elasticity of lungs and impedes gas exchange.

FIFRA – See Federal Insecticide, Fungicide, and Rodenticide Act

First-pass Effect – The phenomenon of removal of chemicals by the liver before entrance of the blood to the systemic circulation; it is associated with substances that are absorbed from the digestive tract.

Fluorosis – A disease characterized by the replacement of calcium in the bone with fluoride, which results in weakening of the bone.

Follicle-stimulating Hormone – A hormone of the anterior pituitary that, in the female, stimulates the follicles of the ovary, assists in maturation of the follicle, and causes secretion of estrogen from the follicle; in the male, it stimulates the epithelium of the seminiferous tubules and is partially responsible for inducing spermatogenesis.

Food and Drug Administration – A federal agency that regulates food additives, cosmetics, medical devices, and drugs.

Food Toxicologist – A toxicologist involved in the study of toxic substances used in the growing and processing of food.

Footdrop – Flexion of the foot due to weakness or paralysis of the anterior muscles of the lower leg.

Forensic Toxicologist – A toxicologist involved in the study of toxic substances generally in the support of legal activities.

Frank-effect Level – The concentration of a substance at which overt effects are first observed.

FSH – See Follicle-stimulating Hormone

Functional Antagonism – An interaction of two chemicals on the same physiological function, each having the opposite effect.

Gall Bladder – A sac-like structure attached to the liver that stores bile produced by the liver.

Gamma Ray – The highest energy, shortest wavelength electromagnetic radiation released from the nuclei of certain radionuclides; composed of photons with no rest mass.

Gastritis – A chronic or acute inflammation of the stomach, especially of the mucous membrane of the stomach.

Gastroenteritis – An inflammation of the mucous membrane of the stomach and intestines.

Genotoxic Carcinogen – A chemical or physical agent that interacts with the DNA of the cell producing a change in the DNA structure, possibly leading to the development of cancer.

Germinal Mutation – A mutation associated only with the formative cells of the ovaries and testes.

Gingivitis – An inflammation of the gums, characterized by redness and swelling.

Glia – The cells of the nervous system other than neurons; they play a support role in the nervous system. They include astrocytes, oligodendrocytes, and Schwann cells; sometimes referred to as neuroglia.

Globin – The protein portion of the hemoglobin molecule located in the erythrocyte.

Globulin – Any of a group of albuminous proteins, insoluble in water.

Glomerular Nephritis – An inflammation of the kidney that affects primarily the glomeruli.

Glomerulus – The mass of capillary loops at the beginning of each nephron, nearly surrounded by Bowman's capsule.

Glucouria – The excretion of glucose in urine, especially in elevated amounts.

Glucuronidation – The binding of glucuronic acid with a toxic compound or its metabolite.

Glycogen – Also known as animal starch, it is a carbohydrate reserve – especially in muscle and liver tissue – that is composed of polymerized glucose molecules.

Glycoprotein – An organic molecule composed of a protein and a carbohydrate.

GnRH – See Gonadotropin-releasing Hormone

Gonadotropin-releasing Hormone – A hormone released from the hypothalamus region of the brain that stimulates the secretion of FSH and LH from the anterior pituitary.

Gonococcal Opthalmitis – An inflammation of the eye caused by the bacterium *Neisseria gonorrhoeae*.

Gout – A metabolic disease that is a form of acute arthritis and is marked by inflammation of the joints and an excess of uric acid in the blood.

Graafian Follicle – A mature follicle containing a single egg; it develops in response to the secretion of FSH from the anterior pituitary.

Granulocyte – A polymorphonuclear white blood cell with granule-containing cytoplasm; neutrophils, eosinophils, and basophils are granulocytes.

Granulocytopenia – A decrease in the number of granulocytes (neutrophils, basophils and eosinophils).

Granuloma – A term applied to a mass or nodule of chronically inflamed tissue with granulations that is usually associated with an infective process.

Granulosa Cells – A single layer of flat cells that surround a primary oocyte in the ovary.

Gray (Gy) – The international unit that is used as a measure of the amount of radiation absorbed per gram of tissue.

HAV – See Hepatitis A Virus

Hazardous Substances Data Bank – A source of chemical information regarding emergency handling procedures, environmental data, regulatory status, and human exposure.

HBV – See Hepatitis B Virus

HCV – See Hepatitis C Virus

HDV – See Hepatitis D Virus

Hematocrit – The percentage of the volume of a blood sample composed of cells.

Hematology – The science concerned with blood and blood-forming tissues.

Hematopoiesis – The production of blood cells.

Hematuria – Blood in the urine; color of the urine may be slightly smoky, reddish, or very red.

Heme – The portion of the hemoglobin molecule of erythrocytes that contains iron.

Hemocytoblast – A primordial cell capable of differentiating into any type of blood cell.

Hemodialysis – A procedure for removing metabolic waste products or toxic substances from the bloodstream by dialysis.

Hemolysis – The breakup of red blood cells.

Hemolytic Anemia – An anemia resulting from the abnormal destruction of red blood cells, in response to certain toxic infectious agents and in certain inherited blood disorders.

Hemophilia – A hereditary blood disease characterized by a prolonged coagulation time.

Hemophiliac – A person afflicted with hemophilia.

Heparin – A glycoprotein that prevents blood from clotting.

Hepatic Artery – One of the major sources of blood to the liver; it supplies oxygen-rich blood to liver cells.

Hepatic Sinusoid – The large intercellular spaces (10 to 40 micrometers in diameter) located in the liver, through which blood passes.

Hepatitis – An inflammation and enlargement of the liver, usually caused by a virus.

Hepatitis A Virus – Blood-borne virus that exists in the blood for only two weeks; infection occurs by contact with contaminated food or water, and is also sometimes contracted via blood transfusions.

Hepatitis B Virus – Also referred to as serum hepatitis; caused by a DNA virus that persists in the blood and is characterized by a long incubation period. Usually transmitted as a result of contact with infected blood or other body fluids; the most obvious symptom is the development of jaundice.

Hepatitis C Virus – A form of hepatitis with clinical effects similar to those of hepatitis B, caused by a blood borne retrovirus that may be of the hepatitis non-A, non-B type.

Hepatitis D Virus – This virus usually occurs with one of the other hepatitis viruses, their co-existence being necessary for HDV to survive; cirrhosis of the liver often develops.

Hepatocyte – A cell of the liver.

Hepatomegaly – The abnormal enlargement of the liver.

Heptachlor – An organochlorine insecticide.

Histamine – An amine released from basophils, that causes dilation of capillaries, contraction of smooth muscle, stimulation of gastric acid secretion, and that promotes the inflammatory response. It is a neurotransmitter that is released during allergic reactions.

Histone – A simple water-soluble protein that is rich in the basic amino acids lysine and arginine and is complexed with DNA in the nucleosomes of eukaryotic chromatin.

Homeostasis – The existence and maintenance of a relatively constant environment within the body with respect to functions and composition of fluids and tissues.

HSDB – See Hazardous Substances Data Bank

Huntington's Chorea – An inherited disease of the central nervous system that usually has its onset between 30 to 50 years of age; disease characterized by decline in mental capabilities, and irregular, spasmodic contraction of limb and facial muscles.

HVAC – An acronym for Heating, Ventilation, and Air Conditioning.

Hydrocephaly – A condition in which there is an excess accumulation of fluid on the brain.

Hydrolysis – The process in which a chemical compound is split into smaller molecules by reacting with water.

Hydrophobic – Antagonistic to water; any substance that is incapable of dissolving in water.

Hydroxyapatite – The complex crystal structure that makes up the lamellae of the bone.

Hyperkeratosis – The excessive growth of the horny tissue (outer layers) of the skin.

Hyperpigmentation – An excessive amount of pigment in a tissue or body part.

Hyperplasia – An abnormal increase in the number of cells.

Hypocalcemia – A low blood-calcium level.

Hypoplastic Left Heart Syndrome – An abnormality characterized by the left ventricle being smaller than normal and composed primarily of connective tissue instead of muscle.

Hypothalamus – A portion of the brain located beneath the thalamus; it is important in the control of the autonomic and endocrine control centers.

Hypoxia – A condition in which inadequate amounts of oxygen are delivered to the tissues.

IARC – An acronym for the International Agency for Research on Cancer.

IDLH – See Immediately Dangerous to Life or Health

Ileum – The terminal portion of the small intestine, extending between the jejunum and the large intestine.

Immediately Dangerous to Life or Health – The maximum concentration of a toxic substance from which one could escape within 30 minutes without any impairing symptoms or irreversible health effects.

Incidence Rate – The number of new cases that are identified during a fixed period of time such as a week, month, or year.

Induced Mutation – A change in the structure of a gene or chromosome of an organism as a result of exposure to an exogenous chemical or physical agent.

Infectious – Capable of being transmitted with or without contact.

Inflammation – A general body response to tissue injury, characterized by swelling, redness, pain, heat, and infiltration of leukocytes.

Interstitial Fluid – The fluid between cells or body parts.

Intracellular Transport – The movement of substances within the cell; function performed in part by endoplasmic reticulum.

Intradermally – Into the skin, as with an injection.

Intramuscularly – Into the muscle, as with an injection.

Intraperitoneally – Into the peritoneal (body) cavity; this is the cavity that contains the organs of the digestive tract and related structures.

Intravenously – Into a vein, as with an injection.

Ion – An atom or a group of atoms that has acquired a net electric charge by gaining or losing one or more electrons.

Iris – The circular, pigmented tissue of the eye located directly in front of the lens that gives color to eye.

Isomer – One of two or more chemical substances that have the same molecular formula, but different arrangement of the atoms in the molecules, resulting in different chemical and physical properties.

Isoproterenol – A drug used to relieve breathing difficulties associated with asthma.

Jaundice – A yellowish pigmentation of the skin, tissues, and body fluids caused by the deposition of bile pigments.

Jejunum – The second portion of the small intestine located between the duodenum and the ileum.

Kaposi's Sarcoma – A neoplastic disease associated especially with AIDS, affecting the skin and mucous membranes, and characterized usually by the formation of pink to reddish-brown or bluish plaques, macules, papules, or nodules.

Keratin – An insoluble protein produced in the cells that comprise the outer layers of the skin.

Keratitis – An inflammation of the cornea.

Keratocyte – A cell found in the corneal stroma responsible for producing connective tissue fibers.

Klinefelter's Syndrome – An abnormal condition in males characterized by two X and one Y chromosomes, infertility, long legs, subnormal intelligence, and smallness of the testicles.

Krebs Cycle – A series of enzymatic reactions involving oxidative metabolism of acetyl units and producing high-energy phosphate compounds (ATP) that serve as the main source of cellular energy.

Kupffer Cells – Phagocytic cells found in the liver; they are involved in the destruction and removal of foreign substances circulating in the blood; macrophages.

Labile – Unsteady; apt to rapidly change or shift.

Lacrimation – The secretion and discharge of tears.

Lamella (Lamellae, pl.) – A thin sheet or layer of bone.

Large Mutation – Exchanges in the sequence of a large number of bases in the DNA; it is associated with inversion, deletion, or translocation.

Larynx – A short passageway that connects the pharynx to the trachea.

Lavage – A washing, especially of a hollow organ such as the stomach, with repeated injections of water.

LC_{50} – See Lethal Concentration

LD_{50} – See Lethal Dose

Legionnaire's Disease – A severe, often fatal disease characterized by pneumonia and dry cough; as the disease progresses it can lead to cardiovascular collapse. It is caused by the bacterium *Legionella pneumophila*.

Lens – The transparent organ lying posterior to the iris of the eyeball and anterior to the vitreous humor.

Lethal Concentration – The dose of a particular substance in air that – when administered to all animals in a test – is lethal to 50 percent of the animals.

Lethal Dose – The dose of a particular substance that – when administered, either by injection, ingestion, or dermally, to all test animals – is lethal to 50 percent of the animals.

Leukemia – A disease of the bloodstream that results in an increase in the number of white blood cells.

Leukocyte – A white blood cell; it is part of the immune system that circulates in the blood and lymph and participates in reactions against invading microorganisms or foreign particles.

LH – See Luteinizing Hormone

Lipid – Any of various substances that are soluble in nonpolar organic solvents, such as chloroform and ether. Along with proteins and carbohydrates, lipids constitute the principal structural components of living cells.

Lipophilic – Substances that have a strong affinity for, or are soluble in, lipids.

Lipoprotein – A protein bound with a lipid such as cholesterol, phospholipid, or triglyceride.

LOAEL – See Lowest Observed Adverse Effect Level

Lobule – A small, functional unit of the liver.

LOEL – See Lowest Observed Effect Level

Loop of Henle – A U-shaped portion of the nephron extending from the proximal to the distal convoluted tubule and consisting of the descending and ascending limbs.

Lowest Observed Adverse Effect Level – The concentration at which adverse effects observed at higher concentrations are first seen.

Lowest Observed Effect Level – Concentration where minor effects are observed, but are not related to the response measured.

Lumen – The cavity of a tubular organ, such as a blood vessel.

Luteinizing Hormone – A hormone of the anterior pituitary that, in the female, initiates final maturation of the follicles, their rupture to release the oocyte, the conversion of the ruptured follicle into the corpus luteum, and the secretion of progesterone; in the male, it stimulates the secretion of testosterone in the testes.

Lyme Disease – A recurrent inflammatory disease characterized by the appearance of red splotches on the skin and development of fever, fatigue, headache, and swelling of the joints; there may also be associated effects on the cardiovascular and nervous systems. It is caused by a spirochete, *Borrelia burgdorferi*.

Lymph – A clear or yellowish fluid derived from interstitial fluid and found in lymph vessels.

Lymph Node – A round, oval, or bean-shaped mass of lymphoid tissue; it produces white blood cells that fight bacteria and filter the blood. Lymph nodes are located throughout the body.

Lymph Nodule – A localized area within lymph nodes that is comprised primarily of lymphocytes.

Lymphadenopathy – A disease of the lymph nodes.

Lymphatic Capillary – A small vessel, one single cell thick, that carries lymph toward the larger lymphatic vessels.

Lymphocyte – A type of white blood cell characterized by a large, spherical nucleus surrounded by a thin layer of nongranular cytoplasm. There are several types of lymphocytes involved in the immune system including B lymphocytes and natural killer cells (T cells).

Lymphocytopenia – A decrease in the number of lymphocytes in the body; it results in increased susceptibility to illnesses.

Lymphoma – A cancer associated with the lymph system.

Macrophage – Any large, mononuclear, phagocytic cell.

Malaria – An acute and sometimes chronic infectious disease caused by four protozoan, *Plasmodium falciparum*, *Plasmodium vivax*, *Plasmodium ovale*, and *Plasmodium malariae*, transmitted to the human blood by mosquito bite (*Anopheles*); symptoms include chills, fever, and sweats.

Malathion – An organic phosphate compound used as an insecticide.

Malignant Mesothelioma – A disease associated with exposure to asbestos; cancerous growths develop in the tissue lining the exterior of the lungs and the interior walls of the rib cage.

Malignant Tumor – An uncontrolled growth of cells invasive to surrounding tissue; it has the ability to metastasize.

Material Safety Data Sheet – A document produced by chemical manufacturers that describes physical and chemical properties of chemicals, first aid information, personal protective equipment to use, and suggested means of disposal.

Mechanistic Toxicology – The study of the mechanism of how a toxic substance causes its effects.

Medical Transportation Emergency Center – A national database that provides medical information for chemical exposures.

MEDTREC – See Medical Transportation Emergency Center

Medulla – The inner core or deep part of an animal or plant structure; e.g. of the kidney.

Medulla Oblongata – A portion of the brain that connects the spinal cord with the brain; it contains the autonomic control centers for heart rate and respiration.

Mee's Lines – The transverse white lines that appear in the fingernails about five weeks after exposure to arsenic.

Megakaryocyte – A large cell that has a lobulated nucleus, is found especially in the bone marrow, and is the source of blood platelets.

Melanosis – An unusual deposit of black pigment in different parts of the body.

Menstrual Cycle – A series of changes that occur in sexually mature, nonpregnant females; it specifically includes the cyclical changes that occur in the uterus and ovaries.

Mercurialism – A chronic poisoning by mercury as a result of working with metallic mercury; symptoms include soreness of gums and teeth, increased salivation, and diarrhea.

Mesothelioma – A specific type of lung cancer that is primarily associated with exposure to asbestos.

Messenger RNA – A type of ribonucleic acid involved in the transcription (reading) of chromosomal DNA; the end result is protein synthesis.

Metabolism – The sum of all physical and chemical changes that take place in an organism; includes the breakdown of substances, formation of new substances (synthesis), and changes in the energy content of cells.

Metabolite – Any intermediate substance produced as a result of metabolism.

Metal Fume Fever – A syndrome resembling influenza produced by inhalation of metallic oxide fumes such as zinc oxide; symptoms include chills, weakness, sweating, and anorexia.

Metallothionein – A protein, found in cells, that binds to toxic substances such as heavy metals.

Metaplasia – The transformation of one cell type to another; for example, transformation of non-keratinized epithelial cell to a keratinized epithelial cell.

Metastasis – The ability of tumor cells to break off and enter the bloodstream and be transported to other locations in the body where they can begin to grow; this is characteristic of malignant tumors.

Methemoglobin – A compound formed when iron in the hemoglobin molecule is oxidized from the ferrous state (Fe^{2+}) to the ferric state (Fe^{3+}).

Methemoglobinemia – A condition in which more than one percent of the iron in hemoglobin has been oxidized from Fe^{2+} to Fe^{3+} and is therefore unable to carry oxygen.

Methyl Isocyanate – A chemical used in the production of pesticides; exposure causes irritation of mucous membranes, coughing, and increased saliva production; it can be acutely toxic when inhaled.

Methylation – A Phase II metabolic reaction that plays a minor role in the transformation of toxic substances to a different toxic substance, which is retained in the body longer.

MIC – See Methyl Isocyanate

Microcephaly – An abnormal smallness of head often seen in mental retardation.

Microcytic Anemia – An anemia caused by a decrease in the size of the erythrocytes.

Microvilli – Finger-like projections of the cell membrane of certain cells such as those found in the epithelium of the small intestine.

Minamata Disease – A neurological disease caused by the consumption of fish contaminated with methylmercury.

Mitochondrion – A cell structure composed of two membranes, an outer membrane that defines the external boundary and an inner membrane that exhibits many folds that project into the interior; source of ATP production.

Monocyte – A type of leukocyte that transforms to become a macrophage.

Morphine Sulfate – A drug used to relieve pain or used as a sedative.

mRNA – See Messenger RNA

MSDS – See Material Safety Data Sheet

Mucociliary Escalator – A mechanism found in the respiratory tract, it is made up of ciliated and mucous cells, and assists in the removal of particulates from the respiratory tract.

Mucosa – A lubricating membrane that lines a body cavity or organ that opens to the exterior such as the digestive tract, respiratory or genitourinary canals.

Mucous Cell – Modified epithelial cells that secrete a viscous, watery substance; these cells can be found in the respiratory and digestive tract.

Mucous Membrane –See Mucosa

Multiple Sclerosis – A neurological disorder characterized by numbness, tingling, blurred vision, weakness of the limbs, ataxia.

Muscarinic – A type of postganglionic receptor that interacts with the substance muscarine; effects of stimulation are similar to stimulation produced by acetylcholine.

Muscularis – A layer of muscle tissue located in the outer layers of the digestive tract; contraction of this layer propels food along the tract.

Mustard Gas – An oily, volatile liquid (dichlorodiethyl sulfide) used as a poison gas because of its extremely irritating, blistering, and disabling effects.

Mutagen – Any agent that causes genetic mutations.

Mutation – A permanent change in the genetic material; abnormalities that manifest themselves as altered morphology or altered ability to direct the synthesis of proteins.

Myelin Sheath – A lipoprotein envelope made by the Schwann cell or oligodendrocyte cell membrane wrapping around an axon.

Myelinated Nerve – A nerve in which the axon is surrounded by oligodendrocytes in the CNS and Schwann cells in the PNS forming several concentric layers.

Myelinopathy – A disorder in which the myelin sheath surrounding the axon is disrupted; transmission of the action potential is impaired.

Myocarditis – An inflammation of the myocardium (heart muscle).

Myocardium – The middle layer of the walls of the heart, composed of cardiac muscle.

Myometrium – The muscular wall of the uterus, composed of smooth muscle.

NADPH – See Nicotinamide Adenine Dinucleotide Phosphate

Nanometer – A unit of measure equal to 10^{-9} meters.

Nasal Cavity – The space bounded by the structures that compose the external and internal nose cavity lined by epithelial cells, some of which are specialized to produce mucose or have hairs associated with them.

National Center for Environmental Health – An organization of the CDC that addresses hazards associated with chemical exposure outside the workplace.

National Institute for Occupational Safety and Health – Part of the CDC involved in evaluating chemical hazards in the workplace.

National Library of Medicine – The world's largest biomedical library and source of chemical information, medical history, and biotechnology located in Bethesda, MD (toll free number: 1-888-FIND-NLM).

NCEH – See National Center for Environmental Health

NCRP – National Council of Radiation Protection.

Neisseria gonorrhoeae – The bacterium responsible for causing gonorrhea.

NEL – See No-effect Level

Neoplasm – A new growth of tissue serving no physiological function; characterized primarily by rapid, uncontrolled cellular division.

Nephron – The functional unit of the kidney, consisting of the renal corpuscle, the proximal convoluted tubule, the Loop of Henle, and the distal convoluted tubule.

NER – See Nucleotide Excision Repair

Nettle – Plants with hairs on their leaves that secrete a fluid that irritates the skin; class of gases used as chemical warfare agent that produces effects similar to lying in a bed of nettles.

Neuroglia – See Glia.

Neuron – A grayish or reddish granular cell with specialized processes that is the fundamental functional unit of nervous tissue; a nerve cell.

Neuronopathy – A disorder in which the entire neuron is destroyed.

Neuropsychiatric – Of behavioral or functional abnormalities associated with abnormal nerve cell functioning in the brain.

Neurosyphilis – An infection of the central nervous system by *Treponema pallium*.

Neurotransmitter – A chemical substance released from the end of an axon resulting in a stimulatory or inhibitory effect on the target tissue.

Neutron – A neutrally charged particle that may be emitted from the nuclei of atoms.

Neutrophil – A leukocyte with granules; phagocytic white blood cell.

NFPA – National Fire Protection Association.

Nicotinamide Adenine Dinucleotide Phosphate – The reduced form of nicotinamide adenine dinucleotide phosphate, as opposed to the NADP, which is a co-enzyme present in most living cells and interacts with cellular metabolites, including toxic metabolites.

Nicotinic Receptors – A subset of acetylcholine receptors that bind nicotine.

NIOSH – See National Institute for Occupational Safety and Health

NLM – See National Library of Medicine

Nm – See Nanometer

No Observed Adverse Effect Level – Lowest concentration where some toxicant-related effects are observed, but the observed effect seen at higher levels is still absent.

NOAEL – See No Observed Adverse Effect Level

No-effect Level – Concentration at which no effect is observed; similar to NOEL.

NOEL – See No Observed Effect Level

Norepinephrine – A neurotransmitter substance released from neurons of the sympathetic nervous system.

NRC – See Nuclear Regulatory Commission

NTP – National Toxicology Program; it is located in the National Institute of Environmental Health Sciences and provides information to government agencies on the adverse effects of various toxic substances.

Nuclear Regulatory Commission – A government agency responsible for regulating nuclear material use, handling, and storage; it replaces and has assumed the functions of the Atomic Energy Commission since 1975.

Nucleic Acid – A molecule composed of many nucleotides chemically bound together, such as deoxyribonucleic acid (DNA) and ribonucleic acid (RNA).

Nucleotide – Any of the various compounds consisting of the bases (purines and pyrimides), sugar (deoxyribose and ribose), and phosphate group forming the basic constituents of DNA and RNA.

Nucleotide Excision Repair – Like base excision repair, but NER is the more common mechanism used to repair DNA. Its steps involve creating a gap by removal of the damaged DNA, use of the undamaged strand as a template for synthesis of a new segment, and filling the gap by joining the new segment to the existing DNA molecule.

Nucleotide Excision Repair – One of the many cellular mechanisms involved in DNA repair.

Observation Bias – A misidentification of an individual based upon incomplete information, resulting in an individual being placed in the wrong study group; this may lead to identifying an association between the causative agent and the measured effect that may be either weaker or stronger than what is really there.

Occupational Safety and Heath Administration – A government agency involved in evaluating exposure of workers to toxic substances in the workplace.

o-Ethyl o,p-nitrophenyl phenylphosphorothionate – An organophosphate pesticide that is a cholinesterase inhibitor and used for insect control in growing cotton.

Office of Research and Development – An organization within EPA involved in scientific research to support development of environmental policy as it relates to toxic chemicals.

Olestra – A fat substitute approved by the Food and Drug Administration that can be used in certain food types, i.e., manufacturing potato chips; it has been shown to affect normal gastrointestinal functioning in such ways as diarrhea.

Oligodendrocytes – Neuroglial cells with multiple cell processes that form myelin sheaths around axons in the central nervous system.

One-hit Theory – See Single-hit Theory.

Oocyte – A female sex cell.

Opacity – The measure or indicator of how much light can pass through a substance.

Optic Nerve – The nerve-carrying impulses for the sense of sight, formed by the axons of nerve cells located in the retina.

ORD – See Office of Research and Development

Ortho-chlorobenzylidene malononitrile – See CS Gas

OSHA – See Occupational Safety and Heath Administration

Osmosis – The passage of water through a selectively permeable membrane from the area of lesser solute concentration to an area of greater solute concentration.

Osteocyte – A mature bone cell that is surrounded by bone matrix; bone cell.

Osteomalacia – A disease that results from a deficiency of vitamin D or calcium and is characterized by a softening of the bones with accompanying pain and weakness.

Osteoporosis – A bone disorder in which bone and cartilage become dense, hardened, and brittle.

Ovarian Follicle – A general name for an ovum in any stage of development, along with its surrounding group of cells.

Ovary – The female reproductive glands (2) located in the pelvic cavity; produces the oocyte, estrogen, and progesterone.

Ovulation – The release of an oocyte from the mature follicle in the female reproductive system.

Oxidation – The combination of a substance with oxygen; the loss of one or more electrons, or the loss of hydrogen.

Palliative – A substance that seems to relieve or alleviate symptoms without curing.

Pallor – The lack of color especially in the face; paleness.

Pancreas – A gland located near the stomach and small intestine that regulates blood sugar levels.

Pancytopenia – A decrease in the number of all white blood cells.

Papule – A small, circumscribed, solid elevation on the skin.

Parasympathetic Nervous System – Subdivision of the autonomic nervous system; it is involved in involuntary functions such as digestion, defecation, and urination.

Parenteral – The introduction of a substance into the body other than by the digestive system, such as by subcutaneous or intravenous injection.

Paresthesia – A skin sensation, such as burning, prickling, itching, or tingling, with no apparent cause.

Parkinson's Syndrome – A neurological disorder characterized by tremors.

Passive Transport – The movement of molecules, ions, or atoms across the cell membrane without the expenditure of energy; protein carrier molecules may be involved in the transport.

Penicillamine (β,β-Dimethylcysteine) – A chelating agent used in the treatment of lead poisoning.

Perfusion Rate – The amount of blood delivered to tissues or organs over a specific period of time.

Peripheral Nervous System – The part of the nervous system not including the brain and spinal cord.

Peritoneum – The thin connective tissue membrane that covers the organs and the inner wall of the abdominal cavity.

Peritubular Capillary – The capillary network located in the kidneys; associated with the distal and proximal convoluted tubules.

Phagocytosis – The process by which a cell engulfs solid particles; characterized by the formation of cellular extensions called pseudopodia that surround the particle.

Pharyngitis – An inflammation of the pharynx, in acute cases there is a pain in the throat.

Pharynx – The vertebrate alimentary canal that extends down the neck, starting from the posterior region of the nasal cavity and opening into the esophagus.

Phenobarbital, Sodium – A drug used as a sedative or as an anticonvulsant to stop uncontrolled muscle spasms.

Phocomelia – The developmental abnormality characterized by an abnormal shortening of the long bones of the arms and legs.

Phosgene – A colorless gas formed by the reaction of carbon dioxide and chlorine; it is used in the synthesis of isocyanates, polyurethane resins, carbamates and dyes; used also as a chemical warfare agent that causes lung irritation.

Phospholipid – A lipid containing phosphorus, fatty acids, and a nitrogen-containing substance; it is the main component of the lipid bilayer that forms the structural backbone of all cell membranes.

Photon – A quantitative unit of electromagnetic energy with varying wavelengths, generally regarded as a discrete particle having zero rest mass, no electrical charge, and an indefinitely long lifetime.

Photophobia – An unusual intolerance to light.

Pica – An eating disorder characterized by a craving to ingest any material not fit for food, including starch, clay, crayons, grass, metals, or plaster.

Picrotoxin – A toxic substance obtained from the seed of a shrub, *Anamirta cocculas*, which can produce uncontrolled muscle spasms.

Pinocytosis – The engulfing and absorption of droplets of liquids by cells.

Pituitary Gland – The small, gray, rounded gland located at the base of the brain; it secretes a number of chemical substances (hormones) that regulate body processes such as growth, reproduction, and metabolism.

Plasma – The fluid portion of blood, lymph, or milk as distinguished from suspended material; blood minus the formed elements (i.e., blood cells).

Plasma Cell Membrane – The selectively permeable outer layer of the cell; it is the barrier through which all substances entering or leaving the cell must pass; it controls the rate of absorption.

Platelet – The minute fragments of cells derived from megakaryocytes; they play an important role in blood clotting.

Pleural – Referring to the thin membrane (pleura) that covers the outside of lungs and the inner wall of the chest.

Plicae Circularis – A permanent, deep, transverse fold in the mucosa and submucosa of the small intestine that increases the surface area for absorption.

Pneumoconiosis – A condition of the respiratory tract due to inhalation of dust particles.

Pneumonitis – An inflammation of the lungs.

PNS – See Peripheral Nervous System

Point Mutation – A change in a single base in the DNA structure.

Polyneuritis – An inflammation of several nerves at one time, marked by paralysis, pain, and muscle atrophy.

Polyneuropathy – A generalized disorder of peripheral nerves.

Postsynaptic Membrane – A cell membrane of the effector tissue to which the neurotransmitter substance from the presynaptic terminal binds.

Post-traumatic Stress Disorder – The development of characteristic symptoms after a psychologically traumatic event that is generally outside the range of usual human experience; symptoms include re-experiencing the traumatic event, insomnia; recurrent nightmares.

Potentiation – A condition whereby one substance is made more potent in the presence of another substance that alone produces no adverse effect.

Preening – To clean, primp, or in the case of birds to dress or trim feathers.

Presynaptic Terminal – The end of an axon that releases a neurotransmitter, the latter binding to the postsynaptic membrane.

Prevalence Rate – A measure of the number of individuals who have a particular illness or disease at a specific point in time; it is a snapshot in time.

Primary Bronchi – The two primary divisions of the trachea that lead into the right and left lungs.

Primary Oocyte – The oocyte that is present at birth and that has begun the meiotic division.

Primary Spermatocyte – A male stem germ cell (spermatogonia) that has undergone mitotic division producing a pair of haploid cells that give rise in turn to spermatids.

Primordial Follicle – The primary oocyte together with the granulosa cells.

Process – A projection or outgrowth of bone or tissue.

Progesterone – A hormone secreted primarily by the corpus luteum; aids in growth and development of female reproductive organs.

Progressor Agent – A chemical or physical substance that causes the transition of cells from the promotion stage to the progression stage of cell growth.

Prospective Study – An epidemiological study that examines the development of disease in a group of persons determined to be free of the disease at that moment; a forward-looking study.

Prostate Gland – A gland that surrounds the beginning of the urethra in the male; secretion of the gland is a milky fluid that comprises part of the semen.

Prostration – An extreme weakness or exhaustion.

Protein – A molecule consisting of one or more polypeptide chains; each chain is composed of many amino acids.

Proteinurea – An excessive amount of protein in urine.

Prothrombin – A plasma protein needed for blood clotting.

Proximal Convoluted Tubule – The convoluted portion of the nephron that extends from Bowman's capsule to the descending limb of Henle's Loop.

Pseudopodia – A temporary infolding or extension of the cell membrane; usually associated with phagocytosis (formation of a vacuole to engulf a foreign particle) or in cell locomotion.

PSNS – See Parasympathetic Nervous System

Psychoneurosis – A mental or behavioral disorder of mild or moderate severity; it leads to disturbances in thought, feelings, and attitudes.

Pulmonary Artery – The artery that carries blood from the right ventricle to the lungs.

Pulmonary Circulation – The blood flow through the system of blood vessels that carry blood from the right ventricle of the heart to the lungs and back from the lungs to the left atrium.

Pulmonary Edema – An excessive accumulation of fluid in the lungs.

Pulmonary Vein – The blood vessels that collect oxygenated blood from the lungs and that deliver it to the left atria.

Pupillary Light Reflex – A constriction of the pupil upon stimulation of the retina by light.

Purkinje Fiber – A specialized cardiac muscle fiber that conducts action potentials through cardiac muscle; part of the conduction system of the heart.

Pustule – A small inflamed skin swelling that is filled with pus; a small swelling similar to a blister or pimple.

Quinine – A chemical used in the treatment of malaria.

Rad – A unit of energy absorbed from ionizing radiation per gram of irradiated material.

Radiation – The transfer of electromagnetic waves (radiant energy) from a source, across a media, and absorbed by a body.

Radioactive – A substance capable of giving off atomic emissions in the form of particles or rays.

Radioactivity – The spontaneous release of atomic emissions (alpha particles, beta particles, gamma rays, protons, etc.), either directly from an unstable atomic nuclei or as a consequence of a nuclear reaction.

Radionuclide – A nuclide (atom) that is of either artificial or natural origin and exhibits radioactivity.

RCRA – See Resource Conservation and Recovery Act

Receptor Antagonism – When two chemicals compete for the same target tissue receptor and the resulting observed effect is less than the sum of the separate effects for each chemical; chemicals that are receptor antagonists are commonly referred to as blockers.

Reduction – The gaining of electrons by a molecule; a reaction in which hydrogen is combined with or oxygen is removed from a compound.

Reference Concentration/Dose – Estimates of a daily exposure to an agent by inhalation (RfC) or by skin contact or ingestion (RfD) that is assumed to be without an adverse health impact on the human population.

Registry of Toxic Effects of Chemical Substances – A database for chemical information.

Regulatory Toxicologist – Toxicologist involved in establishing "safe" exposure levels for toxic substances based on the evaluation of toxicological data.

Relative Risk – An estimate of the probability that a given individual will develop a particular effect when exposed to a specific etiological factor compared to a nonexposed individual.

Rem – The amount of ionizing radiation required to produce the same biological effect as one rad of high-penetration x-ray.

Renal Agenesis/Dysgenesis – The lack of formation of the kidneys or imperfect development of some portion of the kidneys.

Renal Corpuscle – The structure composed of a Bowman's capsule and its glomerulus.

Renal Pelvis – Large funnel-like structure located in the medulla of the kidneys that collects urine from the collecting ducts, which then passes out of the kidneys through the ureter.

Renal Pyramid – The cone-shaped structure extends from the renal sinus into the cortex of the kidneys.

Renin – An enzyme secreted by the kidneys that interacts with angiotensin.

Resource Conservation and Recovery Act – A federal act that identifies and regulates hazardous waste handling, storage, treatment, and disposal.

Resting Membrane Potential – The charge difference across the membrane of a resting cell, which is a cell that has not been stimulated to produce an action potential.

Reticulocyte – A red blood cell that retains remnants of the endoplasmic reticulum.

Retina – The innermost coat of the posterior part of the eyeball that receives the image produced by the lens, is continuous with the optic nerve, and consists of several layers, one of which contains the rods and cones that are light-sensitive.

Retinoblastoma – A cancer associated with the retina of the eye.

Retrospective Study – An epidemiological study that compares diseased individuals with non-diseased individuals and looks back in time to determine exposures.

RfC – See Reference Concentration/Dose

RfD – See Reference Concentration/Dose

Rhinitis – An inflammation of the lining of the nasal cavity; in acute cases it increases the secretion of mucus.

Ribonucleic Acid – A type of nucleic acid containing the sugar ribose; involved in protein synthesis.

Ribosome – An intracellular structure associated with protein synthesis.

Risk – The probability or chance that an adverse effect will occur.

Risk Assessment – The process of gathering all available information on toxic effects of a chemical and evaluating it to determine the possible risk.

Risk Management – The process of integrating the results of the risk assessment with economic, social, and legal issues to arrive at an acceptable course of action.

RNA – See Ribonucleic Acid

RTECS – See Registry of Toxic Effects of Chemical Substances

SA – See Sinoatrial Node

Saccharin – A sweet, white, powdered synthetic product derived from coal tar; 300 times sweeter to the taste than sucrose.

Safe Drinking Water Act – A federal act that establishes standards for contaminants in drinking water; regulates discharges to underground injection wells, sole source aquifers, and public drinking water systems.

Sarcoma – A malignant tumor that originates in mesodermal tissue such as bone or muscle; for example, a malignant tumor of the bone is referred to as an osteosarcoma.

Sarin – An extremely toxic nerve gas composed of the isopropyl ester of methylphosphonofloride acid.

Saturnine Gout – A disease with gout-like symptoms produced by lead poisoning.

SBS – See Sick Building Syndrome

Schwann Cell – A neuroglial cell forming myelin sheaths around axons in the peripheral nervous system.

Sclera – The white coat of fibrous tissue that forms the outer protective covering over the eyeball except in the area of the cornea.

SDWA – See Safe Drinking Water Act

Sebaceous Gland – A gland in the dermis of the skin – almost always associated with a hair follicle – that secretes sebum (oil).

Secondary Spermatocytes – The primary spermatocytes that undergo meiosis.

Secretion – The passage of material formed by a cell from the inside to the outside of its plasma membrane.

Selection Bias – Occurs when individuals assigned to a group are not representative of the group being studied; therefore, there may be a weaker or stronger association identified than actually exists.

Selectively Permeable – The property of cell membranes that selectively allows certain molecules to pass while others cannot.

Seminal Vesicle – One of two glandular structures whose secretion is a component of semen.

Seminiferous Tubule – The tubule in the testes in which sperm cells develop.

Sensitization – A condition of being made sensitive to a specific substance such as protein or pollen, usually as a result of repeated exposure.

Serosa – A thin membrane of connective tissue lining the outer surface of organs.

Sertoli Cell – A cell in the wall of the seminiferous tubules to which spermatogonia and spermatids are attached.

Serum Hepatitis – A term previously used for what is now known to be caused by either hepatitis B or hepatitis C viruses.

Sevin – An insecticide (1-naphthyl-N-methyl carbamate) used to control fleas on cats and dogs; it is also used in mosquito control programs.

Sharps – The medical articles that may cause punctures or cuts; includes all broken medical glassware, syringes, needles, scalpel blades, suture needles, and disposal razors.

Sick Building Syndrome – An illness caused by the emission of odors, vapors, or gases from the materials used to construct buildings or its furnishings; symptoms include irritation of eyes, throat, and nose; dry skin; dizziness and nausea; lack of concentration and fatigue.

SIDS – See Sudden Infant Death Syndrome

Sievert – An international unit that is used for measuring the biologic dose of ionizing radiation. Like the rem, the sievert takes into account the relative biological effectiveness (RBE) of each form of ionizing radiation on living tissue. A sievert is roughly equivalent in RBE to 1 gray (Gy) or 100 rads of gamma radiation. One millisievert (mSv) equals 10 ergs of gamma radiation energy transferred to 1 gram of living tissue.

Silicosis – A lung disease caused by inhalation of dust-containing silicon dioxide; damage to the alveoli results in cell death; dead alveoli are replaced with connective tissue in localized area (nodules).

Simple Asphyxiant – A simple asphyxiant is a substance – particularly a gas – that results in a decrease in the amount of oxygen supplied to the body; asphyxia occurs because oxygen is displaced by the asphyxiant resulting in its decreased availability.

Single-hit Theory – The theory that describes a worst-case scenario in which a single unrepaired event involving a critical portion of the genome leads to a malignancy.

Sinoatrial Node – A mass of specialized cardiac muscle fibers located in the right atrium; acts as the "pacemaker" of the cardiac conduction system.

Sinusitis – An inflammation of the membrane that lines the nasal sinus.

Sleeping Sickness – An infectious disease characterized by drowsiness, lethargy, muscular weakness, and tremors.

SNS – See Sympathetic Nervous System

Sodium-potassium Pump – An active transport mechanism associated with the cell membrane that is responsible for maintaining sodium and potassium concentrations inside and outside the cell.

Soma – The body of a neuron containing the nucleus and other cellular structures such as the endoplasmic reticulum, ribosomes, etc.

Soman – An extremely toxic nerve gas composed of methylphosphonofluoridic acid-1,2,2-trimethylpropylester.

Somatic – It refers to those cells that make up tissues, organs, and parts of the individual, other than germ cells.

Somatic Cell Mutation Theory – The theory that describes how cancer may be caused as a result of damage to cellular DNA.

Somatic Motor – A type of efferent motor neuron of the peripheral nervous system that innervates skeletal muscle.

Somatic Mutation – Any of the mutations that occur in cells of the body, except the germinal (sex) cells of the reproductive system.

Spermatids – The secondary spermatocytes that undergo meiosis.

Spermatogenesis – The formation and development of sperm cells.

Spermatogonia – The germ cells in the seminiferous tubules scattered between Sertoli cells that divide and form primary spermatocytes.

Spermatozoan – The male sex cell; sperm cell.

Spina Bifida – An abnormality caused by incomplete development and joining of vertebrae in the spinal column resulting in the spinal cord protruding through this area.

Spleen – A large lymphatic organ located in the upper left side (between the stomach and diaphragm) of the abdominal cavity; serves as a reservoir for blood; plays a role in the destruction of worn-out red blood cells and the formation of mature lymphocytes, which respond to foreign substances in the blood.

Spontaneous Mutation – A mutation that arises naturally and not as the result of exposure to an exogenous mutagenic agent.

Squamous Cell Carcinoma – A carcinoma that arises from the squamous epithelium, which is the most common form of skin cancer.

Staphylococcus aureus – A bacterium commonly present on skin and mucous membranes; some strains produce toxins that cause food poisoning and toxic shock syndrome.

Stem Cell – A non-specialized cell capable of differentiating into a specialized cell; in the case of blood cells the stem cell can differentiate into a number of different types of specialized cells.

Stereoisomer – A molecule containing the same number and kind of atoms as another, but differing in spatial arrangement and, therefore, having differing properties.

Stomatitis – An inflammation of the mucous membrane of the mouth.

Stratum Corneum – The outermost layer of epidermal cells; these cells are dead and are continually shed and replaced; they contain the protein keratin, making the epidermis impermeable to most substances.

Strychnine Sulphate – A toxic substance that can cause uncontrolled contraction of muscles and subsequent death.

Subchronic – Of toxicity studies with a duration of 5-90 days.

Subcutaneous – Of the fatty tissue that lies under the skin.

Submucosa – A layer of connective tissue located beneath the mucosa and the muscularis layer in the gastrointestinal tract; it binds the mucosa to the muscularis layer.

Sudden Infant Death Syndrome – Death of an apparently healthy infant usually before one year of age that is of unknown cause and occurs especially during sleep; also called crib death.

Sulfanilamide – A drug used in the treatment of various bacterial diseases.

Sulfation – A Phase II metabolic reaction in which a sulphate group is reacted with a toxic molecule; the end product is highly water soluble.

Sv – See Sievert

Sympathetic Nervous System – A subdivision of the autonomic nervous system; it is generally involved in preparing the body for immediate physical activity.

Synapse – The junction between a nerve cell and any effector tissue that could be another nerve cell, a muscle cell, or a gland cell.

Synaptic Cleft – The space between the presynaptic and the postsynaptic terminal.

Synergistic – The interaction between two or more chemicals that results in an effect greater than the sum of their individual effects.

Syphilis – A chronic contagious disease caused by a spirochete, *Treponema pallidum*; if left untreated it produces chancres, rashes, and systemic lesions. It is usually a venereal disease, but may involve any organ or tissue.

Syrup of Ipecac – Made from the dried root of *Cephaelis ipecacuanha*, it contains emetine, a toxic alkaloid; it is sold as an over-the-counter syrup used to induce vomiting in response to ingestion of toxic substances.

Systemic Circulation – The flow of blood through the system of blood vessels that carries it from the left ventricle to the tissues of the body and back from the body to the right atrium.

Tabes Dorsalis – A disease of the spinal cord associated with syphilis; symptoms include staggering walk, pain, and paresthesias.

Tabun – An extremely toxic nerve gas containing the ethyl ester of dimethylphysphoramidocyanidic acid.

Tachycardia – An excessively increased heartbeat.

Tachypnea – An excessively rapid respiration rate.

Target Tissue – The tissue in which a toxic substance exerts its effects.

TBS – See Tight Building Syndrome

TEPP – See Tetraethyl Pyrophosphate

Teratogen – A chemical or physical agent that causes abnormal development in the fetus resulting in the appearance of birth defects in newborn children.

Testis – One of two male reproductive glands; it produces testosterone and sperm cells.

Testosterone – A steroid hormone secreted primarily by the testes, it aids in spermatogenesis and controls maintenance and development of male reproductive organs.

Tetraethyl Pyrophosphate – An organophosphate pesticide used for aphids, mites, and as a rodenticide.

Tetrodotoxin – A nerve poison found in eggs of the California newt and certain puffer fish; in concentrated form it is more toxic than cyanide.

Thalidomide – A drug initially used to treat nausea associated with pregnancy; it has been shown to cause severe birth defects.

Threshold – The lowest dose of a chemical at which a specific measurable effect is observed and below which it is not observed.

Threshold Limit Value – A set of standards established by the American Conference of Governmental Industrial Hygienists for concentrations of airborne substances in the workplace; they are time-weighted averages based on conditions to which workers may be repeatedly exposed day after day without adverse effects.

Thrombocyte – A cell fragment involved in blood clot formation.

Thrombocytopenia – A decrease in the number of thrombocytes.

Thymus – A glandular structure of largely lymphoid tissue that functions especially in the development of the body's immune system (T cells). It is present in the young of most vertebrates typically in the upper anterior chest or at the base of the neck, and tends to atrophy in the adult.

Thyroid – A gland located on both sides of the neck just below the voice box; involved in regulating normal body metabolism.

Tight Building Syndrome – An illness caused by reduced ventilation in well insulated buildings; symptoms similar to those associated with sick building syndrome (SBS).

TLV – See Threshold Limit Value

TOCP – See Tri-Ortho-Cresyl Phosphate

Tonic – Of a prolonged muscular contraction; suited to produce a healthy muscular condition.

Total Leucocyte Count – A test to determine the number of white blood cells present in the circulatory system; it is calculated by counting the number of white blood cells in a milliliter of blood.

Toxaphene – A waxy white chlorinated camphene used as an organochlorine pesticide.

Toxic – A substance that produces or causes an adverse effect on an organism.

Toxic Shock Syndrome – A rare and sometimes fatal disease caused by a toxin or toxins produced by certain strains of the bacterium *Staphylococcus aureus*.

Toxic Substances Control Act – A federal act that requires testing and reporting of chemicals prior to manufacture, distribution, and use; restricts the use of chemicals that pose a threat to human health and the environment.

Toxicin – The specific agent in a toxic substance that causes the observed adverse effect.

Toxicological Study – The controlled toxicological investigation on animals designed to establish a direct cause-effect relationship.

Toxicology – The study of the adverse effects of chemicals on living organisms.

Toxicology Data Network – A chemical database maintained by the National Library of Medicine.

TOXNET – See Toxicology Data Network

Toxoplasmosis – A disease due to infection with the protozoa Toxoplasma gondii; symptoms include muscle pain, fever, and – in severe cases – pneumonitis, hepatitis, and encephalitis.

Trachea – A tubular air passageway located in front of the esophagus and extending from the larynx to the primary bronchi.

Transfer RNA – A type of ribonucleic acid, attached to a single amino acid, which transcribes the nucleic acid sequence on the mRNA resulting in protein synthesis.

Transferrin – A beta globulin in blood plasma capable of transporting dietary iron to the liver, spleen, and bone marrow.

Transitional Point Mutation – A change in a single base such a cytosine for guanine, a thymine for adenine, or vice versa.

Transport – See Facilitated Diffusion

Transversion Point Mutation – A replacement of one of the purine bases (adenine or guanine) by one of the pyrimidine bases (thymine, cytosine, or uracil), for example adenine replaced by cytosine, guanine by thymine, etc.

Treponema pallium – A bacterium that causes syphilis in humans.

Tri-Ortho-Cresyl Phosphate – Used as plasticizer for polyvinyl rubber, vinyl plastics, and polystyrene, and targets nervous system tissue.

tRNA – See Transfer RNA

True Skin – See Dermis

TSCA – See Toxic Substances Control Act

TSS – See Toxic Shock Syndrome

Tumor – An abnormal mass of tissue, the growth of which exceeds and is uncoordinated with that of normal tissue; the two types are benign and malignant.

Turner's Syndrome – A congenital hormone disorder caused by failure of the ovaries to respond to pituitary hormone stimulation; individuals are short in stature and do not attain sexual maturation.

Typhus – Any of a group of acute infectious diseases characterized by severe headache, a rash, and sustained high fever.

Ulcer – An injury of the skin or of a mucous membrane – such as the one lining the stomach – that is accompanied by formation of pus and necrosis of surrounding tissue.

Ultraviolet Radiation – Electromagnetic radiation with a wavelength between 10 and 390 nm; a nonionizing form of radiation used for microscopy and hospital air sterilization.

Unmyelinated Nerve – Axons that simply lie within an indentation of an oligodendrocyte cell in the CNS or a Schwann cell in the PNS.

Ureter – Tube conducting urine from the kidneys to the urinary bladder.

Urethra – A duct leading from the bladder that discharges urine externally.

Uric Acid – A crystalline acid occurring as an end product of purine metabolism; normally excreted by the body; decreased excretion is associated with kidney disease and lead poisoning.

Uterus – A hollow muscular organ in which the fertilized oocyte develops into a fetus.

UV – See Ultraviolet Radiation

Vagus Nerve – The tenth cranial nerve and part of the parasympathetic nervous system. It is involved in regulating heart rate and its stimulation slows the rate.

Ventricle – A cavity of a bodily part or organ, such as the chamber of the heart, which receives blood from a corresponding atrium and from which blood is forced into the arteries.

Ventricular Fibrillation – A fast, irregular contraction of the ventricles.

Venule – A small vein that collects blood from the capillaries; several venules unite to form larger veins.

Vertigo – A sensation of irregular or whirling motion, either of oneself or of external objects.

Vesicant – An agent that induces blistering, also called blister gas (i.e. mustard gas and lewisite).

Vesicle – A small sac inside a cell containing liquid or gas.

Villus (Villi, pl.) – A minute, finger-like projection from the surface of the small intestine mucosa, into the lumen of the intestine, that increases the surface area for absorption.

Vital Capacity – The volume of air that can be expelled from the lungs following full capacity.

Vitreous Humor – A transparent, jelly-like substance located between the lens and the retina of the eyeball.

Wristdrop – A condition in which the hand is flexed at the wrist and cannot be extended; due to injury to the nerve or paralysis of extensor muscles of wrist and hand.

Xeroderma Pigmentosum – A disease characterized by extreme sensitivity to sunlight and increased susceptibility of individuals to skin cancer; it is associated with mutation of the DNA molecule caused by UV radiation.

X-ray – Extremely high energy, short wavelength (0.006-12 nm), electromagnetic radiation emitted as the result of the electron transitions in the inner orbits of heavy atoms bombarded by cathode rays in a vacuum tube.

Yellow Fever – An acute infectious disease characterized by jaundice, vomiting, and hemorrhaging.

Yersinia pestis – A bacterium responsible for causing bubonic plague.

Bibliography

Abelson, P.H., "Testing for Carcinogens with Rodents," Science 249:1357, 1990.

Agency for Toxic Substances and Disease Registry, "Toxicology Profile for Chromium," USPHS, ATSDR/TP-88/10.

Agency for Toxic Substances and Disease Registry, "Toxicological Profile for Nickel," U.S. Environmental Protection Agency, October 1987.

Aldridge, W.N., Bares, J.M. et al., "Studies on Delayed Neurotoxicity Produced by Some Organophosphorus Compounds," Annals of the New York Academy of Sciences 160:314, 1969.

Amdur, M.O. "The Response of Guinea Pigs to Inhalation of Formaldehyde and Formic Acid along with a Sodium Chloride Aerosol," Journal of Air Pollution 3:201-220, 1960.

Amdur, M.O. "Aerosols Formed by Oxidation of Sulfur Dioxide: Review of Their Toxicology," Archives of Environmental Health 23:459-468, 1971.

Ames, B.N., Shigenaga, M.K. et al., "DNA Lesions, Inducible DNA Repair, and Cell Division: Three Key Factors In Mutagenesis and Carcinogenesis," Environmental Health Perspectives 93:35-44, 1993.

Anders, M.C. editor, *Bioactivation of Foreign Compounds*, New York, Academic Press, 1985.

Antila, A., Sallmén, M. et al., "Carcinogenic Chemicals in the Occupational Environment," Pharmacology and Toxicology 72:69-76, 1993.

Araki, S., Honma, T., "Relationships Between Lead Absorption and Peripheral Nerve Conduction Velocities in Lead Workers," The British Journal of Industrial Medicine 39:157-160, 1982.

Ariens, E.J., Simonis, A.M. et al., *Introduction to General Toxicology*, San Francisco, Academic Press, 1976.

Astrand, I., "Uptake of Solvents from the Lungs," The British Journal of Industrial Medicine, 42:217-218, 1985.

Auerbach, C., "The Chemical Production of Mutations," Science 158:1141, 1967.

Bakalkin, G. et al., "P53 Binds Single-stranded DNA Ends and Catalyzes DNA Renaturation and Strand Transfer," Proceedings of the National Academy of Sciences, USA 91:413, 1994.

Baker, E., Fine, L., "Solvent Neurotoxicity, the Current Evidence," Journal of Occupational Medicine 28:126-129, 1986.

Balls, M., Riddell, R.J. et al. editors, *Animals and Alternatives in Toxicity Testing*, San Diego, Academic Press, 1983.

Banaszak, E.F., Fink, J.J. et al., "Hypersensitivity Pneumonitis Due to Contamination of an Air Conditioner," New England Journal of Medicine 283:271-274, 1970.

Bell, A., "Toxicity of Lead at Low Dose," The British Journal of Industrial Medicine 47:288, 1990.

Bennet, J.V., Brachman, P.S. editors, *Hospital Infections*, 3rd edition, Boston, Little, Brown and Company, 1992.

Beral, V., Robinson, N. et al., "The Relationship of Malignant Melanoma, Basal and Squamous Skin Cancers to Indoor and Outdoor Work," The British Journal of Cancer 44:886, 1981.

Bigalow, J.E., "The Effects of Ionizing Radiation on Mammalian Cells," Chemical Education 58:144-156, 1981.

Blair, A., Steward, P.A. et al., "Mortality for Lung Cancer among Workers Employed in Formaldehyde Industries," American Journal of Industrial Medicine 17(6):683-699, 1990.

Blank, I.H., Schereplein, R.J. et al., "Transport into and Within the Skin," The British Journal of Dermatology 81 (suppl 4): 4-10, 1969.

Blot, W.J., "Alcohol and Cancer," Cancer Research 52:2119s-2123s, 1992.

Bogen, K.T., "Cancer Potencies of Heterocyclic Amines Found in Cooked Foods," Food and Chemical Toxicology 32:505-515, 1994.

Bogerd, W., *The Bhopal Tragedy: Language, Logic and Politics in the Production of a Hazard*, Boulder, CO, Westview Press, 1989.

Brash, D.E., Haseltine, W.A. et al., "UV-induced Mutation Hotspots Occur at DNA Damage Hotspots," Nature 298:189, 1982.

Brent, R.L., Beckman, D.A. et al., *Principles of Teratology*, In Evans, M.I. editor, *Reproductive Risks and Prenatal Diagnosis*, Norwalk, CT, Appleton and Lange, 1992.

Brotons, J.A., Olea-Seranno, M.F. et al., "1995 Xenoestrogens Released from Lacquer Coatings in Food Cans," Environmental Health Perspective 103: 608-612.

Brown, D.P., Kaplan, S.D. et al., "Retrospective Cohort Mortality Study of Dry Cleaner Workers Using Perchloroethylene," Journal of Occupational Medicine 29:535-541, 1987.

Bullivant, C.M., "Accidental Poisoning by Paraquat: Report of Two Cases in Man," British Medical Journal 1:1272, 1966.

Burrows, C. editor, *Chromium: Metabolism and Toxicity*, Boca Raton, FL, CRC Press, 1983.

Calabrese, E.J., *Multiple Chemical Interactions*, Lewis Publishers, Michigan, 1991.

Camner, P., Clarkson, T.W. et al., *Routes of Exposure, Dose and Metabolism of Metals* in Friberg, L., et al. editors, *Handbook on the Toxicology of Metals*, New York, Elsevier/North-Holland and Biomedical Press, 1979.

Carson, R., *Silent Spring*, Boston, Houghton Press, 1962.

Casarett and Doull's, Toxicology: the Basic Science of Poisons, 5th edition, Klaasen, C.D., Amdur, M.D. et al. editors, New York, MacMillan Publishing Company, 1996.

Cavanaugh, J.B., "Long Term Persistence of Mercury in the Brain," The British Journal of Industrial Medicine 45(10):649-651, 1990.

Cavanaugh, J.B., "Peripheral Neuropathy Caused by Chemical Agents," Critical Reviews of Toxicology 2:365, 1973.

Chissick, S.S., Vevricott, R. editors, *Asbestos: Properties, Applications and Hazards*, Vol 2, New York, John Wiley and Son, 1983.

Clarkson, T.W., "Recent Advances in the Toxicology of Mercury with Emphasis on the Alkylmercurials," CRC Critical Reviews of Toxicology 1:203-234, 1972.

Clarkson, T.W., "Metal Toxicity in the Central Nervous System," Environmental Health Perspectives 75:59-64, 1987.

Clayman, C. editor, *The Human Body: An Illustrated Guide to Its Structure, Function and Disorders*, New York, Dorling Kindersley, 1995.

Clayton, G.D., Clayton, F.E., *Patty's Industrial Hygiene and Toxicology*, Volume II, Parts A, B, C, John Wiley and Son, Inc., New York.

Coburn, R.F. editor, "Biological Effects of Carbon Monoxide," Annals of the New York Academy of Sciences, USA 174:1-40, 1970.

Cohen, G.M., *Target Organ Toxicity*, Vol. I, Boca Raton, Florida, CRC Press, Inc., 1986.

Cohen, S.M., Ellwein, L.B., "Genetic Errors, Cell Proliferation, and Carcinogenesis," Cancer Research 51:6493-6505, 1991.

Steven Rose editor, *Conference on Chemical and Biological Warfare*, Boston, Beacon Press, 1968.

Corbett, T., "Cancer and Congenital Anomalies Associated with Anesthetics," Annals of the New York Academy of Sciences, USA 271:58-66, 1976.

Corcoran, G.B., Fix, L. et al., "Apoptosis: Molecular Control Point in Toxicity," Toxicology and Applied Pharmacology 128:169-181, 1994.

Craighead, J.E., *Pathology of Environmental and Occupational Disease*, St. Louis., Mosby, 1995.

Cremer, J.E., "Toxicology and Biochemistry of Alkyl Lead Compounds," Occupational Health Reviews 17:14, 1965.

Cullen, M.R., Cherniak, M.G. et al., "Occupational Medicine," New England Journal of Medicine 322:675-682, 1990.

Curran, W.J., "Dangers for Pregnant Women in the Workplace," New England Journal of Medicine 312:164-165, 1985.

Curren, R.D., Yang, L.L. et al., "Genotoxicity of Formaldehyde in Cultured Human Bronchial Fibroblasts," Science, pp. 89-90, 1985.

Dean, B.J., "Genetic Toxicology of Benzene, Toluene, Xylenes, and Phenols," Mutation Research 47:75-97, 1978.

DeFlora, S., Serra, D. et al., "Mechanistic Aspects of Chromium Carcinogenicity," Archives of Toxicology (suppl 13):28-39, 1989.

Dekant, W. et al., "Glutathione-dependent Bioactivation of Zenobiotics," Xenobiotica 23:873-887, 1993.

Dekant, W., Vamvakas, S. et al., "Bioactivation of Haloalkenes by Glutathione Conjugation: Formation of Toxic and Mutagenic Intermediates by Cysteine Conjugate β-lyase," Drug Metabolism Reviews 20:43-89, 1989.

Dekant, W., Neumann, H.C. et al., *Tissue-Specific Toxicity: Biochemical Mechanisms*, New York, Academic Press, 1992.

Derban, L.K.A., "Poisoning Due to Alkylmercury Fungicide," Archives of Environmental Health 28:49, 1974.

Derbyshire, M.K. et al., "Nonhomologous Recombination in Human Cells," Molecular and Cellular Biology 14:156, 1994.

DeRobertis, E.D.P., Nowinski, W.W. et al., *Cell Biology*, 8th edition, Philadelphia, W. B. Saunders, 1987.

DeSesso, J.M., Goeringer, G.C., "Ethoxyquin and Nordihydroguaiaretic Acid Reduce Hydroxyurea Developmental Toxicity," Reproductive Toxicology 4:267, 1990.

Doll, R., Morgan, L.G. et al., "Cancer of the Lung and Nasal Sinuses in Nickel Workers," The British Journal of Industrial Medicine 24:623-632, 1970.

Doll, R., "Cancer of the Lung and Nose in Nickel Workers," The British Journal of Industrial Medicine 15:217-223, 1958.

Doll, R., Mathews, J.D. et al., "Cancers of the Lung and Nasal Sinuses in Nickel Workers: A Reassessment of the Period of Risk," The British Journal of Industrial Medicine 34:102-105, 1977.

Donovan, A., Wood, R.D. et al., "Identical Defects in DNA Repair in Xeroderma Pigmentosum Group G and Rodent ERCC Group 5," Nature 363:185, 1993.

Dorian, C., Gattone, V.H. II et al., "Renal Cadmium Deposition and Injury as a Result of Accumulation of Cadmium-metallothionein (CdMT) by Proximal Convoluted Tubules – A Light Microscope Autoradiographic Study with ^{109}CdMT," Toxicology and Applied Pharmacology 114:173-181, 1992.

Dreisbach, R.H., Robertson, W.D. et al., *Handbook of Poisoning*, 12th edition, Appleton & Lange, Norwalk, CT, 1987.

Durante, M. et al., "Non-random Alkylation of DNA Sequences Induced in Vivo by Chemical Mutagens," Carcinogenesis 10:1357, 1989.

Eastman, A., Barry, M.A., "The Origins of DNA Breaks: A Consequence of DNA Damage, DNA Repair, or Apoptosis?" Cancer Investigations 10:229-240, 1992.

Ecobiochen, D.J. et al., "Pharmacodynamic Study of DDT in Cockerels," Canadian Journal of Physiology and Pharmacology 46:785-794, 1968.

Ecobiochen, D.J., *The Basis of Toxicity Testing*, Ann Arbor, CRC Press, 1992.

Elinder, G.G., "Zinc" In Frigers, L., et al. editors, *Handbook on the Toxicology of Metals*, 2nd edition, Elsevier Science Publishers, Amsterdam, 1986.

Ellis, K.J., Yuen, K. et al., "Dose-response Analysis of Cadmium in Man: Body Burden vs. Kidney Dysfunction," Environmental Research 33(1):216-226, 1984.

Encyclopedia of Occupational Health and Safety, 3rd edition, L. Parmeggiani editor, Vol I and II, International Labour Office.

Enterline, P.E., Henderson, V.L. et al., "Exposure to Arsenic and Respiratory Cancer: A Reanalysis," American Journal of Epidemiology 125:929-938, 1987.

Environmental Protection Agency, *Toxicology Handbook*, Government Institutes, Inc., Rockville, MD, 1986.

Environmental Risk Analysis for Chemicals, Conway, R. A. editor, New York, Van Nostrand Reinhold, 1982.

Elements of Toxicology and Chemical Risk Assessment, Environs, Washington, D.C., Environ Corporation, 1986.

Evans, A.S., Brachman, P.S., *Bacterial Infections of Humans, Epidemiology and Control*, 2nd edition, New York and London, Plenum Medical Book Company, 1991.

Evans, A.S., *Viral Infections of Humans, Epidemiology and Control*, 3rd edition, New York and London, Plenum Medical Book Company, 1989.

Fabro, S., McLachlan, J.A. et al., "Chemical Exposure of Embryos During the Preimplantation Stages of Pregnancy: Mortality Rate and Intrauterine Development," American Journal of Obstetrics and Gynecology 148:929, 1984

Falwell, J.K., Hunt, S. et al., *Environmental Toxicology, Organic Pollutants*, John Wiley & Sons, New York, 1988.

Ferm, V.H., Carpenter, S.J., "Teratogenic Effect of Cadmium and its Inhibition by Zinc," Nature 216:1123, 1967.

Fidler, A., Baker, E. et al., "Neurobehavioral Effects of Occupational Exposure to Organic Solvents among Construction Painters," The British Journal of Industrial Medicine 44:292-308, 1987.

Field, B., Kerr, C., "Herbicide Use and Incidence of Neural-tube Defects," Lancet 1342-1342, 1979.

Filov, V.A., Bandman, A.L. et al. editors, *Harmful Chemical Substances*, Vol I, New York, Ellis Harwood, 1993.

Finnegan, M.J., Pickering, C.A.C. et al., "The Sick Building Syndrome: Prevalence Studies," British Medical Journal 289:1573-1575, 1984.

Fishburn, C., Zenz, C., "Metal Fume Fever," Journal of Occupational Medicine 11:142-144, 1969.

Franzblau, A., Lilis, R., "Acute Arsenic Intoxication from Environmental Arsenic Exposure," Archives of Environmental Health 44(6):385-390, 1989.

Freedman, L.S., Clifford, C. et al., "Analysis of Dietary Fat, Calories, Body Weight, and the Development of Mammary Tumors in Rats and Mice: A Review," Cancer Research 50:5710-5719, 1990.

Friberg, L., Nordberg, G.F. et al. editors, *Handbook of Toxicology of Metals*, Vol I, II ed 2, Amsterdam, Elsevier, 1986.

Friedberg, E.C., "Xeroderma Pigmentosum, Cockayne's Syndrome, Helicases, and DNA Repair: What's the Relationship?" Cell 71:887, 1992.

Friedman, G.D., *Primer of Epidemiology*, 4th edition, McGraw-Hill, Inc., New York, 1994.

Fristedt, B., Sterner, N. et al., "Warfarin Intoxication from Cutaneous Absorption," Archives of Environmental Health 11:205, 1965.

Frobisher, M., Fuerst, R. et al., *Microbiology in Health and Disease*, 15th edition, Philadelphia, W.B. Saunders, 1983.

Frost, D.V., "Arsenicals in Biology – Retrospect and Prospect," Federation Proceedings 26:194-208, 1967.

Fullerton, P.M., Kermer, M., "Neuropathy after Intake of Thalidomide," British Medical Journal 2:855-858, 1961.

Gage, J.C., "The Distribution and Excretion of Inhaled Mercury Vapor," The British Journal of Industrial Medicine 18:287, 1961.

Gajewski, E. et al., "Modification of DNA Bases in Mammalian Chromatin by Radiation-generated Free Radicals," Biochemistry 29:7876, 1990.

Galli, C.L., Murphy, S.D., et al. editors, *The Principles and Methods in Modern Toxicology*, Elsevier/North-Holland Biomedical Press, New York, 1980.

Gibbs, L.M., *Love Canal: My Story*, Albany, State University of New York Press, 1981.

Girman, J.R., "Volatile Organic Compounds and Building Bake-out," Occupational Medicine 4:695-712, 1989.

Glass, A.G., Hoover, R.N. et al., "The Emerging Epidemic of Melanoma and Squamous Cell Skin Cancer," Journal of the American Medical Association 262:2097, 1989

Goering, P.L., Waalkes, M.P. et al., *Toxicology of Cadmium*, in Goyer, R. A. et al. editors, *Toxicology of Metals: Biochemical Aspects*, Berlin, Springer-Verlag, 1995.

Goering, P.L., "Lead-protein Interactions as a Basis for Lead Toxicity," Neurotoxicology 14:45-60, 1993.

Goldfrank, L.R., Flomenbaum, N.E. et al., *Goldfrank's Toxicologic Emergencies*, 5th edition, Appleton-Century-Crofts, Norwalk, CT., 1994.

Goldsmith, J.R., Landau, S.A., "Carbon Monoxide and Health," Science 162:1352-1359, 1968.

Gorbach, S.L., Bartlett, J.G. et al., *Infectious Diseases*, W.B. Saunders Company, 1992.

Goyer, R.A., "Lead Toxicity: A Problem in Environmental Pathology," The American Journal of Pathology 64:167, 1971.

Grandjean, P., Andersen, O. et al., "Carcinogenicity of Occupational Nickel Exposures: An Evaluation of the Epidemiological Evidence," American Journal of Industrial Medicine 13:193-209, 1988.

Guengerich, F.P., *Mammalian Cytochrome*, Boca Raton, Fl, CRC, 1987.

Habibi, K., "Characterization of Particulate Lead in Vehicle Exhaust – Experimental Techniques," Environmental Science and Technology 4:239, 1970.

Haddad, L.M., Winchester, J.F., *Clinical Management of Poisoning and Drug Overdose*, 2nd edition W. B. Saunders Company, New York, 1990.

Hafen, B.Q., Peterson, B., *First Aid for Health Emergencies*, West PublishingCompany, New York, 1977.

Haley, T. J., Berndt, W.D., *Handbook of Toxicology*, Washington Hemisphere Publishing Corporation.

Hammond, P.B., Aronsen, A.L., "Lead Poisoning in Cattle and Horses in the Vicinity of a Smelter," Annals of the New York Academy of Sciences 111:595, 1964.

Hammond, E.C., "Smoking in Relation to the Death of One Million Men and Women," *Epidemiological Studies of Cancer and Other Chronic Diseases*, National Cancer Institute Monograph 19, 1966.

Handbook of Pesticide Toxicology, Vol I, General Principles, W.J. Hayes, Jr., et al., Jr. Academic Press, Inc., New York, 1991.

Hanninen, H., "Psychological Picture of Manifest and Latent Carbon Disulphide Poisoning," The British Journal of Industrial Medicine 28:374-381, 1971.

Hascheck, W.M., Rousseaux, C.G. editors, *Handbook of Toxicologic Pathology*, New York, Academic Press, Inc., 1991.

Hawkins, M.M., Smith, R.A. et al., "Pregnancy Outcomes in Childhood Cancer Survivors: Probable Effects of Abdominal Radiation," International Journal of Cancer 43:399, 1989.

Hayes, W.J., *Insecticides, Rodenticides and Other Economic Poisons*, in Di Palma, J.R. editor, *Drill's Pharmacology in Medicine*, 4th edition, New York, McGraw-Hill, 1971.

Hearne, F., Pifer, J. et al., "Absence of Adverse Mortality Effects in Workers Exposed to Methylene Chloride: An Update," Journal of Occupational Medicine 32:234-240, 1990.

Heath, D.F., Vandekar, M., "Toxicity and Metabolism of Dieldrin in Rats," The British Journal of Industrial Medicine 21:269-279, 1964.

Hecht, S.S., "Chemical Carcinogenesis: An Overview," Clinical Physiology and Biochemistry 3:89-97, 1985.

Herrmann, H., Speck, L.B. et al., "Interaction of Chromate with Nucleic Acids in Tissues," Science 119:221, 1954.

Herskowitz, A., Ishii, N. et al., "N-hexane Neuropathy," New England Journal of Medicine 285:82-85, 1971.

Hine, C.H., "Methyl Bromide Poisoning," Journal of Occupational Medicine 11:1, 1969.

Hipolito, R.M., "Xylene Poisoning in Laboratory Workers: Case Reports and Discussion," Laboratory Medicine 1:593-595, 1980.

Hodgkins, D.G. et al., "A Longitudinal Study of the Relation of Lead in Blood to Lead in Air Concentrations among Battery Workers," The British Journal of Industrial Medicine 49:241, 1992.

Hruza, L.L., Pentland, A., "Mechanisms of UV-induced Inflammation," The Journal of Investigative Dermatology 100:35S, 1993.

Hunziker, P.E., Kägi, J.H.R. et al., "Metallothionein" in Harrison, P. editor, *Metalloproteins, Part 2: Metal Proteins with Non-redox Roles*, Deerfield Beach, Fl, 1985.

Industrial Toxicology: Safety and Health Applications in the Workplace, Williams, P.L. and Burson, J.L. editors, New York, Van Nostrand Reinhold Company, 1985.

Jakoby, W.B., Bend, J.R. et al., *Metabolic Basis of Detoxification: Metabolism of Functional Groups*, New York, Academic Press, 1982.

Jensh, R.P., Brent, R.L., "Effects of Prenatal X-irradiation on the 14th-18th Days of Gestation on Postnatal Growth and Development in the Rat," Teratology 38:431, 1988.

Jobling, S.T., Reynolds, R. et al., "A Variety of Environmentally Persistent Chemicals, Including Some Phthalate Plasticizers, Are Weakly Estrogenic," Environmental Health Perspectives, 103:582-587, 1995.

Johnson, M.K., "The Delayed Neuropathy Caused by Some Organophosphorus Esters: Mechanism and Challenge," Critical Reviews of Toxicology 3:289, 1975.

Johnson, E.S., "Human Exposure to 2,3,7,8-TCDD and Risk of Cancer," Critical Reviews of Toxicology 21:451-462, 1992.

Johnson, M.K., "A Phosphorylation Site in Brain and the Delayed Neurotoxic Effect of Some Organophosphorus Compounds," The Biochemical Journal 111:487, 1969.

Jones, R.R., "Ozone Depletion and Cancer Risk," Lancet 2:443, 1987.

Kamrin, M.A., *Toxicology: A Primer on Toxicology Principles and Applications*, Chelsea, MI, Lewis Publishers, 1988.

Kato, R., Eastabrook, R.W. et al. editors, *Xenobiotic Metabolism and Disposition*, London, Taylor and Francis, 1989.

Keane, W.T., Zavon, M.R., "The Total Body Burden of Dieldrin," Bulletin of Environmental Contamination and Toxicology 4:1-16, 1969.

Kelly, D., Koshal., K., "The Effect of Milk Diet on Lead Metabolism in Rats," Environmental Research 6:355-360, 1973.

Kessel, M., Balling, R., "Variations of Cervical Vertebrae after Expression of a Hox-1,1 transgene in Mice," Cell 61:301-308, 1990.

Klaasen, C.D., "Immaturity of the Newborn Rat's Hepatic Excretory Function for Ouabain," The Journal of Pharmacology and Experimental Therapeutics, 183:520-526, 1972.

Klaasen, C.D., "Comparison of the Toxicity of Chemicals in Newborn Rats to Bile Duct-ligated and Sham-operated Rats and Mice," Toxicology and Applied Pharmacology 24:37-44, 1973.

Klaasen, C.D., Shoeman, D.W. et al., "Biliary Excretion of Lead in Rats, Rabbits and Dogs," Toxicology and Applied Pharmacology 29:434-446, 1974.

Klaasen, C.D., "Biliary Excretion of Mercury Compounds," Toxicology and Applied Pharmacology 33:356-365, 1975.

Klaasen, C.D., "Hepatic Excretory Function in the Newborn Rat," The Journal of Pharmacology and Experimental Therapeutics 184:721-728, 1973.

Kleinfeld, M., "Acute Pulmonary Edema of Chemical Origin," Archives of Environmental Health 10:942, 1965.

Kleinsmith, L. J., Kish, V.M. et al., *Principles of Cell and Molecular Biology*, 2nd edition, New York, Harper Collins College Publishers, 1995.

Lodish, H., Baltimore, D. et al., *Molecular Cell Biology*, 3rd edition, New York, W. H. Freeman and Co., 1995.

Kornberg, A., Baker, T. et al., *DNA Replication*, 2nd Edition, New York, W.H. Freeman and Company, 1992.

Kunkel, T.A., "DNA Replication Fidelity," Journal of Biological Chemistry 267:18251, 1992.

Kragh-Hansen, U., "Molecular Aspects of Ligan Binding to Serum Albumin," Pharmacological Reviews 333:17-53, 1981.

Krishna, D.R. and Klotz, U., Extrahepatic Metabolism of Drugs in Humans, Clinical Pharmacokinetics 26:144-160, 1994.

Kunz, J.R.M. et al., *The American Medical Association Family Medical Guide*, Random House, New York, ISBN 0-934-55582-1, 1987.

LaDou, J. editor, *Occupational Medicine*, Norwalk, CT., Appleton & Lange 1990.

Lagerwerff, J.V., Specht, A.W., "Contamination of Roadside Soil and Vegetation with Cadmium, Nickel, Lead and Zinc," Environmental Science and Technology 4:583, 1970.

Landrigan, P.J., "Toxicity of Lead at Low Dose," The British Journal of Industrial Medicine 46:593-596, 1989.

Lane, B.P., Mass, M.J., "Carcinogenicity and Carcinogenicity of Chromium Carbonyl in Heterotopic Tracheal Grafts," Cancer Research 37:1476-1479, 1977.

Lauwerty, R.R., Bucket, J.P. et al., "Occupational Exposure to Mercury Vapors and Biological Action," Archives of Environmental Health 27:65, 1973.

Lee, W.R., Moore, M.R. et al., "Low Level Exposure to Lead: The Evidence for Harm Accumulates," British Medical Journal 301:504, 1990.

Lefevre, M.J., *First Aid Manual for Chemical Accidents*, Van Nostrand Reinhold Company, New York, 1989.

Lehman-McKeeman, L.D., Caudell, D. et al., "µ2U-Globulin Is the Only Member of the Lipocalin Protein Superfamily That Binds to Hyaline Droplet Inducing Agents," Toxicology and Applied Pharmacology 116:170-176, 1992.

Levine, A., *Love Canal: Science, Politics and People*, Lexington, MA, Lexington Books, 1982.

Lewis, L., "Mercury Poisoning in Tungsten-molybdenum Rod and Wire Manufacturing Industry," Journal of the American Medical Association 129:123, 1945.

Lilienfield, D. E., Stolley, P.D. et al., *Foundations of Epidemiology*, 3rd edition, New York, Oxford University Press, 1994.

Lillienfeld, D.E., Gallo, M.A. et al., "2,4-D, 2,4,5-T and 2,3,7,8-TCDD: An Overview," Epidemiologic Reviews 11:28-58, 1989.

Livingston, J.M., Jones, C.R., "Living Area Contamination by Chlordane Used for Termite Treatment," Bulletin of Environmental Contamination and Toxicology 27:406-411, 1981.

Lodish, H., Baltimore, D. et al., *Molecular Cell Biology*, 3rd edition, New York, W. H. Freeman and Co., 1995.

Loomis, T.A., *Essentials of Toxicology*, 3rd edition, Philadelphia, Lea and Febiger, 1976.

Lu, F.C., *Basic Toxicology: Fundamentals, Target Organs, and Risk Assessment*, New York, Hemisphere Publishing Company, 1996.

Luckey, T.D., Venugopal., B., *Metal Toxicity in Mammals, Vol. 1: Physiologic and Chemical Basis for Metal Toxicity*, New York, Plenum Press, 1977.

Lutz, W.K., Schlatter, J., "Chemical Carcinogens and Overnutrition in Diet-related Cancer," Carcinogenesis 13:2211-2216, 1992.

Lutz, W.K., Schlatter, J. et al., "The Relative Importance of Mutagens and Carcinogens in the Diet," Pharmacology and Toxicology 72:s104-s107, 1993.

Mackenzie, J.N., "Some Observations on the Toxic Effects of Chrome on the Nose, Throat and Ear," Journal of the American Medical Association 3:601-603, 1984.

Mahaffey, K.R., "Exposure to Lead in Childhood: The Importance of Prevention," New England Journal of Medicine 327:1308, 1992.

Maibach, H.I., Mark, D. et al., "Percutaneous Penetration of Benzene and Benzene Contained in Solvents Used in the Rubber Industries," Archives of Environmental Health 36:256-260, 1981.

Malaka, T., Kodama, A.M. et al., "Respiratory Health of Plywood Workers Occupationally Exposed to Formaldehyde," Archives of Environmental Health 45(5):288-294, 1990.

Manahan, S.E., *Toxicological Chemistry*, 2nd edition, Lewis Publishers, Chelsea, MI, 1992.

Mandell, G.L., Bennett, J.E. et al. editors, *Principles and Practices of Infectious Diseases*, Vol. 2, 4th edition, New York, Churchill Livingstone, 1995.

Marzulli, F.N., Cellahan, J.F. et al., "Chemical Structure and Skin Penetrating Capacity of a Short Series of Organic Phosphates and Phosphoric Acid," The

Journal of Investigative Dermatology 44:339-344, 1965.

Mathews, H.B., Matsumura, F., "Metabolic Fate of Dieldrin in the Rat," Journal of Agricultural and Food Chemistry 17:845-852, 1969.

May, G., "Chloracne from the Accidental Production of Tetrachlorodibenzodiozin," The British Journal of Industrial Medicine 30:276, 1973.

Mayer, L., Guelish, J. et al., "Hydrogen Fluoride (HF) Inhalation and Burns," Archives of Environmental Health 7:445, 1963.

McBride, W.G., "Thalidomide and Congenital Anomalies," Lancet 2:1358, 1961.

McCarthy, T.B., Jones, R. et al., "Industrial Gassing Poisonings Due to Trichloroethylene, Perchloroethylene, and 1,1,1-Trichloroethane, 1961-80," The British Journal of Industrial Medicine 40:450-455, 1983.

McCunney, R.J., *A Practical Approach to Occupational and Environmental Medicine*, 2nd edition, New York, Little, Brown & Co., 1994.

Medical Microbiology, An Introduction to Infectious Diseases, J.C. Sherris editor, New York, Elsevier, 1984.

Mettler, F.A., Upton, A.C., *Medical Effects of Ionizing Radiation*, 2nd edition, New York, W.B. Saunders Company, 1995.

Michaels, L., Chissick, S.S. et al. editors, *Asbestos: Properties, Applications and Hazards*, Vol I, New York, John Wiley and Sons, 1979.

Milby, T.H., "Prevention and Management of Organophosphate Poisoning," Journal of the American Medical Association 216:2131, 1971.

Milham, S., Strong, T., "Human Arsenic Exposure in Relation to a Copper Smelter," Environmental Research 7:176-182, 1974.

Modrich, P., "Mismatch Repair, Genetic Stability, and Cancer," Science 266:1959-1960, 1994

Mohsenin, V., "Airway Responses to Nitrogen Dioxide in Asthmatic Subjects," Journal of Toxicology and Environmental Health 2:371-380, 1987.

Monks, T.J., Anders, M.W. et al., "Contemporary Issues in Toxicology: Glutathione Conjugate Mediated Toxicities," Toxicology and Applied Pharmacology 106:149, 1990.

Morehouse, W., Subramanian, M.A., "The Bhopal Tragedy: What Really Happened and What it Means for American Workers and Communities at Risk," New York, Council of International and Public Affairs, 1986.

Morrison, W.L., "Effects of Ultraviolet Radiation on the Immune System in Humans," Photochemistry and Photobiology 50:515, 1989.

Mulvihill, J.J., et al., "Pregnancy Outcomes in Cancer Patients. Experience in a Large Cooperative Group," Cancer 60:1143, 1987.

Muscat, J.E., Wynder, E.L., "Cigarette Smoking, Asbestos Exposure, and Malignant Mesothelioma," Cancer Research 51:2263-2267, 1991.

National Institute for Occupational Safety and Health, "Occupational Exposure to Chromium VI, Criteria for a Recommended Standard," U.S. Department of Health, Education, and Welfare, PHS, Center for Disease Control, 1975.

National Institute for Occupational Safety and Health, "A Recommended Standard for Occupational Exposure: Hydrogen Cyanide and Cyanide Salts," U.S. Department of Health, Education, and Welfare, PHS, Center for Disease Control, 1975.

National Institute for Occupational Safety and Health, U.S. Department of Health, Education, and Welfare, "Criteria for a Recommended Standard: Occupational Exposure to Sulfur Dioxide," Washington, D.C., 1974, U.S. Government Printing Office.

National Institute for Occupational Safety and Health/Occupational Safety and Health Administration, "Occupational Health Guide for Chemical Hazards: Inorganic Mercury," DHHS (NIOSH) Publication No. 81-123, 1981.

National Research Council, *Disposal of Chemical Munitions and Agents*, National Academy Press, Washington, D.C., 1984.

National Institute for Occupational Safety and Health, "Criteria for a Recommended Standard: Occupational Exposure to Oxides of Nitrogen (Nitrogen Dioxide and Nitric Oxide)," Cincinnati, U.S. Department of Health, Education and Welfare, PHS, Center for Disease Control, HEW (NIOSH) Publ. 76-149, 1976.

National Research Council (U.S.), *Biosafety in the Laboratory: Prudent Practices for the Handling and Disposal of Infectious Materials*, Washington, D.C., National Academy Press, 1989.

Needleman, H.L., Gatsonis, C.A., "Low-level Lead Exposure and the IQ of Children," Journal of the American Medical Association 263:673-678, 1990.

Nomiyama, K., Nomiyamma, H., *Mechanism of Urinary Excretion of Cadmium: Experimental Studies in Rabbits*, in Nardberg, G.F. editor, *Effects and Dose-Response Relationships of Toxic Metals*, New York, Elsevier Scientific Publishing Company, 1976.

Norseth, T., "The Carcinogenicity of Chromium and its Salts," The British Journal of Industrial Medicine 43:649-651, 1986.

Norton, M.E. et al., "Neonatal Complications after the Administration of Indomethacin for Preterm Labor," New England Journal of Medicine 329:1602, 1993.

Osebold, J.W., Gershwin, L.J. et al., "Studies on the Enhancement of Allergic Lung Sensitization by Inhalation of Ozone and Sulfuric Acid Aerosol," Journal of Environmental Pathology and Toxicology 3:221-234, 1980.

Otake, M., Schull, W.J., "In Utero Exposure to A-bomb Radiation and Mental Retardation: A Reassessment," British Journal of Radiology 57:409, 1984

Ott, M.G., Holder, B.B. et al., "Respiratory Cancer and Occupational Exposure to Arsenicals," Archives of Environmental Health 29:250-255, 1974.

Ottoboni, M.A., *The Dose Makes the Poison*, 2nd edition, New York, Van Nostrand Reinhold, 1991.

Pati, S., Helmbrecht, G.D., "Congenital Schizencephaly Associated with Warfarin Exposure," Reproductive Toxicology 8:115, 1994.

Peak, M.J., Peak, J.G. et al., "Induction of Direct and Indirect Single-strand Breaks in Human Cell DNA by Far- and Near-ultraviolet Radiations: Action Spectrum and Mechanisms," Photochemistry and Photobiology 45:381, 1987

Pfeiffer, C.J., *Gastroenterologic Responses to Environmental Agents - Absorption and Interactions*, in Lee, D.H. ed, *Handbook of Physiology*, Section 9: Reactions to Environmental Agents, Bethesda, MD, American Physiological Society, 1977.

Pitot, H.C., "Endogenous Carcinogenesis: The Role of Tumor Promotion," Proceedings of the Society of Experimental Biology and Medicine 198:661-666, 1991.

Poisoning and Drug Overdose, 2nd edition, K.R. Olson editor, Appleton & Lange, Norwalk, CT, 1994.

Polednak, A.P., "Mortality among Welders, Including a Group Exposed to Nickel Oxide," Archives of Environmental Health 36:235-242, 1981.

Prescott, D.M., Cells: *Principles of Molecular Structure and Function*, Boston, Jones and Bartlett Publishers, 1988.

Preston, D.S., Stern, R.S., "Nonmelanoma Cancers of the Skin," New England Journal of Medicine 327:1649, 1992.

Price, A., "The Repair of Ionizing Radiation-induced Damage to DNA," Cancer Biology 4:61, 1993.

Principles and Methods of Toxicology, 2nd edition, Hayes, A.W. editor, New York, Raven Press, 1989.

Rabinowitz, M.B., Wetremill, G.W. et al., "Absorption, Storage and Excretion of Lead by Normal Humans," The Journal of Clinical Investigation 58:260-270, 1976.

Reinhardt, P.A., Gordon, J.G., *Infectious and Medical Waste Management*, Chelsea, MI, Lewis Publishers.

Renkin, E.M., *Capillary Permeability*, in Mayerson, H.S. editor, *Lymph and the Lymphatic System*, Springfield, IL, Thomas, 1968.

Restani, P., Galli, C.L., "Oral Toxicity of Formaldehyde and its Derivatives," CRC Critical Reviews in Toxicology 21(5):315-328, 1991.

Rinsky, R. A., Smith, A.B. et al., "Benzene and Leukemia: An Epidemiological Risk Assessment," New England Journal of Medicine, 316:1044-1050.

Rodricks, J.V., "Risk Assessment, the Environment, and Public Health," Environmental Health Perspectives, Vol. 102:258-263, 1994.

Rom, W.N., "Accelerated Loss of Lung Function and Alveolitis in a Longitudinal Study of Non-smoking Individuals with Occupational Exposure to Asbestos," American Journal of Industrial Medicine 21:835-844, 1992.

Rosenstock, L., Cullen, M.R., *Textbook of Clinical Occupational & Environmental Medicine*, W.B. Saunders Co., Philadelphia, PA, 1994.

Rothschild, J.H., *Tomorrow's Weapons. Chemical and Biological*, New York, McGraw-Hill Book Company, 1964.

Rugh, R., *Radiology and the Human Embryo and Fetus*, in Dalrymple, G.V. et al., editors, *Medical Radiation Biology*, Philadelphia, W.B. Saunders Company, 1973.

Russell, L.B., "X-ray Induced Developmental Abnormalities in the Mouse and Their Use in the Anlysis of Embryological Patterns," Journal of Experimental Zoology 114:545, 1950.

Safe, S.H., "Environmental and Dietary Estrogens and Human Health: Is There a Problem?" Environmental Health Perspectives, 103: 346-351, 1995.

Saleh, M.A., Turner, W.A. et al., "Polychlorobornane Components of Toxaphene: Structure-toxicity Relations and Metabolic Reductive Dechlorination," Science 198:1256-1258, 1977.

Sasser, L.B., Jarboe, G.L. et al., "Intestinal Absorption and Retention of Cadmium in Neonatal Rats," Toxicology and Applied Pharmacology 41:423-431, 1977.

Schottelius, B. A., Schottelius, D.D. et al., *Textbook of Physiology*, 18th edition, St. Louis, C. V. Mosby Company, 1978.

Schroeder, H.A., Tipton, I.H., "The Human Body Burden of Lead," Archives of Environmental Health 17:965, 1968.

Schull, W.J., Otake, M. et al., "Genetic Effects of the Atomic Bombs: A Reappraisal," Science 213: 1220, 1981.

Seeley, R.R., Stephens, T.D. et al., *Essentials of Anatomy and Physiology*, 2nd edition, New York, C. V. Mosby, 1996.

Selby, C.P., Sancar, A., "Molecular Mechanism of Transcription-repair Coupling," Science 260:53, 1993.

Sellakumar, A. et al., "Carcinogenicity of Formaldehyde and Hydrogen Chloride in Rats," Toxicology and Applied Pharmacology 81:401-406, 1985.

Senanayake, N., Albert, R. et al., "Neurotoxic Effects of Organophosphorus Insecticides," New England Journal of Medicine 316:761-763, 1987.

Seppäläinen, A.M., Lindstrom, K. et al., "Neurophysiological and Psychological Picture of Solvent Poisoning," American Journal of Industrial Medicine 1:31-42, 1980.

Setlow R.B., Woodhead, A.D. et al., "Animal Model for Ultraviolet Radiation-induced Melanoma; Playfish-swordfish Hybrid," Proceedings of the National Academy of Sciences, USA 86:8922, 1989.

Shapira, J., *Radiation Protection. A Guide for Scientists and Physicians*, 3rd edition, Cambridge, MA., Harvard University Press, 1990.

Shastri, L., *Bhopal Disaster: An Eyewitness Account*, New Delhi: Criterion Publications, 1985.

Sim, V.M., Pattle, R.E. et al., "Effect of Possible Smog Irritants on Human Subjects," Journal of the American Medical Association 165:1908-1913, 1957.

Sjostrom, H., Nilsson, R., *Thalidomide and the Power of the Drug Companies*, Marmondsworth, Penguin Press, 1972.

Smith, D.C., "Lead Ingestion, Ceramic Glaze – Alaska, 1992," Journal of the American Medical Association 268:2498, 1992.

Soto, A.M. et al., "The Pesticides Endosulfan, 1994 Toxaphene, and Dieldrin Have Estrogenic Effects on Human Estrogen-Sensitive Cells," Environmental Health Perspectives, 102: 380-383.

Sterling, T.D., Collett, C.W. et al., "Environmental Tobacco Smoke and Indoor Air Quality in Modern Office Work Environment," Journal of Occupational Medicine 28(1):57-62, 1987.

Streiner, D. L., Norman, G.R. et al., *PDQ Epidemiology*, Philadelphia, B. C. Decker, Inc., 1989.

Streissguth, A.P., Clarren, S.K. et al., "Natural History of the Fetal Alcohol Syndrome: a 10-year Follow-up of Eleven Patients," Lancet 2:85, 1985

Sunderman, F.W. Jr., "The Treatment of Acute Nickel Carbonyl Poisoning with Sodium Diethyldithiocarbonate," Annals of Clinical Research 3:182-185, 1971.

Timbrell, J.A., *Principles of Biochemical Toxicology*, Taylor and Francis, London, 1982.

Timbrell, J.A., *Introduction to Toxicology*, Taylor and Francis, New York, 1989.

Tortora, G. J., *Principles of Human Anatomy*, San Francisco, Canfield Press, 1977.

Toskulkao, C., Nhongsaeng, J. et al., "Potentiation of Carbon Tetrachloride Induced Hepatotoxicity by Thinner Inhalation," Journal of Toxicological Science 15:75-86, 1990.

Sperling, F., *Toxicology Principles and Practices*, Vol 2, John Wiley and Sons, New York, 1984.

Tsuchija, K., "Lead" in Friberg, L., et al. editors, *Handbook of Toxicology of Metals*, Amsterdam, Elsevier, 1979.

U.S. Department of Health and Human Services, Agency for Toxic Substances and Disease Registry, "Case Studies in Environmental Medicine: Lead Toxicity," 1990.

U.S. Environmental Protection Agency, "Maximum Contaminant Level Goals and National Primary Drinking Water Regulations for Lead and Copper," Federal Register 56:26469-26470, 1991.

U.S. Environmental Protection Agency, *Risk Assessment Guidelines and Information Directory*, Rockville, MD, Government Institutes, Inc., 1988.

Vainio, H., Wilbourn, J., "Cancer Etiology: Agents Causally Associated with Human Cancer," Pharmacology and Toxicology 72:4-11, 1993.

Vigliani, E.C., Saita, G. et al., "Benzene and Leukemia," New England Journal of Medicine 271:872, 1964.

Vigliani, E.C., Forni, A. et al., "Benzene and Leukemia," Environmental Research 11:122-127, 1976.

Volans, G.N. et al. editors, V International Congress of Toxicology, *Basic Science in Toxicology*, Taylor and Francis, New York, 1989.

Volfe, J.J., "Effects of Cocaine on the Fetus," New England Journal of Medicine 327:399-407, 1992.

Wallace, R. A., *Biology: The World of Life*, 6th edition, New York, Harper-Collins Publishers, 1992.

Wallén, M., et al., "Coexposure to Toluene and P-xylene in Man: Uptake and Elimination," The British Journal of Industrial Medicine 42:111-116, 1985.

Wang, H.H., MacMahon, B., "Mortality of Workers Employed in the Manufacture of Chlordane and Heptachlor," Journal of Occupational Medicine 21:11, 1979.

Waterhouse, J., "Cancer among Chromium Platers," The British Journal of Cancer 2:262, 1975.

Waterman, M.R., Johnson, E.F. editors, *Cytochrome P450. Methods in Enzymalogy*, Vol. 206, New York, Academic Press, 1991.

Weber, L.W.O., Ernst, S.W. et al., "Tissue Distribution and Toxicokinetics of 2,3,7,8-tetrachloro-dibenzo-p-dioxin in Rats after Intravenous Injection," Fundamentals of Applied Toxicology 21:523-534, 1993.

White, R., Feldman, R. et al., "Neurobehavioral Effects of Toxicity Due to Metals, Solvents, and Insecticides," Clinical Neuropharmacology 13:392-412, 1990.

Wilcox, A.J. et al., "Incidence of Early Pregnancy Loss," New England Journal of Medicine 31:189, 1988.

Wilson, J.C., *Current Status of Teratology* in Wilson, J.G., et al. editors, *Handbook of Teratology*, New York, Plenum Press, 1977.

Wilson, C., *Chemical Exposure and Human Health*, McFarland and Company, Inc., North Carolina, 1993.

Zenz, C., Diskerson, O.B. et al., *Occupational Medicine*, 3rd edition, St. Louis, Mosby, 1994.

Zheng, P., Kligman, L.H., "UVA-induced Ultrastructural Changes in Hairless Mouse Skin: A Comparison to UVB-induced Damage," The Journal of Investigative Dermatology 100:194, 1993.

Zimmerman, S., Groehler, K. et al., "Hydrocarbon Exposure and Chronic Glomerulonephritis," Lancet 2:199-201, 1975.

Index

A

AA concentration ranges 345
AAF 77, 78, 292
Absorption 36, 46, 49, 86
   active transport 39
   carrier-mediated transport 88
   enterohepatic circulation 120
   factors affecting 38, 41, 43, 44, 46, 49, 52, 86
      physical and chemical factors 73
      routes of exposure 73
   lung absorption 90
   passive diffusion 87
   passive transport 39
   phagocytosis 89
   pinocytosis 89
   skin absorption 89
Acceptable Daily Intake (ADI) 67, 73, 74, 294
Acetaldehyde 157, 166
Acetaminophen 116
Acetone 74, 89, 153, 172
Acetylaminofluorene, 2 (AAF) 77, 78, 292
Acetylation 105, 113

Acetylcholine 132, 134, 137, 152, 216
Acetylcholinesterase 134, 137, 215
ACGIH 7. *See also* American Conference of Governmental Industrial Hygienists
Acidosis 133
Acids 44, 54
Acquired Immunodeficiency Syndrome (AIDS) 3, 243
Acrolein 157
Acrylamide 134
Actinomycin 66, 67
Action potential 130
Active transport 39, 41, 88, 90, 91, 93, 96, 100, 120, 136
   adenosine triphosphate (ATP) 38
Acute exposure 3, 11
Acute toxicity 29, 58
Additive effects 80
Adenosine triphosphate (ATP) 38, 88, 126, 135, 153, 162, 202, 222, 224
ADI 67, 73, 74, 294
Adulterant 224
Aerosols 90

Afferent nerves 127
Aflatoxin 275
Agency for Toxic Substances and Disease Registry (ATSDR) 4, 7
Agent Orange 221
Agent Q 238
Agranulocytes 139
AHH 79
AIDS 3, 243
Air pollution 184
   indoor air pollution 184, 189
      building-associated illnesses 189
      building-related illnesses 189
      sick-building syndrome 189
      tight-building syndrome 189
   outdoor air pollution 184
      carbon dioxide 186
      carbon monoxide 185
      nitrogen oxides 186
      ozone 188
      particulates 189
      sulfur oxides 188
Albumin 94
Aldicarb 137
Aldrin 81, 110, 218, 219
Alopecia 222

Alpha particles 190, 231, 280
ALS 137
Aluminum 133
Alveoli 45, 46, 155, 157
    fibrosis 158
    pulmonary edema 157
        ammonia 158
        chlorine 158
        hydrogen fluoride 158
        oxides of nitrogen 158
        ozone 158
        perchloroethylene 158
Alveolitis 325
Alzheimer's disease 133
Amelia 271
American Conference of Governmental Industrial Hygienists (ACGIH) 7, 178, 279
Amines 75
Aminonaphthalene, 2- 113
Aminonicotinamide, 6- 270
Ammonia 156, 157, 158, 171, 207
    bronchitis 207
    conjunctivitis 207
    cyanosis 207
    pneumonia 207
    pulmonary edema 207
Amosite 209
Amyotrophic lateral sclerosis (ALS) 137
Anemia 141, 194, 195, 202, 245
    aplastic anemia 142
    benzene 142
    carbon tetrachloride 142
    chlordane 142
    ethanol 141
    Fanconi's anemia 265
    hemolytic anemia 324
    lead 141
    microcytic anemia 141
    nephrotoxicity-induced anemia 142
    radiation 142
Angiotensin 162
Aniline 77, 113
Animal-derived toxins 251
    Red tide 251
Anorexia 197
Anoxia 69
ANS 127
Antagonistic effects 80, 81
    chemical antagonism 81
    dispositional antagonism 81
    functional antagonism 81
    receptor antagonism 81
Anterior chamber 53

Anterior pituitary 166
Anthracene 148
Anthrax 239
    Bacillus anthracis 239
Antibody 81, 140, 146, 148, 149, 156
Antibody-mediated immunity 146
Anticholinesterase agents 135, 137
Aplastic anemia 142
Apostosis 163
Aqueous humor 53
Aromatic hydrocarbons 193
Arrhythmia 153, 192, 193, 194, 197, 212, 219, 224, 245
Arsenic 45, 111, 113, 119, 148, 159, 161, 172, 197, 238, 271, 276
Arsenic acid 197
Arsenic disulfide 197
Arsenic trifluoride 197
Arsenic trioxide 197
Arsine 197
Arterioles 91
Aryl hydrocarbon hydroxylase (AHH) 79
Asbestos 20, 23, 46, 75, 80, 90, 156, 158, 159, 189, 190, 209, 273, 276, 277
    amosite 209
    asbestosis 158
    chrysotile 209
    crocidolite 209
    lung cancer 209
    malignant mesothelioma 158
    mesothelioma 209
    pleural fibrosis 209, 210
    pulmonary fibrosis 209, 210
Asbestosis 158, 210
Aspartame 14
Asphyxia 186, 211, 238
Asthma 188, 189
Astrocytes 102, 128
Ataxia 133, 201, 203, 221, 223
ATP 38, 88, 126, 135, 153, 162, 202, 222, 224
Atria 91
Atrial fibrillation 153
Atrioventricular (AV) node 150
Atrioventricular bundle 150
ATSDR 4, 7
Attributable risk 26
Autoimmune response 148
Autonomic nervous system (ANS) 127
AV 150
Axon 128
Axonopathy 133, 134, 216

B

B cell lymphocytes 146
Bacillus anthracis 239
Bacteremia 245
Bacteria-derived toxins 249
    botulism 249
    Lyme disease 249
    tetanus 250
    Toxic Shock Syndrome 250
Bacterial agents 239
BAI 189
BAL 311
Barium 153
Baroreceptor reflex 152
Baroreceptors 152
Basal cell carcinoma 197, 284
Base-excision repair (BER) 265
Bases 44, 45, 54
Basophils 139
β,β-dimethylcysteine 311
BEI 7
Benign tumor 273
Benzanthracene 148
Benzene 19, 20, 46, 54, 90, 105, 112, 116, 137, 141, 142, 143, 149, 172, 193, 273, 276
Benzene hexachloride 218, 219
Benzo[a]pyrene 79, 111, 114, 148, 156, 159
Benzoyl peroxide 276
BER 265
Beryllium 158, 159
Beta particles 231, 280
Bias 26
    observation bias 26, 27
    selection bias 26, 27
Bile 55
Bilirubin 160
Biliverdin 161
Bioaccumulation 87, 100
Bioactivation 114
    acetaminophen 116
    benzo[a]pyrene 114
    bromobenzene 116
    malathion 114
    nitrosamines 115
    parathion 114
Biological Exposure Indices (BEI) 7
Biological warfare agents 236, 239
    anthrax 239
    Bacillus anthracis 239
    bacterial agents 239
    Brucella abortus 239
    Brucella melitensis 239
    Brucella suis 239

brucellosis 239
bubonic plague 239
Coccidioides immitis 242
dengue fever 242
fungal agents 242
Rickettsias 242
viral agents 242
yellow fever 242
Yersinia pestis 239
Biotransformation 54, 105, 156
   1,1-dichloroethylene 76
   2-acetylaminofluorene 77
   2-aminonaphthalene 113
   acetaminophen 116
   aldrin 110
   aniline 77, 113
   arsenic 111, 113
   benzene 105, 112, 116
   benzo[a]pyrene 111, 114, 156
   bromobenzene 116
   carbon tetrachloride 112, 156, 193
   catabolic reactions 105
   chloroform 112
   chromium 200
   cis-1,2-dichloroethylene 76
   DDT 110, 111
   dichlorvos 217
   dieldrin 110
   dinitrotoluene 80
   ethanol 110, 113
   ethylene glycol 77
   factors affecting metabolism 117
   heptachlor 110
   liver 105
   malathion 77, 111, 114, 217
   methanol 110
   methyl mercury 113
   nicotine 113
   nitrosamines 115
   parathion 79, 106, 114, 217
   phenols 113
   polycyclic aromatic hydrocarbons 156
   pyridine 160
   selenium 113
   tetraethyl pyrophosphate 217
   toxaphene 111
   trans-1,2-dichloroethylene 76
Bipyridyl derivatives 221
   diquat 221
   paraquat 221
Blocking agents 135
Blood 139
   erythrocytes 139
   leukocytes 139
   macrophages 140
   plasma 140
   platelets 139
   stem cells 139
   thrombocytes 139
Blood-borne pathogens 243
   AIDS (Acquired Immunodeficiency Syndrome) 3, 243
   bacteremia 245
   hepatitis 244
   Human Immunodeficiency Virus 243
   Rocky Mountain Spotted fever 245
   syphilis 244
   toxoplasmosis 245
Blood-brain barrier 102, 133, 135, 136, 172
Bone 100, 103, 148, 198
   cadmium 198
   fluorosis 102
   hypocalcemia 212
   osteomalacia 198
   osteoporosis 101, 198
Bone cancer 235, 283
Boric acid 79
Borrelia burgdorferi 249
Botulism 67, 135, 249
Bowman's capsule 97
Bradycardia 137, 153
BRI 189
Bromoacetone 236
Bromobenzene 116
Bronchioles 45
Bronchitis 157, 198, 203, 207, 211, 213, 238
Bronchogenic carcinoma 277
Bronchopneumonia 187, 188
Brucella abortus 239
Brucella melitensis 239
Brucella suis 239
Brucellosis 239
   Brucella abortus 239
   Brucella melitensis 239
   Brucella suis 239
Buboes 239
Bubonic plague 19, 239
Building-associated illnesses (BAI) 189
Building-related illnesses (BRI) 189
Bulbourethral gland 165
Bundle branches 150
Bundle of His 150
BZ 239

C

CAA 5, 184
CaBP 88
Cadmium 44, 49, 52, 54, 79, 88, 90, 97, 100, 119, 120, 142, 148, 153, 158, 159, 162, 167, 168, 189, 198, 270, 271, 273, 275
Cadmium chloride 198
Cadmium oxide 198
Caffeine 136
Calcium chromate 200
Calcium cyanide 211
Calcium-binding protein (CaBP) 88
Cancer
   basal cell carcinoma 197, 284
   benign tumor 273
   bone cancer 235, 283
   bronchogenic carcinoma 277
   carcinoma 273
   leukemia 19, 143, 265, 281, 282, 283
   lung cancer 80, 159, 204, 277, 283
   lymphoma 243
   malignant tumor 273
   mechanisms of action 274
     one-hit theory 274
     single-hit theory 274
     somatic cell mutation theory 274
   mesothelioma 20, 158, 209, 277
   nasal cancer 157, 204
   neoplasm 273
   ovarian cancer 288
   sarcoma 273
   skin cancer 235, 283, 284
   squamous cell carcinoma 197, 284
   stages of development 274
     initiation stage 274
     progression stage 274, 276
     promotion stage 274, 275
   ultraviolet radiation 262, 284
Cannula 246
Carbamate insecticides 2, 82, 215
   carbaryl 2, 137, 215, 218
   Sevin 2, 218
Carbaryl 2, 137, 215, 218
Carbohydrates 37
Carbon dioxide 156, 186
Carbon disulfide 45, 134, 172
Carbon monoxide 46, 75, 81, 90, 134, 142, 153, 156, 185, 190
Carbon tetrachloride 3, 45, 46, 73, 80, 82, 89, 112, 120, 136, 142, 143, 149, 156, 161, 191, 192, 193
Carbonic acid 3

Carboxyhemoglobin 142, 185
Carcinogenic 14, 202, 222, 273
    1,1-dichloroethylene 76
    1,4-dichlorobenzene 80
    2-acetylaminofluorene 77, 292
    2,2,4-trimethylpentane 276
    2,5,2',5'-tetrachlorobiphenyl 276
    acetaldehyde 157
    acrolein 157
    aflatoxin 275
    anthracene 148
    arsenic 159, 276
    asbestos 20, 159, 273, 276, 277
    benzanthracene 148
    benzene 19, 20, 273, 276
    benzo[a]pyrene 79, 148, 159
    benzoyl peroxide 276
    beryllium 159
    cadmium 198, 273, 275
    chromium 159, 200, 275
    cigarette 19, 159
    cyclamates 14
    D-Lomolene 80
    dimethyl sulfate 275
    dimethylbenz[a]anthracene 148
    epigenetic carcinogens 274
    estrogen 276
    ethanol 276
    formaldehyde 78, 157, 208
    fungicides 222
    genotoxic carcinogens 274
    heterocyclic amines 275
    isophorone 80
    lead 202
    methylcholanthrene 148
    nickel 159, 204, 275
    nitrosamines 115, 157, 275
    nitrosureas 275
    organophosphate pesticides 279
    phenobarbital 276
    polycyclic aromatic hydrocarbons (PAH) 79, 159, 275
    progesterone 276
    radiation 273
    saccharin 14, 276
    sources of 276
    strontium 102
    tetrachlorodibenzo-p-dioxin 276
    tobacco 277
    trimethylpentane 100
    unleaded gasoline 80
    vinyl chloride 157, 279
Carcinoma 273
Cardioregulatory center 141, 153

Cardiotoxicity 150
    arrhythmias 153
        atrial fibrillation 153
        bradycardia 153
        tachycardia 153
        ventricular fibrillation 153
    carbon monoxide 153
    metals 153
        barium 153
        cadmium 153
        cobalt 153
        lead 153
        manganese 153
        nickel 153
    organic solvents 153
        acetone 153
        halogenated hydrocarbons 153
        ketones 153
        methylene chloride 153
        toluene 153
Carotid body 212
Carrier molecules 38, 40, 41, 93, 96
    active transport 41
    carrier-mediated transport 88
    facilitated diffusion 40
Carrier-mediated transport 88
Catabolic reactions 105
    hydrolysis 105
    oxidation 105
    reduction 105
Catalysts 126
Cataracts 283, 285
Catecholamines 136
Cathartics 311
Cause-effect relationship 20, 29, 30, 62, 64
CDC 4, 6
Cells of Leydig 165
Centers for Disease Control and Prevention (CDC) 4, 6
Central nervous system (CNS) 127
Central vein 95
CERCLA 5
Cerebellum 203
Cerebral cortex 201
Cerebral palsy 203
CFR 5
Chancre 244
Chelating agents 311
    2,3-dimercapto-1-propanesulfonic acid 311
    2,3-dimercaptopropanol 311
    penicillamine 311
Chemical antagonism 81
Chemical asphyxiant 212

Chemical interactions 80
    additive effects 80
    antagonistic effects 80, 81
    potentiation effects 80, 81
    sensitization effects 80, 81
    synergistic effects 80
Chemical Manufacturers Association 279
Chemical Transportation Emergency Center (CHEMTREC) 7
Chemical warfare agents 3, 236
    arsenical 238
    bromoacetone 236
    BZ 239
    chlorine gas 236, 238
    CN gas 236
    CS gas 236
    dichloroformoxime 238
    diphenyl chlorasine 236
    mustard gas 3, 236, 238
        Agent Q 238
    nerve gas 239
    phosgene 3
    phosgene gas 238
    sarin 74, 239
    soman 11, 239
    tabun 11, 239
Chemoreceptor 143, 212
CHEMTREC 7
Chloracne 220
Chlordane 54, 81, 95, 142, 218, 219
Chlordecone 167
Chlorine 156, 157, 158, 210
    bronchitis 211
    pneumonia 211
    pulmonary edema 211
Chlorine gas 236, 238
Chloroacetophenone gas (CN) 236
Chloroform 3, 46, 47, 73, 80, 90, 112, 120, 122, 136, 163
Chloroquinone 172
Cholera 19
Choroid 53
Chrome ulcers 200
Chromic acid 199
Chromium 44, 45, 49, 90, 156, 157, 159, 161, 198, 210, 275
    calcium chromate 200
    chromic acid 199
    chromium oxychloride 199
    chromium trioxide 200
    sinusitis 200
    thallium dichromate 200
Chromium oxychloride 199

Chromium trioxide 200
Chromosomes 163
Chronic exposure 11
Chronic toxicity 29, 61
Chrysotile 209
Cigarette 19, 114, 157, 159, 185, 187, 198, 204
　　toxic effects 46
　　　lung cancer 19, 80
Ciliary bodies 53
Ciliated epithelial cells 46
Ciliostasis 157
Circulatory system
　　anatomy of 91
　　pulmonary circulation 91
　　systemic circulation 91
Cirrhosis 117, 161, 244
Cis-1,2-dichloroethylene 76
Clean Air Act (CAA) 5, 184
Clean Water Act (CWA) 5
Clinical toxicologists 15
Clinical toxicology 15
Clostridium botulinum 15, 67, 135, 249
CN gas 236
CNS 127
Coal dust 158
Cobalt 153
Cocaine 136, 272
Coccidioides immitis 242
Cockayne's syndrome 265
Code of Federal Regulations (CFR) 5
Codon 257
Colic 202
Collecting duct 97
Comedomes 221
Common bile duct 120
Comprehensive Environmental Response, Compensation, and Liability Act (CERCLA) 5
Concentration gradient 39
Confounding factors 22, 26, 30, 31
Conjugation 105, 111
Conjunctivitis 207
Consumer Product Safety Commission (CPSC) 10
Contraindications 310
Copper 271
Cornea 2, 53, 283
Corneal stroma 171
Corpus luteum 168
Cortex 97
Cotton dust 157
CPSC 10
Creosote 222
Crocidolite 209

Cryptococcal meningitis 243
CS gas 236
Curare 252
CWA 5
Cyanosis 69, 185, 187, 207, 208, 219, 239
Cyclamates 14
Cytoplasm 105
Cytotoxic hypoxia 143
　　hydrogen cyanide 143
　　hydrogen sulfide 143

D

D-Lomolene 80
DBCP 166, 167
DDE 81, 167, 169
DDOH 110
DDT 3, 13, 43, 54, 79, 81, 89, 95, 100, 110, 111, 135, 143, 149, 167, 218, 219, 288, 297
Deactivation 114
　　malathion 114
　　parathion 114
Deamination 260
DEHP 166
Dementia 223
Dendrites 128
Dengue fever 242
Deoxyribonucleic acid (DNA) 112, 232, 256, 257
Department of Transportation (DOT) 306
Depolarizing agents 135
Depressants 135, 136
Dermatitis 191, 192, 194, 195, 197, 199, 204, 223, 251
Dermis 43, 44
DES 272
Descriptive toxicologists 11
Descriptive toxicology 11
Dialysis 244
Diaphragm 137
Diazinon 137
Dibenzofurans 147
Dibromochloropropane 168
Dichlorobenzene, 1,4- 80
Dichlorodiphenyltrichloroethane (DDT) 3. See also DDT
Dichloroethane, 1,1- 191, 192
Dichloroethane, 1,2- 191, 192
Dichloroethylene, 1,1- 76
Dichloroethylene, 1,2- 76
Dichloroformoxime 238
Dichloromethane 191, 193
Dichlorophenoxyacetic acid, 2,4- 220

Dichlorvos 215, 217
Dicumarol 224
Dieldrin 94, 95, 110, 218, 219
Diethylstilbestrol (DES) 272
Diffusion 39, 90, 91, 120
Digestive tract 47
　　absorption 47, 49
　　　factors affecting 49, 52
　　anatomy of
　　　duodenum 49
　　　esophagus 47
　　　ileum 49
　　　jejunum 49
　　　lumen 47
　　　mucosa 47
　　　muscularis 47
　　　serosa 47
　　　submucosa 47
Dimercapto-1-propanesulfonic acid, 2,3- 311
Dimercaptopropanol, 2,3- (BAL) 311
Dimethyl sulfate 275
Dimethyl sulfoxide 172
Dimethylbenz[a]anthracene 148
Dinitrobenzene, 1,3- 166
Dinitrophenol, 2,4- 126, 143, 172
Dinitrotoluene, 2,4- 80, 149
Dioxins 103
Diphenyl chlorasine 236
Diquat 172, 221
Dispositional antagonism 81
Distal convoluted tubule 97
Distribution of toxic substances 91
Divinyl ether 87
DMPS (2,3-dimercapto-1-propanesulfonic acid) 311
DNA 112, 232, 256, 257
　　codons 257
　　histones 256
　　nucleotides 256
DNA repair 265
　　base-excision repair 265
　　nucleotide excision repair 265
DNP 172
Dopamine 136
Dose 3, 11
Dose-response curve 65
　　FEL 67
　　LOAEL 67
　　LOEL 67
　　NEL 67, 73
　　NOAEL 67
　　NOEL 67
　　relative toxicity 65
　　threshold 65, 66
Dose-response relationship 11, 275, 276

Dose-time relationship 72
Ductus deferens 165
DUI (driving under the influence) 13
Duodenum 49
Dyspnea 188

E

Ebola 250
EDB 166
Edema 157
EDS 166
EDTA (ethylene diaminetetraacetic acid) 311
Efferent ductules 165
Efferent nerves 127
Electromagnetic fields (EMF) 23
Emergency Planning and Community Right to Know Act (EPCRA) 5
EMF 23
Emphysema 157, 158, 188, 198, 213
    cadmium 159
    cigarette smoke 159
    nitrogen dioxide 159
    ozone 159
Encephalitis 215
Encephalopathy 201
Endometrium 168
Endoplasmic reticulum (ER) 37, 38, 105, 133
Endosulfan 167
Endrin 218, 219
Enterocolitis 233
Enterohepatic circulation 120
Environmental Protection Agency (EPA) 4, 5, 13, 279, 289
    CAA 5
    CERLA 5
    CWA 5
    EPCRA 5
    FIFRA 5
    RCRA 5
    SDWA 5
    TSCA 5
Environmental toxicologists 13
Environmental toxicology 13
Enzymes 38
Eosinophils 139
EP 9
EPA 4. *See also* Environmental Protection Agency (EPA)
EPCRA 5

Epidemiological studies 5, 11, 18, 19, 20, 21, 30
    advantages of 23, 24, 30
    attributable risk 26
    attributes of 20
    bias 26
    confounding factors 22
    disadvantages of 22, 23, 24, 30
    incidence rate 24, 25
    prevalence rate 24
    prospective epidemiological studies 21, 23, 31
    relative risk 25
    retrospective epidemiological studies 21, 31
Epidemiology 4, 19
Epidermis 43, 89
Epididymis 165
Epigenetic carcinogen 274
Epilation 233
Epilepsy 201
Epinephrine 136
Epithelial cells 41
EPN 80, 81
Equilibrium 40
ER 37
Erythema 195, 282
Erythrocyte protoporphyrin (EP) 9
Erythrocytes 139
Erythropoietin 142
Escherichia coli 15
Esophagus 47
Estrogen 168, 276
Ethanol 80, 89, 102, 110, 113, 137, 141, 166, 272, 276
Ether 3
Ethyl chloride 191, 192
Ethylene 46, 47, 87, 90, 122
Ethylene dibromide (EDB) 166
Ethylene dimethane sulfonate (EDS) 166
Ethylene glycol 77, 163
Ethylene oxide 269
Etiological factor 18
Etiology 18
Excretion 54, 55, 120
    cadmium 120
    kidneys 120
    lead 120
    liver 120
    lungs 120, 122
    mercury 120
Exencephaly 271
Exposure pathways 43, 53, 78
    dermal absorption 43, 74, 75, 76

    eye 43
    ingestion 43, 75, 79
    inhalation 43, 45, 75
    intradermally 53
    intramuscularly 53
    intraperitoneally 53
    intravenously 53
Extravascular space 95
Extremely-low-frequency radiation 285
Exudate 187
Eye 52, 170
    anatomy of 52
        anterior chamber 53
        aqueous humor 53
        choroid 53
        ciliary bodies 53
        cornea 53
        iris 53
        lens 53
        retina 53
        sclera 52
        vitreous humor 53
    toxic effects 171
        2,4-dinitrophenol (DNP) 172
        2,4,6-trinitrotoluene 172
        acetone 172
        ammonia 171
        benzene 172
        carbon disulfide 172
        chloroquinone 172
        dimethyl sulfoxide 172
        diquat 172
        heptachlor 172
        hydrofluoric acid 171
        hydroxyquinone 172
        lead 172
        lime 171
        mercury 172
        methanol 77, 172
        naphthalene 172
        silver 172
        sodium hydroxide 171
        sulfuric acid 171
        toluene 172

F

Facilitated diffusion 39, 40, 88, 89
Facilitated transport 90, 93
Fallopian tubes 167
Fanconi's anemia 265
FAS 272
FDA 4, 13

Federal Insecticide, Fungicide, and Rodenticide Act (FIFRA) 4, 5
FEL 67
Female reproductive system 167
    anatomy 167
        fallopian tubes 167
        ovaries 167
        uterine tubes 167
        uterus 167
    physiology 168
    toxic effects 168
        cadmium 168
        DDE 169
        dibromochloropropane 168
        lead 168
        manganese 168
        mercury 168
        polycyclic aromatic hydrocarbons 168
        polystyrene 168
        styrene 168
        tin 168
Fenestrae 97
Fetal alcohol syndrome (FAS) 272
Fiberglass 46, 90
Fibrinogen 140
Fibrosis 158
    asbestos 158
    beryllium 158
    cadmium 158
    coal dust 158
    paraquat 158
    silica 158
FIFRA 4, 5
First-pass effect 120
Fluorosis 102
Follicle-stimulating hormone (FSH) 166
Food additives 14
    aspartame 14
    cyclamates 14
    olestra 14
    saccharin 14
Food and Drug Administration (FDA) 4, 289
Food toxicologists 13
Food toxicology 13
Footdrop 201
Forensic toxicologists 13
Forensic toxicology 13
Formaldehyde 78, 156, 157, 190, 208
    pulmonary edema 208
Frank-Effect Level (FEL) 67
FSH 166
Fumigants 223
    methyl bromide 223

phosphine 223
Functional antagonism 81
Fungicides 222
    creosote 222
    hexachlorobenzene 222
    organomercurial compounds 222
    pentachlorophenol 222

G

Gall bladder 160
Gamma rays 231, 280
Gasoline, unleaded 80
Gastroenteritis 327
Gastrointestinal tract
    absorption
        factors afffecting 90
Genotoxic carcinogen 274
Genotoxicity 269
Gingivitis 203
Glia 128
Globin 160
Globulin 94, 100
Glomerular nephritis 148, 203
Glomerulus 97
Glucuronidation 105, 112
Glycogen 160
Glycol ethers 166
Glycoprotein 160
GnRH 166
Gonadotropin-releasing hormone (GnRH) 166
Gonococcal ophthalmitis 172
Graafian follicle 168
Granulocytes 139
Granulocytopenia 141, 143
Granulomas 167
Granulosa cells 168
Gray (Gy) 232
Gy 232

H

HAH 147
Halogenated aromatic hydrocarbons (HAH) 147
Halogenated hydrocarbons 153, 191
Hantavirus 250
HAV 244
Hazardous Substances Data Bank (HSDB) 7
HBV 244
HCV 244

HDV 244
Heart 91, 150
    structure and function of 150
        atrioventricular (AV) node 150
        atrioventricular bundle 150
        bundle branches 150
        bundle of His 150
Heating, ventilation, and air conditioning systems 189
Hematopoiesis 140
Hematoxicity 139, 141
    anemia 141
    benzene 141, 142, 143
    carbon dioxide 186
    carbon monoxide 142, 185
    carbon tetrachloride 142, 143
    carboxyhemoglobin 142, 185
    chlordane 142
    DDT 143
    dinitrophenol 143
    ethanol 141
    granulocytopenia 141, 143
    hydrogen cyanide 143
    hydrogen sulfide 143
    hydroxylamine hydrochloride 142
    hypoxia 142
    jaundice 161
    lead 141
    leukemia 143
    lymphocytopenia 141, 143
    methemoglobin 69
    methemoglobinemia 69, 142
    methylene chloride 141
    naphthalene 141
    nitrates 67, 68
    nitrites 67, 68
    nitrobenzene 141, 142
    pancytopenia 141
    phenylhydrazine 79
    primaquine 79
    radiation 142
    sodium nitrite 142
    strychnine 224
    thrombocytopenia 141, 143
    trinitrotoluene 143
    warfarin 223
Hematuria 197
Heme 160
Hemocytoblast 141
Hemoglobin 62
Hemolysin 250
Hemolysis 79, 197
Hemolytic anemia 324, 325
Hemophiliacs 244
Heparin 140
Hepatic artery 95

Hepatic sinusoids 95
Hepatitis 161, 244, 249, 250
Hepatitis A (HAV) 244
Hepatitis B (HBV) 244
Hepatitis C (HCV) 244
Hepatitis D (HDV) 244
Hepatocytes 96
Hepatomegaly 222
Hepatotoxicity 160, 161
   1,1-dichloroethane 192
   1,1,1-trichloroethane 191, 192
   1,1,1,2-tetrachloroethane 191, 192
   1,1,2-trichloroethane 191, 192
   1,1,2,2-tetrachloroethane 191, 192
   1,2-dichloroethane 191, 192
   acetaminophen 116
   arsenic 119, 161
   bromobenzene 116
   carbon tetrachloride 73, 161, 191, 192
   chloroform 73
   chromium 161
   cirrhosis 117, 161
   dichloromethane 191, 193
   diquat 221
   ethyl chloride 191, 192
   ethylene dichloride 191
   hepatitis 161
   hepatomegaly 222
   hexachlorobenzene 222
   hexachloroethane 191, 192
   methyl chloroform 96, 191, 192
   methylene chloride 191
   monochloroethane 191, 192
   organochlorine pesticides 219
   perchloroethylene 191
   tetrachlorethylene 191, 193
   trichloroethane 96, 193
   trichloroethylene 191
Heptachlor 110, 172
Herbicides 220
   2,4-dichlorophenoxyacetic acid 220
   2,4,5-trichlorophenoxyacetic acid 220
   bipyridyl derivatives 221
   chlorophenoxy herbicides 220
   diquat 221
   paraquat 221
Heterocyclic amines 275
Hexachlorobenzene 222
Hexachloroethane 191, 192
Hexachlorophene 134
Hexane 89, 134

Histamine 140, 282
Histones 256
HIV 243
Homeostasis 98
HSDB 7
Human Immunodeficiency Virus (HIV) 243
HVAC 189
Hydrazine 89
Hydrocephaly 245
Hydrochloric acid 156
Hydrofluoric acid 156, 171, 212
   arrhythmia 212
   hypocalcemia 212
Hydrogen cyanide 75, 134, 143, 211
Hydrogen fluoride 158, 207
   cyanosis 207
   pneumonia 207
   pulmonary edema 207, 213
Hydrogen sulfide 143, 211
   pulmonary edema 211
Hydrolysis 105, 111
Hydrophobic 103
Hydroxyapatite 101
Hydroxylamine hydrochloride 142
Hydroxyquinone 172
Hydroxyurea 269
Hyperpigmentation 197
Hyperplasia 156
Hypocalcemia 212
Hypotension 250
Hypothalamus 166
Hypoxia 142
   carbon monoxide 142
   carboxyhemoglobin 142
   cytotoxic hypoxia 143
   hydrogen cyanide 143
   hydrogen sulfide 143
   methemoglobinemia 142

I

IARC 178, 279
IDLH 179
Ileum 49
Immediately Dangerous to Life or Health (IDLH) 179
Immune system 144
   antibody 81, 140, 146, 148, 149, 156
   B cell lymphocytes 146
   lymph nodes 144
   lymphocytes 144, 145
   macrophages 144, 145
   natural killer (NK) cells 146

   neutrophils 146
   spleen 145
   T cell lymphocytes 146
   thymus 145
Immunotoxicity 144
   glomerular nephritis 148
   halogenated aromatic hydrocarbons 147
      dibenzofurans 147
      PBBs 147
      PCBs 147
      TCDD 147
   heavy metals 148
      arsenic 148
      cadmium 148
      lead 148
      mercury 148
   organic solvents 149
      2,4-dinitrotoluene 149
      benzene 149
      carbon tetrachloride 149
      nitrobenzene 149
      toluene 149
   polycyclic aromatic hydrocarbons 148
      anthracene 148
      benzanthracene 148
      benzo[a]pyrene 148
      dimethylbenz[a]anthracene 148
      methylcholanthrene 148
   pesticides 149
      DDT 149
      malathion 149
      parathion 149
Incidence rate 24, 25
Induced mutation 261
Infectious diseases 18
Infectious wastes 247
Inflammation 140
Infrared radiation 284
Ingestion 47
Inhalation 43, 45, 75
International Agency for Research on Cancer (IARC) 178, 279
Interspecies variability 76, 77
Intracellular 38
Intradermally 53
Intramuscularly 53
Intraperitoneally 53
Intraspecies variability 76, 78
Intravenously 53
Ionizing radiation 231, 280
   effects of 281
      cellular effects 281
      tissue effects 282
   sources of 281

types of 280
 alpha particles 280
 beta particles 280, 281
 gamma rays 280, 281
 neutrons 280, 281
 x-rays 280, 281
Ions 38
Iris 53
Isophorone 80
Isoproterenol 58

J

Jaundice 161, 192, 195, 197, 244, 245
Jejunum 49

K

Kaposi's sarcoma 243
Keratin 43
Keratitis 222
Keratocytes 171
Keratosis 197
Kerosene 194
Ketones 153
Kidneys 95, 120, 162, 198, 201, 203, 221
 1,1-dichloroethane 191, 192
 1,1,1-trichloroethane 191, 192
 1,1,1,2-tetrachloroethane 191, 192
 1,1,2-trichloroethane 191, 192
 1,1,2,2-tetrachloroethane 191, 192
 1,2-dichloroethane 191, 192
 arsine 197
 cadmium 119, 142, 198
 carbon tetrachloride 191, 192
 dichloromethane 191, 193
 diquat 221
 erythropoietin 142
 ethyl chloride 191, 192
 ethylene dichloride 191
 excretion 120
 footdrop 201
 function of 97, 162
 glomerular nephritis 148
 hexachloroethane 191, 192
 mercury 119, 142, 203
 methyl chloroform 191, 192
 methylene chloride 191
 monochloroethane 191, 192
 nephrotoxicity-induced anemia 142
 organochlorine pesticides 219
 perchloroethylene 191
 sodium fluoroacetate 224
 storage in 100
 structure of 97
  cortex 97
  medulla 97
  nephron 97
  renal pelvis 97
  renal pyramids 97
  ureter 97
 tetrachlorethylene 191
 tetrachloroethylene 193
 toxicity of
  cadmium 120, 162
  chloroform 163
  ethylene glycol 163
  lead 120, 162
  mercury 120, 162
 trichloroethylene 191, 193
 wristdrop 201
Krebs cycle 224
Kupffer cells 160

L

Labile 203
Lacrimation 208
Lamellae 100
Larynx 45
Lasers 285
Lavage 311
LC_{50} 61
LD_{50} 58
Lead 8, 44, 49, 52, 54, 73, 75, 79, 88, 90, 95, 96, 97, 101, 102, 103, 120, 134, 141, 148, 153, 162, 166, 168, 172, 189, 200, 270, 271
 anemia 202
 ataxia 201
 encephalopathy 201
 epilepsy 201
 footdrop 201
 lead line 202
 saturnine gout 202
 wristdrop 201
Lead arsenate 200
Lead arsenite 200
Lead chromate 200
Lead fluoroborate 200
Lead line 202
Lead nitrate 200
Lead thiocyanate 200
Legionella pneumophila 189
Legionnaire's disease 4, 189
Lens 53
Lethal concentration (LC) 61
Lethal dose (LD) 61
Leukemia 19, 116, 143, 193, 235, 265, 281, 282, 283
 benzene 143
Leukocytes 139
Leukopenia 282
LH 166
Lime 171
Lindane 218, 219
Lipids 38
Lipophilic 103
Lipoproteins 94
Liver 95, 103, 105, 110, 120, 160, 221
 excretion 120
 functions of 160
 storage in 96
 structure of 95
  central vein 95
  hepatic artery 95
  hepatic sinusoids 95
  hepatocytes 96
  lobules 95
LOAEL 67
Lobules 95
Local effects 54, 89
LOEL 67
Loop of Henle 97
Love Canal 4
Lowest-Observed-Adverse-Effect Level (LOAEL) 67
Lowest-Observed-Effect Level (LOEL) 67
Lumen 47
Lung cancer 80, 159, 204, 209, 277, 283
 arsenic 159
 asbestos 159
 benzo[a]pyrene 159
 beryllium 159
 bronchogenic carcinoma 277
 chromium 159
 cigarette smoke 159
 nickel 159
 polycyclic aromatic hydrocarbons 159
Lungs 122
 excretion 120
Luteinizing hormone (LH) 166
Lyme disease 4, 249
Lymph 144
Lymph nodes 144
Lymph nodules 144
Lymphadenopathy 245
Lymphatic capillaries 144

Lymphatic system 144
Lymphocytes 140, 144, 145, 155
   B cell lymphocytes 146
   natural killer (NK) cells 146
   T cell lymphocytes 146
Lymphocytopenia 141, 143
Lymphoma 243

M

Macrophages 140, 144, 145, 148, 149, 155, 210
Malaria 3
Malathion 3, 43, 77, 79, 80, 111, 114, 137, 149, 215, 217
Male reproductive system 165
   anatomy 165
      bulbourethral gland 165
      cells of Leydig 165
      ductus deferens 165
      efferent ductules 165
      epididymis 165
      prostate gland 165
      seminal vesicles 165
      seminiferous tubules 165
      sertoli cells 165
      testes 165
      urethra 165
   physiology 166
   toxic effects 166
      1,2-dibromo-3-chloropropane (DBCP) 166
      1,3-dinitrobenzene 166
      acetaldehyde 166
      cadmium 167
      chlordecone 167
      DDE 167
      DDT 167
      di-(2-ethylhexyl)phthalate (DEHP) 166
      dibromochloropropane (DBCP) 167
      endosulfan 167
      ethanol 166
      ethylene dibromide (EDB) 166
      ethylene dimethane sulfonate (EDS) 166
      glycol ethers 166
      lead 166
      methoxychlor 167
      methylene chloride 167
      toxaphene 167
Malignant tumor 273
Manganese 90, 133, 153, 168
Material Safety Data Sheets (MSDS) 10

Maternal toxicity 270
Mechanistic toxicology 11
Medical Transportation Emergency Center (MEDTREC) 7
Medical wastes 247
MEDTREC 7
Medulla 97
Medulla oblongata 152
Mees' lines 197
Megakaryocytes 140
Melanosis 197
Menstrual cycle 168
Mercurialism 203
Mercuric chloride 163, 202, 203
Mercury 12, 52, 65, 66, 67, 96, 97, 100, 119, 120, 142, 148, 162, 168, 172, 202
   (acetate) phenyl mercury 202
   ataxia 203
   bronchitis 203
   cerebral palsy 203
   glomerular nephritis 203
   mercurialism 203
   mercuric chloride 202, 203
   metallic mercury 202, 203
   methyl mercury 72, 113, 133
   Minamata Disease 12
   paresthesia 203
   pneumonitis 203
Mercury compound 202
   liquid 202
   solid 202
Mesothelioma 20, 158, 209, 277
Messenger RNA (mRNA) 257
Metabolism 62
Metabolism of toxic substances 105
   1,1-dichloroethylene 76
   2-acetylaminofluorene 77
   2-aminonaphthalene 113
   acetaminophen 116
   aldrin 110
   aniline 77, 113
   arsenic 113
   benzene 105, 112, 116
   benzo[a]pyrene 111, 114, 156
   bromobenzene 116
   carbon tetrachloride 112, 156
   catabolic reactions 105
   chloroform 112
   cis-1,2-dichloroethylene 76
   DDT 110, 111
   dieldrin 110
   dinitrotoluene 80
   ethanol 110, 113
   ethylene glycol 77
   factors affecting metabolism 117
   heptachlor 110
   liver 105

   malathion 77, 111, 114
   methanol 110
   methyl mercury 103, 113
   nicotine 113
   nitrosamines 115
   parathion 79, 106, 114
   Phase I (catabolic reactions) 105
   Phase II reactions 105
   phenols 113
   polycyclic aromatic hydrocarbons 156
   selenium 113
   toxaphene 111
   trans-1,2-dichloroethylene 76
Metabolite 79
Metal fume fever 206
Metallic mercury 74, 203
Metallothionein 97, 100, 162
Metals 196
   arsenic 197. *See also* Arsenic
   arsenic acid 197
   arsenic disulfide 197
   arsenic trifluoride 197
   arsenic trioxide 197
   arsine 197
   cadmium 198. *See also* Cadmium
   cadmium chloride 198
   cadmium oxide 198
   calcium chromate 200
   chromic acid 199
   chromium 198. *See also* Chromium
   chromium oxychloride 199
   chromium trioxide 200
   lead 200. *See also* Lead
   lead arsenate 200
   lead arsenite 200
   lead chromate 200
   lead fluoroborate 200
   lead nitrate 200
   lead thiocyanate 200
   mercuric chloride 202
   mercury 202. *See also* Mercury
   mercury compound 202
   nickel 203. *See also* Nickel
   nickel carbonyl 204
   nickel nitrate 204
   nickel sulfate 204
   phenyl mercury 202
   tetramethyl lead 200
   thallium dichromate 200
   zinc 206
   zinc chlorate 206
   zinc chloride 206
Metaplasia 157
Metastasize 273
Methanol 77, 110, 133, 137, 172

Methemoglobin 69, 78
Methemoglobinemia 69, 142
  hydroxylamine hydrochloride 142
  nitrobenzene 142
  sodium nitrite 142
Methoxychlor 167
Methyl bromide 223
Methyl chloroform 96, 192
Methyl isocyanate (MIC) 2
Methyl mercury 72, 103, 113, 133, 203, 271
Methylation 105, 113, 261
Methylcholanthrene 148
Methylene chloride 52, 75, 90, 136, 141, 153, 167, 297
MIC 2
Microcephaly 245
Microcytic anemia 141
Microvilli 49
Microwave radiation 285
Minamata Disease 12
Mineral spirits 195
Mitochondria 37, 38, 105
Mixed wastes 231, 235
Monitoring and sampling
  difference between 67
Monochloroethane 191, 192
Monocytes 140
mRNA 257
MSDS 10
Mucociliary escalator 90, 155
Mucosa 47
Mucous cells 45
Mucous membranes 2
Multiple sclerosis 134
Muscarinic 217
Muscularis 47
Mustard gas 3, 89, 236, 238
  Agent Q 238
Mutagen 256, 260
  5-bromouracil 261
  Actinomycin D 262
  aldehydes 262
  chloroethylnitros-ureas 262
  epoxides 262
  fungicides 222
  nitrogen mustards 262
  ultraviolet radiation 262
Mutation 112, 233, 256, 284
  addition 261
  Cockayne's syndrome 265
  deamination 260
  deletion 261
  Fanconi's anemia 265
  germinal mutation 263
  induced mutation 261
  large mutation 263

methylation 261
mutagen 256, 260
point mutation 263
somatic mutation 263
spontaneous mutation 260
teratogen 256
transitional point mutation 263
transversion point mutation 263
Xeroderma pigmentosum 265
Myelin sheath 128
Myelinated nerve 128
Myelinopathy 133, 134
Myocarditis 245
Myometrium 168

N

NADPH 110
Naphthalene 141, 172
Nasal cancer 157, 204
Nasal cavity 45
Nasopharyngeal region 156
  ammonia 156
  chlorine 156
  chromium (VI) 156
  formaldehyde 156
  hydrochloric acid 156
  hydrofluoric acid 156
National Center for Environmental Health (NCEH) 6
National Fire Protection Association (NFPA) 178, 306
National Institute for Occupational Safety and Health (NIOSH) 6, 178
National Library of Medicine (NLM) 7
National Toxicology Program 279
National Toxicology Program (NTP) 178
Natural killer (NK) cell 146, 148, 149, 156, 210
NCEH 6
NCRP 231
Necrosis 163
Neisseria gonorrhoeae 172
NEL 67, 73, 74
Neoplasm 273
Nephron 97
Nephrotoxicity 162
  heavy metals 162
    cadmium 162
    lead 162
    mercuric chloride 163
    mercury 162
  organic substances 163
    1,1-dichloroethane 191, 192

    1,1,1-trichloroethane 191, 192
    1,1,1,2-tetrachloroethane 191, 192
    1,1,2-trichloroethane 191, 192
    1,1,2,2-tetrachloroethane 191, 192
    1,2-dichloroethane 191, 192
    carbon tetrachloride 191, 192
    chloroform 163
    dichloromethane 191, 193
    ethyl chloride 191, 192
    ethylene dichloride 191
    ethylene glycol 163
    hexachloroethane 191, 192
    methyl chloroform 191, 192
    methylene chloride 191
    monochloroethane 191, 192
    perchloroethylene 191
    tetrachlorethylene 191
    tetrachloroethylene 193
    trichloroethylene 191, 193
Nephrotoxicity-induced anemia 142
NER 265
Nerve gas 239
  sarin 239. See also Sarin
  soman 239. See also Soman
  tabun 239. See also Tabun
Nervous system 127
  central nervous system (CNS) 127
  neuron 127
  peripheral nervous system (PNS) 127
Neuroglia 128
  astrocytes 128
  oligodendrocytes 128
  Schwann cells 128
Neuron 127
Neuronopathy 133
Neuropathy 172, 193, 197, 221, 222, 223
  polyneuropathy 195
Neuropsychiatric 133
Neurosyphilis 172
Neurotoxicity 127, 133
  acrylamide 134
  axonopathy 133
  benzene 137
  botulism 135
  caffeine 136
  carbamate insecticides
    aldicarb 137
    carbaryl 2, 137
    Sevin 2, 137
  carbon disulfide 134
  carbon monoxide 134

carbon tetrachloride 73, 136
chloroform 73, 136
chlorophenoxy herbicides 220
  2,4-dichlorophenoxyacetic acid 220
  2,4,5-trichlorophenoxyacetic acid 220
cocaine 136
ethanol 137
halogenated hydrocarbons 191
  1,1-dichloroethane 191
  1,1,1-trichloroethane 191
  1,1,1,2-tetrachloroethane 191
  1,1,2-trichloroethane 191
  1,1,2,2-tetrachloroethane 191
  1,2-dichloroethane 191
  carbon tetrachloride 191
  dichloromethane 191
  ethyl chloride 191
  ethylene dichloride 191
  hexachloroethane 191
  methyl chloroform 191
  methylene chloride 191
  monochloroethane 191
  perchloroethylene 191
  tetrachlorethylene 191
  trichloroethylene 191
hexachlorophene 134
hexane 134
hydrogen cyanide 134
metals
  aluminum 133
  lead 8, 134
  manganese 133
  mercury 12
  methyl mercury 133
methanol 133, 137
methyl bromide 223
methylene chloride 136
myelinopathy 133
nerve gases
  soman 11
  tabun 11
neuronopathy 133
nicotine 135
organochlorine pesticides 3
  DDT 3, 135
organomercurial fungicides 222
organophosphate pesticides 3, 45, 134
  diazinon 137
  malathion 3, 137
  parathion 137
pyrethroid insecticides 135
toluene 137
xylene 137

Neurotransmitter 130, 152
  acetylcholine 132, 137, 152
  anticholinesterase agents 135, 137
  blocking agents 135
  depolarizing agents 135
  depressants 135, 136
  dopamine 136
  epinephrine 136
  norepinephrine 132, 136, 152
  stimulants 135
Neutrons 231, 280
Neutrophils 139, 146, 155
NFPA 178
Nickel 153, 158, 159, 189, 203, 204, 210, 275
  dermatitis 204
  nickel carbonyl 204
  nickel nitrate 204
  nickel sulfate 204
  pneumonitis 204
Nickel carbonyl 204
Nickel nitrate 204
Nickel sulfate 204
Nicotinamide adenine dinucleotide phosphate (NADPH) 110
Nicotine 74, 113, 135
Nicotinic receptors 135
NIOSH 6, 178
Nitrates 67, 68
Nitric acid 44, 89, 212
  pneumonitis 212
  pulmonary edema 212
Nitrites 67, 68
Nitrobenzene 78, 141, 142, 149
Nitrogen dioxide 46, 157, 159
Nitrogen oxide 90, 156, 158, 186, 188
Nitrosamines 115, 157, 275
Nitrosureas 275
Nitrous oxide 87
NLM 7
Nm (nanometers) 102
No-Effect Level (NEL) 67, 73, 74
No-Observed-Adverse-Effect Level (NOAEL) 67
No-Observed-Effect Level (NOEL) 67
NOAEL 67
NOEL 67
Noninfectious diseases 18, 19
Nonionizing radiation 280, 283
  types of 284
    extremely-low-frequency radiation 285
    lasers 285

    microwave radiation 285
    solar radiation 284
Norepinephrine 132, 134, 136, 152
NRC 4, 7
NTP 178
Nuclear Regulatory Commission (NRC) 4, 7
Nucleic acids 112
Nucleotide excision repair (NER) 265
Nucleotides 256

O

O-ethyl O-(p-nitrophenyl) phenylphosphorothionate 80
Observation bias 26, 27
Occupational Safety and Health Act 6
Occupational Safety and Health Administration (OSHA) 4, 13, 279, 289
Office of Research and Development (ORD) 6
Olestra 14
Oligodendrocytes 128
Oligouric 250
Oliguria 192
One-hit theory 274
Oocytes 168, 283
Oogenesis 282
Optic nerve 171
ORD 6
Organic solvents 44, 149, 153, 163
Organochlorine pesticides 3, 70, 73, 218
  aldrin 218, 219. *See also* Aldrin
  benzene hexachloride 218, 219
  chlordane 218, 219. *See also* Chlordane
  DDT 3, 43, 54, 135, 218, 219. *See also* DDT
  dieldrin 218, 219. *See also* Dieldrin
  endrin 218, 219
  Lindane 218, 219
Organomercurial compounds 222
Organophosphate pesticides 3, 45, 54, 70, 74, 80, 112, 134, 215, 217, 239, 279
  dichlorvos 215, 217
  EPN 80, 81

malathion 3, 43, 215, 217. *See also* Malathion
parathion 106, 215, 217. *See also* Parathion
tetraethyl pyrophosphate 215, 217
Ortho-chlorobenzalmalononitrile gas (CS) 236
OSHA 4. *See also* Occupational Safety and Health Administration (OSHA)
Osmosis 39, 40
Osteocytes 100
Osteomalacia 198
Osteoporosis 101, 198
Ovarian follicles 168
Ovaries 167
Ovulation 168
Oxidation 105, 110
Ozone 156, 157, 158, 159, 188

P

PAH 79. *See also* Polycyclic aromatic hydrocarbon (PAH)
Palliative 233
Pancytopenia 141
Paracelsus 3
Paraquat 158, 221
Parasympathetic nervous system (PSNS) 127
Parathion 79, 81, 89, 106, 114, 137, 149, 215, 217
Parenterally 311
Paresthesia 195, 197, 203, 216, 219, 221, 222
Parkinson's syndrome 133
Particulates 46, 74, 89, 90, 155, 156, 185, 188, 189
    asbestos 46, 156
    fiberglass 46
    silica 46, 156
Passive diffusion 87, 89, 155
Passive transport 39, 91, 93, 100, 120
    diffusion 39
    facilitated diffusion 39
    osmosis 39
    passive diffusion 87
PBBs 147, 148
PCBs 100, 147, 148
Penicillamine (β,β-dimethylcysteine) 311
Pentachlorophenol 126, 222
Perchloroethylene 158
Perfusion rate 98

Peripheral nervous system (PNS) 127
Peritoneum 210
Peritubular capillaries 97
Pesticides 149, 215
    carbamate insecticides 215
    fumigants 223
    fungicides 222
    herbicides 220
    organochlorine pesticides 218
    organophosphate insecticides 215
    rodenticides 223
Petroleum distillates 191, 194
Petroleum ether 195
Phagocytosis 41, 89, 90, 146, 155, 160, 210
Pharyngitis 208
Pharynx 45
Phase I reactions 105, 106
    catabolic reactions 105
Phase II reactions 105, 106, 111
    acetylation 105
    conjugation 105
    glucuronidation 105
    methylation 105
    sulfation 105
Phenobarbital 276
Phenols 113
Phenyl mercury 202
Phenylhydrazine 79
Phocomelia 271
Phosgene 3, 207, 238
    cyanosis 208
    pulmonary edema 208
phosgene 3
Phosgene gas 238
Phosphine 223
Phospholipid 37
Pica 8
Pinocytosis 41, 89, 90, 93
Placental toxicity 270
Plant-derived toxins 251
Plasma 94, 140
Plasma cell membrane 37, 38
    carrier molecules 38
    carrier-mediated transport 88
    function of 38
    structure of
        carbohydrates 37
        phospholipid 37
        proteins 37
Plasma proteins 94
    albumin 94
    globulin 94
    lipoproteins 94
    transferrin 94

Platelets 139
Pleural fibrosis 209, 210
Plicae circulares 49
Pneumoconiosis 189
Pneumonia 207, 211, 245
Pneumonitis 188, 193, 195, 198, 203, 204, 212, 283
PNS 127
Poison ivy 44
Polybrominated biphenyls (PBBs) 147, 148
Polychlorinated biphenyls (PCBs) 100, 147, 148
Polychlorinated dibenzo-p-dioxins (TCDD) 147
Polycyclic aromatic hydrocarbon (PAH) 79, 110, 114, 120, 148, 156, 159, 168, 275
Polyneuritis 327
Polyneuropathy 195, 216
Polystyrene 168
Post-traumatic stress disorder 2
Postsynaptic membrane 132
Potassium cyanide 211
Potassium hydroxide 213
Potentiation effects 80, 81
Presynaptic terminal 132
Prevalence rate 24
Primaquine 79
Primary bronchi 45
Primary oocytes 168
Primary spermatocytes 165
Primordial follicles 168
Processes 128
Progesterone 168, 276
Prospective epidemiological studies 21, 23, 31
    advantages of 24
    disadvantages of 24
Prostate gland 165
Proteins 37
Prothrombin 160
Proximal convoluted tubule 97
Pseudopodia 41
PSNS 127
Psychoneurosis 223
Pulmonary artery 91
Pulmonary circulation 91
Pulmonary edema 157, 187, 188, 193, 194, 198, 206, 207, 208, 211, 212, 213, 221, 223, 238, 245, 283
    ammonia 158, 207
    chlorine 158
    fibrosis 158
    formaldehyde 208
    hydrogen fluoride 158, 207
    oxides of nitrogen 158

ozone 158
perchloroethylene 158
phosgene 3, 207
Pulmonary fibrosis 209, 210
Pulmonary toxicity 155
    alveoli 156
    ammonia 157, 158, 207
    asbestos 156, 158, 209
    beryllium 158
    bronchitis 157
    cadmium 158
    carbon dioxide 156
    carbon monoxide 156
    chlorine 157, 158, 210
    chromium 157
    cigarette smoke 157
    ciliostasis 157
    coal dust 158
    cotton dust 157
    emphysema 157, 158
    fibrosis 158
    formaldehyde 208
    hydrogen fluoride 207
    hydrogen sulfide 211
    hyperplasia 156
    metaplasia 157
    methyl isocyanate (MIC) 2
    nasopharyngeal region 156
    nickel 158
    nitrogen oxides 156, 157, 158, 186
    ozone 156, 158, 188
    paraquat 158
    particulates 188, 189
    perchloroethylene 158
    phosgene 207
    phosphine 223
    pulmonary edema 157
    silica 156, 158
    silicosis 158
    sulfur dioxide 156, 157, 188
    sulfur oxides 188
    tracheobranchiole region 156
Pulmonary veins 91
Purkinje fibers 141
Pustules 221
Pyrethroid insecticides 135
Pyridine 160
Pyrogenic 250

Q

Quinine 3

R

Rad 232
Radiation 3, 142, 231, 271, 273, 280
    alpha particles 231. *See also* Alpha particles
    beta particles 231. *See also* Beta particles
    gamma rays 231. *See also* Gamma rays
    gray 232
    ionizing radiation 231, 280
    neutrons 231. *See also* Neutrons
    nonionizing radiation 280, 283
    rad 232
    rem 231
    Sievert 232
    x-rays 231. *See also* X-rays
Radioactive 102, 280
Radioactive materials 231
Radionuclides 231
Radon 190, 281, 283
RCRA 5
Receptor antagonism 81
Red tide 251
Reduction 105, 111
Reference concentration (RfC) 294
Reference dose (RfD) 294
Registry of Toxic Effects of Chemical Substances (RTECS) 7
Regulatory toxicologists 13
Regulatory toxicology 13
Relative risk 25
Relative toxicity 65, 69, 71
Rem 231
Renal corpuscle 97
Renal pelvis 97
Renal pyramids 97
Renin 162
Resource Conservation and Recovery Act (RCRA) 5
Respiratory tract 45, 155
    anatomy of
        alveoli 45
        bronchioles 45
        ciliated epithelial cells 46
        larynx 45
        mucous cells 45
        nasal cavity 45
        pharynx 45
        primary bronchi 45
        trachea 45
    factors affecting absorption 46, 90
    functions of 155
        defense mechanisms 155
        gas exchange 155
Resting membrane potential 128, 131, 135, 136
Reticulocytes 160
Retina 53
Retrospective epidemiological studies 21, 31
    advantages of 23
    disadvantages of 22, 23
RfC 294
RfD 294
Rhinitis 208
Ribonucleic acid (RNA) 112
Ribosomes 38
Rickettsias 242
Risk 4, 25
Risk assessment 288
Risk management 288
RNA 112
Rocky Mountain Spotted fever 245
Rodenticides 223
    sodium fluoroacetate 224
    strychnine 224. *See also* Strychnine
    warfarin 223. *See also* Warfarin
RTECS 7

S

Saccharin 14, 72, 276
Safe Drinking Water Act (SDWA) 5
Safe human dose 66
Sarcoma 273
Sarin 74, 89, 239
Saturnine gout 202
SBS 189
Schwann cell 128
Sclera 52
SDWA 5
Sebaceous glands 43, 44
Secondary spermatocytes 165
Secretion 38
Selection bias 26, 27
Selectively permeable 38
Selenium 113
Seminal vesicles 165
Seminiferous tubules 165
Sensitization effects 80, 81
Serosa 47
Sertoli cells 165
Serum hepatitis 244
Sevin 2, 137, 218

Sharps 230
Sick-building syndrome (SBS) 189
SIDS 272
Sievert (Sv) 232
Silica 46, 156, 158
Silicosis 158, 189
Silver 172
Simple asphyxiant 186
Single-hit theory 274
Sinoatrial (SA) node 141
Sinusitis 200
Skin 43
    absorption 44, 89
      factors affecting 44, 89
      dermis 43
      epidermis 43
      sebaceous glands 43
      subcutaneous fatty tissue 43
Skin cancer 235, 283, 284
Sleeping sickness 172
Small intestine 49
    duodenum 49
    ileum 49
    jejunum 49
    microvilli 49
    plicae circulares 49
    villi 49
SNS 127
Sodium cyanide 211
Sodium fluoroacetate 224
Sodium hydroxide 44, 89, 171, 213
Sodium nitrite 142
Sodium-potassium pump 135
Solar radiation 284
    infrared radiation 284
    ultraviolet radiation 284
    visible radiation 284
Soma 128
Soman 11, 239
Somatic cell mutation theory 274
Somatic motor 127
Sources chemical information
    ACGIH 7
    ATSRD 7
    CHEMTREC 7
    HSDB 7
    MEDTREC 7
    NLM 7
    NRC 7
    RTECS 7
    TOXNET 7
Sperm 283
Spermatids 165
Spermatogenesis 165, 282
Spermatogonia 165
Spermatozoan 165

Spleen 95, 145
Spontaneous mutations 260
Squamous cell carcinoma 197, 284
Staphylococcus aureus 15
Stem cell 139, 141, 282
Stereo isomer 219
Stimulants 135
Stoddard solvent 195
Storage of toxic substances 95
Stratum corneum 43
Strontium 88, 101, 102
Strychnine 74, 79, 224
Styrene 168
Subchronic 63
Subcutaneous fatty tissue 43, 44
Submucosa 47
Sudden Infant Death Syndrome (SIDS) 272
Sulfanilamide 3
Sulfation 105, 113
Sulfur dioxide 90, 156, 157, 188
Sulfuric acid 44, 89, 171, 213
    bronchitis 213
    emphysema 213
    pulmonary edema 213
Sv 232
Sympathetic nervous system (SNS) 127
Synapse 132
    postsynaptic membrane 132
    presynaptic terminal 132
    synaptic cleft 132
Synaptic cleft 132
Synergistic effects 80
Syphilis 244
Syrup of ipecac 311
Systemic circulation 91
Systemic effects 54, 89, 233

T

T cell lymphocytes 146
Tabes dorsalis 244
Tabun 11, 239
Tachycardia 153, 187
Tachypnea 187
Target tissues 55
TBS 189
TCDD 147, 221, 276
TCLP regulated contaminants 345
TEPP 81, 215, 217
Teratogen 256, 267
Teratogenic agents 133, 222
    6-aminonicotinamide 270
    arsenic 271
    aspirin 269

    cadmium 270, 271
    cocaine 272
    copper 271
    diethylstilbestrol 272
    ethanol 272
    ethylene oxide 269
    folic acid 270
    hexachlorobenzene 222
    hydroxyurea 269
    lead 270, 271
    methyl mercury 133, 271
    radiation 271
    stress 270
    thalidomide 3, 271
    Toxoplasma gondii 271
    viral infection 270
    vitamin A 270
    x-rays 269
Testes 165
Testosterone 165
Tetanus 250
Tetrachlorobiphenyl, 2,5,2',5'- 276
Tetrachlorodibenzo-p-dioxin (TCDD) 276
Tetrachlorodibenzo-p-dioxin, 2,3,7,8- 147
Tetrachlorodibenzo-para-dioxin, 2,3,7,8- 221
Tetrachloroethane, 1,1,1,2- 191, 192
Tetrachloroethane, 1,1,2,2- 191, 192
Tetrachloroethylene 120, 191, 193
Tetraethyl pyrophosphate (TEPP) 81, 215, 217
Tetramethyl lead 200
Thalidomide 3, 271
Thallium dichromate 200
The National Council of Radiation Protection (NCRP) 231
Threshold 65, 66
Threshold Limit Values (TLV) 7, 178
Thrombocytes 139
Thrombocytopenia 141, 143, 194
Thymus 145, 148
Tight-building syndrome (TBS) 189
Tin 168
TLV 7, 178
TOCP 77
Toluene 137, 149, 153, 172, 194, 215
Toxaphene 111, 167
Toxic 3
Toxic Shock Syndrome (TSS) 250
Toxic Substances Control Act (TSCA) 5

Toxicin 3
Toxicity
    factors affecting absorption 86
        first-pass effect 120
        plasma proteins 94
        redistribution of toxic substances 103
        storage of toxic substances 95
    factors afffecting absorption
        distribution of toxic substances 91
    genotoxicity 269
    maternal toxicity 270
    placental toxicity 270
Toxicological studies 5, 11, 18, 29, 30
    acute toxicity 29, 58
    advantages of 30
    chronic toxicity 29, 61
    disadvantages of 30
    dose-time relationship 72
    interspecies variability 77
    intraspecies variability 78
    physical and chemical factors 73
    variables influencing toxicity 72
Toxicology 3, 11
    clinical toxicology 15
    descriptive toxicology 11
    environmental toxicology 13
    food toxicology 13
    forensic toxicology 13
    mechanistic toxicology 11
    regulatory toxicology 13
Toxicology Data Network (TOXNET) 7
TOXNET 7
Toxoplasma gondii 271
Toxoplasmosis 243, 245
Trachea 45
Tracheobronchiole region 157
    ammonia 157
    chlorine 157
    chromium (VI) 157
    cigarette smoke 157
    cotton dust 157
    nitrogen dioxide 157
    ozone 157
    sulfur dioxide 157
Trans-1,2-dichloroethylene 76
Transfer RNA (tRNA) 257
Transferrin 94
Transport 38
Treponema pallium 172, 244
Tri-ortho-cresyl phosphate (TOCP) 77
Trichloroethane 96, 191
Trichloroethane, 1,1,1- 191, 192
Trichloroethane, 1,1,2- 191, 192
Trichloroethylene 74, 120, 193
Trichlorophenoxyacetic acid, 2,4,5- 220
Trimethylpentane 100
Trimethylpentane, 2,2,4- 276
Trinitrotoluene 143
Trinitrotoluene, 2,4,6- 172
tRNA 257
TSCA 5
TSS 250
Tumors 100
Typhus 19

U

Ulcer 244
Ultraviolet radiation (UV) 262, 284
    Cockayne's syndrome 265
    Xeroderma pigmentosum 265
Unmyelinated nerve 128
Uranium dioxide 90
Ureter 97
Urethra 165
Uric acid 202
Uterine tubes 167
Uterus 167
UV 262

V

Vagus nerve 150
Ventricles 91
Ventricular fibrillation 153
Venules 91
Vertigo 194
Vesicants 238
Vesicle 41
Villi 49
Vinyl chloride 157, 279
Virus-caused illnesses 250
    Ebola 250
    Hantavirus 250
Visible radiation 284
Vital capacity 213
Vitreous humor 53

W

Warfarin 79, 94, 223
Wristdrop 201

X

X-rays 158, 231, 269, 280, 283
Xeroderma pigmentosum 265, 274
Xylene 54, 74, 137, 194, 215

Y

Yellow fever 242
Yersinia pestis 239

Z

Zinc 206
    metal fume fever 206
    pulmonary edema 206
Zinc chlorate 206
Zinc chloride 206